Trends in Stem Cell Biology and Technology

Hossein Baharvand
Editor

Trends in Stem Cell Biology and Technology

 Humana Press

Editor
Hossein Baharvand, Ph.D.
Department of Stem Cells
Cell Science Research Center
Royan Institute ACECR, Tehran, Iran
and
Department of Developmental Biology
University of Science and Culture, ACECR
Tehran, Iran

ISBN 978-1-60327-904-8   e-ISBN 978-1-60327-905-5
DOI 10.1007/978-1-60327-905-5
Springer Dordrecht Heidelberg London New York

Library of Congress Control Number: 2008944052

Printed on acid-free paper

Springer is part of Springer Science+Business Media (www.springer.com)

*To the memory of Dr. Saeid Kazemi Ashtiani,
a wonderful colleague, a great stem cell
biologist, and an inspirational advocate
of stem cell research in Iran.*

And

To my family:
*My wife (Parvaneh), daughter (Fatemeh),
father (Ali), mother (Fatemeh), brothers
(Hassan, Abbas, Mohammad, Amir), and
sister (Afsaneh)*

Preface

Stem cells, characterized by the ability to both self-renew and to generate differentiated functional cell types, have been derived from the embryo and from various sources of the postnatal animals and human. The recent advances in stem cell research have led to a better understanding of self-renewal, maintenance, and differentiation of both embryonic and somatic stem cells. This has significantly increased our knowledge of cellular and developmental biology in general and will certainly continue to do so for a long time to come. Moreover, given their role in maintaining and replenishing tissues, stem cells represent a potential means of restoring tissue function and thereby treating the root cause of degenerative disease. Therefore, in parallel, we need to improve our cognizance of the challenges involved in applying stem cells in clinical settings. The current chapters highlight both of these aspects: that of understanding the "actual" and that of developing the "possible."

In recognition of the growing excitement and potential of stem cells as models for both the advancement of basic science and future clinical applications, I felt it timely to edit this book in which forefront investigators would provide new findings for the use of stem cells to study various lineages and tissue types and some applications. We are pleased to provide *Trends in Stem Cell Biology and Technology*, a broad-scaled series of cutting-edge chapters that have already been shown to have, or will soon have, tremendous utility with stem cells and their differentiated progeny. The authors have put together recent advances and perspectives in important fields of stem cell research: embryonic stem (ES) cells, somatic stem cells, and stem cell therapy, which deal with embryonic and somatic stem cells and their potential therapeutic applications.

Embryonic stem cells are pluripotent cells with the capacity to give rise to every somatic cell type. The nature, characteristics, and potentials of human ES cells are described in the article by Bongso and Fong. In addition, Eckardt and McLaughlin describe the generation of ES cells from gamete-derived uniparental embryos, which can be patient-derived and potentially histocompatible with the gamete donor. They also address evaluation of the integrity of the lines generated, an essential criterion in interpreting differentiation assays in vivo and in vitro. Also, Ragina and Cibelli explain the derivation of parthenogenetic embryonic stem (PGES) cells from the inner cell mass of parthenogenetic embryo at the blastocyst stage. These pluripotent stem cells offer an easily obtainable pool of stem cells that can be used as a source for derivation of autologous tissues, albeit limited to females in reproductive age. PGES cells' derivation does not require destruction of a viable embryo and therefore bypasses the ethical debates surrounding the use of naturally fertilized embryos. Moreover, Zuccotti and coworkers summarize the advancement in nuclear reprogramming and in cell reprogramming by cell fusion, using amphibian eggs or egg extracts, with cell extracts, with synthetic molecules, or by induced expression of specific genes and production of induced pluripotent stem cells. In contrast, Mardanpour et al. describe general considerations regarding molecular and cellular aspects of reprogramming of germ cells at different developmental stages to stem cells compared with their counterpart, ES cells. Moreover, epigenetic modifications, such as covalent modifications of histones and DNA methylation, are extremely important control mechanisms for self-renewal, cell fate, and cloning which describe by Andollo et al. and Balbach et al. Production of genetically manipulated mice by genetic manipulation of mouse ES cells is one of the premier tools for the study of genetic diseases. Matthaei describes his methods to produce these animals that have proven to be highly reliable as well as give exceptionally high rates of germline transmission with all strains of ES cells that he has used. Moreover, in just the past few years amazing progress has been made in germ cells differentiation from stem cells in vitro, which is review by Marqués-Mari et al.

Several chapters summarize the current state of knowledge in the somatic stem cell field. De Rooij reviews recent developments in the field of spermatogonial stem cells (SSCs). These cells are important for male fertility and recently it has been shown that at least mouse SSC are able to transform into multipotent stem cells capable of differentiation into various other cell lineages. Moreover, Olive and coworkers describe recent experimental results, including data from their laboratory, regarding gene expression profile of the SSC population. The chapter focuses on both up- and down-regulated protein coding transcripts and several differentially expressed microRNAs, which are increasingly being implicated in stem cell functions, such as pluripotency. In their article, Abdallah et al. describe mesenchymal stem cells, which occur in bone marrow stroma and in the stroma of diverse organs. They can give rise to, for example, osteoblasts, adipocytes, and chondrocytes and are currently being introduced into the clinic for the treatment of a variety of diseases.

Stem cells and their application in therapeutic replacement strategies are described in six articles focusing on heart failure, deafness, diabetes, and corneal injury. Stamm and coworkers summarize the basic research background of cardiac regenerative medicine and give a critical appraisal of the current efforts to translate the experimental approaches into the clinical setting. Moreover, Saric et al. critically review the current literature on use of fully undifferentiated ES cells for cardiac repair, elaborate on the tumorigenic risk of ES cells and pluripotent cells in general, and summarize strategies for elimination of this threat as an important step toward translation of ES cell–based therapies to clinic. This discussion is also highly relevant for clinical applicability of newly developed autologous ES cell–like stem cells, so-called induced pluripotent stem (iPS) cells, which circumvent ethical and, to some extent, immunological concerns linked to the use of blastocyst-derived ES cells, but still possess high tumorigenic potential. Trachoo and Rivolta review several protocols used to generate neural precursors from human ES cells, including initial attempts to establish otic placodal precursors. They discuss their potential application in the development of a new therapy for deafness.

Franceschini and coworkers describe recent experimental results, including data from their laboratory, regarding the first evidence that transplanted stem cell that migrate to the neurolfactory mucosa may contribute to neuroepithelium structure restoration with resumption of the sensorineural olfactory loss. Moreover, diabetes is a degenerative pathology that has different causes. Roche et al. summarize the key work that has been performed in the bioengineering of both ES cells and adult stem cells toward insulin-secreting cells to treat diabetes. The adult corneal epithelium is continuously regenerated from stem cells under both normal conditions as well as following injury and is located at the basal layer of the corneoscleral limbus. These stem cells simultaneously retain their capacity for self-renewal and maintain a constant cell number by giving rise to fast-dividing progenitor cells. Kolli and coworkers discuss corneal epithelial anatomy, corneal epithelial stem cell biology, and the application of this biology in the field of regenerative medicine.

Moreover, in an interesting review, Hosseinkhani and Hosseinkhani review the application of scaffolding materials together with stem cell technologies for applications in tissue regeneration. Conventional in vitro models to study differentiation of stem cells are freshly isolated cells grown in two-dimensional cultures. Clinical trails using in vitro stem cell culture can be expected only when the differentiated stem cells mimic the tissue regeneration in vivo. Therefore, the design of an in vitro three-dimensional model of biodegradable scaffolds that mimics the in vivo environment is needed to effectively study its application for regenerative medicine. Tissue engineered scaffolds have a significant effect on stem cells proliferation and differentiation. Moreover, Wolf and Mofrad describe the significance of processes that convert mechanical signals into a cascade of biochemical signals that affect the phenotype of stem cells, a process called cellular mechanotransduction. Mechanotransduction, in combination with other experimental techniques, may provide new insights into the operations that occur at the cellular level. Understanding cellular mechanotransduction can also prove useful in understanding the overall effect on biological systems resulting from a change in just a few small variables. To elucidate the particular roles that stem cells play in healing during the adult stages, a role for stem cells that is still poorly understood as compared to what is known about them in an embryonic environment, experimental approaches must combine both mechanical and biochemical observations.

Collectively, these chapters should prove a useful resource not only to those who are using or wish to use stem cells to study specific applications, but also to those interested in stem cell biology advances. We hope this book will also serve as a catalyst to spur others to use stem cells for both the fundamental understanding of stem cells and their potential utility.

I am extremely grateful to the contributors for their commitment, dedication, and promptness with submissions! I am also grateful to Dr. Hamid Gourabi, Dr. Abdolhossein Shahverdi, Dr. Ahmad Vosough Dizaj and Dr. Mohsen Gharanfoli, for having faith in and supporting me throughout this project. I wish also to acknowledge the great support provided by many at Humana Press, specifically Mindy Okura-Marszycki and Vindra Dass. A special thank you goes to my dedicated coworkers at the Department of Stem Cells who, with their tireless commitment, became a crucial factor in the editing and completion of the book. I am grateful to Zahra Maghari for her help with collecting the chapters and in follow-up.

Finally, I hope that the book will achieve the intent that I had originally imagined: that it will prove to be a book with something for both experts and novices alike, and that it will serve as a launching point for further developments in stem cells.

Hossein Baharvand Ph.D
Department of Stem Cells
Royan institute
and
Department of Developmental Biology
University of Science and Culture
Tehran, Iran

Contents

Contributors

Basem M. Abdallah
Endocrinology Research Laboratory (KMEB), Department of Endocrinology, Odense University Hospital and Medical Biotechnology Centre, University of Southern Denmark, Odense, Denmark

Sajjad Ahmad
North East Stem Cell Institute, New Castle University, Central Parkway, New Castle upon Tyne NEI 3BZ, UK

Noelia Andollo
Laboratory of Developmental Biology and Cancer, Department of Cell Biology and Histology, Faculty of Medicine and Odontology, University of the Basque Country, 48940 Leioa, Vizcaya, Spain

Marcos J. Araúzo-Bravo
Max Planck Institute for Molecular Biomedicine, Röntgenstrasse 20, 48149 Münster, Germany

Juan Aréchaga
Laboratory of Developmental Biology and Cancer, Department of Cell Biology and Histology, Faculty of Medicine and Odontology, University of the Basque Country, 48940 Leioa, Vizcaya, Spain

Maribel Arribas
Instituto de Bioingenieria, Universidad Miguel Hernandez, Alicante, Spain

Sebastian T. Balbach
Max Planck Institute for Molecular Biomedicine, Röntgenstrasse 20, 48149 Münster, Germany

Simone Bettini
Department of Evolution and Experimental Biology, University of Bologna, Via Selmi 3, 40126 Bologna, Italy

Michele Boiani
Max Planck Institute for Molecular Biomedicine, Röntgenstrasse 20, 48149 Münster, Germany

Ariff Bongso
Department of Obstetrics and Gynaecology, Yong Loo Lin School of Medicine,
National University of Singapore, Kent Ridge, Singapore 119074

M. Dolores Boyano
Laboratory of Developmental Biology and Cancer, Department of Cell Biology
and Histology, Faculty of Medicine and Odontology, University of the Basque
Country, 48940 Leioa, Vizcaya, Spain

Oliver Brüstle
Institute of Reconstructive Neurobiology, Life and Brain Centre, University of
Bonn and Hertie Foundation, Bonn, Germany

F. M. Cavaleri
Max Planck Institute for Molecular Biomedicine, Röntgenstrasse 20, 48149
Münster, Germany

Jose Bernardo Cibelli
Department of Animal Science and Physiology, Michigan State University, B270
Anthony Hall, East Lansing, MI 48824, USA
Programa Andaluz de Terapia Celular y Medicina Regenerativa, Andalucía, Spain

N. Crosetto
Institute of Biochemistry II, J. W. Goethe University Hospital, Theodor-Stern-Kai 7,
60590 Frankfurt am Main, Germany

François Cuzin
Inserm U636, 06108 Nice, France
Laboratoire de Génétique du Développement Normal et Pathologique, Université
de Nice-Sophia Antipolis, 06108 Nice, France
Equipe Labellisée Ligue Nationale Contre le Cancer, 06108 Nice, France

M. del Mar Zalduendo
Laboratory of Developmental Biology and Cancer, Department of Cell Biology
and Histology, Faculty of Medicine and Odontology, University of the Basque
Country, 48940 Leioa, Vizcaya, Spain

Dirk G. de Rooij
Center for Reproductive Medicine, AMC, 1105 AZ Amsterdam, The Netherlands
Department of Endocrinology, Utrecht University, 3584 CH Utrecht, The
Netherlands

Sigrid Eckardt
Center for Animal Transgenesis and Germ Cell Research, New Bolton Center,
University of Pennsylvania, Kennett Square, PA 19348, USA

Wolfgang Engel
Institute of Human Genetics, University of Göttingen, Göttingen, Germany

Azra Fatima
Medical Center, Institute for Neurophysiology, University of Cologne, Cologne, Germany

Francisco Figueiredo
North East Stem Cell Institute, and Department of Ophthalmology, Royal Victoria Infirmary, Queen Victoria Road, Newcastle upon Tyne NE1 4LP, UK

Chui-Yee Fong
Department of Obstetrics and Gynaecology, Yong Loo Lin School of Medicine, National University of Singapore, Kent Ridge, Singapore 119074

Valeria Franceschini
Department of Evolution and Experimental Biology, University of Bologna, Via Selmi 3, 40126 Bologna, Italy

Lukas P. Frenzel
Medical Center, Institute for Neurophysiology, University of Cologne, Cologne, Germany

Silvia Garagna
Dipartimento di Biologia Animale, Centro Interdipartimentale di Ingegneria Tissutale, Piazza Botta 9, 27100 Pavia, Italy

Luca Gentile
Max Planck Institute for Molecular Biomedicine, Röntgenstrasse 20, 48149 Münster, Germany

Tamara Glaeser
Institute of Reconstructive Neurobiology, Life and Brain Centre, University of Bonn and Hertie Foundation, Bonn, Germany

Kaomei Guan
Department of Cardiology and Pneumology, University of Göttingen, Göttingen, Germany

Manoj K. Gupta
Medical Center, Institute for Neurophysiology, University of Cologne, Cologne, Germany

Gerd Hasenfuss
Department of Cardiology and Pneumology, University of Göttingen, Göttingen, Germany

Jürgen Hescheler
Medical Center, Institute for Neurophysiology, University of Cologne, Cologne, Germany

Roland Hetzer
Deutsches Herzzentrum Berlin, Cardiothoracic Surgery, Augustenburger Platz 1, 13353 Berlin, Germany

Hossein Hosseinkhani
Department of Cardiovascular Medicine, Graduate School of Medicine, Kyoto University Hospital, Kyoto 606-8507, Japan

Mohsen Hosseinkhani
International Center for Young Scientists (ICYS), National Institute for Materials Science (NIMS), Tsukuba, Ibaraki 305-0044, Japan

Moustapha Kassem
Endocrinology Research Laboratory (KMEB), Department of Endocrinology, Odense University Hospital & Medical Biotechnology Centre, University of Southern Denmark, Odense, Denmark

Sai Kolli
North East Stem Cell Institute, and Institute of Human Genetics, Newcastle University, Central Parkway, Newcastle upon Tyne NE1 3BZ, UK
Department of Ophthalmology, Royal Victoria Infirmary, Queen Victoria Road, Newcastle upon Tyne NE1 4LP, UK

Majlinda Lako
North East Stem Cell Institute, and Institute of Human Genetics, Newcastle University, Central Parkway, Newcastle upon Tyne NE1 3BZ, UK

Jae Ho Lee
North East Institute of Stem Cell Biology, Institute of Human Genetics, International Centre for Life, Central Parkway, University of Newcastle upon Tyne, Newcastle upon Tyne NE1 3BZ, UK

Parisa Mardanpour
Department of Cardiology and Pneumology, University of Göttingen, Göttingen, Germany

Ana Isabel Marqués-Marí
Valencia Stem Cell Bank, Prince Felipe Research Center (CIPF), Valencia, Spain

Klaus I. Matthaei
Stem cell and Gene Targeting Laboratory, The Division of Molecular Bioscience, The John Curtin School of Medical Research, The Australian National University, Canberra ACT Australia 0200

K. John McLaughlin
Center for Animal Transgenesis and Germ Cell Research, New Bolton Center, University of Pennsylvania, Kennett Square, PA 19348, USA

José Vicente Medrano
Valencia Stem Cell Bank, Prince Felipe Research Center (CIPF), Valencia, Spain

Mohammad R.K. Mofrad
Molecular Cell Biomechanics Laboratory, Department of Bioengineering, University of California, Berkeley, CA 94720, USA

Boris Nasseri
Deutsches Herzzentrum Berlin, Cardiothoracic Surgery, Augustenburger Platz 1,
13353 Berlin, Germany

Karim Nayernia
North East Institute of Stem Cell Biology, Institute of Human genetics,
International Centre for Life, Central Parkway, University of Newcastle upon
Tyne, Newcastle upon Tyne NE1 3BZ, UK

Jessica Nolte
Institute of Human Genetics, University of Göttingen, Göttingen, Germany

Virginie Olive
Inserm U636, 06108 Nice, France
Laboratoire de Génétique du Développement Normal et Pathologique, Université
de Nice-Sophia Antipolis, 06108 Nice, France
Equipe Labellisée Ligue Nationale Contre le Cancer, 06108 Nice, France

Beatriz Paredes
Instituto de Bioingenieria, Universidad Miguel Hernandez, Alicante, Spain

Neli Petrova Ragina
Department of Animal Science, Michigan State University, B270 Anthony Hall,
East Lansing, MI 48824, USA

Minoo Rassoulzadegan
Inserm U636, 06108 Nice, France
Laboratoire de Génétique du Développement Normal et Pathologique, Université
de Nice-Sophia Antipolis, 06108 Nice, France
Equipe Labellisée Ligue Nationale Contre le Cancer, 06108 Nice, France

Carlo Alberto Redi
Fondazione IRCCS Policlinico San Matteo, 27100 Pavia, Italy

Roberto P. Revoltella
Foundation "Stem Cells and Life" onlus, Via Zamenhof 8, 56127 Pisa, Italy

Marcelo N. Rivolta
Centre for Stem Cell Biology and Department of Biomedical Sciences, University
of Sheffield, Sheffield, UK

Enrique Roche
Instituto de Bioingenieria, Universidad Miguel Hernandez, Alicante, Spain

Riccardo Saccardi
Bone Marrow Transplantation Center, Department of Hematology, University
Hospital of Careggi, Via Morgagni 85, 50134 Firenze, Italy

Hamid Saeed
Endocrinology Research Laboratory (KMEB), Department of Endocrinology, Odense University Hospital and Medical Biotechnology Centre, University of Southern Denmark, Odense, Denmark

Tomo Saric
Medical Center, Institute for Neurophysiology, University of Cologne, Cologne, Germany

Hans R. Schöler
Max Planck Institute for Molecular Biomedicine, Röntgenstrasse 20, 48149 Münster, Germany

Carlos Simón
Valencia Stem Cell Bank, Prince Felipe Research Center (CIPF), and Fundación IVI, University of Valencia, Valencia, Spain

Christof Stamm
Deutsches Herzzentrum Berlin, Cardiothoracic Surgery, Augustenburger Platz 1, 13353 Berlin, Germany

Objoon Trachoo
Centre for Stem Cell Biology and Department of Biomedical Sciences, University of Sheffield, Sheffield, UK

Nestor Vicente-Salar
Instituto de Bioingenieria, Universidad Miguel Hernandez, Alicante, Spain

Christopher B. Wolf
Molecular Cell Biomechanics Laboratory, Department of Bioengineering, University of California, Berkeley, CA 94720, USA

Gerald Wulf
Department of Haematology and Oncology, University of Göttingen, 37073 Göttingen, Germany

Maurizio Zuccotti
Dipartimento di Medicina Sperimentale, Sezione di Istologia ed Embriologia, Università degli Studi di Parma, Via Volturno 39, 43100 Parma, Italy

Human Embryonic Stem Cells: Their Nature, Properties, and Uses

Ariff Bongso and Chui-Yee Fong

Abstract The ability to grow human embryos in vitro to the blastocyst stage via coculture or sequential culture media led to the isolation and growth of human embryonic stem cells (hESCs) from blastocysts left over from in vitro fertilization programs. These cells being pluripotent can be differentiated into almost all the tissues types of the human body and therefore offers promise in the treatment of a variety of incurable diseases by transplantation therapy. They also provide an ideal screening tool for potential drugs in the pharmaceutical industry and allow the study of early human development and infant cancers. Although all National Institutes of Health (NIH)–registered hESC lines are research grade, having been derived and propagated on xenosupports and with xenoproteins, clinical grade hESC lines derived and propagated in xenofree culture conditions and under current good manufacturing practices (cGMP) facilities are now available. hESCs have been differentiated in vitro into pancreatic islets, neurons, and cardiomyocytes, and transfer of such hESC-derived tissues into diseased animal models have shown successful engraftment. However, two hurdles are delaying hESC-derived cell therapy reaching human clinical trials: (1) possible immunorejection of hESC-derived tissues and (2) the concern of teratoma formation. To overcome immunorejection, attempts are being made to customize tissues for patients via nuclear transfer and other reprogramming methods. Recently, human skin fibroblasts were reprogrammed to the pluripotent embryonic state by transfection with four genes (induced pluripotent stem cells). This approach not only allows tissue customization but is also an embryo-free method that overcomes ethical sensitivities. The development of several hESC banks worldwide containing a diverse range of clinical grade hESC lines that could be HLA typed and tissue matched for treatment is also a practical approach to preventing immunorejection. Several approaches to eliminating teratoma formation from undifferentiated renegade hESCs residing in transferred hESC-derived tissues are in progress. It is hoped that this hurdle will be circumvented soon, then allowing the application of current successful animal validated transplantation studies in the human.

A. Bongso (✉) and C.-Y. Fong
Department of Obstetrics and Gynaecology, Yong Loo Lin School of Medicine,
National University of Singapore, Kent Ridge, Singapore 119074
e-mail: obgbongs@nus.edu.sg

H. Baharvand (ed.) *Trends in Stem Cell Biology and Technology*,
DOI 10.1007/ 978-1-60327-905-5_1,
© Humana Press, a Part of Springer Science+Business Media, LLC 2009

1

Keywords Embryonic stem cells • Derivation • Culture • Differentiation

Introduction

Certain lower animals possess remarkable regenerative qualities that have fasci-
nated man for many years. Regeneration is a physiological process where lost body
parts are replaced over time with no human intervention. The household gecko can
drop its tail at will to protect itself from its predator and within days the remaining
tissues at the base of the tail can organize itself to reproduce the missing body part *(1)*.
Skinks from the Outback of Australia drop their tails, and a barrage of tails are
reproduced instead of one. If the head of a flatworm is removed, a completely new
head will be formed, and if a flatworm is cut into ten pieces, each cut piece can
produce a completely new flatworm. Interestingly, although the molecular machinery
for such regenerative abilities is present in mammals, they have lost the regenerative powers in major organs except for the liver. However, nature has provided two
tradeoffs for this loss of regenerative power: (1) efficient wound healing and (2) the
presence of a very mysterious cell, the stem cell, which gets involved in repair during
tissue injury. Stem cells behave as blank slates that can not only assist in immediate
repair but can also differentiate into a variety of cell types, each with its own functions.
It is known today that most tissue repair events in mammals are dedifferentiation-
independent events resulting from activation of preexisting stem or progenitor
cells *(2)*.

Today the field of stem cell biology has gained tremendous importance and has
drawn a lot of publicity, with several reports showing the promise of this science in the
future cure of a variety of diseases by transplantation therapy. The successful translation of this science from bench to bedside will change the quality of life of millions of
people worldwide who suffer from illnesses where current approaches in medicine
have not been able to take full control. The field of human embryonic stem cell (hESC)
biology unfortunately is fraught with many ethical controversies as human embryos
need to be destroyed to derive such cells. Additionally, since the transplanted hESC-
derived tissues originally come from donor embryos, there is concern of immunorejection, so customization of tissues to the sick patient by reprogramming the patient's own
cells is necessary to circumvent this and this in itself involves the equally sensitive area
of nuclear transfer (NT) or therapeutic cloning *(3)*. However, rapid progress is being
made in this field and hopefully such issues will be circumvented.

Definition of a Stem Cell

The term "stem cell" actually originated from botanical monographs where the
word "stem" was used for cells in the apical meristem, which is responsible for the
continued growth of plants *(4)*. In mammals, given the vast variety of stem cells
isolated from preimplantation embryos, the fetus, umbilical cord, and adult organs,
it becomes necessary to provide a more general definition for the term "stem cell"

and a more specific definition based on the type of stem cell. In general, stem cells can be defined as specialized or undifferentiated cells that can self-renew and be differentiated into other cell types, each new cell type possessing a different function *(5)*. The degree of differentiation of stem cells to various other tissue types varies with the different types of stem cells, and this phenomenon is referred to as plasticity. Plasticity can range from totipotency to pluripotency to multipotency to unipotency. Mammalian blastomeres from early cleaving embryos are considered totipotent as they have the potential to produce complete organisms, while embryonic stem cells are considered pluripotent as they can differentiate into almost all 210 tissue types of the mammalian body but cannot produce a whole individual. Multipotency is restricted to those mesenchymal stem cell types that can differentiate into a small variety of tissues, while unipotency is generally restricted to stem cell sources that can be differentiated only into one lineage *(2)*.

More recently, hESCs have been defined more specifically as cells that must have the following properties: the ability to (1) self-renew, (2) differentiate into cells of all three primordial germ layers (ectoderm, mesoderm, and endoderm), and (3) pass through a full battery of stem cell characterization tests, such as (a) morphological characteristics, (b) surface marker antigens (e.g., SSEA-1, -3, -4; Tra-1-60, -80), (c) Oct-3 and -4, (d) alkaline phosphatase, (e) karyotype, (f) genomic markers for the three primordial germ layers, (f) telomerase, and (g) the production of teratomas in severely combined immunodeficient (SCID) mice. Such a definition becomes necessary for (1) registration of hESC lines on the National Institutes of Health (NIH) registry, (2) exchange of cell lines between institutions, and (3) the reliable differentiation to tissue types given the fact that several differences exist between derived hESC lines *(6)*, which may be intrinsic, based on the quality of embryos used or due to different derivation protocols. In fact, transcriptome profiling has clearly illustrated several properties that are common to all hESCs at the molecular level, but certain gene differences do exist between some cell lines *(7, 8)*. Such essential attributes of "stemness" have been proposed in detail by Ramalho-Santos et al. *(9)*.

Classification of Stem Cells

Stem cells in the human can be classified into many types based on their source of origin. More recently, they have been classified based on the presence or absence of a battery of CD and embryonic stem cell (ESC) markers (Fig. 1). The male and female gonads contain stem cells referred to as spermatogonia and oogonia, respectively. Through their self-renewal and subsequent meiosis they are responsible in producing the cells of the germ line and eventually spermatozoa and oocytes. These two haploid gametes eventually fertilize to establish diploidy and produce the zygote. The zygote remains at the top of the hierarchical stem cell tree, being the most primitive cell, and the germ cells therefore possess the unique feature of developmental totipotency *(10, 11)*. The zygote undergoes cleavage in the human through a period of 5–6 days, producing two to four blastomeres (two- to four-cell stage) on day 2, eight blastomeres (eight-cell stage) on day 3, fusing or completely fused blastomeres (compacting or compacted stage) on day 4, and blastocyst stages

STEM CELLS

	hHSC (Blood, Bone marrow)	hMSC (Bone marrow, Blood, Organs)	hESC (Embryos)
Stem cell markers:	CD ++ ESC −	CD ++ ESC −	CD − ESC ++
Uses:	Tissue transplantation	Tissue transplantation	Tissue transplantation Pharmaceutical screening Human development
Plasticity:	Unipotent	Multipotent	Pluripotent
Teratomas:	No	No	Yes
Rejection:	No	Yes	? (Customize)
Cell lines:	No	No	Yes
Scaling up:	+	++	+
Telomerase:	−	−	++

Other embryonic stem cell types (epiblast)

Amniotic membrane: CD++, No terat, multi
Umbilical cord blood: CD++, No terat, multi
Umbilical cord matrix: CD++, No terat, multi

Amniotic fluid: CD++, ESC+, No terat, multi, telo+
Abortus (<12 wks): CD++, ESC+, HLA -, multi, telo+
Wharton's jelly: CD++, ESC+, No terat, multi, telo+
CBE: CD++, ESC+, No terat, multi, telo+

Testis/Ovary ??
Endometrium ??

Fig. 1 Classification of stem cells according to characteristics, uses, and plasticity. Note (in *lower part* of figure) that besides embryonic stem cells several other stem cells of the reproductive system that possess characteristics in between adult and embryonic stem cells have been isolated, opening a new area of reproductive stem cell science. *Abbreviations: hHSC* human hematopoietic stem cells; *hMSC* human mesenchymal stem cells; *hESC* human embryonic stem cells

on days 5 and 6 *(12)*. Each of the blastomeres is considered totipotent because it has the potential to produce a complete organism, as demonstrated when blastomeres are placed in the uterus of rabbits or mice. In the strictest sense of the definition of a stem cell, however, such blastomeres cannot be called stem cells because they do not self-renew.

The first bona fide stem cell to be produced in the mammal is in the inner cell mass (ICM) of the 5-day-old blastocyst. These cells self-renew and eventually produce two cell layers: the hypoblast and epiblast. The hypoblast generates the yolk sac, which degenerates in the human, and the epiblast produces the three primordial germ layers (ectoderm, mesoderm, and endoderm). These germ layers produce all the various tissues of the organism. Transmission electron microscopy studies have shown in the 9-day-old human embryo the transition of ICM to hESCs *(13)*. Thus hESCs are considered pluripotent and not totipotent because they cannot produce complete human beings but have the potential to produce all the 210 tissues of the human body. During embryogenesis and fetal growth such embryonic stem cells that have not participated in organogenesis remain as adult stem cells in organs during adulthood. It can thus be hypothesized that the adult stem cells residing in specific organs are already differentiated cells, and their function is to be dedifferentiated and be recruited for repair of injury incurred by the specific organ. Unfortunately, such adult stem cells in the organs are few in number and inadequate to complete repair with a subsequent breakdown of disease of that specific organ. It has been shown that fetal and adult stem cells could cross boundaries by transdifferentiating into other tissue types and are thus referred to as multipotent *(14–17)*. Those stem cells that are unable to transdifferentiate but differentiate into one specific lineage are referred to as unipotent. An example of such unipotency is the differentiation of bone marrow hematopoietic stem cells to blood. Thus as embryogenesis shifts to organogenesis, infancy, and then adulthood, stem cell plasticity shifts from pluripotency to multipotency.

Recently there has been tremendous interest in the derivation of stem cells from other embryonic tissues that arise from the epiblast, such as the amniotic membrane, amniotic fluid, and umbilical cord *(18, 19)*. The amniotic membrane, amniotic fluid, and some stem cell types in the umbilical cord possess both CD and some ESC markers, and although considered multipotent, some of them have certain properties in between pluripotency and multipotency and as such are useful cells for transplantation therapy *(19)*. The umbilical cord, for example, has three types of stem cells: (1) in cord blood, (2) in the Wharton's jelly (Fig. 2), and (3) in the perivascular matrix around the umbilical blood vessels within the cord itself *(20)*.

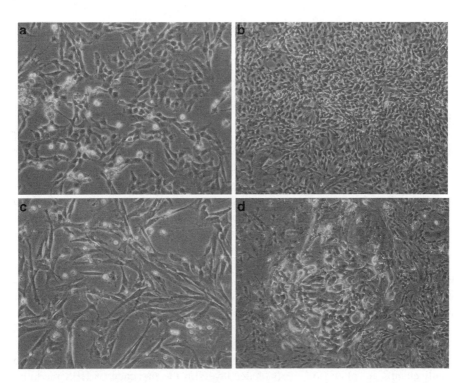

Fig. 2 Human Wharton's jelly stem cells (hWJSC). (**a**) Primary culture of hWJSC grown in the presence of human embryonic stem cell (HES) culture medium showing epitheliod-like cell growth. (**b**) Same culture maintaining epitheliod-like morphology when confluent after 7 days. (**c**) Primary culture of hWJSC grown in umbilical cord matrix stem cell (UCMSC) medium showing fibroblast-like morphology. (**d**) hWJSCs showing human embryonic stem cells (hESC)-like colony formation when grown in HES medium

Embryonic stem cells have the advantages of possessing pluripotent markers, producing increased levels of telomerase, and being coaxed into a whole battery of tissue types and thus remain as the hallmark of stem cell biology with the greatest potential for cell-based therapy. However, they have the disadvantages of potential teratoma production, their derived tissues have to be customized to patients to prevent immunorejection, and their numbers have to be scaled up in vitro for clinical application. Adult bone marrow stem cells and stem cells from the Wharton's jelly have the advantages of availability in large numbers and do not produce teratomas, but have the limitations of being multipotent or unipotent and yield low levels of telomerase.

Genuine hESCs have the following characteristics: (1) self-renewal in an undifferentiated state for very long periods of time with continued release of large amounts of telomerase, (2) maintenance of "stemness" or pluripotent markers, (3) teratoma formation in SCID mice that contains tissues from all three primordial germ layers, (4) maintenance of a normal stable karyotype, (5) clonality, (6) OCT-4, and other genomic (e.g., NANOG) expression, and (7) ability to produce chimeras when injected into blastocysts in the mouse model. Many of the multipotent stem cells from fetal, cord, and adult tissues that are usually positive for only CD markers are mesenchymal stem cells (MSCs).

Derivation and Propagation of hESCs

The ability to grow human embryos to the day-5 blastocyst stage (blastocyst culture) in in vitro fertilization (IVF) programs *(21)* set the stage for the first isolation of hESCs *(22, 23)*. Thereafter, several methods of hESC derivation have been reported with success *(24, 25)*. These include (1) the whole embryo method, (2) immunosurgery, and (3) mechanical separation of the ICM. In the whole embryo method *(23)*, the zona pellucida of the blastocyst is first removed by enzymatic treatment with a protease (pronase). The zona-free blastocyst is then plated on mouse or human fibroblast feeder layers previously treated with mitomycin C or irradiated to stop their own growth. The culture medium (human embryonic stem cell [HES] medium) used is a mixture of Dulbecco's modified Eagle's (DMEM) medium, supplemented with fetal calf serum, human serum or a knockout (KO) serum supplemented with basic fibroblast growth factor (bFGF), mercaptoethanol and insulin–transferrin–selenium (ITS) supplement. After approximately 2 weeks, the ICM grows as a raised clump of cells on the feeder layer, while the peripheral trophectodermal (TE) cells spread out as a patch of large cells. The ICM clump is then carefully dissected out with fine needles, disaggregated into smaller clumps of cells, and the small clumps are plated on fresh feeder layers in the presence of HES medium. After about 7–8 days, each cluster forms a small colony of hESCs and each hESC has high nuclear-cytoplasmic ratios with prominent nucleoli. Once the colonies reach a reasonably sized diameter, they are exposed to a brief treatment of dispase for slight detachment from the feeder, then dissected into smaller pieces, and each

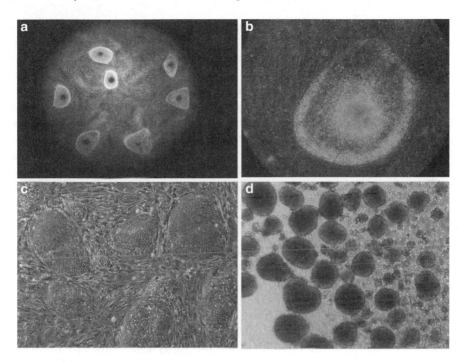

Fig. 3 (**a**) Stereo microphotograph showing hESC colonies growing on human feeder cells (fetal skin). Note *oblong thin flat* colonies with minimal differentiation. (**b**) Large human embryonic stem cells (hESC) colony growing on murine embryonic fibroblast (MEF) feeder. Note *circular* shaped colony with thick edges and slight differentiation at center and periphery of colony. (**c**) hESC colonies on MEFs grown by the enzymatic bulk culture method. Note many colonies of different sizes. (**d**) hESC embryoid bodies (hESC-EB) of different sizes

piece is then plated onto fresh feeders. Dissection is carefully carried out to avoid transfer of any differentiated hESCs present at the center and periphery of the colonies. This method is commonly referred to as the "cut and paste" method (Fig. 3a, b).

hESCs could also be derived from the zona-free blastocyst by first separating the ICM from the TE and culturing only the ICM. The ICM can be separated from the TE either mechanically with pointed needles or by immunosurgery. The immunosurgery approach is more efficient and reliable *(24, 25)*. For immunosurgery, the zona is first removed with pronase and the zona-free blastocyst is exposed to anti-human antibodies in the presence of guinea pig complement and monitored under phase contrast optics to observe the gradual lysis of the TE cells keeping the ICM intact. The ICM is then washed in DMEM medium to remove the antibodies and complement, and then seeded onto fresh inactivated mitomycin-C murine embryonic fibroblast (MEFs) or human feeders in the presence of HES medium and incubated at 37°C in a 5% CO_2 atmosphere. The rest of the protocol is the same as the cut and paste method described above. hESC numbers can be scaled up to some extent using the enzymatic bulk culture method (Fig. 3c). In this method at the time of passaging the colonies and feeder cells are enzymatically treated and seeded together onto fresh feeders. Several small, medium, and large colonies sprout up from clumps of hESCs that attach to the new feeder.

To avoid any risk of contamination to the hESC from the animal xenosupport system (MEFs) and xenoproteins (guinea-pig complement, porcine transferrin, bovine insulin), xenofree protocols using sterile human feeders and recombinant or human-based reagents have been reported for derivation and propagation of hESCs *(26–28)*. Feeder-free protocols using Matrigel have recently replaced feeder cells for propagation of hESCs but not used for hESC derivation *(29)*. Based on the xenofree protocol, the first six proprietary clinical grade hESC lines developed under current good manufacturing practices (cGMP) conditions were produced by ESI Singapore (http://www.escellinternational.com). Xenofree cryopreservation methods (vitrification or slow programmable freezing) using human-based freezing reagents have also been reported to allow storage of hESCs in closed embryo straws in the vapor phase of liquid nitrogen (−180°C) *(30)*.

Donation of Embryos

Embryos that are used for derivation of hESCs are usually those in surplus that are donated by informed consent from IVF patients. Such patients have a choice of (1) donating their spare embryos to other childless couples, (2) disposal, or (3) donating for research with informed consent. Interestingly, some centers have also provided IVF patients a fourth choice of using their spare frozen embryos to derive and store hESC lines for their own use later on. This approach which may be ethically controversial is being practiced in some states in the USA (http://www.stemlifeline.com). It is important to note that a vast number of surplus IVF embryos are currently frozen in centers worldwide, and if consent can be sought to derive hESC lines from these embryos, repositories containing large numbers of hESC lines can be established on a diverse ethnic basis, which may be useful later for HLA tissue matching for transplantation therapy. Currently, there are only two major hESC line repositories: one based with the NIH in the United States and the Medical Research Council (MRC) bank in the United Kingdom. Tremendous differences appear to occur between the currently stored hESC lines in terms of gene differences and method of derivation *(6)*, and this is all the more reason that many more hESC lines must be derived and stored. It has been estimated that about 150 hESC lines may give a reasonably good tissue match to avoid hESC-derived tissue rejection, although some workers claim that many more cell lines may be required *(31)*.

Embryoid Body Formation

hESCs have the unique ability of producing embryoid bodies (EBs) (Fig. 3d). These are circular sphere-like structures that contain cells from all three primordial germ layers. EBs are usually produced by the hanging drop method, where hESC clusters are discouraged from attaching to specially coated plastic culture plates in the presence of special ingredients in the culture medium, thus allowing them to round up to form the spherical EBs. EB production is usually the first step in attempting to differentiate hESC into desirable cell lineages for transplantation therapy. The spontaneous development of hESCs to EBs in vitro and then mechanical separation and enrichment of neuronal cells from the rest of the lineages within the EB without the need for the use of differentiating agents has been successfully carried out *(25)*.

Applications of hESCs

hESCs have many applications in human medicine. The first is the production of hESC-derived tissues for transplantation therapy for the treatment of a variety of incurable diseases. A list of the potential diseases that may be treated with hESC-derived tissues is shown in Table 1. Future interaction of such tissues with scaffolds made from polymers may find their use in organ transplantation. After transplantation of the scaffolded tissue, the scaffold could be broken down in vivo with depolymerizing agents, allowing the tissue to take the place of the organ. This has opened a whole new field referred to as tissue engineering, involving both bioengineers and medical researchers. Second, hESCs and their derived tissues also serve as useful tools in the screening of potential drugs for the pharmaceutical industry. For example, a potentially new heart drug could be tested on a hESC-derived cardiomyocyte cell line in vitro. Currently, pharmaceutical companies use animal cell lines for drug testing, which gives tremendous variability in the results in response to different drugs. Third, hESCs can be used to study early human development (e.g., congenital anomalies) and the pathogenesis of infant cancers. Fourth, hESCs can serve as ideal vehicles for gene therapy.

Scientific Hurdles to hESC Application

Tremendous progress has been made thus far in the conversion of hESC into desirable cell lineages. Additionally, successful functional outcome has also been demonstrated when such hESC-derived tissues are transplanted into animal models.

Table 1 Some potential diseases treatable with human embryonic stem cells–derived tissues

Human embryonic stem cells–derived tissue	Disease
Cardiomyocytes	Myocardial infarction
Neuronal cells	Parkinson's, Alzheimer's, spinal cord injuries
Pancreatic islets	Diabetes
Keratinocytes	Burns, cosmetic surgery
Hepatocytes	Cirrhosis, hepatitis
Bone, cartilage	Cartilage injuries, osteoarthritis
Blood	Leukemias, thalassemias
Skeletal muscle	Muscular dystrophy
Retinal, corneal cells	Macular degeneration, corneal diseases

The transplanted tissues engraft successfully, enter the in vivo stem cell niche, integrate with the host microenvironment, and improve cell function of malformed organs or tissues (32–34). However, there are still some obstacles that are delaying taking hESC-derived tissues to human clinical trials. These obstacles are: (1) the fear that such transplanted tissues may be rejected as they originate from donor embryos, (2) the concern that any residual rogue undifferentiated pluripotent hESCs in the hESC-derived tissue (after the differentiation process) may produce teratomas at the transplanted site if the cells are injected directly into the site or in extratransplanted sites if the cells are administered systemically, and (3) the number of cells available for treatment are inadequate, and methods to scale up cell numbers rapidly are urgently required.

To overcome the issue of immunorejection several approaches are being investigated. Many laboratories are attempting to customize hESC-derived tissues to patients by NT. This involves electrofusing the somatic nucleus of the patient, requiring tissue customization with an enucleated donor human oocyte. The fused product undergoes cleavage to yield a blastocyst from which customized hESCs and hESC-derived tissues for the patient could be derived and propagated (Fig. 4). Thus far this approach has not been successful in the production of cloned human embryos but has recently

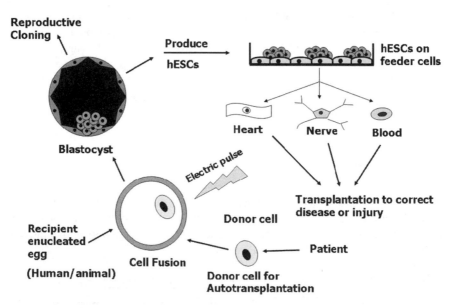

Fig. 4 Cartoon illustrating the customization of tissues by the method of nuclear transfer (NT) or therapeutic cloning. A somatic cell nucleus from the sick patient is inserted into an enucleated human or animal egg, the product then electrofused to produce a blastocyst. Such blastocysts may have two uses: (**a**) the production of a reproductive clone if placed in a surrogate mother or (**b**) the derivation of human embryonic stem cells (hESC) lines from which customized tissues for the patient could be generated because the genome of such hESC-derived tissues would be the same as the donor nucleus of the same patient that generated the blastocyst. Theoretically, when such customized tissues are transplanted back to the patient, immunorejection could be avoided

been successful for the nonhuman primate. Cattle, sheep, and other domestic animal embryos are routinely produced by the NT method, and the birth of live offspring from such NT embryos has helped increase the genetic merit of livestock industries. There have been recent claims that the use of rabbit or bovine oocytes to reprogram human somatic cells by the NT method may be more practical, given the paucity of human oocytes. Although ethically sensitive, the United Kingdom recently approved research on human–animal chimeras for this purpose. Several scientific hurdles need to be overcome before NT becomes a routine useful approach to customize hESC-derived tissues for sick patients (Table 2).

Other embryo-free and reprogramming methods have also been successful (Table 3). Recently, fetal and adult somatic skin fibroblasts were reprogrammed to the embryonic state by transfection with four pluripotent genes (induced pluripotent stem cells [iPSCs]) *(35–37)*. The ensuing cell lines were confirmed as pluripotent and were similar to hESC lines derived from surplus embryos. In one of these reports, an adult patient's skin fibroblasts were reprogrammed to produce iPSCs *(35)*, demonstrating that iPSC-derived tissues could thus be customized to a specific patient, preventing immunorejection. The iPSC approach has the added advantage of not requiring embryos to derive hESC lines, bypassing the ethical sensitivities of using surplus IVF embryos and creating embryos by NT *(38)*.

It has also been shown that lymphocytes and other somatic cell types can be reprogrammed by whole intact hESCs, hESC karyoplasts, and hESC cytoplasts by chemical fusion. It was claimed that the hESC cytoplast has powerful reprogramming powers similar to the ooplasm of oocytes ("stembrids") *(39–41)*. The production of parthenogenetic hESC lines that are pluripotent has also been successful for

Table 2　Problems with nuclear transfer

Parthenogenesis?
Efficiency: 0.57–6%
Faulty faithful epigenetic reprogramming
Eggs needed for each patient to customize human embryonic stem cells (hESCs)
Availability of human eggs?
Animal eggs useful, but unethical?
Disharmony between nuclear and mitochondrial genes
Meiotic spindle retention after egg enucleation: Implications!
Disarrayed mitotic spindles after nuclear transfer
Misaligned chromosomes after nuclear transfer
Aneuploid embryos after nuclear transfer
Stricter molecular requirements for mitotic spindle assembly in primate nuclear transfer

Table 3　Methods to derive human embryonic stem cells (hESC) without destroying embryos

From single blastomeres via blastomere biopsy for preimplantation genetic diagnosis (PGD)
Reprogramming adult fibroblasts to embryonic state by ectopic expression of transcription factors (POU5F1, SOX2, c-MYC, KLF4, LIN28)
Reprogramming adult fibroblasts with hESC karyoplast and cytoplasts using cell fusion
hESC chromosome transfer into arrested zygotes

Table 4 Alternative approaches to preventing rejection

Producing panels of xeno-free human embryonic stem cells (hESC) lines from surplus IVF
    embryos for tissue matching (~450 hESC lines)
Modifying the histocompatability locus: universal donor hESC lines
Encapsulating hESC derived cells with immunoprivileged membranes
hESCs being embryonic: will they be immunoprivileged?

Table 5 Prevention of teratoma formation

Purification and lineage selection
FACS (flow cytometric fluorescence-activated cell sorting)
MACS (magnetic-activated cell sorting)
Directed expression of suicide or apoptosis-controlling genes in graft tissues (ceramide analogues
    prevent hESC-induced teratoma formation)
Alginate encapsulated cell delivery
Selection against undifferentiated hESCs by cytotoxic antibodies
Separation of undifferentiated hESCs from hESC-derived cells using discontinuous density gradients

customizing hESC-derived tissues for the females only *(42)*. However, the most
logical and easiest approach to preventing immunorejection may be the development of several stem cell banks worldwide containing large numbers of fully characterized diverse clinical grade hESC lines derived from surplus IVF embryos that are
HLA typed that can then be tissue matched for treatment (Table 4).

Several approaches are being attempted to eliminate the concern of teratoma
production in transplanted hESC-derived tissues *(43)* (Table 5). These include
methods to separate residual rogue undifferentiated hESCs present in differentiated
tissues by flow cytometric fluorescence-activated cell sorting or magnetic-activated
cell sorting (FACS/MACS), the use of density gradients, the selective induction of
apoptosis in such residual hESCs, the encapsulation of hESC-derived tissues with
immunoprivileged membranes at the time of delivery *(44)*, and the administration
of hESC conditioned medium (hESC-CM) or hESC extracts (membrane disrupted
cells) rather than administration of whole intact hESC-derived cells *(45)*. It is also
definitely not known whether the injection of differentiated tissues containing some
renegade undifferentiated hESCs will actually induce teratomas in specific transplanted sites besides the hind limb and kidney capsule where such teratomas have
been demonstrated in animal models after injection of clusters of hESCs only.

Currently, hESC can be grown in bulk by the enzymatic culture method. However,
cell numbers are still inadequate to provide for patient treatment as it is estimated
that at least one to five million hESC-derived cells need to be administered at each
site for successful functional outcome. Also, repeated injections may be necessary.
As such, methods are being investigated (e.g., the use of bioreactors) in attempting
to scale up numbers for future treatment. An alternate approach to cell-based therapy
would be to only prime differentiation of hESCs in vitro along a specific lineage for
about 36–48 h and then inject the primed hESC-derived cells rather than inject
terminally differentiated tissues. It is hoped that the host's damaged organ itself or

its stem cell niche will trigger the continuation of the differentiation process of the transplanted hESC-derived cells in vivo along the injured organ's lineage, thus increasing cell numbers to help in repair. Additionally, it is also possible that the mere presence of the transplanted primed hESC-derived cells in the diseased organ will also help to mobilize important growth factors from the transplanted or extratransplanted sites to the injured site to assist in repair via a paracrine effect. The delivery of hESC-CM or hESC extracts may also provide improved functional outcome via similar mechanisms.

Strategies for Differentiation of hESCs Along Desirable Cell Pathways

Differentiation is a biological phenomenon where an unspecialized cell acquires the properties of a specialized cell. For example, in vivo, bone marrow stem cells differentiate into blood. Differentiation in vitro can either be spontaneous or controlled. In high density culture, hESCs differentiate spontaneously into cells of the three primordial germ layers with preference via a default pathway into neuronal cells (25). The desired differentiated cell type can be mechanically separated, enriched, and a pure culture of that specific cell type obtained. Neuronal cells secreting dopamine and serotonin in vitro have been produced in this way (32, 46). Controlled differentiation can be achieved in three ways. The first way is the coculture of hESCs with companion cells (preferably fetal). The companion cells release certain factors that entice the hESCs to differentiate along a desired lineage. For example, hESCs have been cocultured in direct contact with visceral endodermal cells in vitro, and within 10–14 days the hESCs were differentiated into beating cardiomyocytes (33). Second, certain growth factors and biochemical agents can be added into the culture medium that bathes the hESCs, helping them to differentiate. For example, retinoic acid is well known to differentiate hESCs into neurons, and dimethylsulfoxide (DMSO) differentiates hESCs into bone (47). The third way is transfection of the hESCs with specific gene constructs that can induce differentiation along a desired lineage. The cardiomyosin gene, when transfected into murine embryonic stem cells (mESCs), can convert mESCs into functional cardiomyocytes (48). Either undifferentiated hESCs or hESC-derived EBs could be used for differentiation by the various methods outlined above.

Current State of the Art with Respect to Clinical Application of hESC-Derived Tissues

Tremendous progress has been made in the field of hESC biology, although the tissues produced by these cells have not been used to date clinically in patients. hESCs have been successful differentiated into cardiomyocytes in vitro using a visceral endodermal coculture protocol and then transplanted into ischemic mice whose coronary arteries were either tied or occluded by freezing. The transplanted cardiomyocytes were tracked by a reporter gene construct linked to it (Lac Z or GFP protein) and engraftment confirmed by the integration of the colored tagged cells. Cardiac function was also compared against controls by patch-clamp electro-physiology, echocardiography, magnetic resonance imaging, and the use of tachy-cardiac and bradycardiac drugs in the mouse model. Similarly, insulin-secreting pancreatic islets have been generated by transfecting hESCs with the PDX-1 gene construct and/or growth factors and when transplanted into diabetic mice glycemic levels have been controlled. hESC-derived neuronal cells have been transplanted into the brains of parkinsonian mice, with successful migration and integration of the transplanted neurons into the circuitry of the brain *(32)*. hESCs have also been successfully differentiated into other tissues such as bone *(47)*. However, the appli-cation of all these approaches in human trials is being held back until the issues of teratoma formation and immunorejection are resolved.

Unanswered Questions when Translating hESC-Derived Tissues to the Clinic

Several issues remain unanswered when attempting to take hESC-derived tissues to the clinic (Table 6). Currently, doctors administer fluids (injections), solids (pills), or carry out surgical intervention to correct disease. Regenerative medicine hopes to bring a new dimension to such treatment by using a cell-based therapy involving the use of autologous or differentiated donor stem cells. As a result several safety measures need to be taken because a whole cell is administered and a variety of impurities may be administered with it, both from within the cell and outside the cell. The cells must be clinically compliant and generated under cGMP or (cGTCP) current good tissue culture

Table 6 Unanswered questions

Targeted
    Delivery
    Homing
In vivo somatic stem cell niche
    How do niche cells influence stem cell fate decisions?
Dosage and long-term stability
Side effects
In vivo function
Safety
What should we be transplanting?
    Undifferentiated stem cells?
    Terminally differentiated tissues/short induction of differentiation?
    Progenitor cells?
    Conditioned medium?

practice conditions using xenofree protocols that are preferably U.S. Food and Drug Administration (FDA) approved to prevent the risk of transmission of adventitious agents and rogue undifferentiated hESCs that may induce teratomas. Cell-based therapy requires the availability of large numbers of cells. Thus large-scale cell production strategies utilizing bioreactors and perfusing systems need to be developed *(49)*.

It is still not known what the best route of administration is and how many cells are required for each administration. The frequency of administration has also not been worked out with both autologous and donor stem cells and their derived tissues. Direct cell injections into the malfunctioned organ would be preferred to peripheral or portal vein administration to prevent the cells homing in unwanted sites, thus inducing cancers. The administration of hESC-derived progenitor cells, hESC-derived conditioned medium, or hESC extracts may prevent teratoma production and improve function by making use of the host's stem cell niche to trigger normal tissue development rather than the use of hESC-derived whole cells that has its own associated problems. The stem cell niche is a combination of cells and extracellular matrix components in the host's local tissue environment that govern cell behavior. Stem cell niches have been identified in the pericryptal myofibroblasts at the base of the crypts in the intestine, in the bulge beneath the sebaceous glands of hair follicles, and in other locations in the human body *(50)*.

Long-term in vivo functional outcome after hESC-derived tissue transplantation also needs to be properly worked out. Validation of such information using the non–human primate model rather than the mouse or rat model would be the ideal approach. The more complicated the organ such as the brain, the greater is the likelihood of the presence of many progenitor cell types for different parts of the brain, and hence the more difficult to generate a single specific hESC-derived tissue for treatment. Preparing insulin-secreting islets from hESCs appears to be less challenging given the size of the pancreas and limited number of cell types. Future treatments for the variety of incurable diseases in the human via hESC-directed stem cell therapy are expected in the short term for diabetes and heart diseases.

References

1. Brockes JP. Amphibian limb regeneration: rebuilding a complex structure. Science 1997;276:81–7.
2. Bongso A, Richards M. History and perspective of stem cell research. Best Pract Res Clin Obstet Gynaecol 2004;18:827–42.
3. Bongso A, Lee EH. Stem Cells: From Bench to Bedside. World Science, Singapore, 2005.
4. Kiessling AA, Anderson SC. Human Embryonic Stem Cells. Jones and Bartlett, Boston, MA, 2003.
5. Alison MR, Poulsom R, Forbes S, Wright NA. An introduction to stem cells. J Pathol 2002;197:419–23.
6. Adewumi O, Aflatoonian B, Ahrlund-Richter L, et al. Characterization of human embryonic stem cell lines by the International Stem Cell Initiative. Nat Biotechnol 2007;25:803–16.
7. Richards M, Tan SP, Tan JH, Chan WK, Bongso A. The transcriptome profile of human embryonic stem cells as defined by SAGE. Stem Cells 2004;22:51–64.

8. Richards M, Chan WK, Bongso A, Tan SP. Reverse SAGE characterization of orphan SAGE tags from human embryonic stem cells identifies the presence of novel transcripts and antisense transcription of key pluripotency genes. Stem Cells 2006;24:1162–73.

9. Romalho-Santos M, Yoon S, Matsuzaki Y, Mulligan RC, Melton DA. 'Stemness': transcriptional profiling of embryonic and adult stem cells. Science 2002;298:597–600.

10. Yoshimizu T, Sugiyama N, De Felice M, et al. Germline-specific expresión of the Oct-4/green fluorescent protein (GFP) transgene in mice. Dev Growth Differ 1999;41:675–84.

11. Pesce M, Wang X, Wolgemuth DJ, Scholer H. Differential expression of the Oct-4 transcription factor during mouse germ cell differentiation. Mech Dev 1998;71:89–98.

12. Bongso A, Tan S. Human blastocyst culture and derivation of embryonic stem cell lines. Stem Cell Rev 2005;1:87–98.

13. Fong CY, Sathananthan AH, Wong PC, Bongso A. Nine-day-old human embryo cultured in vitro: a clue to the origins of embryonic stem cells. Reprod Biomed Online 2004;9:321–5.

14. Bjornson CR, Rietze RL, Reynolds BA, Magli MC, Vescovi AL. Turning brain into blood: a hematopoietic fate adopted by adult neural stem cells in vivo. Science 1999;283:534–7.

15. Jackson KA, Mi T, Goodell MA. Hematopoietic potential of stem cells isolated from murine skeletal muscle. Proc Natl Acad Sci U S A 1999;96:14482–6.

16. Clarke DL, Johansson CB, Wilbertz J, et al. Generalised potential of adult neural stem cells. Science 2000;288:1660–3.

17. Krause DS, Theise ND, Collector MI, et al. Multi-organ, multi-lineage engraftment by a single bone marrow derived stem cell. Cell 2001;105:369–77.

18. McGuckin CP, Forraz N, Baradez MO, et al. Production of stem cells with embryonic characteristics from human umbilical cord blood. Cell Prolif 2005;38:245–55.

19. Fong CY, Richards M, Manasi N, Biswas A, Bongso A. Comparative growth behaviour and characterization of human Wharton's jelly stem cells. Reprod Biomed Online 2007;15:708–18.

20. Sarugaser R, Lickorish D, Baksh D, Hosseini MM, Davies JE. Human umbilical cord perivascular (HUCPV) cells: a source of mesenchymal progenitors. Stem Cells 2005;23:220–9.

21. Bongso A, Fong CY. The effect of coculture on human zygote development. Curr Opin Obstet Gynecol 1993;5:585–93.

22. Bongso A, Fong CY, Ng SC, et al. The growth of inner cell mass cells from human blastocysts. Theriogenology 1994;41:161.

23. Bongso A, Fong CY, Ng SC, et al. Isolation and culture of inner cell mass cells from human blastocysts. Hum Reprod 1994;9:2110–7.

24. Thomson JA, Itskovitz-Eldor J, Shapiro SS, et al. Embryonic stem cell lines derived from human blastocysts. Science 1998;282:1145–7.

25. Reubinoff BE, Pera MF, Fong CY, Trounson A. Bongso A. Embryonic stem cell lines from human blastocysts: somatic differentiation in vitro. Nat Biotechnol 2000;18:399–404.

26. Richards M, Fong CY, Chan WK, Wong PC, Bongso A. Human feeders support prolonged undifferentiated growth of human inner cell masses and embryonic stem cells. Nat Biotechnol 2002;20:933–6.

27. Richards M, Tan S, Fong CY, Biswase A, Chan WK, Bongso A. Comparative evaluation of various human feeders for prolonged undifferentiated growth of human embryonic stem cells. Stem Cells 2003;21:546–56.

28. Hovatta O, Mikkola M, Gertow K, et al. A culture system using human foreskin fibroblasts as feeder cells allows production of human embryonic stem cells. Hum Reprod 2003;18:1404–9.

29. Xu C, Inokuma MS, Denham J, et al. Feeder-free growth of undifferentiated human embryonic stem cells. Nat Biotechnol 2001;19:971–4.

30. Richards M, Fong CY, Tan S, Chan WK, Bongso A. An efficient and safe xeno-free cryopreservation method for the storage of human embryonic stem cells. Stem Cells 2004;22:779–89.

31. Rao MS, Auerbach JM. Estimating human embryonic stem cell numbers. Lancet 2006; 367:650–1.

32. Ben-hur T, Idelson M, Khaner H, et al. Transplantation of human embryonic stem cell-derived neural progenitors improves behavioral deficit in parkinsonian rats. Stem Cells 2004;22:1246–55.
33. Mummery C, Ward-van Oostwaard D, Doevendans P, et al. Differentiation of human embryonic stem cells to cardiomyocytes: role of coculture with visceral endoderm-like cells. Circulation 2003;107:2733–40.
34. Miyazaki S, Yamato E, Yasui Y. Regulated expression of Pdx-1 promotes in vitro differentiation of insulin-producing cells from embryonic stem cells. Diabetes 2004;53:1030–7.
35. Takahashi K, Tanabe K, Ohnuki M, et al. Induction of pluripotent stem cells from adult human fibroblasts by defined factors. Cell 2007;131:1–12.
36. Yu J, Vodyanik MA, Smuga-Otto K, et al. Induced pluripotent stem cell lines derived from human somatic cells. Science 2007;315:1917–20.
37. Wernig M, Meissner A, Foreman R, et al. In vitro reprogramming of fibroblasts into a pluripotent ES cell-like state. Nature 2007;448:318–24.
38. Meissner A, Jaenisch R. Generation of nuclear transfer-derived pluriotent ES cells from cloned Cdx2-deficient blastocysts. Nature 2006;439:212–5.
39. Verslinsky Y, Strelchenko N, Rechitsky S, et al. Human embryonic stem cells with genetic disorders. Reprod Biomed Online 2005;10:105–10.
40. Cowan CA, Atienza J, Melton DA, Eggan K. Nuclear reprogramming of somatic cells after fusion with human embryonic stem cells. Science 2005;309:1369–73.
41. Do JT, Scholer HR. Nuclei of embryonic stem cells reprogram somatic cells. Stem Cells 2004;22:941–9.
42. Revazova ES, Turovets NA, Kochetkova OD, et al. Patient-specific stem cell lines derived from human parthenogenetic blastocysts. Cloning Stem Cells 2007;9:1–18.
43. Hentze H, Graichen R, Colman A. Cell therapy and the safety of embryonic stem cell-derived grafts. Trends Biotechnol 2006;25:24–32.
44. Dean SK, Yulyana Y, Williams G. Differentiation of encapsulated embryonic stem cells after transplantation. Transplantation 2006;82:1175–84.
45. Qin M, Tai G, Colla P, Polak JM, Bishop AE. Cell extract-derived differentiation of embryonic stem cells. Stem Cells 2005;23:712–8.
46. Zhang S, Wernig M, Duncan ID, Brustle O, Thomson JA. In vitro differentiation of transplantable neural precursors from human embryonic stem cells. Nature 2001;19:1129–33.
47. Buttery LD, Bourne S, Xynos JD. Differentiation of osteoblasts and in vitro bone formation from embryonic stem cells. Tissue Eng 2001;7:89–99.
48. Klug MG, Soonpaa MH, Koh GY, Field LJ. Genetically selected cardiomyocytes from differentiating embryonic stem cells form stable intracardiac grafts. J Clin Invest 1996;98:216–24.
49. Thomson H. Bioprocessing of embryonic stem cells for drug discovery. Trends Biotechnol 2007;25:224–30.
50. Scadden DT. The stem cell niche as an entity of action. Nature 2006;441:1075–9.

Production of Uniparental Embryonic Stem Cell Lines

Sigrid Eckardt and K. John McLaughlin

Abstract Embryonic stem cells, or induced pluripotent cells derived from somatic cells, can yield differentiated progeny with potential applicability for tissue repair. This chapter describes the generation of embryonic stem cells from gamete-derived uniparental embryos. These embryonic stem cells can be patient-derived and potentially histocompatible with the gamete donor. The production of uniparental embryos followed by derivation of embryonic stem cells can be accomplished without producing fertilized zygotes, an advantage that avoids some ethical issues. We describe methods for the generation of uniparental embryonic stem cells from mouse uniparental embryos. We also address evaluation of the integrity of the lines generated, an essential criterion in interpreting differentiation assays in vivo and in vitro.

Keywords Embryonic stem cells • Uniparental • Parthenogenetic • Androgenetic • Derivation • Pronuclear transfer

Introduction

Uniparental Embryonic Stem Cells: A Source of Pluripotent Stem Cells

Pluripotent embryonic stem (ES) cells have stimulated considerable interest for their potential use in therapeutic cell and tissue replacement (*1–3*). Aside from graft efficacy and safety concerns, one major hurdle for the potential application of differentiated ES cells in tissue repair is how to manage the requirement for immune

S. Eckardt and K.J. McLaughlin (✉)
Center for Animal Transgenesis and Germ Cell Research, University of Pennsylvania,
382 West Street Road, Kennett Square, PA 19348, USA,
e-mail: kjmclaug@vet.upenn.edu

H. Baharvand (ed.), *Trends in Stem Cell Biology and Technology*,
DOI 10.1007/ 978-1-60327-905-5_2,
© Humana Press, a Part of Springer Science+Business Media, LLC 2009

compatibility *(4–7)*. Potential approaches for the generation of patient-derived and autologous pluripotent stem cells include the derivation of ES cells from somatic cell nuclear transfer (SCNT) embryos, from uniparental embryos *(8)*, and the induction of pluripotency in somatic cells (iPSCs) *(9)*. This chapter describes the production of uniparental embryos and the derivation of embryonic stem cells thereof.

Mammalian uniparental embryos with only maternally (oocyte) or paternally (sperm) derived genomes fail early in development *(10, 11)*; however, they typically reach at least the blastocyst stage and can give rise to ES cell lines *(12, 13)*. Autologous uniparental ES cells derived using the gametes of a patient could be a potential source of tissue for cell replacement therapy. Unlike the extreme inefficiency observed with SCNT, murine uniparental embryos and ES cells can be obtained at rates similar to those of fertilized embryos *(14, 15)* (48% and 43% from one-cell stage, respectively), and human parthenogenetic (PG) ES cell lines have been derived *(16–20)*. PG human ES cell lines are frequently major histocompatibility complex (MHC) matching to the oocyte donor *(17)*. As uniparental conceptuses occur spontaneously but are not viable and can produce neoplasms (ovarian teratoma, hydatidiform moles) and even aggressive choriocarcinomas *(21)*, using them to generate ES cells subjectively side steps some of the ethical perspectives associated with the destruction of potentially viable fertilized and SCNT embryos *(22–24)*. iPSCs potentially address the concerns of destroying viable embryos. However, this approach requires considerable refinement to reduce the potential side effects of using genetic manipulation.

Origin and Generation of Uniparental Embryos and ES Cells

The most commonly known type of uniparental embryo is the PG embryo. Parthenogenesis is a type of gynogenesis, or generation of an organism with exclusively maternal genomes. It can be initiated by spontaneous or experimental activation of an unfertilized oocyte (Fig. 1a; top row). Diploidy of PG embryos can be achieved by suppressing extrusion of the second polar body during activation. Mouse experiments investigating the role of maternal and paternal genomes, including those requiring specific genotypes, often apply the transfer of maternal and paternal pronuclei between zygotes (pronuclear transfer, PNT; *(25)*; Fig. 1b), producing gynogenetic (GG) embryos with two maternal genomes from different oocytes (Fig. 1b, top row), and androgenetic (AG) embryos with two paternal genomes (AG; Fig. 1b, bottom row). GG embryos develop similarly to PG embryos and are often used to study the consequences of two maternal genomes *(26–28)*.

To avoid the ethical concerns of destroying a viable fertilized embryo, uniparental embryos by definition, and in practice, can be generated using only the genetic material of one individual of reproductive age of either sex, without fertilizing an

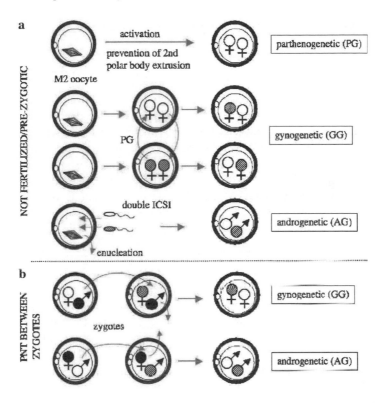

Fig. 1 Strategies for the production of uniparental embryos. (**a**) Approaches not involving fertilized embryos, by either activating a female's oocyte (parthenogenetic, PG), by exchange of pronuclei between PG embryos to produce gynogenctic (GG) embryos, or by injecting two sperm into an enucleated donor oocyte (androgenetic, AG). (**b**) Pronuclear transfer between zygotes *(25)*, an approach often used in murine experiments, producing GG embryos or AG embryos with two maternal or paternal genomes from different zygotes, respectively.

intact oocyte (Fig. 1a). This includes PG embryos, as well as GG embryos that would result from the transfer of a pronucleus from one haploid or diploid PG embryo to another (if diploid PG are used, less oocytes are needed; Fig. 1a, middle row). For embryos derived only from paternal genomes, two sperm would be introduced into an enucleated donor oocyte by intracytoplasmic sperm injection (ICSI) or in vitro fertilization (IVF) at high sperm concentrations (AG; Fig. 1a, bottom row); this method has been established in the mouse and the bovine *(29, 30)*. Based on the clinical success of ICSI, human AG embryos should be technically feasible. Although production of human uniparental embryonic stem cells would require the manipulation of human gametes, i.e., oocytes and sperm, it could be accomplished without the manipulation of zygotes (oocyte containing both sperm and egg genome).

Developmental Potential of Uniparental Embryos and Uniparental ES Cells

Androgenetic embryos rarely develop to or past early somite stages, with abundant trophoblast but retarded development of the embryo proper *(10)*. Diploid partheno-genetic or gynogenetic embryos can occasionally develop to later somite stages but typically lack extraembryonic tissues *(26, 31)*. This limited developmental potential of uniparental embryos *(10, 11)* is due to the lack of, or overexpression of, imprinted genes, i.e., genes that are preferentially expressed from only one parental allele *(32)*. When combined with normal cells in composite animals (chimeras), uniparental embryonic cells and uniparental ES cells can contribute to various tissues, including the germline, although AG and PG derived cells exhibit bias in their differentiation into, and exclusion from, certain lineages *(33–36)*. AG cells in developmental chimeras are frequently found in tissues derived from mesodermal lineages. Conversely, PG cell derivatives are often present in brain but rarely in mesodermal tissues such as skeletal muscle *(12, 33, 34, 36–38)*. This contribution bias manifests later in development with marked differential in contribution to some tissues or elimination thereof at later fetal stages *(33, 38)*. AG cells, even at low levels of contribution, cause severe defects and often lethality in chimeras *(12, 33, 39)*, and readily transform in vitro *(40)*. PG-derived cells contrast in that they exhibit a deficit in their capacity to proliferate at normal rates *(40, 41)*. The basis for at least some of these developmental and proliferative defects arises from abnormal expression levels of imprinted genes that are involved in fetal growth regulation *(14, 40, 42)*.

Therapeutic Potential of Uniparental ES Cell Derivatives

The therapeutic applicability of uniparental ES cells depends on their functional equivalence to normal ES cells, i.e., those that are derived from fertilized embryos. Consequences of genomic imprinting are potentially a major restriction to the therapeutic utility of uniparental ES cells. While PG ES cells have been proposed as a source of patient-derived therapeutic material *(8)*, AG ES cells have not been similarly considered, in part because mouse experiments have shown that unlike PG cells, AG cells cause severe defects and frequent lethality when combined with normal cells in chimeras *(12, 33, 39)*. AG ES cells, however, are pluripotent including germline transmission *(12, 14, 33, 35, 43)*. Therapeutic utility of AG ES cell derivatives would include males of reproductive age into the patient pool, more than doubling the potential number of patients that could benefit from uniparental cell transplants. In contrast to developmental stages, the relevance of genomic imprinting for normal adult tissue function is largely unresolved *(44)*. Our studies on hematopoietic transplantation of uniparental fetal liver cells and in vitro derivatives of uniparental ES cells suggest that genomic imprinting does not preclude a potential therapeutic applicability of uniparental cells in certain tissues *(45)*.

We observed differentiation of AG and GG fetal liver derived cells into major blood lineages in adult recipients as defined by expression of CD4, B220, Gr-1, and Ter-119, and recipient mice were phenotypically normal when maintained in a specific pathogen-free (SPF) facility. We also observed engraftment and contribution of in vitro derivatives of uniparental ES cells in adults, using ectopic expression of the homeodomain protein HoxB4 in differentiating ES cells as an approach to promote formation of cells with a definitive hematopoietic phenotype (46). Contribution levels of uniparental ES cell derivatives in adult hosts were similar to those of normal (46) and nuclear-transfer–derived (47) ES cell derivatives, but due to the constitutive HoxB4 expression approach taken, we observed predominantly myeloid lineage differentiation (45).

Genomic Imprinting in Uniparental ES Cells

Genomic imprinting, defined as a bias in allele-specific expression of imprinted genes, is regulated in association with parent-specific imprinting marks that are set in the germ line, some of which involve differential methylation of regulatory regions. Uniparental ES cell lines are useful for studying genomic imprinting, however, with some limitations. Developmental potential, phenotypes of, and gene expression in uniparental ES cell chimeras, particularly those generated with AG ES cell lines (12, 48, 49), show that uniparental ES cells retain many aspects of genomic imprinting. The derivation and extended in vitro culture of ES cells can, however, change epigenetic marks and regulation of imprinted gene expression (50). ES cells, both those derived from normal and uniparental embryos, do not have an identified developmental state, are largely defined by their characteristics in in vitro culture, and can be epigenetically unstable with consequences on gene expression and thus differentiation potential (51, 52). For uniparental ES cell lines, AG lines appear to retain imprinting more faithfully than PG ES cell lines. For the latter, chimeras often exhibit phenotypic differences compared to PG aggregation chimeras, including the absence of growth deficits (49) and more widespread tissue contribution of PG ES cells (53). Consistent with our own observations that demonstrate less faithful conservation of gametic methylation marks in GG compared to AG ES cell lines (45), reactivation of paternally expressed genes in PG ES cell lines and changes in imprinted gene expression have been reported for PG ES cell lines and their derivatives in chimeras (54, 55).

Methods for the Derivation of Uniparental ES Cells

The methods for the derivation of uniparental ES cells include protocols for obtaining murine uniparental embryos, i.e., parthenogenetic activation for PG embryos, and pronuclear transfer for AG and GG embryos, culture to the blastocyst stage, and the derivation of ES cell lines from blastocysts. This chapter emphasizes the description of protocol components relevant for uniparental ES cell production. For general techniques in embryo manipulation and ES cell culture, we refer to existing extensive protocols *(56, 57)*.

Parthenogenetic Activation of Oocytes

To obtain diploid parthenogenetic embryos for the derivation of ES cell lines, oocytes are activated in the presence of cytochalasin B. The activation protocol is based on culture of oocytes in medium supplemented with strontium chloride in the absence of calcium chloride. This protocol works well for oocytes obtained from C57Bl6 × 129 or C57Bl6 × C3H F1 hybrid mice. Common alternative activation protocols include activation using ethanol exposure *(57, 58)*.

Equipment and Reagents

- Dissection microscope; CO_2 incubator (5% CO_2, 37°C)
- Mouth or syringe controlled pipetting device for embryo handling
- Pasteur pipettes pulled to 0.3 mm outer diameter
- Dissection tools, fine forceps, 27-gauge syringes
- 35-mm suspension dishes (Corning 430588)
- 35-mm tissue culture dishes with four inner rings (Greiner Cellstar 627170)
- M2 medium (commercial supplier; or as described in *(56)*)
- CZB culture medium (see below)
- Activation medium (see below): CZB culture medium without $CaCl_2$, supplemented with 10 mM $SrCl_2$ and cytochalasin B (5 µg/mL)
- Light mineral oil (Sigma); washed four times with millipore water (add twofold amount of sterile water to oil in sterile bottle/tissue culture flask, shake vigorously by hand several times, or on a shaker for 10–15 min, remove water; after last wash, add fresh water to oil in bottle).
- Hyaluronidase (Sigma H3884, 100 µg/mL in M2 medium; filter sterilize; store in 2–3 mL aliquots at −20°C)

Preparation of Culture and Activation Medium

Prepare media fresh from the following stock solutions on the day before activation:

Solution (concentration, volume)	Chemical	Grams
Solution 1 (10×; g in 500 mL)	KH_2PO_4	0.80
(add in the following order)	$MgSO_4 \times 7H_2O$	1.45
	NaCl	23.85
	KCl	1.80
	Na lactate	
	(60% syrup; Sigma L 4263)	29.24mL
	D-glucose	5.00
Solution 2 (10×; g to 48.5 mL)	$NaHCO_3$	1.055
	Add 1 mL of 5 mg/mL solution phenol red	
Solution 3 (100×; g to 50 mL)	Sodium pyruvate	0.145
Solution 4 (100×; g in 100 mL)	$CaCl_2 \times 2H_2O$	2.51
Solution 5 (200×)	Glutamine (200 mM; Gibco 25030-081)	
Solution 6 (100×)	Pen/Strep (100×; Gibco 15140-122)	

Storage: 1 and 4, filter sterilize, store at 4°C for 8 weeks; 2 and 3, make fresh; 5 and 6, store aliquots at −20°C.

Stock	Component	10 mL
	Water	7.65
1	Salts/sugars	1
2	HCO_3	1
3	Pyruvate	0.1
4	Ca	0.1
5	Glutamine	0.05
6	Pen/Strep	0.1

Preparation of CZB Culture Medium (10 mL):

Add 0.04 g BSA (Serologicals 81-003-2) per 10 mL, then filter sterilize. Equilibrate medium in tube or in culture drops overnight in CO_2 incubator. For culture drops, place 20 μL drops onto 35-mm suspension dish and cover with light mineral oil.

Preparation of CZB Medium for Activation (10 mL)

Combine water and stock solutions 1, 2, 3, 5, and 6 as described for CZB culture medium. Instead of stock solution 4, add 0.1 mL of a 1 M stock solution of $SrCl_2$ in water (10 mM final concentration of $SrCl_2$). Add BSA (0.04 g/10 mL), filter sterilize and equilibrate overnight in CO_2 incubator. On the day of activation, add 0.05 mL of 1 mg/mL cytochalasin B in dimethylsulfoxide (DMSO) (prepare sterile, store in aliquots at −80°C). Prepare culture drops as described above.

Parthenogenetic Activation of Unfertilized Oocytes

Induce ovulation in female mice (6–12 weeks old) by intraperitoneal injection of pregnant mare serum gonadotropin (PMSG; 5 IU) followed 48 h later by intraperitoneal

injection of human chorionic gonadotropin (hCG; 5 IU). Sacrifice females 17 h
post–hCG injection, dissect oviducts, place in large (0.5 mL drop) of hyaluronidase
solution in the lid of a 35-mm culture dish and release oocyte–cumulus complexes
from oviduct by opening the ampulla with a 27-guage needle while pinching it
with fine forceps. Incubate for approximately 5–10 min at room temperature.
Pick up oocytes using a mouth or syringe controlled fine Pasteur pipette and
wash twice in a large drop of M2 medium (easiest in about 150 µL in a four-ring
Greiner dish). Wash oocytes twice in 0.15 mL activation medium in a four-ring
Greiner dish (dish preequilibrated in CO_2 incubator), then twice in 50 µL drops
of activation medium covered with mineral oil. Place into preequilibrated
culture drops made with activation medium and culture for 5–6 h in CO_2 incubator.
Wash thoroughly in CZB culture medium (use large, preequilibrated culture
drops, incubate several minutes in each drop) and discard fragmented oocytes,
oocytes lacking a perivitelline space or of other unusual morphology or color.
Place in preequilibrated CZB culture drops and culture for 3 days to the blastocyst stage. On the next day, score the number of embryos that have cleaved to the
two-cell stage to assess activation. Remove fragmented embryos from the culture
drops and discard.

Production of Androgenetic and Gynogenetic Embryos by Pronuclear Transfer

This approach involves the reciprocal exchange of male and female pronuclei
between zygotes, resulting in embryos with two paternal (AG) or two maternal
(GG) genomes from different zygotes. One pronucleus is removed with a micromanipulation pipette without disruption of the plasma membrane of the zygote, and
the membrane-surrounded pronucleus (karyoplast) is subsequently fused with the
recipient zygote, a method referred to as pronuclear transplantation (25). The
experimental design for this approach needs to include genetic markers that allow
it to distinguish AG or GG embryos after manipulation from the zygotes used for
their production, such that errors (i.e., culture of nonmanipulated zygotes or false
identification of pronuclei) become apparent. Such markers can include enhanced
green fluorescent protein (eGFP; for example ubiquitously expressed transgene
(59)) and/or the intracellular biochemical marker glucose-phosphate-isomerase 1
(Gpi1). For example, when producing AG embryos with a desired genetic background of 129S1 (homozygous for the a allele of Gpi1), zygotes could be from the
intercross of 129S1 male mice with C57Bl6 female mice (homozygous for the b
allele of Gpi1). Only AG embryos and the resulting ES cell lines would be
homozygous for the a allele of Gpi1, whereas those derived from fertilized embryos
or manipulated embryos with a falsely introduced female pronucleus would be
heterozygous for the a and b alleles, and embryos with two maternal genomes

would be homozygous for the *b* allele (see also characterization of newly derived ES cell lines in the section "Derivation of ES Cell Lines").

Equipment and Reagents

- Micromanipulation setup (see below)
- Glass capillaries (Clark Electromedical GC100-15 for holding, and Clark Electromedical GC100T-15 for enucleation pipettes)
- Pipette puller (P80 or better; Brown-Flaming)
- Capillary grinder (Bachhofer)
- Microforge (Defonbrune style)
- Micromanipulation chamber (i.e., 6 cm glass bottom Petri dish)
- Electrofusion setup: modified AC function generator or commercial electrofusion device
  Cell fusion chamber: two parallel electrodes with a distance that will sustain a field strength of 1.5 kV/mm
- HEPES-buffered medium for manipulation/washes (M2)
- Nocodazole (0.3 mg/mL in DMSO; 1,000×)
- Cytochalasin B (see the section "Preparation of Culture and Activation Medium")
- Fusion medium (see below)
- CZB culture medium (see the section "Preparation of Culture and Activation Medium")
- Silicon oil (200 fluid/20 centistokes)

The micromanipulation setup requires two three-dimension movement micromanipulators attached to an inverted microscope. To control meniscus movement in pipettes, 2-μm syringes are attached to holding and enucleation pipette instrument holders via thick-walled plastic tubing, and the system is filled with silicon oil.

Production of Media

Enucleation Medium

M2 supplemented with 0.3 μg/mL nocodazole and 5 μg/mL cytochalasin B. Make fresh on day of use.

Fusion Medium

0.3 M mannitol
0.1 mM $MgSO_4$
0.05 mM $CaCl_2$

Adjust pH at 7.4 and osmolarity to 280 mOsm. Filter, sterilize, and store at 4°C for up to 3 months.

Production of Pipettes

Pipettes are pulled from glass capillary tubing (150 mm long) on a pipette puller. Holding pipettes are pulled from thick walled tubing (Clark Electromedical GC100-15), are cut on a microforge to an O.D. of 150–180 µm and are polished with the microforge filament to an I.D. of 80–100 µm. Enucleation manipulation pipettes are made from thin walled tubing (Clark Electromedical GC100T-15), cut to an O.D. of 20–25 µm and ground on a capillary grinder to produce a 45° bevel. Pipettes are then washed in 25% hydrofluoric acid to sharpen the bevelled edge followed by spiking of the distal tip of the bevel on the microforge filament. All pipettes are bent on the microforge to facilitate positioning into the micromanipulation chamber.

Pronuclear Transfer

This procedure involves the following steps: identification of pronuclei in zygotes, removal of one pronucleus from a zygote, transfer of a karyoplast with a pronucleus from a different zygote under the zona pellucida, and subsequent electrofusion. For production of AG embryos, the female pronuclei are removed from zygotes and replaced with a second male pronucleus from a different zygote. For GG embryos, male pronuclei are replaced with female pronuclei. Depending on the desired genotype of AG or GG embryos, zygotes from the same or different intercrosses are used.

To recover zygotes, superovulate female mice as described above (see the section "Parthenogenetic Activation of Unfertilized Oocytes") and after hCG administration mate with male mice. On the next morning, check for the presence of copulatory plugs, and 15–17 h post-hCG, dissect oviducts from females with a copulatory plug and recover putative zygotes from the ampulla as described for oocytes (see the section "Parthenogenetic Activation of Unfertilized Oocytes").

Prior to manipulation, incubate presumptive zygotes for 20 min in enucleation medium. Visualize pronuclei and verify that both the male and female pronucleus are visible: The male pronucleus is initially smaller and peripheral in the oocyte cytoplasm, where the sperm entered the egg. Subsequently, the male pronucleus migrates toward the cortex and becomes larger. In contrast, the female pronucleus is typically proximal to the polar body. To simplify the manipulation phase, only zygotes with both pronuclei visible on a dissection microscope are selected for manipulation (Fig. 2a).

Manipulation is performed in groups of approximately 20 zygotes (estimating about 45–60 min for manipulation exposure). Position the zygote with the holding

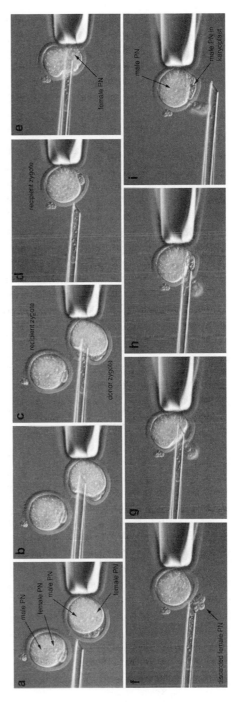

Fig. 2 Micromanipulation procedures to produce androgenetic (AG) embryos by pronuclear transfer. (**a**) Two zygotes at the pronuclear stage, arrows indicate male and female pronuclei (PN). (**b**) Positioning of the enucleation pipette for removal of the male PN from one zygote. (**c**) Removal of the karyoplast with the male pronucleus from the donor zygote by suctioning into the pipette. (**d**) Positioning of the recipient zygote for removal of the female PN. (**e**) Removal of the female PN from the recipient zygote. (**f**) The female pronucleus is expelled from the pipette. (**g, h**) Insertion and placement of the karyoplast with the male pronucleus from the donor zygote subzonally of the recipient zygote from which the female pronucleus has been removed. (**i**) AG construct with two male PN, one subzonally, ready for electrofusion.

pipette, and using a 20–25 μM O.D. pipette beveled to 45 degrees, remove male for production of AG (Fig. 2b, c), or female pronucleus for production of GG embryos. With this pronucleus in the pipette, the opposite parental pronucleus is removed from another zygote (Fig. 2d, e), expelled from the pipette (Fig. 2e), and the transplant pronucleus inserted subzonally immediately afterward (Fig. 2g, h). The successfully manipulated zygotes can be readily identified within the batch at the end of each group manipulation based on the presence of the karyoplast subzonally (Fig. 2i).

To fuse subzonal karyoplasts with the manipulated zygotes, the constructs are first equilibrated in electrofusion medium for 2 min. Constructs are then placed in an AC field to polarize and align karyoplasts and to achieve close contact (AC stimulus at 500–1,000 kHz, sine wave, 0–20 V pp). Once alignment and juxtaposition are achieved, a DC pulse is generated (1–1.5 kV/cm, 1–2 pulses, 50–100 ms interval, 20–100 ms duration). Optimal fusion parameters vary between chambers and electrodes and can be determined by fusing two-cell stage embryos and assessing fusion rate versus blastocyst development (60).

After fusion treatment, constructs are immediately washed in CZB culture medium (as described above for oocytes post activation in the section "Parthenogenetic Activation of Unfertilized Oocytes") and cultured in CZB culture drops to the blastocyst stage.

Derivation of ES Cell Lines

Equipment and Reagents

- 96, 48, 24, 12, 6-well tissue culture dishes (Falcon)
- Pulled Pasteur pipettes, lightly flame polished for embryo handling
- Feeder cells for ES cell derivation: STO fibroblasts; available from ATCC (CRL-1503). STO cells that have been stably transfected with a neo^r vector and an LIF expression vector (SNL cells) are courtesy of Allan Bradley and Elizabeth Robertson
- M2 medium (see above)
- Tyrode's solution, acidic
- Media and solutions described in the section "Preparation of Solutions and Media"

Preparation of Solutions and Media

ES Cell Medium

- 500 mL DMEM (Specialty Media/Chemicon EmbryoMax SLM-220-B; without l-glutamine and Na-pyruvate; with 4,500 mg/L glucose, 2,250 mg/L Na Bicarb)
- 6 mL nonessential amino acids (100×; Gibco 11140-050)

- 6 mL Pen/Strep (100×; Gibco 15140-122)
- 6 mL L-glutamine (100×; Gibco 25030-081)
- 0.6 mL β-mercaptoethanol (1,000×; Gibco 21958-023)
- 75 mL fetal bovine serum (Hyclone defined FBS; SH30070.03)

Store at 4°C. After 3 weeks, replenish glutamine and 2-mercaptoethanol from the respective stock solution according to the amount of medium left in bottle. If using mouse primary embryonic fibroblasts (MEF) or STO cells as feeder layers, add LIF (Chemicon Esgro® LIF ESG1106 or 1107) to the medium (500 U/mL final).

Medium for Feeder Cells

Same composition as ES cell growth medium, but lower concentration of FBS (35 mL FBS per 500 mL, i.e., 7%), also use DMEM from Gibco (11965). Store at 4°C.

1× Dulbecco's Phosphate Buffered Saline (DPBS)

Dilute from 10× stock (Gibco 14200-075; without calcium or magnesium).

Mitomycin C (MMC)

Dissolve 2 mg of mitomycin C (Sigma M 0503) in 5.0 mL of 1× DPBS (40× stock). Store 200-μL aliquots in sterile tubes at –80°C; add one aliquot to 10 mL medium for treatment (8 μg/mL final).

PBS/Gelatin

Add 1 g gelatin (Sigma G 2500) to 1,000 mL 1× DPBS in glass bottle, then autoclave. Store at 4°C after opening; can keep for 2 months.

Trypsin/EDTA

0.25% Trypsin (Sigma T 4799), 1 mM EDTA, 1× DPBS.

To prepare 1,000 mL	Trypsin	2.5 g
	EDTA 500 mM stock	2 mL
	10 × DPBS stock	100 mL

Add water ad 200 mL, filter sterilize, aliquot 10 mL of the resulting 5× stock into 50 mL tubes and store at −20°C. To produce 50 mL working solution, add 39.5 mL water and 0.5 mL of filter sterilized 5% (w/v) BSA (Sigma A 9647 in water) to thawed stock. Store the working solution at 4°C for 1–2 weeks.

Cell Freezing Solutions

–   Solution I: 50% (v/v) FBS in DPBS+ (1×; Gibco 14287-080).
–   Solution II: mixture of 20 mL 1× DPBS and 5 mL DMSO (Sigma D 2650).
–   Store at 4°C. Discard after 14 days.

Derivation of ES Cell Lines

The method described has been adapted from various published protocols, including *(58, 61)*. An overview of the procedure is outlined in Fig. 3. Blastocyst stage embryos from which the zona pellucida has been removed, are placed in individual wells of a 96-well plate and are cultured for 3 days to form outgrowths. To disaggregate these outgrowths (passage 0), the whole well is treated with trypsin, and all cells from each well of the 96-well plate are transferred to a well of a 48-well plate. An alternative method to this is picking and disaggregating only the inner cell mass of the outgrowth *(58, 61)*. Culture expansion is continued by treating entire wells with trypsin and transferring them into gradually larger dishes. Again, an alternative method at passage 1 is picking individual ES cell colonies rather than transferring the whole well. All ES cell derivation and culture is performed on wells covered with feeder layers.

Preparation of Feeder Layers

Grow STO or SNL on 15 cm dish to confluency. Aspirate medium and replace with feeder medium containing MMC (1×) and incubate for 2 h at 37°C in CO_2 incubator. Wash plate several times with PBS, trypsinize cells, and determine cell count. Plate cells on gelatin-treated culture dishes (resuspend cells at a concentration of 1×10^7 cells/mL, plate 3.5 µL per well of a 96-well plate, 8.75, 17.5, 35, and 70 µL per well of a 48-, 24-, 12-, and 6-well plate, respectively). For gelatin-treatment of dishes, cover bottom of wells with gelatin/PBS and incubate at 4°C for 1 h or at room temperature for 20 min. Aspirate gelatin completely (tilt plates when aspirating) and add feeder medium to the well.

Zona Removal

Place blastocyst stage embryos into drop of M2 medium. Transfer into large drop (200 µL) of Tyrode's solution and observe dissolving of the zona pellucida through a dissection microscope. Remove embryos from Tyrode's as soon as zona is dissolved, wash several times in M2 medium.

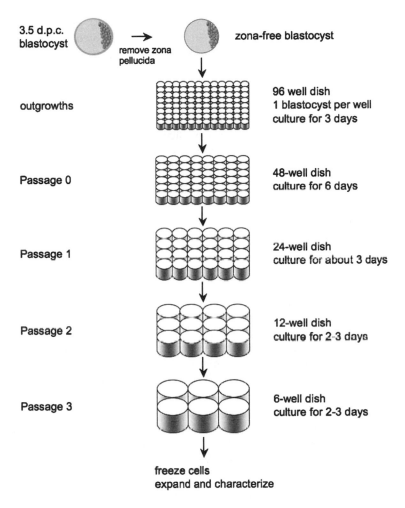

Fig. 3 Overview of embryonic stem (ES) cell derivation. Passaging between stages is performed by trypsin treatment, and the whole contents of each well transferred to a fresh well with feeder cells. After passage 1, subsequent passaging may have to be adjusted to the density of the ES cell colonies. Wells with only few colonies should be passaged onto the same size well (24 well) to increase colony density.

Outgrowths and Early Passage to Establish ES Cell Lines

Place zona-free blastocyst stage embryos individually onto wells of a 96-well plate covered with STO or SNL feeder cells. Culture in ES cell medium for 3 days, until the inner cell mass has formed a large, often mushroom-shaped outgrowth. Trypsinize the whole well (passage zero, p0): aspirate medium, rinse with 1× DPBS, rinse briefly with trypsin, and then add small amount of trypsin such that the bottom of the well is just covered. Place into incubator for approximately 3–5 min. Add ES cell medium and, using Gilson pipette, repeatedly pipette up and down to break up clumps and obtain a cell suspension. Transfer cells from each well onto a fresh well of a 48-well plate with feeder cells. Culture cells in ES cell medium for 6–7 days. Colonies appearing to be ES cell-like can form earlier but should be ignored until the 6–7 days after passage. Trypsinize each well as described (passage 1, p1), and plate onto wells of a 24-well plate. Distinct ES cell colonies will appear in about 2–3 days. Continue passaging, using a 1:2 to 1:3 split to obtain sufficient density at early passages (see Fig. 3). Typically, new lines will be in a well of a 6-well plate at passage 3 or 4 and can be frozen (two vials per well of a 6-well plate) or expanded for further characterization. If the density of colonies is low, for example at passage 1 or 2, passage cells once on the same size well, i.e., perform passage 2 onto a 24 well or passage 3 onto a 12 well. Once lines are established, ES cells are split 1:3 to 1:6 every 2 days, when they are at approximately 70% confluency.

Freezing and Thawing ES Cells

Freezing. ES cells should be at about 70% confluency; change medium 2–3 h before freezing. Dissociate cells by trypsinization, add 2 mL ES cell maintenance medium and collect by brief centrifugation (1,100 rpm, 3 min). Resuspend cell pellet in half of the desired final volume of freezing solution I, then slowly, drop by drop, add an equal volume of freezing solution II while mixing carefully by gently flicking tube. Transfer into cryovial and place into −80°C freezer in a cooling device providing a controlled freezing rate (StrataCooler® or similar); then transfer to liquid nitrogen storage after 24 h. Freeze two vial with 500 µL volume each from one well of a 6-well plate or four vials from one 6-cm dish.

Thawing. Place vial in 37°C water bath to thaw, transfer contents into 15-mL tube. Slowly add ES cell medium to about 3–4 mL, collect cells by brief centrifugation, resuspend in ES cell maintenance medium, and plate onto wells with feeder cells.

Characterizing Newly Derived ES Cell Lines

The characterization of newly derived ES cell should include analysis of chromosome number for all lines, verification of uniparental origin for AG and GG ES cell lines derived from embryos generated by pronuclear transfer between zygotes, and analysis of Y-chromosome status for AG ES cell lines. Other analyses can include but are not limited to verification of genomic integrity (i.e., chromosome duplication or translocation) by G-banding of metaphase spreads; single nucleotide polymorphisms (SNP) analysis to determine the degree of homozygosity, which also serves as verification of uniparental origin, for example, in PG lines *(62)*. The ability of ES cell lines to contribute in chimeras can only be ascertained by performing blastocyst injection (for protocols, see *(56, 63, 64)*), and subsequent analysis of the presence of ES cell derived cells in fetal stage or postnatal chimeras. Markers to identify ES cells in chimeras include eGFP (ES cells derived from eGFP transgenic mouse strains), or the presence of mouse strain-specific glucose-phosphate-isomerase 1 isoforms that can be ascertained by electrophoretic analysis of tissue samples (Gpi-1 isozyme analysis) *(65)*.

Determine chromosome number by making metaphase spreads (for protocol see *(56)*). To verify uniparental origin using Gpi-1 isoforms, collect about five large ES cell colonies with a pulled Pasteur pipette, freeze-thaw twice in 20 µL water and perform Gpi-1 electrophoresis. (If feeders are included in the sample, STO feeder cells are homozygous for the B isoform.) To determine Y-chromosome status, perform polymerase chain reaction (PCR) on genomic DNA: isolate genomic DNA from ES cells, or collect about ten large colonies as described above and digest at 55°C for 1–2 h in a shaker block in 10× volume of digest buffer (100 mM Tris–HCl pH 8.0, 0.5% (v/v) Tween 80, 0.5% (v/v) NP-40 with 100 mg/mL proteinase K). Dilute viscous digest 1:10 and use 2 µL in a 20 µL PCR reaction. Primers for the murine Zfy gene are 5'-CTC ATG CTG GGA CTT TGT GT-3' and TGT GTT CTG CTT TCT TGG TG-3', amplifying a fragment of 406 base pairs length. Verify presence and accessibility of genomic DNA in each sample by also performing PCR for a housekeeping gene such as β-actin.

References

1. Shufaro Y, Reubinoff BE. Therapeutic applications of embryonic stem cells. Best Pract Res Clin Obstet Gynaecol 2004;18:909–27.
2. Keller G. Embryonic stem cell differentiation: emergence of a new era in biology and medicine. Genes Dev 2005;19:1129–55.
3. Zeng X, Rao MS. Human embryonic stem cells: long term stability, absence of senescence and a potential cell source for neural replacement. Neuroscience 2007;145:1348–58.
4. Kaufman DS, Thomson JA. Human ES cells—haematopoiesis and transplantation strategies. J Anat 2002;200:243–8.
5. Boyd AS, Higashi Y, Wood KJ. Transplanting stem cells: potential targets for immune attack. Modulating the immune response against embryonic stem cell transplantation. Adv Drug Deliv Rev 2005;57:1944–69.
6. Drukker M. Immunogenicity of human embryonic stem cells: can we achieve tolerance?. Springer Semin Immunopathol 2004;26:201–13.
7. Taylor CJ, Bolton EM, Pocock S, Sharples LD, Pedersen RA, Bradley JA. Banking on human embryonic stem cells: estimating the number of donor cell lines needed for HLA matching. Lancet 2005;366:2019–25.
8. Cibelli JB, Grant KA, Chapman KB, et al. Parthenogenetic stem cells in nonhuman primates. Science 2002;295:819.
9. Takahashi K, Yamanaka S. Induction of pluripotent stem cells from mouse embryonic and adult fibroblast cultures by defined factors. Cell 2006;126:663–76.
10. Barton SC, Surani MA, Norris ML. Role of paternal and maternal genomes in mouse development. Nature 1984;311:374–6.
11. McGrath J, Solter D. Completion of mouse embryogenesis requires both the maternal and paternal genomes. Cell 1984;37:179–83.

12. Mann JR, Gadi I, Harbison ML, Abbondanzo SJ, Stewart CL. Androgenetic mouse embryonic stem cells are pluripotent and cause skeletal defects in chimeras: implications for genetic imprinting. Cell 1990;62:251–60.
13. Robertson EJ, Kaufman MH, Bradley A, Evans MJ. Isolation, properties, and karyotype analysis of pluripotential (EK) cell lines from normal and parthenogenetic embryos. In: Silver LM, Martin GR, Strickland S, editors. Teratocarcinomal Stem Cells Cold Spring Harbor Conferences on Cell Proliferation, 1983, Cold Spring Harbor Laboratory Press, Cold Spring Harbor, NY, 1983:647–63.
14. McLaughlin KJ, Kochanowski H, Solter D, Schwarzkopf G, Szabo PE, Mann JR. Roles of the imprinted gene Igf2 and paternal duplication of distal chromosome 7 in the perinatal abnormalities of androgenetic mouse chimeras. Development 1997;124:4897–904.
15. Kim K, Lerou P, Yabuuchi A, et al. Histocompatible embryonic stem cells by parthenogenesis. Science 2007;315:482–6.
16. Marchant J. Human eggs supply 'ethical' stem cells. Nature 2006;441:1038.
17. Revazova ES, Turovets NA, Kochetkova OD, et al. Patient-specific stem cell lines derived from human parthenogenetic blastocysts. Cloning Stem Cells 2007;9:432–49.
18. Lin G, OuYang Q, Zhou X, et al. A highly homozygous and parthenogenetic human embryonic stem cell line derived from a one-pronuclear oocyte following in vitro fertilization procedure. Cell Res 2007;17:999–1007.
19. Mai Q, Yu Y, Li T, et al. Derivation of human embryonic stem cell lines from parthenogenetic blastocysts. Cell Res 2007;17:1008–19.
20. Revazova ES, Turovets NA, Kochetkova OD, et al. HLA homozygous stem cell lines derived from human parthenogenetic blastocysts. Cloning Stem Cells 2008;10:11–24.
21. Mutter GL. Role of imprinting in abnormal human development. Mutat Res 1997; 396:141–7.
22. Daley GQ, Ahrlund Richter L, Auerbach JM, et al. Ethics. The ISSCR guidelines for human embryonic stem cell research. Science 2007;315:603–4.
23. Hipp J, Atala A. Tissue engineering, stem cells, cloning, and parthenogenesis: new paradigms for therapy. J Exp Clin Assist Reprod 2004;1:3.
24. Jaenisch R. Human cloning—the science and ethics of nuclear transplantation. N Engl J Med 2004;351:2787–91.
25. McGrath J, Solter D. Nuclear transplantation in the mouse embryo by microsurgery and cell fusion. Science 1983;220:1300–2.
26. Surani MA, Barton SC. Development of gynogenetic eggs in the mouse: implications for parthenogenetic embryos. Science 1983;222:1034–6.
27. Kono T, Obata Y, Wu Q, et al. Birth of parthenogenetic mice that can develop to adulthood. Nature 2004;428:860–4.
28. Kawahara M, Wu Q, Takahashi N, et al. High-frequency generation of viable mice from engineered bi-maternal embryos. Nat Biotechnol 2007;25:1045–50.
29. Lagutina I, Lazzari G, Duchi R, Galli C. Developmental potential of bovine androgenetic and parthenogenetic embryos: a comparative study. Biol Reprod 2004;70:400–5.
30. Kono T, Sotomaru Y, Sato Y, Nakahara T. Development of androgenetic mouse embryos produced by in vitro fertilization of enucleated oocytes. Mol Reprod Dev 1993;34:43–6.
31. Kaufman MH, Barton SC, Surani MA. Normal postimplantation development of mouse parthenogenetic embryos to the forelimb bud stage. Nature 1977;265:53–5.
32. Bartolomei MS, Tilghman SM. Genomic imprinting in mammals. Annu Rev Genet 1997;31:493–525.
33. Barton SC, Ferguson-Smith AC, Fundele R, Surani MA. Influence of paternally imprinted genes on development. Development 1991;113:679–87.
34. Fundele RH, Norris ML, Barton SC, et al. Temporal and spatial selection against parthenogenetic cells during development of fetal chimeras. Development 1990;108:203–11.
35. Fundele R, Barton SC, Christ B, Krause R, Surani MA. Distribution of androgenetic cells in fetal mouse chimeras. Roux's Arch Dev Biol 1995;204:484–93.
36. Nagy A, Sass M, Markkula M. Systematic non-uniform distribution of parthenogenetic cells in adult mouse chimaeras. Development 1989;106:321–4.

37. Paldi A, Nagy A, Markkula M, Barna I, Dezso L. Postnatal development of parthenogenetic in equilibrium with fertilized mouse aggregation chimeras. Development 1989;105:115–8.
38. Fundele R, Norris ML, Barton SC, Reik W, Surani MA. Systematic elimination of parthenogenetic cells in mouse chimeras. Development 1989;106:29–35.
39. Mann JR, Stewart CL. Development to term of mouse androgenetic aggregation chimeras. Development 1991;113:1325–33.
40. Hernandez L, Kozlov S, Piras G, Stewart CL. Paternal and maternal genomes confer opposite effects on proliferation, cell-cycle length, senescence, and tumor formation. Proc Natl Acad Sci U S A 2003;100:13344–9.
41. Jagerbauer EM, Fraser A, Herbst EW, Kothary R, Fundele R. Parthenogenetic stem cells in postnatal mouse chimeras. Development 1992;116:95–102.
42. Eggenschwiler J, Ludwig T, Fisher P, Leighton PA, Tilghman SM, Efstratiadis A. Mouse mutant embryos overexpressing IGF-II exhibit phenotypic features of the Beckwith–Wiedemann and Simpson–Golabi–Behmel syndromes. Genes Dev 1997;11:3128–42.
43. Narasimha M, Barton SC, Surani MA. The role of the paternal genome in the development of the mouse germ line. Curr Biol 1997;7:881–4.
44. Burns JL, Jackson DA, Hassan AB. A view through the clouds of imprinting. FASEB J 2001;15:1694–703.
45. Eckardt S, Leu NA, Bradley HL, Kato H, Bunting KD, Mclaughlin KJ. Hematopoietic reconstitution with androgenetic and gynogenetic stem cells. Genes Dev 2007;21:409–19.
46. Kyba M, Perlingeiro RC, Daley GQ. HoxB4 confers definitive lymphoid–myeloid engraftment potential on embryonic stem cell and yolk sac hematopoietic progenitors. Cell 2002;109:29–37.
47. Rideout WM, Hochedlinger K, Kyba M, Daley GQ, Jaenisch R. Correction of a genetic defect by nuclear transplantation and combined cell and gene therapy. Cell 2002;109:17–27.
48. Mann JR. Properties of androgenetic and parthenogenetic mouse embryonic stem cell lines, are genetic imprints conserved.? Semin Dev Biol 1992;3:77–85.
49. Allen ND, Barton SC, Hilton K, Norris ML, Surani MA. A functional analysis of imprinting in parthenogenetic embryonic stem cells. Development 1994;120:1473–82.
50. Szabo P, Mann JR. Expression and methylation of imprinted genes during in vitro differentiation of mouse parthenogenetic and androgenetic embryonic stem cell lines. Development 1994;120:1651–60.
51. Dean W, Bowden L, Aitchison A, et al. Altered imprinted gene methylation and expression in completely ES cell-derived mouse fetuses: association with aberrant phenotypes. Development 1998;125:2273–82.
52. Humpherys D, Eggan K, Akutsu H, et al. Epigenetic instability in ES cells and cloned mice. Science 2001;293:95–7.
53. Sturm KS, Berger CN, Zhou SX, Dunwoodie SL, Tan S, Tam PP. Unrestricted lineage differentiation of parthenogenetic ES cells. Dev Genes Evol 1997;206:377–88.
54. Jiang H, Sun B, Wang W, et al. Activation of paternally expressed imprinted genes in newly derived germline-competent mouse parthenogenetic embryonic stem cell lines. Cell Res 2007;17:792–803.
55. Horii T, Kimura M, Morita S, Nagao Y, Hatada I. Loss of genomic imprinting in mouse parthenogenetic embryonic stem cells. Stem Cells 2008;26:79–88.
56. Nagy A, Gertsenstein M, Vintersten K, Behringer R. Manipulating the Mouse Embryo, Third ed., Cold Spring Harbor Laboratory Press, Cold Spring Harbor, NY, 2003.
57. Wassarman PM, DePamphilis ML, editors. Guide to Techniques in Mouse Development, Academic Press, Inc., San Diego, CA, 1993.
58. Mann JR. Deriving and propagating mouse embryonic stem cell lines for studying genomic imprinting. Methods Mol Biol 2001;181:21–39.
59. Okabe M, Ikawa M, Kominami K, Nakanishi T, Nishimune Y. 'Green mice' as a source of ubiquitous green cells. FEBS Lett 1997;407:313–9.
60. McLaughlin KJ. Production of tetraploid embryos by electrofusion. Methods Enzymol 1993;225:919–30.
61. Abbondanzo SJ, Gadi I, Stewart CL. Derivation of embryonic stem cell lines. Methods Enzymol 1993;225:803–23.

62. Kim K, Ng K, Rugg-Gunn PJ, et al. Recombination signatures distinguish embryonic stem cells derived by parthenogenesis and somatic cell nuclear transfer. Cell Stem Cell 2007;1(3):346–52.
63. Stewart CL. Production of chimeras between embryonic stem cells and embryos. Methods Enzymol 1993;225:823–55.
64. Mann JR. Surgical techniques in production of transgenic mice. Methods Enzymol 1993; 225:782–93.
65. Nagy A, Rossant J. Production of completely ES cell derived fetuses. In: Joyner AL, editor. Gene Targeting, First ed., IRL, Oxford;1993:147–78.

Parthenogenetic Embryonic Stem Cells in Nonhuman Primates

Neli Petrova Ragina and Jose Bernardo Cibelli

Abstract Parthenogenesis is a naturally occurring process where an oocyte is activated without sperm contribution. In mammals, parthenogenetic (PG) embryos cannot develop to term. The most commonly used method of artificially making diploid PG embryos is using via chemical activation of the egg and by preventing extrusion of the second polar body. Parthenogenetic embryonic stem (PGES) cells are derived from the inner cell mass of PG embryo at the blastocyst stage. They are pluripotent, i.e., they can differentiate into all three germ layers: ecto-, meso-, and endoderm, and can be propagated as stem cells in culture for prolonged periods of time. PGES cells offer an easily obtainable pool of stem cells that can be used as a source for derivation of autologous tissues, albeit limited to females in reproductive age. PGES cells derivation does not require destruction of a viable embryo and therefore bypasses the ethical debates surrounding the use of naturally fertilized embryos.

Nonhuman primates are the closest species to human in the tree of evolution and therefore are excellent models for studying human development and diseases. PGES cells from nonhuman primate and human parthenogenetically activated oocyte have recently been derived *(1–3)*. These cells offer a valuable tool for studying the developmental, differentiation, and functional potential of the PGES cells in the context of their clinical application in organ and tissue transplantations in humans.

Keywords Parthenogenesis • Stem cells • Imprinting

Introduction

Parthenogenesis is the process by which the oocyte is activated without paternal contribution. It is a natural way of reproduction in some lower organisms.

N.P. Ragina and J.B. Cibelli (✉)
Cellular Reprogramming Laboratory, Michigan State University, B270 Anthony Hall,
East Lansing, MI 48824, USA,
e-mail: cibelli@msu.edu

H. Baharvand (ed.), *Trends in Stem Cell Biology and Technology*,
DOI 10.1007/ 978-1-60327-905-5_3,
© Humana Press, a Part of Springer Science+Business Media, LLC 2009

Parthenogenesis leads to the formation of an embryo consisting of only maternal genome complement. The opposite process is called androgenesis, these are embryos of only paternal genome. Androgenetic embryos can be produced by fertilization of a previously enucleated oocyte by a haploid sperm followed by the duplication of its chromosomes *(1)*. Alternatively, an oocyte can be normally fertilized and the female pronucleus later removed. In mammals, parthenogenetic embryos are not compatible with life and are spontaneously aborted *(2)*. Sometimes, parthenogenetic development in the female reproductive tract can lead to the formation of ovarian teratomas. Ovarian teratomas are benign tumors consisting exclusively of maternal genomic complement *(3)*.

Although parthenogenetic embryos are not viable, in mouse, embryonic stem (ES) cells derived from the inner cell mass (ICM) of parthenogenetically activated oocytes were able to give rise to viable animals when inserted into a fertilized mouse blastocyst. In primates, parthenogenetic embryonic stem (PGES) cells have been derived from *Macaca fascicularis* (called PGES Cyno1 cells) and from rhesus monkey PGES cells. However, their contribution to a chimeric embryo in nonhuman primates has not been tested.

In humans, human PGES cells have also been successfully derived in vitro from human parthenogenetic blastocysts *(4–6)*. Moreover, a child whose peripheral blood leukocytes were entirely parthenogenetic by origin has been reported *(7)*. This phenomena indicates that although parthenogenetic development in utero is aborted and rarely detected, a chimeric individual can develop from a single zygote *(7, 8)*.

Parthenogenetic embryos can be artificially derived by chemical activation of the oocyte using ionomycin/dimethylaminopurine (DMAP). This particular protocol also prevents the extrusion of the second polar body that renders the embryonic genome diploid *(9)*. Since the parthenogenetic embryos are created without sperm contribution, they only contain the maternal genome.

Parthenogenetic (PG) fetuses exhibit severe growth and differentiation defects, fail to establish proper placental growth, and die early in gestation. One of the major reasons for this phenomenon has been attributed to deregulation of imprinted genes *(10–12)* (Fig. 1).

PGES cells are derived from the ICM of PG embryos at a blastocyst stage. PGES can be maintained as stem cells for a prolonged period of time. During the process of derivation and propagation in vitro, however, PGES cells are prone to chromosomal aberrations such as loss of one of the X-chromosomes, which renders the cells aneuploid. This phenomenon is mostly due to the presence of two active XX-chromosomes, which is associated with global reduction of DNA methylation. In the cell there is a selective mechanism against loss of methylation, which may provide the tendency of X-chromosome instability *(13, 14)* (Fig. 2).

Primate PGES cells have been derived for the first time by Cibelli et al. *(15)* in 2002 from *Macaca fascicularis* (Cyno-1) (Fig. 3). These cells exhibit normal ES cell morphology, i.e., small cytoplasmic/nuclear ratio, numerous nucleoli, and cytoplasmic lipid bodies. They stain positive for markers for primate ES cells and

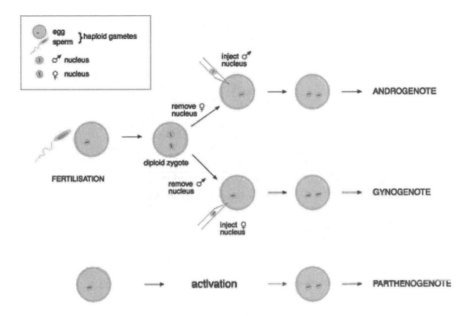

Fig. 1 Construction of diploid embryos with genomes derived from a single sex. In a fertilized egg, the male and female pronuclei can be removed before they fuse, and a diploid egg re-created by injection of another pronucleus. The development of these diploid eggs can then be followed in vitro by culturing or in vivo by transplantation to pseudopregnant mice. Parthenogenetic embryos can be created by the chemical activation of an unfertilized egg *(12)* (*see Color Plates*)

express pluripotency markers such as OCT4. They exhibit high telomerase activity and are karyotypically normal (42XX) *(16)*. The capacity of the Cyno-1 PGES cells to differentiate was tested in vivo by teratoma formation. Teratoma derived from Cyno-1 PGES cells demonstrated that they can give rise to all three germ layers: meso-, ecto-, and endoderm, such as cartilage, muscle, and bone (mesoderm), neurons, melanocytes, and hair follicles (ectoderm), and intestinal and respiratory epithelial (endoderm) with a preference toward ecto- and endoderm derivatives *(15)*. Tumors derived from Cyno-1 PGES cells consisted of mature differentiated tissues with a little contribution of mitotically dividing cells, which indicates their benign origin (Fig. 4). Cyno-1 PGES cells were successfully differentiated in vitro into neurons and were capable of producing neural specific mediators such as dopamine and serotonin. Further characterization of Cyno-1 cell-derived neurons demonstrates that these cells express tyrosine hydroxylase and are electrophysiologically active *(16, 17)*. When dopamine neurons derived from Cyno-1 cells were transplanted in vivo in rodents and in primate models of Parkinson's disease, they were able to maintain stable phenotype, express dopaminergic markers, and show long-term survival *(17, 18)* (Fig. 5).

Primate parthenogenetic stem cells have been derived from rhesus monkey as well *(19)*. Similar to Cyno-1 cells, when injected into immunocompromised mice,

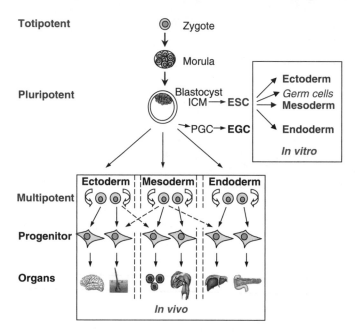

Fig. 2 Stem cell hierarchy. Zygote and early cell division stages (blastomeres) to the morula stage are defined as totipotent because they can generate a complex organism. At the blastocyst stage, only the cells of the inner cell mass (ICM) retain the capacity to build up all three primary germ layers, the endoderm, mesoderm, and ectoderm as well as the primordial germ cells (PGC), the founder cells of male and female gametes. In adult tissues, multipotent stem and progenitor cells exist in tissues and organs to replace lost or injured cells. At present, it is not known to what extent adult stem cells may also develop (transdifferentiate) into cells of other lineages or what factors could enhance their differentiation capability (*dashed lines*). Embryonic stem (ES) cells, derived from the ICM, have the developmental capacity to differentiate in vitro into cells of all somatic cell lineages as well as into male and female germ cells *(14)* (*see Color Plates*)

rhesus PGES cells were able to form teratomas consisting of derivatives of all tree germ layers (ecto-, meso-, and endoderm). The rhesus monkey ES cells were morphologically similar to normal biparental stem cells derived from the same species. They expressed the key pluripotency markers (such as OCT4, SSEA-3 and -4, TRA-1-60, NANAOG, SOX2, TDGF, LEFTYA, and TERT) and were karyotypically normal *(19)*.

A high degree of genomic homozygosity is expected from monoparental stem cell lines such as parthenogenetic and androgenetic embryonic stem cells. Genetic homozygosity is one of the major obstacles to potential implication of the PGES cells for cell replacement therapy due to the possibility of immune rejection. Dighe et al. *(19)* have demonstrated that diploid rhesus monkey PGES cells restored heterozygosity at 64% of the examined loci on average in the five PGES cell lines analyzed by using a microsatellite analysis (STR). Moreover, the rhesus monkey PGES cells demonstrated reestablishment of heterozygosity within 15 analyzed

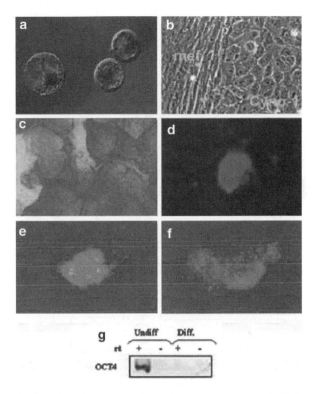

Fig. 3 Characterization of parthenogenetic Cyno-1 embryos and derived cell lines. (**a**) Parthenogenetically activated eggs at day 8 of development before inner cell mass (ICM) isolation. (**b**) Phase contrast of Cyno-1 stem cells growing on top of mitotically inactivated mouse feeder layer (mef). (**c**) Alkaline phosphatase staining. (**d**) Stage-specific embryonic antigen 4. (**e**) Tumor rejection antigen 1–60. (**f**) Tumor rejection antigen 1–81 staining. (**g**) Reverse Transcription Polymerase Chain Rxn (RT-PCR) octamer-binding transcription factor 4 expression in undifferentiated Cyno-1 cells. (Scale bars = 50 μm in (**a**), 10 μm in (**b**) and (**d–f**), and 4 mm in (**c**).) *(16) (see Color Plates)*

microsatellite markers of the major histocompatability (MHC) region in three of the five rhesus monkey PGES cells tested. This rendered the genotype of the MHC region of these three cell lines identical to the egg donors. The ability to generate isogenic MHC rhesus monkey PGES cells suggests the potential application of these cells as a source of a patient's own MHC-matched cells for therapeutic cell replacement therapy. Dighe et al. *(19)* have also genotyped one Cyno-1 PGES cell line for the 15 microsatellite markers of the MHC region. This Cyno-1 PGES cell line was homozygous for the MHC region STR markers.

Human PGES cells have been derived only recently *(4, 6)*. These cells stained positive for the major pluripotency markers, such as SSEA-3, SSEA-4, TRA1-60, TRA1-81, and OCT4, similar to nonhuman primate PGES cells and also exhibit similar morphology characterized by tightly packed colony with prominent

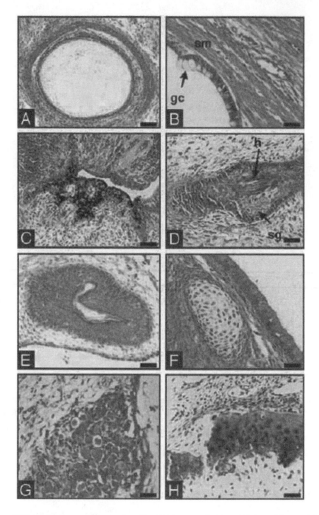

Fig. 4 In vivo differentiation of Cyno-1 cells. Cells were injected intraperitoneal in severe combined immunodeficient mice. Eight and 15 weeks after injection, teratomas 12 and 30 mm in diameter, respectively, were isolated, fixed with 10% paraformaldehyde, and paraffin embedded. Sections were stained with hematoxylin-eosin. The following complex structures were observed: gut (**a**), intestinal epithelium with typical goblet cells (gc), and smooth muscle (sm) (**b**), neuronal tissue with melanocytes (**c**), hair follicle complex with evident hair (h), and sebaceous gland (sg) (**d**), skin (**e**), cartilage (**f**), ganglion cells (**g**), and bone (**h**). (Scale bars = 40 μm in (**a**), 10 μm in (**b**) and (**d–h**), and 20 μm in (**c**).) *(16)* (*see Color Plates*)

nucleoli, and a small cytoplasm to nucleus ratio (Fig. 6). Upon induction of differentiation in vitro, the human PGES cells give rise to derivatives of all three germ layers *(6)*. DNA profiling of the human PGES cells revealed that these cells can also give rise to MHC matched PGES cells, isogeneic to the oocyte donor *(6)*. Recently a derivation of new lines of human (PGES) cells have been reported *(20)*.

Despite the ability to produce karyotypically normal, heterozygous at MHC locus primate and human PGES cells, a major obstacle to a potential successful

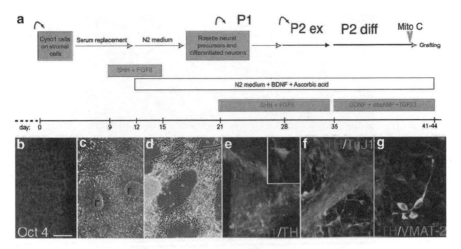

Fig. 5 In vitro differentiation of primate Cyno-1 parthenogenetic embryonic stem (PGES) cells. (a) Schematic representation of the sequential steps designed to induce a dopamine neuronal phenotype from undifferentiated primate embryonic stem (ES) cells. ES cells are first grown in a coculture system on stromal feeders. At the rosette stage, these neuroepithelial structures are replated on coated dishes in a feeder-free system for differentiation. (b–d) Microscopic images illustrating the aspect of the colonies at each passage. Undifferentiated stage Cyno-1 cells grow in colonies and all cells (b) express the transcription factor Oct-4. (c) Soon after the first passage, cells are organized into typical neuroepithelial structures (rosettes, r). (d) At the differentiation stage neurons grow in clusters and extend out neurites and (e) express midbrain genes like Engrailed-1 (En1), which is colocalized in some neurons with TH. (f) TH neurons coexpressed the neuronal marker β-tubulin III (clone TuJ1) and (g) some coexpressed VMAT-2 (yellow coexpression). (Scale bars = 100 μm (b–c), 150 μm (d), 75 μm (e), 50 μm (f), 35 μm (g) and inset in (e).) *Abbreviations: BDNF* brain-derived neurotropic factor; *FGF8* fibroblast growth factor 8; *GDNF* glial-derived neurotropic factor; *TH* tyrosine hydroxylase; *SHH* sonic hedgehog; *TGFβ3* transforming growth factor β3; *TH* tyrosine hydroxylase; *VMAT* vesicular monoamine transporter *(17)* *(see Color Plates)*

application of parthenogenetic stem cells in clinical trials is the severe deregulation of the imprinted genes, which renders these cells prone to tumorigenesis and other abnormalities upon engraftment/transplantation.

Imprinting in Parthenogenetic Primate Stem Cells

The phenomenon of imprinting refers to the differential expression of the maternally or paternally inherited alleles of specific genes (named imprinted genes) *(21)*. Thus a paternally imprinted gene is silenced when inherited from the paternal chromosome and expressed when inherited from the maternal chromosomeand vise versa. The expression pattern changes do not occur as a result of changes of the underlying DNA sequence but as a result of different epigenetic modifications, such as DNA methylation, histone modification, spreading of noncoding RNA molecules, small

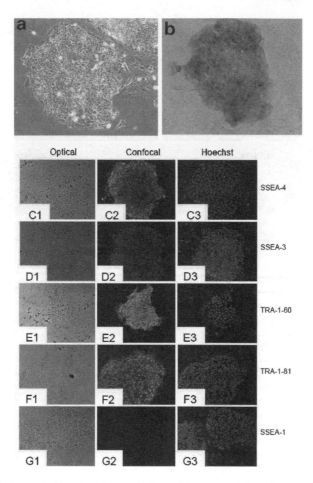

Fig. 6 Morphology, alkaline phosphatase (AP), and immunostaining of stem cell markers for human parthenogenetic embryonic stem-1 (hPES-1) cell line. (**a**) Morphology of hPES-1 colony under invert microscope; (**b**) AP staining of hPES-1; (**c**) SSEA4; (**d**) SSEA3; (**e**) TRA-1-60; (**f**) TRA-1-81; (**g**) SSEA1. (**c–g**) Listed are optical, confocal images and the corresponding Hoechst staining for the hPES-1 cells *(4)* *(see Color Plates)*

interfering RNAs, and other yet unidentified factors *(22)*. The imprinting pattern of the genes is initially erased and then reestablished during germ cells development in a parent-specific manner and is later maintained in all tissues and organs. The expression of the imprinted genes in most cases is under the control of an imprinting control region (ICR), which is differentially methylated at Cytosin-5 in CpG dinucleotide repeats. Imprinted genes are usually found in clusters. In most of the cases the differential methylation of the ICR can regulate the expression of an imprinted gene cluster bidirectionally *(23)*. In humans, loss of imprinting of certain genes such as *IGF2*, *H19*, *SNRPN*, *NDN*, and others, can lead to the development of several congenital disorders and cancer (Table 1). Parthenogenetic embryos and

Table 1 Diseases and syndromes that result from problems to the imprinting mechanisms or from errors in the imprinting of genes (23)

Disorder	Affected genes	Phenotype	Art link?
Angelman syndrome	Chromosome 15: maternal copy, loss of SNRPN imprinting	Mental retardation, ataxic gait, seizures, sociable disposition	Yes
Autism	Unknown X-linked gene (not always connected to imprinting)	Impaired language development, problems with social and motor skills	
Beckwith–Wiedemann syndrome	11p15 region: altered expression of IGF2, H19, and LIT1	Undescended testes, large newborn, seizures, abdominal wall defects	Yes
Cancer	Variable, e.g., IGF2 in lung cancer (not always connected to imprinting	Tumors	
ICF (immunodeficiency, centromeric region instability and facial anomalies syndrome)	DNMT3B	Immune problems, facial anomalies, growth retardation	
Paraganglioma	Paternal mutations SDHA (PGL1) and PGL2	Glomus tumors of the parasympathetic ganglia mainly in the head and neck region, tend to be slow growing and benign	
Prader–Willi syndrome	Chromosome 15: paternal copy	Undescended tests, mental retardation, short stature, obesity, small hands and feet	
Preeclampsia	Not yet defined	Serious complication of pregnancy	
Pseudohypoparathyroidism type IA (Albright hereditary osteodystrophy)	Imprinted GNAS cluster	Parathyroid hormone resistance, short stature, round face, and short hand bones	
Pseudohypoparathyroidism type IB	Imprinted GNAS cluster	Parathyroid hormone resistance localized to renal system, causing hypocalcaemia and hyperphosphatemia	
Rett syndrome	MeCP2	Childhood neurodevelopmental disorder mainly affecting females. Loss of motor function and mental retardation	
Silver–Russell syndrome	Cases which are imprinting related – chromosome 7	Short stature, excessive sweating, triangular face, inward curving fifth fingers and colored spots on the skin	

(continued)

Table 1 (continued)

Disorder	Affected genes	Phenotype	Art link?
Transient neonatal diabetes	An imprinted gene at 6q24. Candidates are ZAC and HYMAI	Growth retardation and diabetes which develops during the first 6 months of life but corrected by 18 months	
Turner syndrome	Complete or partial loss of second X chromosome	Affects females – short stature, social problems and ovarian failure	
Wilms' tumor	IGF2 loses imprinting	Childhood kidney tumor	

embryonic stem cells exhibit severe deregulation of the imprinted genes due to their uniparental origin, i.e., they lack the paternal set of the imprinted genes *(24–26)*. Loss of imprinting is one of the major causes for failure of the parthenogenetic embryos to develop to term and for the limited differentiation potential of the parthenogenetic stem cell *(11, 24–28)*. Additional changes in imprinting can arise in the process of derivation of parthenogenetic embryos due to yet not-identified reasons. Loss of imprinting of the parthenogenetic stem cells can occur during culturing or manipulation of these cells in vitro. This holds true and is extensively investigated in primate PGES embryonic stem and in normal primate biparental embryonic stem cells that are generated from a normally fertilized embryo *(29)*. The expression pattern and methylation status of two of the most extensively studied gene clusters, one on chromosome 11p15 and the other on chromosome 15p11 in primates and humans, have been extensively investigated *(29)*. Chromosome 11p15.5 harbors *IGF2* and *H19* imprinted genes, which expression is regulated by a common ICR, also called differentially methylated region (DMR), which lies between the two genes. In the maternal allele the DMR is not methylated, which allows an insulator protein to bind and prevent action of an enhancer downstream of *H19* to stimulate *IGF2* expression, and thus *H19* is expressed and *IGF2* is silenced. On the paternal allele DMR is methylated on the CpG dinucleotides, and the enhancer stimulates *IGF2* expression while *H19* is silenced.

The region on chromosomes 15q11-q13 is characterized with a set of oppositely imprinted genes, the most studied of which is *SNURF/SNRPN*, *UBE3A*, and *NDN*. *NDN* is paternally expressed gene, i.e., maternally imprinted. *SNURF/SNURPN* region is controlled by an imprinted center located at the 5′ end of the *SURF/SNURPN* gene *(30)*. The *SNURF/SNURPN* gene is expressed from the paternal allele. *UBE3A* is maternally expressed gene, i.e., paternally imprinted in the opposite direction of *SNURF/SNURPN* gene and its imprinted expression is restricted to certain tissues *(30)*.

Loss of imprinting of *IGF2* and *H19* and other imprinted genes has been associated with multiple congenital disorders such as Beckwith–Wiedemann syndrome *(31)* (*IGF2* and *H19*), Prader–Willi and Angelman syndromes *(32)* (*SNURF/SNRPN*, *UBE3A* and *NDN*), and many others *(33)*. In PG embryos and PGES cells there is a severe deregulation of the imprinted genes, with *H19* being expressed biallelically and therefore up-regulated, while *IGF2* is normally barely detectable *(11, 34)*. *H19* is a paternally imprinted gene (i.e., expressed from the maternal allele) and *IGF2* is maternally imprinted gene (i.e., expressed from the paternal allele). The *H19* gene codes for an untranslated mRNA molecule. During embryogenesis *H19* mRNA transcription is activated in a tissue-specific manner at certain developmental periods such as in the extraembryonic cell types at the time of implantation. Later in development, during midgestation, *H19* expression is observed in certain tissues and cells, most of which are derived from the endoderm germ layer, such as the developing liver, gut, muscle, and kidney. In adult tissues, *H19* expression becomes predominantly restricted to the skeletal muscles, thymus, heart, and lungs *(35)*.

The *IGF2* (insulin-like growth factor 2) gene encodes a member of the insulin family of polypeptide growth factors that is involved in development and growth.

IGF2 promotes differentiation and migration, inhibits apoptosis, and is essential for proper embryo and placental growth *(36)*. *IGF2* is expressed by the embryo during early pregnancy and later becomes localized to the cytoplasm of the trophoblast cells, the cells that are going to contribute to the formation of the placenta and other extraembryonic organs *(37)*. *IGF2* is also a potent mitogenic factor. *IGF2* is not only overexpressed in a number of human neoplasms *(38–40)*, but it also leads to malignant transformations, tumor development, and hyperplasia when overexpressed in mouse models *(41–43)*.

Due to deregulation of imprinted genes and *IGF2* and *H19* in particular, PG embryos are defective in ICM population maintenance and differentiation and in generation of multinucleated trophoblastic cells (syncytiotrophoblasts), leading to defective placentation and embryonic lethality *(25, 44)*.

It has been reported that biparental rhesus monkey embryonic stem cells exhibit loss of imprinting of *H19* and *IGF2* due to aberrant methylation of the DMR region *(29)*. Mitalipov et al. *(29)* took advantage of the availability of single nucleotide polymorphisms (SNPs) in the monkey to study the parental-specific methylation pattern of the *H19/IGF2* and *SNURF/SNRPN* imprinted genes in biparental rhesus monkey ES cells. They perform methylation analysis on 17 rhesus monkey ES cell lines for the methylation status of the DMR region by bisulfate sequencing and Southern blot using SNPs genotyping for defining the parent-of-origin region. They found severe deregulation of the methylation of the DMR region, which leads to biallelic expression of *IGF2* and *H19*. The methylation status and imprinting expression of *SNURF/SNRPN* imprinting center (IC) was not affected. These results suggest that the imprinting status of genes is influenced and can be changed during in vitro culturing, and that different genes have different degrees of susceptibility to the epigenetic changes induced in culture. This is specifically critical and of huge concern for the application and use of primate parthenogenetic stem cells (and all parthenogenetic stem cells in general) in clinics since by origin these cells lack the paternal imprints, and additional imprinting defects can accumulate during cell culture.

In a later report, the same group has analyzed the imprinted pattern of rhesus monkey PGES cell lines *(19)*. Expression levels of maternally expressed genes such as *UBE3A* and *H19* and paternally expressed genes such as *IGF2*, *PEG10*, *DIRAS3*, *SGCE*, *PEG3*, *MEST*, *ZIM2*, *PLAGL1*, *MAGEL2*, *MKRN3*, and *SNRPN* were analyzed by reverse transcription-polymerase chain reaction (RT-PCR) and by real time PCR analyses. A significant expression of paternally expressed genes *IGF2*, *SGCE*, and *DIRAS3* were detected in rhesus monkey PGES cells that normally are expected to be silenced. Moreover, transcription from some of the other paternally expressed genes, i.e., maternally imprinted, was detected. To confirm the gene expression data, the methylation status of the *H19/IGF2* and *SNURF/SNURPN* IC regions was evaluated by employing a methylation sensitive Southern blot and bisulfate sequencing analysis *(19)*. The methylation profiling of these two well-defined imprinted centers demonstrated sporadic hypermethylation of the *H19/IGF2* IC region, which normally is expected to be hypomethylated on both maternal alleles and can explain the significant upregulation of *IGF2* gene expression in rhesus monkey PGES cells reported in this study. The methylation status of *SNURF/SNURPN* IC region was, as expected with the region of both maternal alleles, being completely methylated *(19)*. These data reinforce the notion that multiple factors such as stem cell derivation, culture conditions, and manipulation, combined with the intrinsic genetic instability that PGES cells possess, may cause changes in the methylation status and inappropriate expression of the imprinted genes.

The expression pattern of several of the imprinted genes such as *H19*, *IGF2*, *SNRPN*, *NDN*, *PEG10*, and *P57* was analyzed in Cyno-1 cells as well by the means of quantitative real time PCRon cDNA derived from total RNA (Cibelli, personal communication. As in rhesus monkey PGES cells, *H19* mRNA expression levels were significantly higher in the Cyno-1 PGES cells compared to that of normal biparental *Macaca fascicularis* embryonic stem cells. The expression of the paternally expressed gene *PEG10* was low but detectable, while the expression of another set of paternally expressed genes, *SNRPN* and *NDN*, was undetectable, as expected according to their parent of-origin imprinting status. *P57* is a maternally expressed gene, i.e., paternally imprinted, and therefore high levels of *P57* mRNA expression in Cyno-1 PGES cells were detected compared to a normal biparental cell line (Cibelli, personal communication). Significant up-regulation of *IGF2* gene expression, normally imprinted on the maternal allele, was detected in Cyno-1 PGES cells, similar to that observed in rhesus monkey PGES cells. An interesting observation was that upon down-regulation of *H19* in Cyno-1 PGES cells by using a short hairpin RNA (shRNA), the mRNA levels of *IGF2* dropped significantly as well. The same phenomenon was observed upon down-regulation of *H19* in mouse PGES cells. The causes of this unexpected observation are yet to be investigated.

The expression patterns of *H19*, *TSSC5*, *PEG1*, and *SNURPN* have been analyzed in human PGES cells *(6)*. The expression of *PEG1* and *SNRPN*, two paternally expressed genes, was lower but detectable in human PGES cells compared to the expression levels in normal biparental human embryonic stem cells. These data indicate that human PGES cells also exhibit deregulation of some of the imprinted genes analyzed, similar to that observed in nonhuman primate PGES cell. Another set of imprinted genes suggested of being important for development of parthenogenetically derived embryos and for proper differentiation of PGES cells are the oppositely imprinted genes *DLK-DIO3* located on chromosome 14 in human and nonhuman primates. It was recently reported that parthenogenetic mice can develop to term when the *H19* DMR and *Dlk1-Dio3* intergenic germline-derived DMR region are deleted *(45)*.

Thus PGES and ES cells derived from normally fertilized embryos exhibit aberrant expression of the imprinted genes when cultured in vitro. This phenomenon raises a concern regarding the safe use of these cells in clinical trials since the majority of the imprinted genes control cell growth, division, and proliferation and may assist in the development of tumors or disease phenotypes *(38–43)*. Nevertheless, when induced to differentiate or after manipulation of the ICR regions of key imprinted genes such as *H19*, *IGF2*, *DLK-DIO3*, these cells were able, by yet unknown mechanism, to give rise to terminally differentiated tissues or to viable animals capable of producing offspring *(27, 28, 45, 46)*.

Primate Parthenogenetic Stem Cells:
Future Applications and Concerns

The rate of the organ transplantations performed in the United States is increasing *(47)*. In 2006 a record total of 28,923 organ transplant operations have been performed, which is up from 28,112 the year before, as reported by the New York Organ Donor Network. Despite the increasing number of organ transplantations, the pool of patients waiting for organ donations exceeds the supply *(48)*. There is a lack of awareness in the U.S. population for the need to donate organs that must be addressed if this gap between supply and demand is to be closed.

Even when donors are available, a major issue in organ and tissue transplantation continues to be tissue matching. Engraftment of solid organs, tissues, and cells from unrelated donors carries a high risk of rejection due to lack of compatibility between the MHC alleles of the recipient and donor. This obstacle can be surpassed by deriving tissues and organs from a cell pool that contains the patient's own genome. PGES cells can be an alternative source of matched cells for women in reproductive age.

In the context of tissue transplantation, primate PGES cells have not yet been extensively studied. More experiments are needed to determine the extent to which these cells can differentiate. Ideally chimeric primate PGES embryos could be generated by injecting PGES cells into a blastocyst of a fertilized embryo, or more stringent yet, by aggregation of PGES cells with tetraploid embryo (tetraploid complementation experiment) *(49, 50)*. Due to their close relationship to humans, nonhuman primate PGES offer a valuable tool not only as a model to study human disease and epigenetics, but also as a model to study their potential to serve as a future source for cell replacement therapy.

There are two questions still to be answered regarding the plasticity of PGES cells. If parthenogenesis does not lead to full-term development, due to failure to produce all organs and tissues of the embryo and the extra-embryonic membranes, how is it possible that ES cells derived from a parthenogenetic blastocyst can give rise to all embryonic germ layers? Recent data show that one of the leading factors is epigenetic changes in the stem cells genome that occur during in vitro culture conditions *(26)*. Some of these culture-induced epigenetic changes may affect a subpopulation of the PGES cells and confer them with growth and/or differentiation advantages that are preferentially selected for in culture. Therefore, it is of great importance that terminal differentiation of these cells can be achieved before transplanting PGES cells derivatives in order to ensure that no pluripotent, epigenetically abnormal PGES cells exist in the graft that may lead to tumors and/or may have yet unknown consequences for the recipient.

Current efforts focus on optimizing the nonhuman primate and human PGES growth conditions as to minimize the presence of animal proteins and other contaminants in the growth media that have the potential to induce a severe immune response upon transplantation.

The use of feeder-free system for manipulation and propagation of the nonhuman primates and human PGES cells will prevent contamination of the PGES cells intended for clinical application with feeder layer cells or their metabolite products. This is another way to ensure the PGES cell purity and thus ameliorate the risk of rejection upon engraftment into the recipient.

Conclusion

Nonhuman primate PGES cells are a valuable tool to study the potential application of PGES cells for cell replacement therapy in humans. Similar to the embryonic stem cells derived from fertilized embryos, PGES cells stain positive for the major pluripotency markers such as SSEA-3, SSEA-4, TRA1-60, TRA1-81, and OCT4. PGES cells morphology is indistinguishable from that of normal biparental ES cells. Unlike ES cells derived from normal biparental embryos, PGES cells derivation does not require destruction of a viable embryo and therefore bypass the ethical issues that some members of society possess.

Moreover, although epigenetically instable, when injected into immunocompromised mice, primate PGES cells have been able to give rise to teratomas consisting of derivatives of all three germ layers. This phenomenon suggests that primate PGES cells are able to correct for the imprinting misexpression upon induction of differentiation and generate terminally differentiated derivatives of ecto-, endo-, and mesoderm origin. This might be due to culture conditions that ultimately have an effect on the epigenetic status of the cells or might be due to yet unidentified factors.

Despite the need for more extensive studies in vivo and despite their unstable genetic and epigenetic status, primate parthenogenetic stem cells offer an unlimited source of pluripotent stem cells and a valuable avenue for studying the molecular basis of human congenital disorders caused by deregulation of imprinting.

References

1. Matsuda T, Wake N. Genetics and molecular markers in gestational trophoblastic disease with special reference to their clinical application. Best Pract Res Clin Obstet Gynaecol 2003;17:827–36.
2. Mutter GL. Role of imprinting in abnormal human development. Mutat Res 1997;396: 141–7.
3. Surti U, Hoffner L, Chakravarti A, Ferrell RE. Genetics and biology of human ovarian teratomas. I. Cytogenetic analysis and mechanism of origin. Am J Hum Genet 1990;47:635–43.
4. Mai Q, Yu Y, Li T, et al. Derivation of human embryonic stem cell lines from parthenogenetic blastocysts. Cell Res 2007;17:1008–19.
5. Revazova ES, Turovts NA, Kochetkova OD, et al. HLA homozygous stem cell lines derived from human parthenogenetic blastocysts. Cloning Stem Cells 2008 10:11–24.
6. Revazova ES, Turovts NA, Kochetkova OD, et al. Patient-specific stem cell lines derived from human parthenogenetic blastocysts. Cloning Stem Cells 2007;9:432–49.
7. Strain L, Warner JP, Johnston T, Bonthron DT. A human parthenogenetic chimaera. Nat Genet 1995;11:164–9.

8. Surani MA. Parthenogenesis in man. Nat Genet 1995;11:111–3.
9. Mitalipov SM, Nusser KD, Wolf DP. Parthenogenetic activation of rhesus monkey oocytes and reconstructed embryos. Biol Reprod 2001;65:253–9.
10. Sotomaru Y, Katsusava Y, Hatada I, Obata Y, Sasaki H, Kono T. Unregulated expression of the imprinted genes H19 and Igf2r in mouse uniparental fetuses. J Biol Chem 2002;277: 12474–8.
11. Ogawa H, Wu Q, Komiyama J, et al. Disruption of parental-specific expression of imprinted genes in uniparental fetuses. FEBS Lett 2006;580;5377–84.
12. Lyle R. Gametic imprinting in development and disease. J Endocrinol 1997;155:1–12.
13. Zvetkova I, Apedaile A, Ramsahoye B, et al. Global hypomethylation of the genome in XX embryonic stem cells. Nat Genet 2005;37:1274–9.
14. Wobus AM, Boheler KR. Embryonic stem cells: prospects for developmental biology and cell therapy. Physiol Rev 2005;85:635–78.
15. Cibelli JB, Grant KA, Pahapman KB, et al. Parthenogenetic stem cells in nonhuman primates. Science 2002;295:819.
16. Vrana KE, Hipp JD, Goss AM, et al. Nonhuman primate parthenogenetic stem cells. Proc Natl Acad Sci U S A 2003;100 Suppl 1:11911–6.
17. Sanchez-Pernaute, R, Stder L, Ferrari D, et al. Long-term survival of dopamine neurons derived from parthenogenetic primate embryonic stem cells (Cyno-1) after transplantation. Stem Cells 2005;23:914–22.
18. Ferrari D, Sanchez-Pernaute R, Lee H, Studer L, Isacson O. Transplanted dopamine neurons derived from primate ES cells preferentially innervate DARPP-32 striatal progenitors within the graft. Eur J Neurosci 2006;24:1885–96.
19. Dighe V, Clepper L, Pedersen D, et al. Heterozygous embryonic stem cell lines derived from nonhuman primate parthenotes. Stem Cells 2008;26:756–66.
20. Cheng L. More new lines of human parthenogenetic embryonic stem cells. Cell Res 2008;18:215–7.
21. Sapienza C. Parental imprinting of genes. Sci Am 1990;263:52–60.
22. Tang Wy, Ho SM. Epigenetic reprogramming and imprinting in origins of disease. Rev Endocr Metab Disord 2007;8:173–82.
23. Swales AK, Spears N. Genomic imprinting and reproduction. Reproduction 2005;130: 389–99.
24. Fujimoto A, Mitalipov SM, Kuo HC, Wolf DP. Aberrant genomic imprinting in rhesus monkey embryonic stem cells. Stem Cells 2006;24:595–603.
25. Newman-Smith ED, Werb Z. Stem cell defects in parthenogenetic peri-implantation embryos. Development 1995;121:2069–77.
26. Horii T, Kimura M, Morita S, Nagao Y, Hatada I. Loss of genomic imprinting in mouse parthenogenetic embryonic stem cells. Stem Cells 2008;26:79–88.
27. Kono T, Obata Y, Wu Q, et al. Birth of parthenogenetic mice that can develop to adulthood. Nature 2004;428:860–4.
28. Kono T, Sotomaru Y, Katsuzawa Y, Dandolo L. Mouse parthenogenetic embryos with monoallelic H19 expression can develop to day 17.5 of gestation. Dev Biol 2002;243:294–300.
29. Mitalipov S, Clepper L, Sritanaudomchai H, Fujimoto A, Wolf D. Methylation status of imprinting centers for H19/IGF2 and SNURF/SNRPN in primate embryonic stem cells. Stem Cells 2007;25:581–8.
30. Runte M, Kroisel PM, Gillessen-Kaesbach G, et al. SNURF-SNRPN and UBE3A transcript levels in patients with Angelman syndrome. Hum Genet 2004;114:553–61.
31. Reik W, Maher ER. Imprinting in clusters: lessons from Beckwith–Wiedemann syndrome. Trends Genet 1997;13:330–4.
32. Nicholls RD, Saitoh S, Horsthemke B. Imprinting in Prader–Willi and Angelman syndromes. Trends Genet 1998;14:194–200.
33. Hurst LD, McVean GT. Growth effects of uniparental distomes and the conflict theory of genomic imprinting. Trends Genet 1997;13:436–43.

34. Sotomaru Y, Kawase Y, Ueda T, et al. Disruption of imprinted expression of U2afbp-rs/U2af1-rs1 gene in mouse parthenogenetic fetuses. J Biol Chem 2001;276:26694–8.
35. Poirier F, Chan CT, Timmons PM, Robertson EJ, Evans MJ, Rigby PW. The murine H19 gene is activated during embryonic stem cell differentiation in vitro and at the time of implantation in the developing embryo. Development 1991;113:1105–14.
36. Han VKM, Carter AM. Spatial and temporal patterns of expression of messenger RNA for insulin-like growth factors and their binding proteins in the placenta of man and laboratory animals. Placenta 2000;21:289–305.
37. Pringle KG, Roberts CT. New light on early post-implantation pregnancy in the mouse: roles for insulin-like growth factor-II (IGF-II)? Placenta 2007;28:286–97.
38. Minniti CP, Luan D, O'Grady C, Rosenfeld RG, Oh Y, Helman LJ. Insulin-like growth factor II overexpression in myoblasts induces phenotypic changes typical of the malignant phenotype. Cell Growth Differ 1995;6:263–9.
39. Pacher M, Seewald MJ, Mikula M, et al. Impact of constitutive IGF1/IGF2 stimulation on the transcriptional program of human breast cancer cells. Carcinogenesis 2007;28:49–59.
40. Prelle K, Wobus AM, Krebs O, et al. Overexpression of insulin-like growth factor-II in mouse embryonic stem cells promotes myogenic differentiation. Biochem Biophys Res Commun 2000;277:631–8.
41. Weber MM, Fottner C, Schmidt P, et al. Postnatal overexpression of insulin-like growth factor II in transgenic mice is associated with adrenocortical hyperplasia and enhanced steroidogenesis. Endocrinology 1999;140:1537–43.
42. Moorehead RA, Sanchez OH, Baldwin RM, Khokha R. Transgenic overexpression of IGF-II induces spontaneous lung tumors: a model for human lung adenocarcinoma. Oncogene 2003;22:853–7.
43. Petrik J, Pell JM, Arany E, et al. Overexpression of insulin-like growth factor-II in transgenic mice is associated with pancreatic islet cell hyperplasia. Endocrinology 1999;140:2353–63.
44. Newman-Smith E, Werb Z. Functional analysis of trophoblast giant cells in parthenogenetic mouse embryos. Dev Genet 1997;20:1–10.
45. Kawahara M, Wu Q, Takahashi N, et al. High-frequency generation of viable mice from engineered bi-maternal embryos. Nat Biotechnol 2007;25:1045–50.
46. Kono T, Kawahara M, Wu Q, et al. Paternal dual barrier by Ifg2-H19 and Dlk1-Gtl2 to parthenogenesis in mice. Ernst Schering Res Found Workshop 2006;23–33.
47. Sheehy E, Conrad SL, Brigham LE, Hiura H, Obata Y. Estimating the number of potential organ donors in the United States. N Engl J Med 2003;349:667–74.
48. Coombes JM, Trotter JF. Development of the allocation system for deceased donor liver transplantation. Clin Med Res 2005;3:87–92.
49. Nagy A, Gocza E, Diaz EM, et al. Embryonic stem cells alone are able to support fetal development in the mouse. Development 1990;110:815–21.
50. Nagy A, Rossant J, Nagy R, Abramow-Newerly W, Roder JC. Derivation of completely cell culture-derived mice from early-passage embryonic stem cells. Proc Natl Acad Sci U S A 1993;90:8424–8.

Nuclear and Somatic Cell Genetic Reprogramming

Maurizio Zuccotti, Silvia Garagna, and Carlo Alberto Redi

Abstract Nuclear and somatic cell genetic reprogramming has seen a huge improvement during the past 10 years since the cloning of the first mammal, Dolly the sheep. In this chapter we will summarise the advancement in nuclear and cell reprogramming by cell fusion, using amphibian eggs or egg extracts, with cell extracts, with synthetic molecules, or by induced expression of specific genes. The latter method of cell reprogramming represents a novel strategy that could allow the use of the patient's own somatic cells for cell therapy, overcoming the incompatibility of heterologous cell transplantation.

Keywords Nuclear transfer • Cell reprogramming • Cell fusion • Cell extracts • iPS

Introduction

Until a few years ago scientists believed that the genetic programme of a mammalian terminally differentiated somatic cell was not reversible, they thought that cytoplasmic determinants and epigenetic mechanisms locked the cell to a status of permanent differentiation.

Although this remains true for differentiated cells in the tissues of living animals, it is not true when cells are treated under particular experimental and culture conditions or their nuclei are transferred into or fused with the cytoplasm of undifferentiated cells. In this chapter we will summarise the main results that have realised the possibility to change the status of differentiation of a nucleus or of a cell as a whole.

M. Zuccotti (✉), S. Garagna, and C.A. Redi
Dipartimento di Medicina Sperimentale, Sezione di Istologia ed Embriologia, Universita' degli Studi di Parma, Via Volturno 39, 43100 Parma, Italy

H. Baharvand (ed.), *Trends in Stem Cell Biology and Technology*,
DOI 10.1007/ 978-1-60327-905-5_4,
© Humana Press, a Part of Springer Science+Business Media LLC 2009

Nuclear Reprogramming

At the beginning of the twentieth century, Hans Spemann first proposed to isolate the nucleus of an embryonic or adult cell and transfer it into the cytoplasm of an enucleated oocyte. In other words, he suggested the first nuclear transfer experiment to understand if the nucleus of a differentiated cell would still be capable of reprogramming the expressed information and then reassume embryonic development. The inadequacy of the technology limited Spemann in his attempt to carry out this experiment, but others after him tried and succeeded. Amphibians were the first animals to have their nuclei reprogrammed when transferred into an enucleated oocyte. John Gurdon in 1966 (1, 2) demonstrated that nuclei of differentiated hepatocytes when transferred into enucleated oocytes could support full-term development and give birth to live individuals that reached adult life.

Despite this success with amphibians, most researchers thought that genetic reprogramming of terminally differentiated nuclei would have been impossible with mammals, for at least two lines of reasoning. First, full-term mammalian development requires the contribution of both male and female genomes (3, 4) due to the different epigenetic status that they have acquired during gametogenesis, a phenomenon known as genomic imprinting. Genomic imprinting must clearly be in place to support correct and full-term development of the reconstructed embryos. How this happens remains unclear. Genomic imprinting has been studied since the discovery in the early 1970s that the paternal X chromosome was preferentially inactivated in the trophectoderm cell lineage of the female preimplantation mouse blastocyst (4, 5). Large parts of the genome, including an entire chromosome (X-chromosome), are subject to a different epigenetic signature that regulates the expression of the maternal and paternal alleles during development. To date, more than 80 genes are under study for their possible imprinted status in both the human and mouse genomes (see http://www.geneimprint.com). DNA methylation, histone acetylation/methylation, chromatin organisation, and nuclear architecture are epigenetic mechanisms involved in modelling and modulating the expression of imprinted genes. The methylation of CpG islands in the promoter region of imprinted genes is one of the mechanisms most studied and better understood. Imprinted methylation is established during gametogenesis and erased at the next generation in the germ line. Erasure occurs in primordial germ cells (PGCs) after they have settled in the forming gonads and prior to the entrance in meiosis, in the mouse between 10.5 and 12.5 dpc. Imprinting is rebuilt in a gender-specific manner during either spermatogenesis or oogenesis by de novo methyltransferase (DNMT3A) and its cofactor DNMT3-like (DNMT3L) (6, 7). Soon after fertilisation, the male genome replaces protamines with histones and then is actively demethylated (Fig. 1), likely by a demethylase; instead, the female genome is passively demethylated, probably through the exclusion of DNMT1o (a form of DNMT1 present in the oocyte) from the nuclei of preimplantation embryos. Transposons and imprinted genes are protected from the demethylation occurring during preimplantation development. How protection is accomplished is yet unknown, although the protein STELLA has been suggested as a possible candidate, since the deletion of its gene from oocytes results

Cell Reprogramming

Reprogramming by Cell Fusion

One of the most used methods of cell reprogramming is fusion between cells (Fig. 2). Heterokaryons may be obtained by spontaneous or induced fusion. When mouse bone marrow cells are grown on a feeder layer of mitotically inactive fibroblasts,

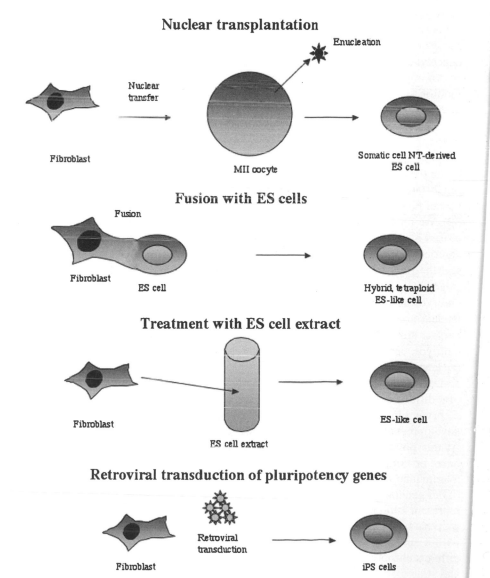

Fig. 2 Different strategies used for cell reprogramming [adapted from (79)]

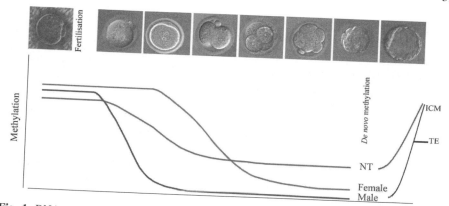

Fig. 1 DNA methylation profile of mouse preimplantation development in embryos obtained after fertilisation or nuclear transfer [adapted from (78)] (*see Color Plates*)

in early preimplantation lethality and loss of methylation of a number of imprinted genes. In cloned embryos, DNA demethylation occurs soon after embryo activation (Fig. 1), but it is less rapid than in the male genome, and several studies have shown aberrant methylation profiles compared to in vitro or in vivo embryonic development (6, 8–10). Once they have reached the blastocyst stage, the cells of the inner cell mass have a methylation similar to that of control embryos, whereas that of trophectoderm cells show aberrant methylation (8).

The proof that the nuclei of terminally differentiated cells are capable of supporting full-term development when transferred into an enucleated oocyte would suggest that developmental genes are correctly expressed and imprinted genes are somehow correctly marked. The epigenetic signature of an adult somatic genome must be erased and replaced with a new developmental blueprint. The demethylating environment that characterises the ooplasm at the very beginning of development may favour this erasing activity. At this stage, we must assume that the genome of a terminally differentiated somatic cell either retains the original biparental imprinting or that the imprinting is built up following nuclear transfer; the latter hypothesis is improbable, as it is unlikely that the ooplasm could build up a biparental imprint. Imprinted genes may still own protective mechanisms that allow them to escape demethylation when transferred into the oocyte (e.g., the protein STELLA), or in the hypothesis that they are demethylated, they may still maintain a target of recognition [i.e., preexisting histone marks (11)] that allows their de novo methylation.

A recent report has challenged our viewpoint of the imprinting phenomenon. This study has shown that imprinted genes regulated by two paternally methylated imprinting-control regions (the H19 differentially methylated region [DMR] and the Dlk1-Dio3 intergenic germ line–derived DMR) are the only paternal barriers that prevent the normal development of bimaternal mouse foetuses to term (12). This experiment suggests that the number of paternally imprinted genes that must be correctly imprinted is lower than expected, and thus perhaps the reprogramming process, as far as the imprinted genes are concerned, is less problematical.

A second hurdle that the reconstructed embryos may encounter during the early stages of preimplantation development is the activation of the embryonic genome (zygotic genome activation, ZGA). Following fertilisation, the zygote changes chromatin organisation from a meiotic to a postmeiotic status with a functional compartmentalisation of the genome capable of influencing gene expression (13). In the mouse, ZGA largely depends on a remodelling of chromatin organisation that leads to a first minor burst of gene activation at the G2 phase of the first cell cycle and a second major burst during the G2 phase of the second cell cycle (14). In other species ZGA occurs later: at the 4-cell stage in rabbit and pig, at the 4/8-cell stage in the human embryo, at the 8/16-cell stage in sheep and cow. It has been understood that ZGA occurs at a precise time after fertilisation, a phenomenon known as "zygotic clock" (15, 16). Approximately five thousand novel genes, whose transcripts are not present in metaphase II oocytes, are expressed in preimplantation development (17–19). Currently, the identity of and the role played by these genes during preimplantation development is unknown; however, lack of or incorrect timing of embryonic gene expression leads to the death of the embryo (15, 20).

In order to proceed with correct development, the reconstituted mouse embryo must initiate and, as far as the ZGA is concerned, complete genome reprogramming before the end of the second cell cycle, when embryonic activation occurs. The analysis of gene expression has shown conflicting results. Some genes seem to be expressed correctly, others are clearly abnormally expressed (21–24). In our study, the genes analysed were activated in nuclear transfer (NT) preimplantation embryos at approximately the correct time compared to control embryos, indicating that the reprogramming phenomenon is developmentally regulated and that the somatic genome is quickly rearranged toward an embryonic type of expression during the early stages of segmentation (25). However, the quantitative profile of gene expression showed high variability in cloned compared to control embryos. Others have found less variability in the profiles of gene expression in cloned embryos than in in vitro fertilisation (IVF) embryos, when compared with those obtained by natural fertilisation (26–28).

As a whole these studies show that reprogramming is not completed by the blastocyst stage. Incomplete reprogramming seems to happen at the time when the trophectoderm first arises, and this is reflected by the high number of phenotypic abnormalities and high rates of embryonic loss (29, 30).

Further reprogramming occurs at the epiblast stage, and this has been shown studying the process of X-chromosome inactivation. In normal female embryos, the paternal X-chromosome is preferentially inactivated at the 4-cell stage, then paternal X-inactivation is maintained in the trophectoderm, whereas within the epiblast the inactive X is reactivated and further random inactivation occurs (31, 32). In female reconstructed embryos, produced with a somatic nucleus with an inactive X-chromosome, there is first an inactivation of both X-chromosomes, then, the reprogramming activity continues with a random reactivation of one X-chromosome in the epiblast (33), suggesting that the reprogramming process is started in the egg and then continues at least until the epiblast stage.

Normal embryonic development proceeds through a series of check points that are also bonds for NT embryonic development. A correct epigenetic reprogramming may represent an important requirement for completion of development of both normal and NT embryos. Besides DNA methylation, other epigenetic mechanisms are probably involved in the reprogramming process. Covalent histone modifications, the introduction of histone variants into nucleosomes, and more in general chromatin remodelling and nuclear architecture have been the topics of intensive studies. Recent work has suggested a link between histone modifications and DNA methylation that may be involved in reprogramming (34–36). When the nuclei of mouse cumulus cells were transferred into enucleated mouse MII oocytes, histones H3 and H4 that are acetylated in these somatic cells were rapidly deacetylated; on the contrary, the same nuclei showed only a slight decrease in histone methylation. The ability to deacetylate the chromatin disappeared rapidly after oocyte activation (37).

An important question in developmental biology is whether epigenetic genome modifications in early embryos represent constraints to the acquisition of a specific nuclear organisation. To answer this question we have recently compared the nuclear architecture of embryos that begin development with nuclei carrying drastically different epigenetic signatures (i.e., preimplantation embryos obtained by IVF or NT) to learn whether they attained common higher-order chromatin arrangements or, on the contrary, whether they followed different patterns of higher-order nuclear organisation. Our research, in addition to that of others', has analysed the localisation of kinetochores in relation to the formation of chromocentres and nucleoli (38, 39). We found that each stage of preimplantation embryonic development is characterised by a stage-specific spatial organisation of nucleoli, kinetochores, and pericentric heterochromatin. Also, we found that, despite differences in the frequencies and the time course of nuclear architecture reprogramming events, by the 8-cell stage NT embryos achieve the same distinct nuclear organisation in the majority of embryos as observed for IVF embryos (39). At this stage the gametic or somatic nuclear architecture of IVF or NT embryos is replaced by a common embryonic nuclear architecture. This finding suggests that the epigenome of the three types of embryos partially acts as a constraint of the nuclear organisation of the three nuclear subcompartments analysed.

During the past 10 years, since the cloning of the first mammal from the nucleus of a differentiated somatic cell, our knowledge of the problems that reconstructed embryos are facing in order to complete preimplantation and postimplantation development has highly improved. However, much still remains to be understood. At this point we have gathered most of the information that could be collected based on our present understanding of the molecular mechanisms that are regulating mammalian development. Further understanding will come from an improvement of our general knowledge of mammalian development and clearly NT embryos will represent an important model study.

What is important is the proof of principle that has been put forward: the idea that within specific environmental conditions a genome can be moulded to express different phenotypes. These results have opened the way to a number of new approaches and strategies to try to reprogramme not only the nucleus, but also the whole cell.

these cells expressed markers from both the cell types and expressed endothelial-specific genes when plated on Matrigel matrix (40). Embryonic stem (ES) cells fuse spontaneously with somatic cells under conditions of coculture (41, 42). Neurophere cells, obtained from the forebrain of a 14.5-dpc mouse, when cultured with ES cells spontaneously fused with them; the heterokaryons obtained expressed markers from both types of cells, they made ES cell-like colonies, held a tetraploid chromosome complement, and contributed to chimeras (42). Another example of spontaneous cell fusion was obtained when hygromycin-sensitive mouse ES cells were cocultured with hygromycin-resistant primary murine brain cells (43). The resulting hybrids formed hygromycin-resistant ES cell-like colonies, expressed the stem-cell specific *Foxd3* gene from the genome of the neural cells, and contributed to all three germ layers in chimeras.

The reprogramming activity that the studies described above have demonstrated in ES cells was shown, through experiments of induced cell fusion, in other pluripotent cell types long before. When embryonal carcinoma (EC) cells are fused with somatic cells, the heterokaryons undergo reactivation of the X-chromosome (44), underlining the reprogramming activity of these pluripotent cells. Fusion of embryonic germ (EG) cells with thymocytes induces demethylation of several genes, including some imprinted genes (45). When hybrids where generated by fusion of male Hprt$^{-/-}$ mouse ES cells with female splenocytes, they formed embryoid bodies with cell types of all three germ layers and contained synchronously replicating X-chromosomes, suggesting that the inactive X-chromosome of splenocyte origin was reactivated (46). In another study, the fusion of female mouse thymocytes containing a green fluorescent protein (GFP) transgene under the control of the *Oct-4* promoter with male mouse ES cells, produced heterokaryons that expressed GFP, contained reactivated X-chromosomes of thymocyte origin, and contributed to all the three germ layers following the formation of chimeras (47). In these hybrids, the methylation of imprinted genes was left unchanged, whereas nonimprinted genes and heterochromatic centromeric regions were demethylated. Other epigenetic marks are modified following the fusion of somatic cells with ES cells. Heterokaryons exhibited histones H3 and H4 hyperacetylation, global lysine 4 and H3 di- and tri-methylation, and hyperacetylation of lysine 4 and H3 in the promoter of the expressed *Oct-4* gene (48).

One major problem for the use of a cell fusion–based therapy is the instability of the cell lines obtained and the loss of chromosomes (49). Some researchers have tried to remove the nucleus of the reprogramming cell, either before or after fusion (the resulting cell is called a cybrid). However, several studies have shown that induction of pluripotency in heterokaryons between ES and somatic cells requires ES nuclear activity. In fact, when prior to cell fusion ES cell nuclei are removed, the somatic nucleus is not reprogrammed to express the *Oct-4* gene (50). These experiments suggest that reprogramming must include nuclear factors of the pluripotent cell, and the production of a cytoplast must take this into account, as a cytoplast without nuclear components will lack the factors needed for the repro-gramming process to be carried out and completed. Another factor that may play an important role in reprogramming is NANOG. NANOG is an activator of pluripotent gene expression in mouse neural stem cells (NSCs). When ES cells overexpressing the *Nanog* gene are fused with NSCs, *Oct-4* is activated and *Olig2* and *Blbp* (two NSC specific genes) are down-regulated (51).

Another cell type that was used for fusion-mediated reprogramming is the PGC. PGCs isolated from male and female genital ridges, undergo imprinting erasure, and express genes of pluripotency. They can be cultured to form embryoid germ cells (EGCs). When ROSA26βgeo mouse PGCs were electrofused with thymocytes, a number of imprinted genes underwent demethylation, assuming a DNA methylation state similar to that of EGCs (45). When these cell hybrids were injected into blastocysts, β galactosidase was widely expressed in chimeric embryos at 9.5 and 10.5 dpc.

Reprogramming by Induced Expression of Specific Genes

A recent study described how the introduction of extra copies of *Oct-4*, *c-Myc*, *Sox-2*, and *Klf-4* genes endowed mouse embryonic fibroblasts with permanent pluripotent characteristics of ES cells, including the capacity to differentiate into cells of all three germ layers in vitro and in vivo, following engraftment into immunocompromised hosts (52). These cells were called induced pluripotent stem (iPS) cells (Fig. 2). This study has given an example of how the introduction and the expression of as few as four inducible genes in differentiated cells can cause dedifferentiation of these cells toward a pluripotent status. This set of four genes seems to be sufficient to promote pluripotency, but they are also related by their association with carcinogenesis (53–56), providing the molecular link between the regulation of pluripotency in ES cells and the acquisition of dedifferentiation in carcinogenesis (57). Other genes, such as *Nanog*, *Fbx15*, *E-Ras*, *Dppa2*, and *Tcl1*, that have also been introduced did not generate the same dedifferentiating effect, although these same genes are involved in the maintenance and sustenance of ES cells pluripotency. These results suggest that *Oct-4*, *c-Myc*, *Sox-2*, and *Klf-4* genes are required to initiate the process of dedifferentiation. Further studies have shown that when iPS cells are injected into mouse blastocysts, they contribute to all tissue types, including sperm and oocytes (58, 59). This method for reprogramming differentiated somatic cells is astonishing; however, its efficiency is still low, around 0.07%. The authors hypothesised that the low number of reprogrammed cells that they have obtained, may have derived from a small population of ~0.067% stem cells that has been found in mammalian skin cells (60, 61).

Reprogramming Using Amphibian Eggs or Egg Extracts

When nuclei of terminally differentiated somatic cells are transferred within an enucleated oocyte, the reconstructed embryo is capable of initiating and completing development. Thus, mature female germ cells have the capacity of removing the status of differentiation of the host genome and establishing novel totipotent/pluripotent characteristics. Some researchers have injected entire somatic cells into

Xenopus oocytes demonstrating a reprogramming capacity. When *Xenopus* oocytes where injected with human lymphocytes, the cell hybrid demethylated and expressed the human *OCT-4* and inactivated the amphibian *Thy-1* (62, 63).

Xenopus eggs have largely been used for preparing cell-free extracts. When human leukocytes where incubated in *Xenopus* egg extracts, these differentiated somatic cells expressed *Oct-4* and alkaline phosphatase (AP) after 7 days of culture and after only 30 min of treatment (64). One important player of the reprogramming process was the chromatin-remodelling factor BRG1 (Brahma-related group 1) (64). *Xenopus* extracts prepared from oocytes or mature eggs showed different nuclear remodelling (65). Oocytes extracts promote transcription, but not DNA replication; on the contrary, mature egg extracts promote replication, but not transcription.

Reprogramming with Cell Extracts

Turning point experiments have shown that cell-free extracts isolated from different types of differentiated somatic cells were capable to reprogramme gene expression in other somatic cell/isolated nuclei types (66–68) (Fig. 2). Fibroblasts exposed to extracts from human T cells or from a transformed T-cell line showed the activation of lymphoid cell–specific genes and expression of T-cell–specific antigens. When fibroblasts were cultured with neuronal precursor cell extracts, they expressed a neurofilament protein and extended neurite-like outgrowths. Human adipose tissue stem cells were reported to take on cardiomyocyte properties following transient exposure to a rat cardiomyocyte extract, indicating the possibility of transpecies efficient reprogramming of somatic cells (68). The functional reprogramming was reported to be stable through many cell divisions over several weeks of cell culture (66, 67). However, the stability of cell reprogramming depended on the source of cell extracts: transient functional cell reprogramming was described after alteration of cell fate when fibroblasts were cultured in the presence of insulinoma cell line extracts (69). These data show the molecular dominance of a certain cell type over another, resulting in the reprogramming of the susceptible cell by the dominant one.

A recent study has tested the hypothesis that pluripotent stem cells could be obtained in vitro by culturing differentiated somatic cells in the presence of cell extracts obtained from ES cells (70). This report showed the evidence that mouse embryonic stem cell extracts elicited the expression of the *Oct-4* pluripotency marker in the majority of NIH-3T3 (3T3) fibroblasts and induced the formation of distinct embryonic stem cell–like colonies. These embryonic stem–like cells formed embryoid-like bodies and differentiated into cells of the three germ layers.

The use of cell extracts for inducing cell transdifferentiation seems to be very promising and could be a powerful system for analysing cell reprogramming events as they occur in vitro and to obtain large quantities of pluripotent cells (Fig. 2). It is of crucial importance that the robustness of this method of cell transdifferentiation is tested by other laboratories before it is advanced to a more ambitious use in cell therapy programmes.

Our laboratory has recently employed the same reprogramming protocol on two lines of immortalised cells: the STO and the 3T3 fibroblasts (71). After ES cell extract treatment, we analysed the expression of a panel of pluripotency markers, including *Oct-4*, *Nanog*, and *Rex-1* genes, OCT-4, Forssman antigen, and SSEA-1 proteins and AP activity. The main results of this study were: first, we confirmed an enduring reprogramming activity of the ES cell extract, although on a very small number of cells that varies from ~0.003 to 0.04% of the total population of fibroblasts and with an effect limited to the induction of *Oct-4* and *Rex-1* gene expression and AP activity. Second, the expression of OCT-4, SSEA-1, and Forssman antigen proteins was never detected. Third, our work has clearly demonstrated that ES cells may survive the procedure of extract preparation, can be source of contamination that is expanded in culture, and can give false-positive results.

These studies demonstrate that in vitro treatment of differentiated somatic cells with ES cell extracts induces the expression of pluripotent marker genes in a small population of cells, providing proof of concept and endorsing the hypothesis that differentiated cells may be reprogrammed to a different state of differentiation and even brought back to a pluripotent status. Interestingly, the frequency (~0.003–0.04%) of reprogrammed fibroblasts that we have found in both populations of extract-treated STO and 3T3 resembles the low frequency (~0.07%) of reprogrammed cells that have been obtained in another type of cell-reprogramming experiment that has been recently carried out by Takahashi and Yamanaka (52) (see also the section "Reprogramming by Induced Expression of Specific Genes"). The authors hypothesise, among other explanations, that the low number of reprogrammed cells they obtained may have derived from a small population of ~0.067% stem cells that has been found in mammalian skin cells (60, 61). The fibroblasts that we used for the reprogramming experiments were immortalised, thus it is unlikely that stem cells are still present; perhaps, there is a small population of cells within the fibroblast culture that has a better responsiveness to the reprogramming treatment and is driven to the expression of *Oct-4*, *Rex-1*, and AP activity. The high chromosome rearrangements that occur in both fibroblast populations may constitute a serious obstacle to the reprogramming treatment of these cells.

Once the reprogramming procedure is firmly established and somatic cells are proved to have permanently acquired full pluripotency characteristics, it will be possible to try to break down the components of the extract that are involved in initiating and maintaining the dedifferentiation process.

Reprogramming with Synthetic Molecules

Recent studies have tried to identify chemical libraries in order to characterise small molecules with biological activities. Combinatorial libraries of over 100,000 heterocyclic compounds were designed around a large number of kinase-directed molecular scaffolds, including substituted purines, pyrimidines, quinazolines, pyrazines, pyrrolopyrimidine, pyrazolopyrimidine, phthalazines, pyridazines, and quinoxalines. These libraries are being screened for molecules that control stem cell fate and self-renewal (embryonic and adult), as well as molecules that induce reprogramming of lineage-committed cells. For example molecules have been identified that induce mouse EC differentiation into cardiomyocytes or neural cells (72, 73). Using libraries of this kind, a molecule called reversine has been used to induce dedifferentiation in myoblasts (74). Reversine is a 2,6-disubstituted purine derivative. Very recently it has been shown that reversine can increase the plasticity of C2C12 myoblasts and that reversine-treated cells gain the ability to differentiate into osteoblasts and adipocytes under lineage-specific inducing conditions (75).

Recently, two independent research groups led by James Thomson and Shinya Yamanaka have published the results of their studies demonstrating that, following retroviral transduction of four genes (*OCT4, SOX2, c-MYC,* and *KLF4* (76); *OCT4, NANOG, SOX2,* and *LIN28* (77)) into human fibroblasts, these cells can be induced to acquire a pluripotent, ES-like status, including the capacity of differentiating into the three germ layers in vitro and to form teratomas. The possibility of obtaining human iPS cells will allow the use of the patient's own somatic cells for cell therapy, overcoming the incompatibility of heterologous cell transplantation.

Acknowledgments We wish to thank the following grant agencies for their contribution to the research that was carried out in our laboratories and that was reported in this review: PRIN 2005, CARIPLO Foundation 2006, Olympus Foundation Science for Life, Millipore, FIRB 2005 (Project RBIP06FH7J); Istituto Superiore di Sanita' (Programma Nazionale Cellule Staminali 2003–4).

References

1. Gurdon JB, Uehlinger V. "Fertile" intestine nuclei. Nature 1966; 210:1240–1.
2. Gurdon JB. From nuclear transfer to nuclear reprogramming: the reversal of cell differentiation. Annu Rev Cell Dev Biol 2006;22:1–22.
3. Surani MA, Barton SC, Norris ML. Development of reconstituted mouse eggs suggests imprinting of the genome during gametogenesis. Nature 1984;308:548–50.
4. McGrath J, Solter D. Completion of mouse embryogenesis requires both the maternal and paternal genomes. Cell 1984;37:179–83.
5. Takagi N, Sasaki M. Preferential inactivation of the paternally derived X chromosome in the extraembryonic membranes of the mouse. Nature 1975;256:640–2.
6. Bourc'his D, Xu GL, Lin CS, et al. Dnmt3L and the establishment of maternal genomic imprints. Science 2001;294:2536–9.
7. Kaneda M, Okano M, Hata K, et al. Essential role for de novo DNA methyltransferase Dnmt3a in paternal and maternal imprinting. Nature 2004;429:900–3.
8. Dean W, Santos F, Stojkovic M, et al. Conservation of methylation reprogramming in mammalian development: aberrant reprogramming in cloned embryos. Proc Natl Acad Sci U S A 2001;98:13734–8.
9. Han YM, Kang YK, Koo DB, et al. Nuclear reprogramming of cloned embryos produced in vitro. Theriogenology 2003;59:33–44.
10. Kang YK, Park JS, Koo DB, et al. Limited demethylation leaves mosaic-type methylation states in cloned bovine pre-implantation embryos. EMBO J 2002;21:1092–100.
11. Jelinic P, Stehle JC, Shaw P. The testis-specific factor CTCFL cooperates with the protein methyltransferase PRMT7 in H19 imprinting control region methylation. PLoS Biol 2006;4:e355.

12. Kawahara M, Wu Q, Takahashi N, et al. High-frequency generation of viable mice from engineered bi-maternal embryos. Nat Biotechnol 2007;25:1045–50.
13. Thompson EM. Chromatin structure and gene expression in the preimplantation mammalian embryo. Reprod Nutr Dev 1996;36:619–35.
14. Schultz RM. The molecular foundations of the maternal to zygotic transition in the preimplantation embryo. Hum Reprod Update 2002;8:323–31.
15. Nothias JY, Majumder S, Kaneko KJ, et al. Regulation of gene expression at the beginning of mammalian development. J Biol Chem 1995;270:22077–80.
16. Schultz RM, Davis W Jr., Stein P, et al. Reprogramming of gene expression during preimplantation development. J Exp Zool 1999;285(3):276–82.
17. Ko MS, Kitchen JR, Wang X, et al. Large-scale cDNA analysis reveals phased gene expression patterns during preimplantation mouse development. Development 2000;127:1737–49.
18. Hamatani T, Carter MG, Sharov AA, et al. Dynamics of global gene expression changes during mouse preimplantation development. Dev Cell 2004;6:117–31.
19. Wang QT, Piotrowska K, Ciemerych MA, et al. A genome-wide study of gene activity reveals developmental signaling pathways in the preimplantation mouse embryo. Dev Cell 2004;6:133–44.
20. Christians E, Campion E, Thompson EM, et al. Expression of the HSP 70.1 gene, a landmark of early zygotic activity in the mouse embryo, is restricted to the first burst of transcription. Development 1995;121:113–22.
21. Inoue K, Ogonuki N, Miki H, et al. Inefficient reprogramming of the hematopoietic stem cell genome following nuclear transfer. J Cell Sci 2006;119:1985–91.
22. Winger QA, Hill JR, Shin T, et al. Genetic reprogramming of lactate dehydrogenase, citrate synthase, and phosphofructokinase mRNA in bovine nuclear transfer embryos produced using bovine fibroblast cell nuclei. Mol Reprod Dev 2000;56:458–64.
23. Daniels R, Hall V, Trounson AO. Analysis of gene transcription in bovine nuclear transfer embryos reconstructed with granulosa cell nuclei. Biol Reprod 2000;63:1034–40.
24. Wrenzycki C, Wells D, Herrmann D, et al. Nuclear transfer protocol affects messenger RNA expression patterns in cloned bovine blastocysts. Biol Reprod 2001;65:309–17.
25. Sebastiano V, Gentile L, Garagna S, et al. Cloned pre-implantation mouse embryos show correct timing but altered levels of gene expression. Mol Reprod Dev 2005;70:146–54.
26. Smith SL, Everts RE, Tian XC, et al. Global gene expression profiles reveal significant nuclear reprogramming by the blastocyst stage after cloning. Proc Natl Acad Sci U S A 2005;102:17582–7.
27. Pfister-Genskow M, Myers C, Childs LA, et al. . Identification of differentially expressed genes in individual bovine preimplantation embryos produced by nuclear transfer: improper reprogramming of genes required for development. Biol Reprod 2005;72:546–55.
28. Somers J, Smith C, Donnison M, et al. Gene expression profiling of individual bovine nuclear transfer blastocysts. Reproduction 2006;131:1073–84.
29. Wilmut I. What regulates developmental plasticity? Cloning Stem Cells 2002;4:177–8.
30. Loi P, Clinton M, Vackova I, et al. Placental abnormalities associated with post-natal mortality in sheep somatic cell clones. Theriogenology 2006;65:1110–21.
31. Mak W, Nesterova TB, de Napoles M, et al. Reactivation of the paternal X chromosome in early mouse embryos. Science 2004;303:666–9.
32. Okamoto I, Otte AP, Allis CD, et al. Epigenetic dynamics of imprinted X inactivation during early mouse development. Science 2004;303:644–9.
33. Bao S, Miyoshi N, Okamoto I, et al. Initiation of epigenetic reprogramming of the X chromosome in somatic nuclei transplanted to a mouse oocyte. EMBO Rep 2005;6:748–54.
34. Santos F, Hendrich B, Reik W, et al. Dynamic reprogramming of DNA methylation in the early mouse embryo. Dev Biol 2002;241:172–82.
35. Arney KL, Bao S, Bannister AJ, et al. Histone methylation defines epigenetic asymmetry in the mouse zygote. Int J Dev Biol 2002;46:317–20.
36. Cowell IG, Aucott R, Mahadevaiah SK, et al. Heterochromatin, HP1 and methylation at lysine 9 of histone H3 in animals. Chromosoma 2002;111:22–36.

37. Fulka H. Changes in global histone acetylation pattern in somatic cell nuclei after their transfer into oocytes at different stages of maturation. Mol Reprod Dev 2008;75:556–64.
38. Martin C, Beaujean N, Brochard V, et al. Genome restructuring in mouse embryos during reprogramming and early development. Dev Biol 2006;292:317–32.
39. Merico V, Barbieri J, Zuccotti M, et al. Epigenomic differentiation in mouse preimplantation nuclei of biparental, parthenote and cloned embryos. Chromosome Res 2007;15: 341–60.
40. Que J, El Oakley RM, Salto-Tellez M, et al. Generation of hybrid cell lines with endothelial potential from spontaneous fusion of adult bone marrow cells with embryonic fibroblast feeder. In Vitro Cell Dev Biol Anim 2004;40:143–9.
41. Terada N, Hamazaki T, Oka M, et al. Bone marrow cells adopt the phenotype of other cells by spontaneous cell fusion. Nature 2002;416:542–5.
42. Ying QL, Nichols J, Evans EP, et al. Changing potency by spontaneous fusion. Nature 2002;416:545–8.
43. Pells S, Di Domenico AI, Gallagher EJ, et al. Multipotentiality of neuronal cells after spontaneous fusion with embryonic stem cells and nuclear reprogramming in vitro. Cloning Stem Cells 2002;4:331–8.
44. Takagi N, Yoshida MA, Sugawara O, et al. Reversal of X-inactivation in female mouse somatic cells hybridized with murine teratocarcinoma stem cells in vitro. Cell 1983;34:1053–62.
45. Tada M, Tada T, Lefebvre L, et al. Embryonic germ cells induce epigenetic reprogramming of somatic nucleus in hybrid cells. EMBO J 1997;16:6510–20.
46. Matveeva NM, Shilov AG, Kaftanovskaya EM, et al. In vitro and in vivo study of pluripotency in intraspecific hybrid cells obtained by fusion of murine embryonic stem cells with splenocytes. Mol Reprod Dev 1998;50:128–38.
47. Tada M, Takahama Y, Abe K, et al. Nuclear reprogramming of somatic cells by in vitro hybridization with ES cells. Curr Biol 2001;11:1553–8.
48. Kimura H, Tada M, Nakatsuji N, et al. Histone code modifications on pluripotential nuclei of reprogrammed somatic cells. Mol Cell Biol 2004;24:5710–20.
49. Serov OL, Matveeva NM, Kizilova EA, et al. "Chromosome memory" of parental genomes in embryonic hybrid cells. Ontogenez 2003;34:216–27.
50. Do JT, Han DW, Scholer HR. Reprogramming somatic gene activity by fusion with pluripotent cells. Stem Cell Rev 2006;2.257–64.
51. Silva J, Chambers I, Pollard S, et al. Nanog promotes transfer of pluripotency after cell fusion. Nature 2006;441:997–1001.
52. Takahashi K, Yamanaka S. Induction of pluripotent stem cells from mouse embryonic and adult fibroblast cultures by defined factors. Cell 2006;126:663–76.
53. Monk M, Holding C. Human embryonic genes re-expressed in cancer cells. Oncogene 2001;20:8085–91.
54. Adhikary S, Eilers M. Transcriptional regulation and transformation by Myc proteins. Nat Rev Mol Cell Biol 2005;6:635–45.
55. Rowland BD, Peeper DS. KLF4, p21 and context-dependent opposing forces in cancer. Nat Rev Cancer 2006;6:11–23.
56. Tsukamoto T, Mizoshita T, Mihara M, et al. Sox2 expression in human stomach adenocarcinomas with gastric and gastric-and-intestinal-mixed phenotypes. Histopathology 2005;46: 649–58.
57. Perry AC. Induced pluripotency and cellular alchemy. Nat Biotechnol 2006;24:1363–4.
58. Okita K, Ichisaka T, Yamanaka S. Generation of germline-competent induced pluripotent stem cells. Nature 2007;448:313–7.
59. Wernig M, Meissner A, Foreman R, et al. In vitro reprogramming of fibroblasts into a pluripotent ES-cell-like state. Nature 2007;448:318–24.
60. Dyce PW, Zhu H, Craig J, et al. Stem cells with multilineage potential derived from porcine skin. Biochem Biophys Res Commun 2004;316:651–8.
61. Toma JG, McKenzie IA, Bagli D, et al. Isolation and characterization of multipotent skin-derived precursors from human skin. Stem Cells 2005;23:727–37.

62. Byrne JA, Simonsson S, Western PS, et al. Nuclei of adult mammalian somatic cells are directly reprogrammed to oct-4 stem cell gene expression by amphibian oocytes. Curr Biol 2003;13:1206–13.
63. Simonsson S, Gurdon J. DNA demethylation is necessary for the epigenetic reprogramming of somatic cell nuclei. Nat Cell Biol 2004;6:984–90.
64. Hansis C, Barreto G, Maltry N, et al. Nuclear reprogramming of human somatic cells by xenopus egg extract requires BRG1. Curr Biol 2004;14:1475–80.
65. Alberio R, Johnson AD, Stick R, et al. Differential nuclear remodeling of mammalian somatic cells by *Xenopus laevis* oocyte and egg cytoplasm. Exp Cell Res 2005;307:131–41.
66. Hakelien AM, Landsverk HB, Robl JM, et al. Reprogramming fibroblasts to express T-cell functions using cell extracts. Nat Biotechnol 2002;20:460–6.
67. Landsverk HB, Hakelien AM, Kuntziger T, et al. Reprogrammed gene expression in a somatic cell-free extract. EMBO Rep 2002;3:384–9.
68. Gaustad KG, Boquest AC, Anderson BE, et al. Differentiation of human adipose tissue stem cells using extracts of rat cardiomyocytes. Biochem Biophys Res Commun 2004;314:420–7.
69. Hakelien AM, Gaustad KG, Collas P. Transient alteration of cell fate using a nuclear and cyto-plasmic extract of an insulinoma cell line. Biochem Biophys Res Commun 2004;316:834–41.
70. Taranger CK, Noer A, Sorensen AL, et al. Induction of dedifferentiation, genomewide transcrip-tional programming, and epigenetic reprogramming by extracts of carcinoma and embryonic stem cells. Mol Biol Cell 2005;16:5719–35.
71. Neri T, Monti M, Rebuzzini P, et al. Mouse fibroblasts are reprogrammed to Oct-4 and Rex-1 gene expression and alkaline phosphatase activity by embryonic stem cell extracts. Cloning Stem Cells 2007;9:394–406.
72. Wu X, Ding S, Ding Q, et al. Small molecules that induce cardiomyogenesis in embryonic stem cells. J Am Chem Soc 2004;126:1590–1.
73. Ding S, Schultz PG. A role for chemistry in stem cell biology. Nat Biotechnol 2004;22:833–40.
74. Chen S, Zhang Q, Wu X, et al. Dedifferentiation of lineage-committed cells by a small molecule. J Am Chem Soc 2004;126:410–1.
75. Chen S, Takanashi S, Zhang Q, et al. Reversine increases the plasticity of lineage-committed mammalian cells. Proc Natl Acad Sci U S A 2007;104:10482–7.
76. Takahashi K, Tanabe K, Ohnuki M, et al. Induction of pluripotent stem cells from adult human fibroblasts by defined factors. Cell 2007;131:861–72.
77. Yu J, Vodyanik MA, Smuga-Otto K, et al. Induced pluripotent stem cell lines derived from human somatic cells. Science 2007;318:1917–20.
78. Dean W, Santos F, Reik W. Epigenetic reprogramming in early mammalian development and following somatic nuclear transfer. Semin Cell Dev Biol 2003;14:93–100.
79. Collas P, Gammelsaeter R. Novel approaches to epigenetic reprogramming of somatic cells. Cloning Stem Cells 2007;9:26–32.

Reprogramming Male Germ Cells to Pluripotent Stem Cells

Parisa Mardanpour, Kaomei Guan, Tamara Glaeser, Jae Ho Lee, Jessica Nolte, Gerald Wulf, Gerd Hasenfuss, Wolfgang Engel, Oliver Brüstle, and Karim Nayernia

Abstract Reprogramming of a differentiated cell into a cell capable of giving rise to many different cell types, a pluripotent cell, which in turn could repopulate or repair nonfunctional or damaged tissue, would present beneficial applications in regenerative medicine. It was shown by different groups that germ cells can be reprogrammed to pluripotent stem cells in all diploid stages of development. Specification of germline lineage is one of the most essential events in development, since this process ensures the acquisition, modification, and reservation of the totipotent genome for subsequent generations. We and other groups have shown that adult male germline stem cells, spermatogonial stem cells, can be converted into embryonic stem cell–like cells that can differentiate into the somatic stem cells of three germ layers. Importantly, cultured germ cells demonstrate normal and stable karyotypes as well as normal patterns of genomic imprinting. Transplantation studies have begun in a variety of models in hopes of defining their potential application of pluripotent stem cells derived from germ cells to treat a wide variety of human conditions, including cardiovascular and neurological disorders. This chapter describes general considerations regarding molecular and cellular aspects of reprogramming of germ cells at different developmental stages to stem cells compared with their counterpart, embryonic stem cells.

Keywords Germ cells • Stem cells • Spermatogonial stem cells • Pluripotency • Regenerative medicine

P. Mardanpour, K. Guan, T. Glaeser, J.H. Lee, J. Nolte, G. Wulf, G. Hasenfuss, W. Engel, O. Brüstle, and K. Nayernia (✉)
North East Institute of Stem Cell Biology, Institute of Human Genetics, International Centre for Life, University of Newcastle upon Tyne, Central Parkway, Newcastle upon Tyne NE1 3BZ, UK
e-mail: karim.nayernia@newcastle.ac.uk

H. Baharavand (ed.), *Trende in Stem Cell Biology and Technology.*
DOI 10.1007/978-1-60327-905-5_05
© Humana Press, a Part of Springer Science+Business Media, LLC 2009

Introduction

Pluripotency is governed by the intricate interplay between genetic and epigenetic factors. Previous studies implicate Oct-4, Sox2, and Nanog as core regulators of the transcriptional circuitry in pluripotent cells. The three transcriptional factors and their downstream target genes promote self-renewal and pluripotency. Reprogramming can be initiated by the introduction of few defined factors. However, the molecular mechanisms driving reprogramming remain largely obscure and await further investigations. Embryonic stem cells (ESCs) and germline stem cells provide exciting models for understanding the underlying mechanisms that make a cell pluripotent. Germline stem cells express factors required for reprogramming fibroblasts to pluripotent stem cells, suggesting that germline stem cells are potentially pluripotent. Knowledge about the pluripotency of germ cells would enable dedifferentiation of any adult somatic cells and their reprogramming in other histocompatible cell types for cell therapy treatments.

In mammalian embryos, germ-cell status is acquired in a democratic way, as a result of interactions between neighbouring cells. In mice, these interactions are initiated when there are only three cell layers in the embryo. In mouse, the inner layer (the epiblast) eventually gives rise to all the cells of the foetus, including the germ cells, while the outer layers are supportive tissues that nevertheless send important early patterning signals to the epiblast. Initially, the epiblast is a cup-shaped sheet, but during the process known as gastrulating, cells move toward one side, drop out of the layer, and give rise to a new cell population, the mesoderm. Most mesoderm cells move into the developing embryo, while the remainder contribute to extraembryonic mesodermal support tissues: the amnion and allantois (1).

The first germ cells that appear in embryo are called primordial germ cells (PGCs) (Fig. 1). In the mouse embryo, PGCs were first identified as alkaline phosphatase–positive cells within the extraembryonic mesoderm at 7.2 days post-coitum (dpc) (2, 3). PGCs require several members of the bone morphogenetic protein (BMP) ligand family for their development (4–7), after which they proliferate and migrate to the genital ridges, where they become enclosed in the seminiferous cords by 11 dpc. The important factors involved in specification of germ cells in vivo were shown in Fig. 2. During this period, the PGCs are proliferating so rapidly that the initial founder population of about 40 PGCs gives rise to about 25,000 germ cells by the time proliferation ceases at 13.5 dpc (8). By 13.5 dpc, the germ cells are also undertaking sexually dimorphic patterns of development.

In human embryos, ~100 PGCs appear between the third and fourth weeks of gestation in the endoderm of the dorsal wall of the yolk sac, near the allantois. PGCs then proceed to migrate through the hindgut during the fourth week and the dorsal mesentery in the fifth week, when ~1,000 PGCs reach the genital ridge (9, 10). Once PGCs enter the gonadal ridge they are generally referred to as gonocytes. In fact, the term gonocyte was originally used in rodents to distinguish the postmigratory PGCs from those in the gonad based on morphological distinctions alone (11, 12). By 7 weeks postfertilization (pF), sexual differentiation of the male

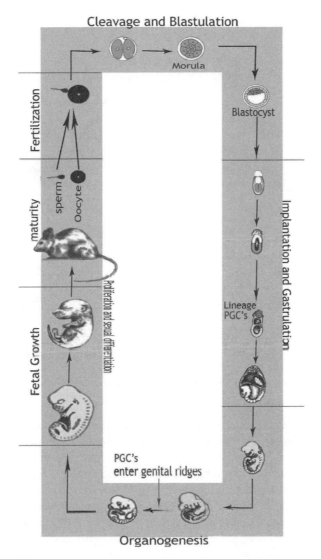

Fig. 1 Reproductive cycle: specification and development of germ cells

gonadal ridge into a testis begins *(13, 14)*, followed by germ cell expansion from approximately 3,000 in week 6 to approximately 30,000 by week 9 pF.

After birth, male germ cells develop to spermatogonial stem cells (SSCs). SSCs are descendants of the PGCs, which migrate from extraembryonic sites to colonize the gonadal ridge early during embryonic life *(15)*. Spermatogenesis depends on SSCs, which have the ability to self-renew and generate a large number of differentiated germ cells in most species. In mammals, millions of spermatozoa are produced every day from SSCs *(16, 17)*. Although SSCs are infrequent in the testis, there are presumably ~1 in 3,000–4,000 cells in adult mouse testis (16, *17)*.

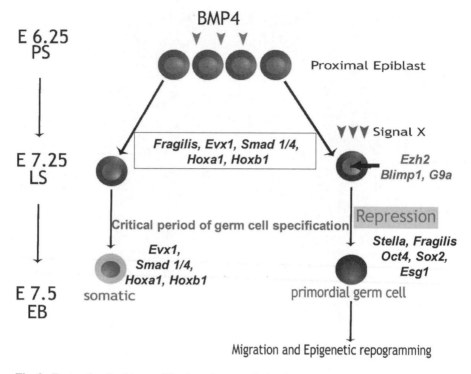

Fig. 2 Factors involved in specification of germ cells in vivo

SSCs can be identified unequivocally by a functional transplantation assay. To maintain spermatogenesis, the processes of self-renewal and differentiation of SSCs must be precisely regulated by intrinsic gene expression in the stem cells and extrinsic signals, including soluble factors or adhesion molecules from the surrounding microenvironment, the stem cell niche *(18)*. Normally, spermatogenesis is controlled by both inter- and intracellular mechanisms. Intercellular signalling pathways must eventually influence spermatogenesis through the intracellular regulatory molecules at important control points. These control points are often revealed by clean-arrest phenotypes, as suggested by extensive genetic analysis in major model organisms.

Evidences of Germ Cell Pluripotency

The male germ cells show pluripotency at different stages of development (Fig. 3). There is evidence for pluripotency of germ cells on the molecular, cellular, and developmental levels. On the molecular level, it has been shown that the factors involved in the pluripotent stem cells are required for germ cell maintenance. Molecular pluripotency programme of stem cells is based on regulation of specific

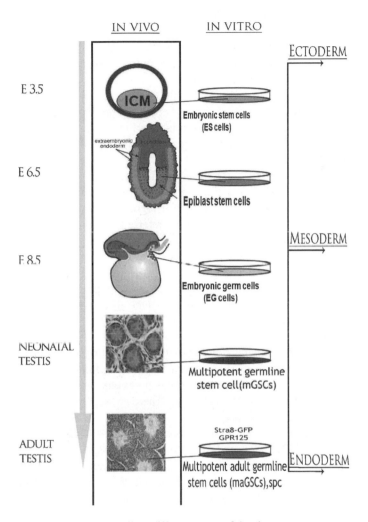

Fig. 3 Pluripotency of male germ cells at different stages of development

transcription factors, chromatin remodelling, regulatory RNA molecules, and defined signalling pathways. Recent studies have provided new insights into the key regulators of ESC pluripotency and their interactions to produce the pluripotent state. Different studies first showed that the homeodomain transcription factors Oct-4 and Nanog are essential regulators of early development and ESC identity *(19–21)*. Disruption of Oct-4 and Nanog causes loss of pluripotency and inappropriate differentiation of inner cell mass (ICM) and ESC to trophectoderm and extraembryonic endoderm, respectively *(19, 22)*. However, recent evidence suggests that Nanog may function to stabilize the pluripotent state rather than being essential for maintaining pluripotency of ESCs *(19)*. Oct-4 can heterodimerize with the HMG-box transcription factor Sox2 in ESCs, and Sox2 contributes to pluripotency, at least in part, by regulating Oct-4 levels. Oct-4 is rapidly and apparently completely silenced during early cellular differentiation *(23)*. The key roles played by Oct-4, Sox2, and Nanog during early development as well as their unique expression pattern make it likely that these regulators are central to the transcriptional regulatory hierarchy that specifies ESC identity *(23)*.

All these factors are expressed in male germ cells during feral and postnatal germ cell development *(24–26)*. However, recent studies have shown that the transcription factor SOX2 is not detected in human PGCs, in contrast to their murine counterparts, highlighting a major interspecies difference relevant to both gamete and stem cell biology *(27)*.

In male embryos, *Oct-4* expression persists in germ cells throughout foetal development. After birth, it is maintained in proliferating gonocytes, prospermatogonia, and later in undifferentiated spermatogonia *(28)*. This highly restricted, cyclic expression pattern in the germ line, referred to as the totipotent cycle, has been taken as an indication that Oct-4 may have a role in maintaining the pluripotency and germline potential of pluripotent embryonic cells. Using germ cell–specific deletion of *Oct-4* in conditional knockout mouse models, it was shown that Oct-4 is required for PGC survival and PGCs undergo apoptosis without Oct-4 *(21)*. Nanog is a divergent homeodomain protein found in mammalian pluripotent cells and developing germ cells. PGCs that lack Nanog fail to mature on reaching the genital ridge *(19)*.

On the cellular level, there is strong evidence for pluripotency of germ cells. In an early embryo a cell has the potential to generate many different cell types. During development cells generally lose this potential, or "potency," and become restricted to making one or a few cell types. There are several lines of evidence that show the potency of germ cells.

Pluripotency of embryonic germ cells derived from PGCs has been shown by several groups *(29–32)*. Cultured PGCs exposed to a specific cocktail of growth factors give rise to embryonic germ cells, pluripotent stem cells that can contribute to all the lineages of chimeric embryos, including the germline. The conversion of PGCs into pluripotent stem cells is a remarkably similar process to nuclear reprogramming in which a somatic nucleus is reprogrammed in the egg cytoplasm. The conversion of PGCs into pluripotent stem cells may be linked in some way with their deregulated proliferation *(33)*.

Other evidence is the pluripotency of testicular germ cell tumours. Testicular teratomas are highly unusual benign tumours containing derivatives of the three primary germ layers. Different studies have shown that embryonal carcinomas are derived from PGCs *(34)*. Assays of developmental potency show that isolated ESCs are pluripotent stem cells but when they lose the ability to differentiate them from malignant teratocarcinomas. Understanding the genetics of embryonal carcinoma cell formation and the growth factor signalling pathways controlling embryonic germ cell derivation could tell us much about the molecular controls on developmental potency in mammals *(35, 36)*.

Recently, the pluripotency of postnatal germline stem cells was clearly demonstrated *(37–41)*. It was reported that mouse SSCs could be maintained in vitro for quite a long period. Mouse spermatogonia were cultured under fairly standard culture conditions with Dulbecco's modified Eagle's medium (DMEM) containing 10% fetal bovine serum (FBS) and several nutrients (pyruvate and lactate). In these experiments STO feeder cells, a cell line derived from mouse fibroblast, were essential for the maintenance of stem spermatogonia. Following incubation for various periods (up to 4 months), mouse stem cells remained alive in culture, and these cells could be successfully transplanted in mice system. Extensive areas of cultured cell-derived spermatogenesis were generated in the seminiferous tubules of the host, and production of mature spermatozoa was observed *(42–44)*. Some of the studies used improved culture system based on a commercial medium (StemPro-34® SFM, Invitrogen) and supplementation with various growth factors and hormones, such as b-estradiol, progesterone, epidermal growth factor (EGF), basic fibroblast growth factor (bFGF), leukemia inhibitory factor (LIF), and glial-cell line-derived, neurotrophic factor (GDNF). Recently, Kanatsu-Shinohara et al. *(26)* demonstrated the pluripotency of a single SSC, providing evidence that a single SSC acquire pluripotency, but its conversion into a pluripotent cell type is accompanied by loss of spermatogenic potential. Also, it was demonstrated that mouse SSC retained the euploid karyotype and androgenetic imprint during the 2-year experimental period and produced normal spermatogenesis and fertile offspring. However, the telomeres in SSC gradually shortened during culture, suggesting that they are not immortal *(15)*. This evidence clearly suggests the pluripotency of germ cells in all stages of development (Fig. 3).

We succeeded in developing a procedure for isolation and purification of SSCs from adult mouse testis *(38)*. In our study we were able to isolate and culture these cells in culture medium containing the precise combination of cellular growth factors needed for the cells to reproduce themselves in vitro. These cells were characterized concerning their molecular profiling and were compared to molecular profiling of ESCs using a stem cell array, which contains relevant genes related to stem cell metabolism. In particular, differentiation of these cells to functional cardiac and neural stem cells was demonstrated (Figs. 4 and 5). The results indicate that SSCs share many molecular characteristics with ESCs. On the cellular level, SSCs resemble ESCs; they form embryoid body structure after 2 weeks of culture. Stem cell potential of isolated SSCs was examined using the transplantation technique. This method allowed SSCs to recolonize the seminiferous tubuli of germ cell–depleted mice and regeneration of spermatogenesis. These cells are able to differentiate into various cell types of the three germ layers in vitro. In contrast to ESCs, use of SSCs for cell transplantation will allow establishment of individual cell-based therapy because the donor and recipient are identical. In addition, the ethical problem is avoided.

Recently a group could confirm our data by different approach. This group discovered a novel orphan G-protein coupled receptor (GPR125) that is restricted to undifferentiated spermatogonia within the testis. GPR125 expression was maintained when the progenitor cells were extracted from the in vivo niche and propagated under growth conditions that recapitulate key elements of the niche. Such

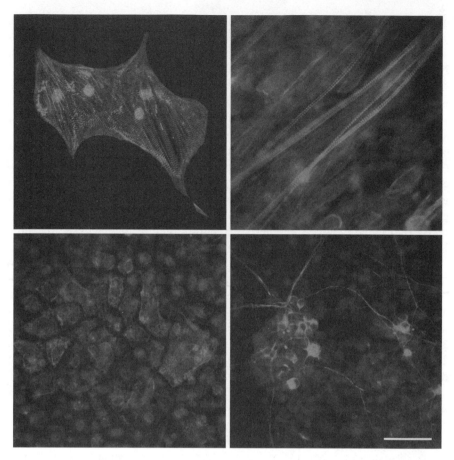

Fig. 4 Differentiation of spermatogonial stem cells into derivatives of three germ layers. (**a**) Organization of sarcomeric protein a-actinine in isolated uninucleate cardiac cells at differentiation day 5 + 7. (**b**) Nebulin-positive myotubes in embryoid body outgrowths at day 5 + 23. (**c**) Pancytokeratin-positive epithelial cells in embryoid body outgrowths at day 5 + 14. (**d**) Neurofilament protein M (NFM)-positive neurons at day 5 + 12. Nuclei are stained with DAPI. (Scale bar, 50 μm)

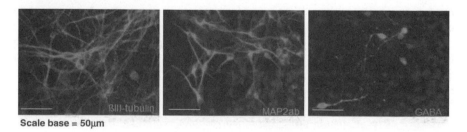

Scale base = 50μm

Fig. 5 Differentiation of spermatogonial stem cells (SSC) to neurons. Expression of different neuron-specific markers (β III-tubulin, MAP2ab, γ-aminobutyric acid [GABA]) could be detected in SSC-derived neurons. (Scale bar, 50 mm)

conditions preserved the ability of the cells to generate multipotent derivatives, known as multipotent adult spermatogonial derived progenitor cells (MASCs) *(39)*.

Our approach provides an accessible in vitro model system for studies of mammalian gametogenesis, as well as for developing new strategies for reproductive engineering, infertility treatment, and establishment of individual cell-based therapy.

Relevance for Regenerative Medicine

Recently, cell-based therapies have developed as a foundation for regenerative medicine. Although most cell-based therapies currently consist of heterogeneous cell populations, it is anticipated that the standard of care needs well-characterized stem cell lines that can be modified to meet the individual needs of the patient. Extensive research in the area of regenerative medicine is focused on the development of cells, tissues, and organs for the purpose of restoring function through transplantation. The general belief is that replacement, repair, and restoration of function is best accomplished by cells, tissues, or organs that can provide the appropriate physiologic or metabolic functions more efficiently than any mechanical devices, recombinant proteins, or chemical compounds.

Several lines of evidence have suggested extensive proliferation activity and pluripotency of germline stem cells, including SSCs. These characteristics provide new and unprecedented opportunities for the therapeutic use of SSCs for regenerative medicine.

Conclusions

Several types of stem cell have been discovered from germ cells. Each of these has promised to revolutionize the future of regenerative medicine through the provision of cell-replacement therapies to treat a variety of debilitating diseases. Stem cell research is politically charged, receives considerable media coverage, raises many ethical and religious debates, and generates a great deal of public interest. The tremendous versatility of ESCs versus the unprecedented reports describing adult stem cell plasticity have ignited debates as to the choice of one cell type over another for future application. If researchers can isolate SSCs from humans, germline stem cells can be reprogrammed to have the ability to differentiate into cells that can be used therapeutically for cell-based regenerative medicine. In addition, development of in vitro culture of SSCs might lead to a new clinically relevant approach in treating male infertility in patients with germ cell arrest or patients at risk of infertility due to chemo- or radiotherapy. Furthermore, these cells could serve as target cells for establishment of germline gene therapeutic approaches.

Acknowledgments This work was supported by The University of Newcastle upon Tyne, ONE North East, and German Research council. We thank Hamed Nayernia for excellent graphical images.

References

1. Hogan, B. Developmental biology: decision, decisions! Nature 2002;418:282–3.
2. Chiquoine, AD. The identification, origin and migration of the primordial germ cells in the mouse embryo. Anat Rec 1954;118:135–46.
3. Ginsburg M, Snow MHL, McLaren A. Primordial germ cells in the mouse embryo during gastrulation. Development 1990;110:521–8.
4. Lawson KA, Dunn NR, Roelen BA, et al. Bmp4 is required for the generation of primordial germ cells in the mouse embryo. Genes Dev 1999;13:424–36.
5. Ying Y, Liu XM, Marble A, et al. Requirement of Bmp8b for the generation of primordial germ cells in the mouse. Mol Endocrinol 2000;14:1053–63.
6. Fujiwara T, Dunn NR, Hogan BL. Bone morphogenetic protein 4 in the extraembryonic mesoderm is required for allantois development and the localization and survival of primordial germ cells in the mouse. Proc Natl Acad Sci U S A 2001;98:13739–44.
7. Ying Y, Zhao GQ. Cooperation of endoderm-derived BMP2 and extraembryonic ectoderm-derived BMP4 in primordial germ cell generation in the mouse. Dev Biol 2001;232:484–92.
8. Tam PP, Snow MH. Proliferation and migration of primordial germ cells during compensatory growth in mouse embryos. J Embryol Exp Morphol 1981;64:133–47.
9. Witschi E. Migration of the germ cells of human embryos from the yolk sac to the primitive gonadal folds. Contrib Embryol 1948;32:67–80.
10. McKay D, Hertig AT, Adams EC, et al. Histochemical observations on the germ cells of human embryos. Anat Rec 1953;117:201–19.
11. Clermont Y, Perey B. Quantitative study of the cell population of the seminiferous tubules in immature rats. Am J Anat 1957;100:241–67.
12. Sapsford C. Changes in the cells of the sex cords and the seminiferous tubules during development of the testis of the rat and the mouse. Aust J Zool 1962;101:178–92.
13. Wartenberg H. Development of the early human ovary and role of the mesonephros in the differentiation of the cortex. Anat Embryol 1982;165:253–80.
14. Bendsen E, Byskov AG, Laursen SB, et al. Number of germ cells and somatic cells in human fetal testes during the first weeks after sex differentiation. Hum Reprod 2003;18:13–8.
15. Bendel-Stenzel M, Anderson R, Heasman J, et al. The origin and migration of primordial germ cells in the mouse. Semin Cell Dev Biol 1998;9:393–400.
16. Olive V, Cuzin F. The spermatogonial stem cells: from basic knowledge to transgenic technology. Int J Biochem Cell Biol 2005;37:246–50.
17. Nayernia K, Li M, Engel W. Spermatogial stem cells. Methods Mol Biol 2004;253:105–20.
18. Kubota H, Avarbock MR, Brinster RL. Growth factors essential for self-renewal and expansion of mouse spermatogonial stem cells. Proc Natl Acad Sci U S A 2004;101:16489–94.
19. Chambers I, Silva J, Colby D, et al. Nanog safeguards pluripotency and mediates germline development. Nature 2007;450:1230–4.
20. Mitsiu K, Tokuzawa Y, Itoh H, et al. The homeoprotein Nanog is required for maintenance of pluripotency in mouse epiblast and ES cells. Cell 2003;113:631–42.
21. Kehler J, Tolkunova E, Koschorz B, et al. Oct4 is required for primordial germ cell survival. EMBO Rep 2004;5:1078–83.
22. Chambers I. The molecular basis of pluripotency in mouse embryonic stem cells. Cloning Stem Cells 2004;6:386–91.
23. Jaenisch R, Young R. Stem cells, the molecular circuitry of pluripotency and nuclear reprogramming. Cell 2008;132:567–82.

24. Kerr CL, Hill CM, Blumenthal PD, Gearhart JD. Expression of pluripotent stem cell markers in the human fetal testis. Stem Cells 2008;26:412–21.
25. Hoei-Hansen CE, Almstrup K, Nielsen JE, et al. Stem cell pluripotency factor NANOG is expressed in human fetal gonocytes, testicular carcinoma in situ and germ cell tumours. Histopathology 2005;47:48–56.
26. Kanatsu-Shinohara M, Lee J, Inoue K, et al. Pluripotency of a single spermatogonial stem cells in mice. Biol Reprod 2008;78:681–7.
27. Perrett RM, Tumpenny L, Eckert JJ, et al. The early human germ cell lineage does not express SOX2 during in vivo development or upon in vitro culture. Biol Reprod 2008;78:852–8.
28. Pesce M, Gross MK, Schoeler HR. In line with our ancestors: Oct-4 and the mammalian germ. Bioassays 1998;20:722–32.
29. Kerr CL, Gearhart JD, Elliott AM, et al. Embryonic germ cells: when germ cells become stem cells. Semin Reprod Med 2006;24:304–13.
30. Kerr CL, Shamblott KJ, Gearhart JD. Pluripotent stem cells from germ cells. Methods Enzymol 2006;419:400–26.
31. Shamblott MJ, Axelman J, Wang S, et al. Derivation of pluripoten stem cells from cultured human primordial germ cells. Proc Natl Acad Sci U S A 1998;95:13726–31.
32. McLAren A, Durcova-Hills G. Germ cells and pluripotent stem cells in the mouse. Reprod Fertil Dev 2001;13:661–64.
33. Cooke JE, Godin I, Ffrench-Constant C, et al. Culture and manipulation of primordial germ cells. Methods Enzymol 1993;225:37–58.
34. Stevens LC. Origin of testicular teratomas from primordial germ cells in mice. J Natl Cancer Inst 1967;38:549–52.
35. Damjanov I. Teratocarcinoma stem cells. Cancer Surv 1990;9:303–19.
36. Bonner AE, Wang Y, You M. Gene expression profiling of mouse teratocarcinomas uncover epigenetic changes associated with the transformation of mouse embryonic stem cells. Neoplasia 2004;6:490–502.
37. Kanatsu-Shinohara M, Inoue K, Lee J, et al. Generation of pluripotent stem cells from neonatal mouse testis. Cell 2004;119:1001–12.
38. Guan K, Nayernia K, Maier LS, et al. Pluripotency of spermatogonial stem cells from adult mouse testis. Nature 2006;440:1199–203.
39. Scandel M, James D, Shmelkov SV, et al. Generation of functional multipotent adult stem cells from GPR125+ germline progenitors. Nature 2007;449:346–50.
40. Kubota H, Brinster RL. Culture of rodent spermatogonial stem cells, male germline stem cells of the postnatal animal. Methods Cell Biol 2008;86:59–84.
41. Falciatori I, Lillard-Wetherell K, Wu Z, et al. Deriving mouse spermatogonial stem cell lines. Methods Mol Biol 2008;450:163–79.
42. Van Saen D, Goossens E, De Block G, et al. Regeneration of spermatogenesis by grafting testicular tissue or injecting testicular cells into the testes of sterile mice: a comparative study. Fertil Steril 2008 Apr. 2.
43. McLaren DJ. Spermatogonial stem cell transplantation, testicular function, and restoration of male fertility in mice. Methods Mol Biol 2008;450:149–62.
44. Kanatsu-Shinohara M, Ogonuki N, Inoue K, et al. Restoration of fertility in infertile mice by transplantation of cryopreserved male germline stem cells. Hum Reprod 2003;18:2660–7.
45. Kanatsu-Shinohara M, Ogonuki N, Iwano T, et al. Genetic and epigenetic properties of mouse male germline stemcells during long-term culture. Development 2005;132:4155–63.

Pluripotent Stem Cell Epigenetics During Development and Cancer

Noelia Andollo, M. Dolores Boyano, M. del Mar Zalduendo,
and Juan Aréchaga

Abstract Studies on biological development and cancer have pointed out the importance of specific epigenetic environments to maintain the equilibrium between repressed and activated genes. It has been possible to establish that this kind of environment induces chromatin structure modification and heritable changes in gene functions without altering primary DNA sequencing. We show here recent results of our laboratory on the expression of two imprinted genes, *U2af1-rs1* and *H19*, in normal and pluripotent male germinal cells and in embryonic stem cell after induction of differentiation and apoptosis by retinoic acid treatments. These experimental observations can shed new light for a better understanding of testis embryonal carcinoma biology.

Keywords Pluripotent stem cells • Embryonic stem cells • Germinal cells • Cancer stem cells • Genomic imprinting • Epigenetics

Introduction

Biological development and cancer are typical phenomena under epigenetic control *(1–4)*. In the widest sense, epigenetic alterations are functional modifications of DNA by external conditions, leading to the modulation of gene expression without altering primary DNA sequences. Until now, the main types of molecular epigenetic changes described are related to chromatin (DNA methylation and histone modification by acetylation, methylation, and phosphorylation). But, from the cell biology point of view, there are plenty of related interactions and signaling molecules to be investigated in relationship with the epigenetic phenomena. Pluripotent stem cell proliferation, differentiation, and apoptosis during development and cancer provide one of the best models to study this *(5–7)*.

N. Andollo, M.D. Boyano, M. del Mar Zalduendo and J. Aréchaga (✉)
Laboratory of Developmental Biology and Cancer, Department of Cell Biology and Histology, Faculty of Medicine and Odontology, University of the Basque Country, 48940, Leioa, Vizcaya, Spain
e-mail: juan.arechaga@ehu.es

H. Baharavand (ed.), *Trende in Stem Cell Biology and Technology.*
DOI 10.1007/978-1-60327-905-5_06
© Humana Press, a Part of Springer Science + Business Media, LLC 2009

As examples of pluripotent stem cell epigenetics approaches, we will analyze in this chapter our recent results on the study of methylation state and DNase-I sensitivity of two genes with opposite genomic imprinting (*U2af1-rs1* and *H19*) in pluripotent germinal cells (EG1), isolated postnatal spermatogonia, and mature sperm cells *(8)*. We will then discuss the epigenetic regulation of imprinting genes during embryonic stem cell (ES cells) differentiation induced by retinoic acid *(9)* and the involvement of epigenetic modifications in the simultaneous apoptosis process *(10)*. These approaches will help increase the understanding of embryonal carcinoma cell (EC cells) biology, mainly because these cancer stem cells closely resemble ES cells *(11)*. Testis embryonal carcinoma is typically a developmental tumor in which we also are currently examining several epigenetic factors like matrix metalloproteinases, insulin-like growth factors, transforming growth factor beta, cadherin/catenin complex, integrins, angiogenic factors, etc. related with its invasive and differentiation properties *(12)*.

Epigenetic Mechanisms of Imprinted Genes in Germinal Cells

Imprinted genes are those genes in which monoallelic expression is governed by the parental origin of the allele. The epigenetic modification that gives rise to their monoallelic expression is known as genomic imprinting *(4, 13, 14)*. Different mechanisms have been proposed as hallmarks of imprinting, such as differences in chromatin structure between the maternal and paternal alleles *(15)*; allele-specific differences in differentially methylated regions (DMRs), which are CpG rich areas subjected to epigenetic modifications *(16, 17)*; the role of regulatory elements at imprinted domains *(18)*; and processes of acetylation and methylation of histones *(19, 20)*.

The occurrence of these processes during the initial stages of development is necessary for correct embryo development. Parental imprinting is conserved in all somatic cells after fertilization. However, in the germ cell line, gene imprinting needs to be reset to establish a new sex-specific imprint that will be inherited by the next generation. The precise timing of the erasure and establishment of imprint for many genes remains to be determined, and the molecular mechanisms underlying genomic imprinting have not yet been fully characterized. We have analyzed the methylation pattern and the chromatin structure of two genes with opposite genomic imprinting (*U2af1-rs1* and *H19* genes) in embryonic germ (EG) cells derived from 8.5 days postcoitum (dpc) mouse embryo primordial germ cells (PGCs), in isolated postnatal spermatogonia cells and in mature sperm cells.

The *U2af1-rs1* gene is a small intronless gene located in the proximal region of mouse chromosome 11, which encodes a protein that shares homology with the U2 small nuclear ribonucleoprotein auxiliary factor *(21, 22)*. The maternal allele of this gene is imprinted, and it is the paternal allele that is expressed in embryonic and adult tissues. The *H19* gene is mapped to a cluster of imprinted genes located in the distal region of mouse chromosome 7. It encodes the RNA of the most abundant ribonucleoprotein particle found during embryo development *(23)*. In contrast to the *U2af1-rs1* gene, the imprinted allele is the paternal one, and expression of this gene takes place from the maternal allele. It is thought that *H19* participates in down-regulating cellular proliferation *(24)*.

Concerning DNA methylation, imprinted genes can be identified in somatic cells as those that present DMR methylation in 50% of their alleles. This percentage corresponds to the maternal alleles in the case of the *U2af1-rs1* gene or the paternal ones in the case of the *H19* imprinted gene. So, a state of complete demethylation for the *U2af1-rs1* gene or complete methylation for the *H19* gene in the male germ line is likely to be indicative of establishment of genomic imprinting. To perform methylation analysis of the *U2af1-rs1* gene, the methylation dependent *Not*I and *Hpa*II restriction enzymes were analyzed in the genomic region limited by the *Bgl*II enzyme, which includes the gene and its DMR (Fig. 1a). We found that the *Not*I and *Hpa*II restriction sites in the *U2af1-rs1* DMR were partially methylated in spermatogonia and sperm cells, indicating that imprint establishment had not yet happened at this stage of male germ cell differentiation. Surprisingly, the level of methylation increased during the development of the male germ cell line (Fig. 2a). In fact, sperm cells presented a complete methylation of the *Not*I and *Hpa*II restriction sites in approximately half of the *U2af1-rs1* alleles, the same as isolated Sertoli cells and MSC-1 Sertoli cell line used as somatic control cells *(8)*. These results indicate that the imprinting of the *U2af1-rs1* gene is not totally established in sperm cells. Previous reports concerning other imprinted genes (*Igf2* and *Igf2r*) *(25, 26)* support the idea that the allelic methylation pattern is not fully established in the gametes, but will be complete after fertilization or even during the initial stages of embryo development. Moreover, it has been shown by immunofluorescence techniques that after fertilization and prior to the first DNA duplication, the paternal genome is selectively demethylated, while the maternal genome displays de novo methylation *(27–29)*.

In relation to the *H19* gene, two DMRs were analyzed: the DMR located 2–4 kb upstream of the *H19* coding sequence, delimited by *Sac*I restriction enzyme, and another small DMR situated upstream near the promoter and included in the *Bst*XI restriction fragment. The methylation-dependent *Hha*I restriction endonuclease was studied in both regions (Fig. 1b). We found a progressive increment in the methylation status of the *H19* gene at the *Hha*I sites in *Sac*I and *Bst*XI fragments for spermatogonia and sperm cells, although methylation was not absolute (Fig. 2b) *(8)*. These results are in agreement with those from other authors *(30)*. In fact, the detected methylation was approximately twofold in the *Sac*I DMR of the *H19* gene in spermatogonia and sperm cells with respect to somatic cells, suggesting that all *H19* alleles are methylated in spermatogonia and gametes. The *Bst*XI fragment presented a similar level of methylation in both spermatogonia and somatic cells, but this degree of methylation increased almost twofold in sperm cells. Therefore, in contrast to the *U2af1-rs1* gene, imprinting of *H19* appears to be completely established in the male gametes.

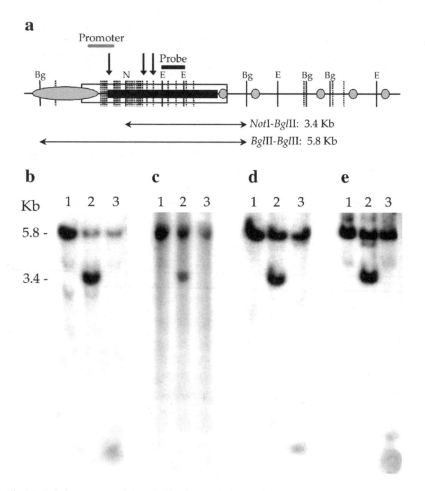

Fig. 1 Methylation pattern of the *U2af1-rs1* gene. (**a**) The *U2af1-rs1* gene (*black box*) and flanking sequences (GenBank accession number AF309654) include the *EcoR*I (E), *Bgl*II (Bg), *Not*I (N), and *Hpa*II (*disrupted lines*) restriction sites. A 630-base pair *EcoR*I fragment was used as a probe. The *open rectangle* denotes the position of the DMR and the *gray rounded* figures indicate the location of repetitive sequences. The *thick line* above the gene represents the promoter. The size of the restriction fragments obtained following digestion with *Bgl*II or *Bgl*II + *Not*I restriction enzymes is also shown. DNase-I hypersensitive sites (*vertical arrows*) are located in a broad region within approximately 1 kb from the *U2af1-rs1* transcription start site. The DNA methylation pattern was analyzed by Southern blotting in spermatogonia (**b**), mature sperm cells (**c**), control isolated Sertoli cells (**d**), and the MSC-1 Sertoli cell line (**e**). To perform methylation assays, samples were incubated with *Bgl*II, *Bgl*II + *Not*I, and *Bgl*II + *Hpa*II restriction enzymes (lanes 1, 2, and 3, respectively)

It is noteworthy that methylation of the *Sac*I DMR was observed during the spermatogonia stage, whereas methylation of the *Bst*XI restriction fragment, which includes the promoter and the 5′-extreme of the *H19* gene, was observed later, during the sperm cell stage of differentiation (Fig. 2b). Thus, the occurrence of methylation

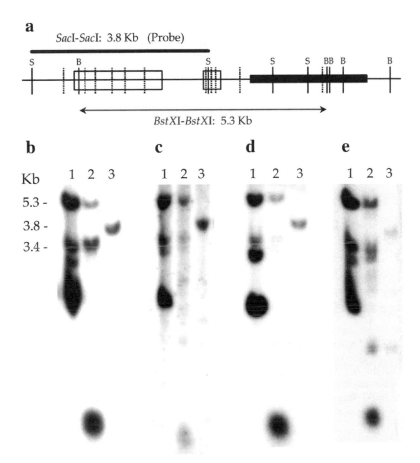

Fig. 2 Methylation pattern of the *H19* gene. (**a**) The *H19* gene (*black box*) and its upstream sequence (GenBank accession number U19619) include the *Bst*XI (Bs), *Sac*I (S), and *Hha*I (*disrupted lines*) restriction sites. A 3.8-kb *Sac*I DNA fragment was used as a probe. The DMRs (*open rectangles*) are located at 2–4 kb and less than 1 kb upstream from the transcription start site. The *Bst*XI restriction fragment is also shown. The DNA methylation pattern was analyzed by Southern blotting in spermatogonia (**b**), mature sperm cells (**c**), control isolated Sertoli cells (**d**), and the MSC-1 Sertoli cell line (**e**). To perform methylation assays, samples were incubated with *Bst*XI, *Bst*XI + *Hha*I, and *Sac*I + *Hha*I restriction enzymes (lanes 1, 2, and 3, respectively)

of the *Bst*XI fragment subsequent to that of the SacI fragment during the course of development corroborates the theory that proximal promoter region is probably involved in the maintenance rather than the establishment of *H19* genomic imprinting *(31)*.

Our methylation results suggest that the establishment of imprinting in the male germ cell line is gene specific, and that this process does not occur during a single stage of development.

Nevertheless, prior to the establishment of methylation, other epigenetic mechanisms may be involved to determine the identification of alleles as future paternal

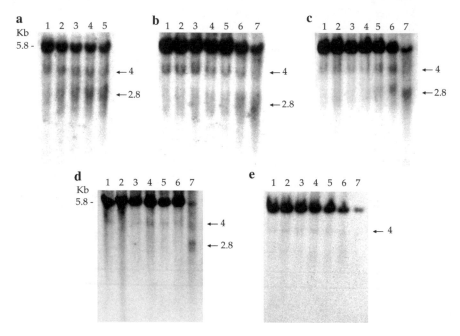

Fig. 3 DNase-I sensitivity of the *U2af1-rs1* gene in male germ differentiation. DNase-I sensitivity was analyzed by Southern blotting in spermatogonia (**a**), control isolated Sertoli cells (**b**), MSC-1 Sertoli cell line (**c**), as well as in male EG-1 (**d**) and female EG-3 (**e**) embryonic germ cell lines. For nuclease sensitivity analysis, after incubation of nuclei with 0, 10, 20, 30, 50, 100, and 250 U/ mL of DNase-I (lanes 1–7), DNA was isolated from nuclei and digested with *Bgl*II. *Arrows* indicate DNase-I digestion products corresponding to DNase-I hypersensitive sites (HSS)

or maternal and subsequently permit the selective methylation of the appropriate maternal or paternal allele of the imprinted gene. We suggest the possibility that modifications in chromatin structure, and more specifically the presence of DNase-I hypersensitive sites (HSS), may be involved in the sex-specific labeling of alleles.

Thus, we analyzed chromatin conformation of the *U2af1-rs1* and *H19* genes by DNase-I sensitivity assay in spermatogonia, Sertoli cells, MSC-1 Sertoli cell line as well as in male EG-1 and female EG-3 embryonic germ cells, which derive from 8.5 dpc mouse embryo PGCs. We detected HSS for the *U2af1-rs1* gene in male germ and somatic cells, but not in female germ cells, which only contain future maternal alleles that will become methylated (Fig. 3) *(8)*. This finding suggests the association of HSS with the *U2af1-rs1* paternal allele. It is possible that the presence of HSS may be related to the joining of specific DNA binding proteins involved in keeping the paternal allele unmethylated, as has been described elsewhere *(32)*.

At the *H19* DMR, we detected several HSS in the different types of cellular nuclei analyzed, with the exception of nuclei of spermatogonia (Fig. 4). Moreover, a close chromatin conformation was detected in EG-1 male embryonic germ cells

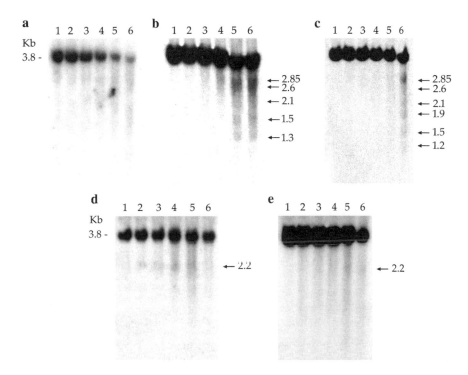

Fig. 4 DNase-I sensitivity of the *H19* gene in male germ differentiation. DNase-I sensitivity was analyzed by Southern blotting in spermatogonia (**a**), control isolated Sertoli cells (**b**), MSC-1 Sertoli cell line (**c**), as well as in male EG-1 (**d**) and female EG-3 (**e**) embryonic germ cell lines. For nuclease sensitivity analysis, after incubation of nuclei with 0, 10, 20, 40, 80, and 160 U/mL of DNase-I (lanes 1–6), DNA was extracted from nuclei and digested with *Sac*I. Arrows indicate DNase-I digestion products corresponding to HSS

in comparison to those of EG-3 female cells *(8)*. Since the spermatogonia will produce male gametes, in which the *H19* gene must be methylated, it is conceivable that the absence of HSS allows these alleles to acquire methylation imprinting (Fig. 5). Consequently, the HSS detected in female EG-3 embryonic germ cells as well as in somatic cells would be associated with maternal alleles. Consistently, other authors have detected HSS in maternal alleles of somatic cells from several tissues with and without gene expression *(33, 34)*, suggesting that the presence of HSS might be constitutively associated with *H19* maternal alleles. Concerning male EG-1 embryonic germ cells, which derive from 8.5 dpc mouse embryo PGCs, the presence of HSS may be due to the fact that the erasure of imprinting of the *H19* gene has not yet happened at this stage of development *(35–37)*.

Therefore, the presence of DNase-I HSS may constitute a molecular marker to identify alleles and subsequently acquire the appropriate methylation imprint. Thus, this identifier would be present or absent for a specific gene according to the sex of the gamete.

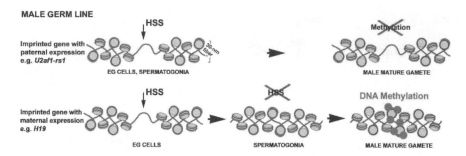

Fig. 5 Hypothetical model for the acquisition of the correct methylation imprint in gametes. Hypersensitive sites (HSS) are discontinuities in the nucleosomal organization of chromatin, which confer an open chromatin conformation. In this model, adapted to the male germ line on the basis of our results, HSS appear as molecular markers to identify alleles. Those alleles that show HSS (*vertical arrow*) for each gene will not acquire methylation, while alleles with no HSS would appear methylated (*dotted*) in that gene

Epigenetic Regulation during Embryonic Stem Cell Differentiation

The embryonic stem cell differentiation pattern closely resembles mouse embryogenesis *(5)*. These undifferentiated and pluripotent cells that are derived from the inner cell mass of blastocysts can differentiate spontaneously in vitro when leukemia inhibitory factor (LIF) is removed from culture media. Differentiation can also be induced with retinoic acid (RA) *(38)*, a molecule extensively associated to the study of the processes of differentiation and morphogenesis during embryo development *(39)*.

Cell type–specific differentiation process implies changes in gene expression. Based on the evidences that gene expression can be modulated by epigenetic modifications *(40)*, together with the suggestion that these epigenetic modifications have a deep importance for the biology of the stem cells *(41)*, we decided to study the involvement of epigenetic mechanisms such as DNA methylation and chromatin structure in RA-induced differentiation of ES cells.

The mechanisms of action of epigenetic modifications have been characterized in imprinted genes, which exhibit a differential expression of the maternal and paternal alleles subjected to modifications in DNA methylation and chromatin structure *(42–44)*. Hence, we chose for this study the maternally imprinted *U2af1-rs1* gene because of its well-characterized methylation pattern *(45)* and its isolated location in the proximal region of mouse chromosome 11 *(46)*, not clustered with other imprinted genes. This intronless gene codes for a protein with homology to the splicing factor U2 small nuclear ribonucleoprotein auxiliary factor *(21, 22)*.

Our differentiation model consisted of treatment of ES cells with 1 μM RA in the absence of LIF during 72 h. In addition to monolayers of differentiated cells, this treatment also produced spontaneous formation of embryoid bodies.

Following the analysis of several differentiation markers, the treatment showed that ES cells differentiated into visceral endoderm along with some cells that remained undifferentiated or with primitive ectoderm phenotypes *(9)*.

To study the control of the *U2af1-rs1* gene expression by epigenetic mechanisms in the differentiation process of ES cells, we started analyzing DNA methylation in relation to gene expression. We observed an increment in the *U2af1-rs1* gene expression during the differentiation process of these cells induced by RA. However, the increase in gene expression did not correlate with a loss of methylation at this gene (Fig. 6) *(9)*. In detail, to perform DNA methylation experiments, several methylation-dependent (*Not*I and *Hpa*II) and methylation-independent (*Msp*I) restriction enzymes were analyzed in the *U2af1-rs1* gene region limited by *Bgl*II restriction enzyme (Fig. 1a). We observed a specific increment of methylation at *Hpa*II sites in differentiated cells and in embryoid bodies, as well as methylation of *Msp*I recognition sites in embryoid bodies (Fig. 6a) *(9)*. Methylation increase does not depend on culture conditions nor of the joining of methyl-CpG-binding proteins (like MeCP2, which cover restriction sites). Indeed, it is due to specific DNA methylation at these regions.

Although methylation of cytosine residues in CpG dinucleotides of CpG islands has been correlated with gene silencing, several exceptions have been described *(15, 47)*, even during differentiation and development *(48–50)*. Furthermore, recent evidence tends to give the main importance to other epigenetic mechanisms such as changes in chromatin structure and covalent modifications of histones in the control of gene expression *(51, 52)*. When we analyzed chromatin structure in the RA-induced process of differentiation of ES cells, we observed that differentiated cells displayed a more open chromatin conformation than undifferentiated cells. Furthermore, together with more sensitiveness of differentiated cells to DNase-I, they showed an HSS *(9)*, which has been thought to be directly associated with *U2af1-rs1* gene expression (Fig. 7). Therefore, as *U2af1-rs1* RNA expression appears to be higher in differentiated cells and in embryoid bodies than in their undifferentiated counterparts, during RA-differentiation of ES cells, the *U2af1-rs1* expression is correlated with an open chromatin conformation. It is not yet clear the mechanism underlying the control of the *U2af1-rs1* gene expression by chromatin structure, but it seems that rather than changes in histone acetylation, it could be associated with repressing chromatin structures analogous to heterochromatin, which would be organized independently from DNA methylation or histone acetylation *(53)*. So our results suggest that the regulation of the *U2af1-rs1* gene expression could be modulated by variations in chromatin structure rather than by changes in the DNA methylation pattern during RA-induced process of differentiation of ES cells.

The epigenetic regulation of different genes such as *Oct-4* or *TERT* during cellular differentiation and development reveals the participation of DNA methylation, modifications on chromatin structure, or both mechanisms on the regulation of gene expression during these processes *(7, 54–57)*. However, the fact that the expression product of the *U2af1-rs1* gene is implicated in RNA splicing could associate epigenetic mechanisms to the control of RNA maturation, a step forward beyond transcription.

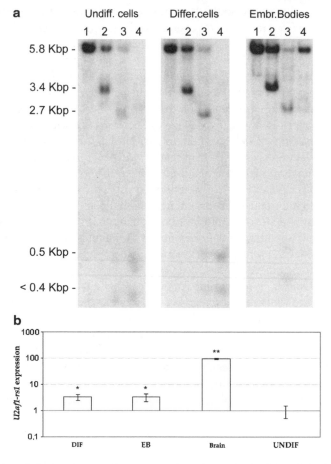

Fig. 6 DNA methylation pattern and gene expression of the *U2af1-rs1* domain in the differentiation process of embryonic stem (ES) cells. (**a**) DNA methylation pattern was analyzed by Southern blot in undifferentiated cells as well as in differentiated cells and embryoid bodies obtained following 72-h retinoic acid (RA) treatment in the absence of leukemia inhibitory factor (LIF). The restriction pattern corresponds to digestions with the following endonucleases: *Bgl*II alone (lane 1) or in combination with *Not*I (lane 2), *Hpa*II (lane 3), and *Msp*I (lane 4). (**b**) Logarithmic representation of the *U2af1-rs1* gene expression by real-time reverse transcription-polymerase chain reaction (RT-PCR). Expression levels were measured in undifferentiated ES cells (UNDIF) as well as in differentiated cells (DIF) and embryoid bodies (EB) obtained following 72-h RA treatment in the absence of LIF. Positive control for the *U2af1-rs1* expression was provided by a brain sample. Quantitative data for the *U2af1-rs1* gene were normalized to that obtained from the combination of three housekeeping genes (*cyclophilin A*, *β-actin*, and *rig/S15*). Each column represents the average of six amplification reactions with samples and controls run in triplicate (error bars represent the standard deviations) performed on three cDNA samples reverse transcribed from RNA prepared from three independent culture assays. Values for undifferentiated ES cells were set to 1. Treated cells and brain sample were compared with untreated cells for statistical significance. *, Significant ($p < 0.05$). **, Very significant ($p < 0.01$). We observed a threefold increase in *U2af1-rs1* gene expression in differentiated ES cells and embryoid bodies as compared with undifferentiated ES cells

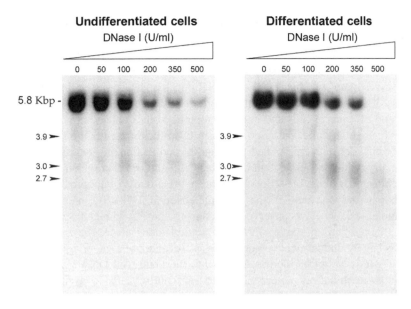

Fig. 7 Southern blot showing DNase-I sensitivity of *U2af1-rs1* gene in undifferentiated embryonic stem (ES) cells and in differentiated ones obtained after 72-h retinoic acid (RA) treatment of cells in the absence of leukemia inhibitory factor (LIF). The assay was performed in isolated nuclei using increasing concentrations of DNase I (0, 50, 100, 200, 350, and 500 U/mL). Afterward, DNA was purified and digested with *Bgl*II restriction enzyme, generating a 5.8- kb DNA fragment. The membrane was hybridized with the *U2af1-rs1* probe. *Arrowheads* point out hypersensitive sites (HSS)

Epigenetic Modifications during Apoptosis of Embryonic Stem Cells

The mechanism of cell death, called apoptosis, plays a crucial role during embryonic development. This physiological process, in which morphological and biochemical events are well characterized, usually finishes with the production of apoptotic bodies. These membrane-bounded and well-preserved cell structures are condemned to degradation. However, little has been studied about their biology. In addition, many aspects of the molecular regulation of the apoptotic process still remain unclear. In this sense, the participation of epigenetic mechanisms in apoptosis has not yet been elucidated, although its importance in the control of gene activation or repression has been extensively demonstrated in genomic imprinting *(44)* and, very recently, in carcinogenesis *(1, 58)*.

We have analyzed the involvement of DNA methylation and chromatin structure in the process of apoptosis of ES cells. In vitro treatment with physiological doses of RA of these cells as well as of EC F9 and P19 cells also included in this study, gave rise to apoptosis associated to the process of cellular differentiation *(59, 60)*.

The chosen treatment consisted of culture of ES cells with 1 µM RA in the absence of LIF during 72 h.

As for the study of epigenetic mechanisms in relation to cellular differentiation *(9)*, we employed the maternally imprinted *U2af1-rs1* gene.

To investigate whether DNA methylation changes could be involved in the process of apoptosis, the methylation pattern of the *U2af1-rs1* gene was analyzed in ES and EC control cells and in their apoptotic bodies obtained by 72-h RA treatment. We used the same restriction enzymes as for the differentiation studies. That is, *Bgl*II (which delimits the studied gene and 5′ and 3′ flanking regions) alone or in combination with the methylation-dependent *Not*I and *Hpa*II or with the methylation-independent *Msp*I enzymes (Fig. 1a). During the process of apoptosis, a characteristic happening consists of the internucleosomal DNA fragmentation due to endonucleases. We have observed that during the RA-induced apoptosis of ES cells, although this massive DNA degradation takes place, the *U2af1-rs1* gene remains largely undigested by endonucleases, revealing the integrity of some copies of this gene in apoptotic bodies. Moreover, the preserved DNA fragments maintained the methylation pattern observed for control cells, consisting of partial methylation for *Not*I and *Hpa*II restriction sites for ES cells (Fig. 8a) *(10)*. Thus, neither methylated nor unmethylated restriction sites were specifically targeted by apoptotic endonuclease activity during chromatin fragmentation, since bands corresponding to each methylated and nonmethylated sequence of the restriction enzymes were evident. So, DNA fragmentation is not dependent on the methylation status of the methylation-dependent recognition sites. The conservation of DNA regions in spite of the common chromatin fragmentation was also observed in the apoptotic process of F9 *(60)* and P19 EC cells *(10)*. Both EC cell lines show completely unmethylated restriction pattern for *Not*I site and for all *Hpa*II recognition sites. This result indicates that the maintenance of the structural integrity of DNA fragments during apoptosis is not linked to a particular cell line.

In addition, densitometric analysis of autoradiographs revealed a significant DNA demethylation of about 15% at *Not*I restriction sites for the *U2af1-rs1* gene in apoptotic bodies with respect to control ES cells, while demethylation at *Hpa*II sites was very slight (Fig. 8a) *(10)*. To determine if the demethylation of these DNA specific sequences in apoptotic bodies was a consequence of RA treatment or, on the contrary, a general event in the apoptosis of ES cells, we cultured cells with 5-aza-2′-desoxycytidine (AZDC), a compound that inhibits general DNA methylation. The fragmentation pattern of DNAs obtained from supernatants of cultures treated at different doses of AZDC revealed their apoptotic nature (Fig. 8b). Moreover, apoptosis was found to be dose dependent, suggesting that modifications on the genomic pattern of methylation in our experimental model are associated with the induction of apoptosis *(10)*. Other authors have previously shown genomic DNA demethylation preceding the execution phase of apoptosis as well *(61–63)*. Recently, cancer has been defined as an epigenetic disease where a wide global hipomethylation is observed, together with hypermethylation of CpG islands located at regulatory gene regions *(1)*. Demethylating compounds and histone deacetylase inhibitors assayed in tumor cell lines restore the expression of some tumor suppressor genes that had suffered repression due to the aberrant

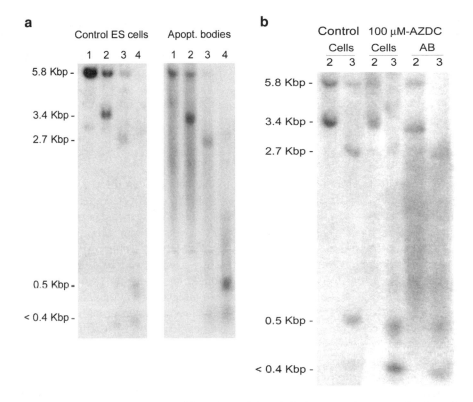

Fig. 8 DNA methylation analysis of the *U2af1-rs1* domain in the apoptotic process of embryonic stem (ES) cells. (**a**) DNA methylation pattern in control cells as well as in apoptotic bodies obtained from supernatants of 72-h retinoic acid (RA) treated cultures of ES cells. The restriction pattern analyzed by Southern blot corresponds to digestions with the following endonucleases: *Bgl*II alone (lane 1) or in combination with *Not*I (lane 2), *Hpa*II (lane 3), and *Msp*I (lane 4). (**b**) Restriction pattern analysis of *U2af1-rs1* gene by Southern blotting after experimental demethylation of ES cells induced by 100 μM 5-aza-2'-desoxicytidine (AZDC) treatment. Untreated ES cells (control) as well as adherent cells and apoptotic bodies (AB) obtained following AZDC treatment during 48 h were analyzed. Lanes 2 and 3 correspond to DNAs digested with *Bgl*II plus *Not*I or *Hpa*II restriction enzymes, respectively

hypermethylation that occurs during transformation. Some tumor suppressor genes, such as *RAR-β2 (34, 35)* or *Apaf-1 (36, 37)*, are directly involved in the apoptotic process. As a result, experimental demethylation of these genes restores the capability to induce apoptosis in tumor cell lines. In a similar way, the variation in the methylation status observed in nontransformed ES cells after treatment with AZDC gives place to the apoptotic process, although the precise genes affected by demethylation that produce apoptosis remain to be found.

Surprisingly, no demethylation at all occurred at *Not*I restriction site of the *U2af1-rs1* gene, even after treatment with high doses (100 μM) of the demethylating agent AZDC during 48 h. Nevertheless, *Hpa*II recognition sites were totally demethylated for both adherent cells and apoptotic bodies (Fig. 8b) *(10)*. These results indicate that demethylation of the *U2af1-rs1* gene at the *Not*I site might be specific for the treatment of ES cells with RA. The *Not*I site is located in the CpG island of the *U2af1-rs1* gene. Although CpG islands are usually in a demethylated state and show an open chromatin structure that promotes gene transcription *(42)*, imprinted genes are an exception and their CpG islands usually show DNA methylation in the repressed allele. The maintenance of the methylation status that we observed for the *Not*I site after AZDC treatment may be a mechanism to preserve the imprinting mark of the gene. According to it, this region of the *U2af1-rs1* gene has been previously shown to exhibit particular resistance to the effect of the highly specific inhibitor of histone deacetylase trichostatin A (TSA) *(19)*.

With regard to chromatin conformation, our data concerning the analysis of sensitivity to DNase-I in apoptotic bodies revealed an unexpected open chromatin structure for the *U2af1-rs1* gene that allows the access and cleavage by this nuclease *(10)*. In fact, genomic DNA obtained from apoptotic bodies showed sensitivity to DNase-I, as we observed by agarose gel electrophoresis. When we performed Southern blotting for the *U2af1-rs1* region, we saw that apoptotic bodies were sensitive to DNase-I at this region. Equivalent results were observed for the *β-ACTIN* gene (Fig. 9). Thus, in spite of the extremely compact chromatin of apoptotic bodies, there is a relatively open chromatin conformation that still makes them accessible to the DNase-I endonuclease. It is feasible that the demethylation observed for this gene might alter the accessibility and action of DNase-I. In this context, similar results have been observed in the highly condensed inactive X-chromosome of murine lung fibroblasts, where demethylation of about 15% induced after AZC (5 azacytidine) treatment produces a dramatic increase in DNase-I sensitivity of the entire chromosome *(64)*. These results show that methylation affects chromatin structure in a model where chromatin is as highly condensed as in apoptotic bodies.

Based on the observation of the relatively open chromatin conformation of apoptotic bodies, we thought that transcriptional capability might be conserved during apoptosis. By agarose gel stained with ethidium bromide, we detected total RNA in apoptotic bodies. Moreover, expression of transcripts of *U2af1-rs1* and *β-ACTIN* genes were clearly detected by Northern blotting, while a faint signal was also detected for the *Gapdh* gene. Furthermore, after performing an in vitro translation assay with purified mRNAs from apoptotic bodies we detected newly synthesized biotin-labeled proteins, revealing the functional state of mRNAs purified from apoptotic bodies *(10)*. Thus, the relatively open chromatin conformation observed in apoptotic bodies and possibly affected by DNA demethylation may facilitate several events that take place in the apoptotic process. For example, it may allow the access of nucleases to DNA to perform oligonucleosomal fragmentation and the access of some nuclear caspases in order to cleave the elements that protect DNA, such as histone H1. Furthermore, decompaction of chromatin in apoptotic bodies might allow the expression of genes required for the proper execution of the apoptotic process, specially considering the large presence of *U2af1-rs1* transcripts observed in apoptotic bodies, which suggests that their RNA maturation machinery should still be able to proceed to a certain extent, as the product of this gene has been implicated in RNA splicing.

U2af1-rs1 gene

β-ACTIN gene

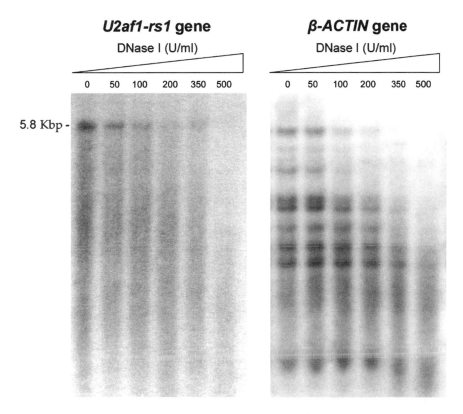

Fig. 9 Southern blot showing DNase-I sensitivity of *U2af1-rs1* and human β-ACTIN genes in apoptotic bodies of embryonic stem (ES) cells obtained after 72-h retinoic acid (RA) treatment of ES cells. The assay was performed in isolated nuclei using increasing concentrations of DNase-I (0, 50, 100, 200, 350, and 500 U/mL). Afterward, DNA was purified and digested with *Bgl*II restriction enzyme, generating a 5.8-kb DNA fragment. The membrane was hybridized with the *U2af1-rs1* probe and with a human *β-ACTIN* probe

Acknowledgments This work was supported by Spanish Ministry of Education and Science grant (BFU 2007–66610/BFI), University of the Basque Country Research Group grant (GIU08/04) to J.A. and Jesús de Gangoiti Barrera Foundation fellowship to N.A.

References

1. Esteller M, Herman JG. Cancer as an epigenetic disease: DNA methylation and chromatin alterations in human tumours. *J Pathol* 2002;196:1–7.
2. Fulka H, St John JC, Fulka J, Hozak P. Chromatin in early mammalian embryos: achieving the pluripotent state. *Differentiation* 2008;76:3–14.
3. Kiefer JC. Epigenetics in development. *Dev Dyn* 2007;236:1144–56.
4. Murrell A. Genomic imprinting and cancer: from primordial germ cells to somatic cells. *Sci World J* 2006;6:1888–910.

5. Rizzino A. Embryonic stem cells provide a powerful and versatile model system. *Vitam Horm* 2002;64:1–42.
6. Spivakov M, Fisher AG. Epigenetic signatures of stem-cell identity. *Nat Rev Genet* 2007;8:263–71.
7. Szutorisz H, Dillon N. The epigenetic basis for embryonic stem cell pluripotency. *Bioessays* 2005;27:1286–93.
8. Boyano MD, Andollo N, Zalduendo MM, Arechaga J. Imprinting in mammalian gametes is gene specific and do not occur at a single stage of differentiation. *Int J Dev Biol* 2008;52(8):1105–11.
9. Andollo N, Boyano MD, Andrade R, Arechaga JM. Epigenetic regulation of the imprinted U2af1-rs1 gene during retinoic acid-induced differentiation of embryonic stem cells. *Dev Growth Differ* 2006;48:349–60.
10. Andollo N, Boyano MD, Andrade R, et al. Structural and functional preservation of specific sequences of DNA and mRNA in apoptotic bodies from ES cells. *Apoptosis* 2005;10:417–28.
11. Harrison NJ, Baker D, Andrews PW. Culture adaptation of embryonic stem cells echoes germ cell malignancy. *Int J Androl* 2007;30:275–81.
12. Diez-Torre A, Silvan U, De Wever O, Bruyneel E, Mareel M, Arechaga J. Germinal tumor invasion and the role of the testicular stroma. *Int J Dev Biol* 2004;48:545–57.
13. Ferguson-Smith AC, Surani MA. Imprinting and the epigenetic asymmetry between parental genomes. *Science* 2001;293:1086–9.
14. da Rocha ST, Ferguson-Smith AC. Genomic imprinting. *Curr Biol* 2004;14:646–9.
15. Feil R, Handel MA, Allen ND, Reik W. Chromatin structure and imprinting: developmental control of DNase-I sensitivity in the mouse insulin-like growth factor 2 gene. *Dev Genet* 1995;17:240–52.
16. Jaenisch R, Bird A. Epigenetic regulation of gene expression: how the genome integrates intrinsic and environmental signals. *Nat Genet* 2003;33 Supp 1: 245–54.
17. Durcova-Hills G, Burgoyne P, McLaren A. Analysis of sex differences in EGC imprinting. *Dev Biol* 2004;268:105–10.
18. Delaval K, Feil R. Epigenetic regulation of mammalian genomic imprinting. *Curr Opin Genet Dev* 2004;14:188–95.
19. Gregory RI, O'Neill LP, Randall TE, et al. Inhibition of histone deacetylases alters allelic chromatin conformation at the imprinted U2af1-rs1 locus in mouse embryonic stem cells. *J Biol Chem* 2002;277:11728–34.
20. Drewell RA, Goddard CJ, Thomas JO, Surani MA. Methylation-dependent silencing at the H19 imprinting control region by MeCP2. *Nucleic Acids Res* 2002;30:1139–44.
21. Hatada I, Sugama T, Mukai T. A new imprinted gene cloned by a methylation-sensitive genome scanning method. *Nucleic Acids Res* 1993;21:5577–82.
22. Hayashizaki Y, Shibata H, Hirotsune S, et al. Identification of an imprinted U2af binding protein related sequence on mouse chromosome 11 using the RLGS method. *Nat Genet* 1994;6:33–40.
23. Brannan CI, Dees EC, Ingram RS, Tilghman SM. The product of the H19 gene may function as an RNA. *Mol Cell Biol* 1990;10:28–36.
24. Bartolomei MS, Tilghman SM. Genomic imprinting in mammals. *Annu Rev Genet* 1997;31:493–525.
25. El-Maarri O, Buiting K, Peery EG, et al. Maternal methylation imprints on human chromosome 15 are established during or after fertilization. *Nat Genet* 2001;27:341–4.
26. Reik W, Walter J. Evolution of imprinting mechanisms: the battle of the sexes begins in the zygote. *Nat Genet* 2001;27:255–6.
27. Mayer W, Niveleau A, Walter J, Fundele R, Haaf T. Demethylation of the zygotic paternal genome. *Nature* 2000;403:501–2.
28. Oswald J, Engemann S, Lane N, et al. Active demethylation of the paternal genome in the mouse zygote. *Curr Biol* 2000;10:475–8.

29. Santos F, Hendrich B, Reik W, Dean W. Dynamic reprogramming of DNA methylation in the early mouse embryo. *Dev Biol* 2002;241:172–82.

30. Ueda T, Abe K, Miura A, et al. The paternal methylation imprint of the mouse H19 locus is acquired in the gonocyte stage during foetal testis development. *Genes Cells* 2000;5:649–59.

31. Davis TL, Yang GJ, McCarrey JR, Bartolomei MS. The H19 methylation imprint is erased and re-established differentially on the parental alleles during male germ cell development. *Hum Mol Genet* 2000;9:2885–94.

32. Feil R, Khosla S. Genomic imprinting in mammals: an interplay between chromatin and DNA methylation? *Trends Genet* 1999;15:431–5.

33. Hark AT, Tilghman SM. Chromatin conformation of the H19 epigenetic mark. *Hum Mol Genet* 1998;7:1979–85.

34. Khosla S, Aitchison A, Gregory R, Allen ND, Feil R. Parental allele-specific chromatin configuration in a boundary-imprinting-control element upstream of the mouse H19 gene. *Mol Cell Biol* 1999;19:2556–66.

35. Hajkova P, Erhardt S, Lane N et al. Epigenetic reprogramming in mouse primordial germ cells. *Mech Dev* 2002;117:15–23.

36. Li JY, Lees-Murdock DJ, Xu GL, Walsh CP. Timing of establishment of paternal methylation imprints in the mouse. *Genomics* 2004;84:952–60.

37. Trasler JM. Gamete imprinting: setting epigenetic patterns for the next generation. *Reprod Fertil Dev* 2006;18:63–9.

38. Mummery CL, Feyen A, Freund E, Shen S. Characteristics of embryonic stem cell differentiation: a comparison with two embryonal carcinoma cell lines. *Cell Differ Dev* 1990;30:195–206.

39. Rohwedel J, Guan K, Wobus AM. Induction of cellular differentiation by retinoic acid in vitro. *Cells Tissues Organs* 1999;165:190–202.

40. Razin A, Kantor B. DNA methylation in epigenetic control of gene expression. *Prog Mol Subcell Biol* 2005;38:151–67.

41. Cerny J, Quesenberry PJ. Chromatin remodeling and stem cell theory of relativity. *J Cell Physiol* 2004;201:1–16.

42. Bird A. The essentials of DNA methylation. *Cell* 1992;70:5–8.

43. Razin A, Cedar H. DNA methylation and genomic imprinting. *Cell* 1994;77:473–6.

44. Tilghman SM. The sins of the fathers and mothers: genomic imprinting in mammalian development. *Cell* 1999;96:185–93.

45. Feil R, Boyano MD, Allen ND, Kelsey G. Parental chromosome-specific chromatin conformation in the imprinted U2af1-rs1 gene in the mouse. *J Biol Chem* 1997;272:20893–900.

46. Nabetani A, Hatada I, Morisaki H, Oshimura M, Mukai T. Mouse U2af1-rs1 is a neomorphic imprinted gene. *Mol Cell Biol* 1997;17:789–98.

47. Feil R, Walter J, Allen ND, Reik W. Developmental control of allelic methylation in the imprinted mouse Igf2 and H19 genes. *Development* 1994;120:2933–43.

48. Vizirianakis IS, Pappas IS, Gougoumas D, Tsiftsoglou AS. Expression of ribosomal protein S5 cloned gene during differentiation and apoptosis in murine erythroleukemia (MEL) cells. *Oncol Res* 1999;11:409–19.

49. Warnecke PM, Clark SJ. DNA methylation profile of the mouse skeletal alpha-actin promoter during development and differentiation. *Mol Cell Biol* 1999;19:164–72.

50. Kaneko KJ, Rein T, Guo ZS, Latham K, DePamphilis ML. DNA methylation may restrict but does not determine differential gene expression at the Sgy/Tead2 locus during mouse development. *Mol Cell Biol* 2004;24:1968–82.

51. Cho KS, Elizondo LI, Boerkoel CF. Advances in chromatin remodeling and human disease. *Curr Opin Genet Dev* 2004;14:308–15.

52. Lee JH, Hart SR, Skalnik DG. Histone deacetylase activity is required for embryonic stem cell differentiation. *Genesis* 2004;38:32–8.

53. El Kharroubi A, Piras G, Stewart CL. DNA demethylation reactivates a subset of imprinted genes in uniparental mouse embryonic fibroblasts. *J Biol Chem* 2001;276:8674–80.

54. Hattori N, Abe T, Suzuki M et al. Preference of DNA methyltransferases for CpG islands in mouse embryonic stem cells. *Genome Res* 2004;14:1733–40.
55. Liu L, Saldanha SN, Pate MS, Andrews LG, Tollefsbol TO. Epigenetic regulation of human telomerase reverse transcriptase promoter activity during cellular differentiation. *Genes Chromosomes Cancer* 2004;41:26–37.
56. Rugg-Gunn PJ, Ferguson-Smith AC, Pedersen RA. Human embryonic stem cells as a model for studying epigenetic regulation during early development. *Cell Cycle* 2005;4:1323–6.
57. Szutorisz H, Canzonetta C, Georgiou A, Chow CM, Tora L, Dillon N. Formation of an active tissue-specific chromatin domain initiated by epigenetic marking at the embryonic stem cell stage. *Mol Cell Biol* 2005;25:1804–20.
58. Jones PA. Cancer Death and methylation. *Nature* 2001;409:141, 143–4.
59. Atencia R, Garcia-Sanz M, Unda F, Arechaga J. Apoptosis during retinoic acid-induced differentiation of F9 embryonal carcinoma cells. *Exp Cell Res* 1994;214:663–7.
60. Asumendi A, Andollo N, Boyano MD, et al. The role of cleavage of cell structures during apoptosis. *Cell Mol Biol (Noisy-le-grand)* 2000;46:1–11.
61. Saitoh F, Hiraishi K, Adachi M, Hozumi M. Induction by 5-aza-2′-deoxycytidine, an inhibitor of DNA methylation, of Le(y) antigen, apoptosis and differentiation in human lung cancer cells. *Anticancer Res* 1995;15:2137–43.
62. Jackson-Grusby L, Beard C, Possemato R, et al. Loss of genomic methylation causes p53-dependent apoptosis and epigenetic deregulation. *Nat Genet* 2001;27:31–9.
63. Stancheva I, Hensey C, Meehan RR. Loss of the maintenance methyltransferase, xDnmt1, induces apoptosis in Xenopus embryos. *EMBO J* 2001;20:1963–73.
64. Jablonka E, Goitein R, Marcus M, Cedar H. DNA hypomethylation causes an increase in DNase-I sensitivity and an advance in the time of replication of the entire inactive X chromosome. *Chromosoma* 1985;93:152–6.

Observing and Manipulating Pluripotency in Normal and Cloned Mouse Embryos

Sebastian T. Balbach, F.M. Cavaleri, Luca Gentile, Marcos J. Araúzo-Bravo, Hans R. Schöler, N. Crosetto, and Michele Boiani

Abstract The mouse ooplasm is the ideal platform to study and compare induced and natural pluripotency because it can support both, after somatic cell nuclear transfer (cloning) and after fertilization, respectively. The amount of pluripotency induced after cloning is variable but always limited compared to fertilization. It can be visualized conveniently if the nucleus donor cells carry a green fluorescent protein (GFP) reporter under control of the pluripotency-associated gene *Oct4* promoter. Thus we produced cloned and fertilized mouse embryos transgenic for *Oct4-GFP* (*GOF18-ΔPE-EGFP*). We also developed and validated a live cell imaging method, whereby we resolve and selectively pick cloned embryos that hold distinct amounts of induced pluripotency as predicted by GFP intensity and measured by embryonic stem cell derivation. Currently we are developing a microinjection method to change the level of Oct4 without modifying the genome of the embryo. Here we discuss our findings in relation to the epigenetic reprogramming of the nucleus transplant and to cell fate decisions in the cloned or fertilized mouse embryo.

Keywords Embryo • Embryonic stem cell • Green fluorescent protein • Nuclear transfer • Oct4 • Pluripotency • Reprogramming

Natural Pluripotency

In mammalian development, pluripotency is the ability of a cell to give rise to a host of tissues belonging to the three primary germ layers: ectoderm, mesoderm, and endoderm, and the germ cells. Totipotency is the ability to give rise to all tissues including extraembryonic ones, e.g., the trophectoderm. It is believed that contribution

S.T. Balbach, F.M. Cavaleri, L. Gentile, M.J. Araúzo-Bravo, H.R. Schöler, N. Crosetto, and M. Boiani (✉)
Max Planck Institute for Molecular Biomedicine, Röntgenstrasse 20, 48149, Münster, Germany
e-mail: mboiani@mpi-muenster.mpg.de

H. Baharvand (ed.), *Trends in Stem Cell Biology and Technology*,
DOI 10.1007/ 978-1-60327-905-5_7,
© Humana Press, a Part of Springer Science + Business Media LLC 2009

of germ cells is sufficient to claim totipotency, since the gametes that form the totipotent zygote arise from germ cells. In mouse development, the zygote and the 2-cell stage blastomeres are totipotent. Isolated 4- and 8-cell stage blastomeres cannot form embryos that are viable *in vivo*, although they can contribute to all tissues of chimeric mouse embryos *(1)*, form the whole mouse when supported with tetraploid blastomeres *(2)*, and give rise to pluripotent cell lines *(3)*. This indicates that cell number is important to implement pluripotency *(4)*. At the blastocyst stage, the first cell lineage decision becomes apparent. The blastocyst is composed of an inner cell mass (ICM) and a trophectoderm (TE). The ICM but not the TE is pluripotent and can be derived in cell lines known as embryonic stem (ES) cells in mouse and human *(5)*. Mouse ES cells cannot give rise to TE except after genetic manipulation *(6)* but can form germ cells both *in vivo* and *in vitro* *(7)*, hence their potency is arguably almost full. Under current protocols, human ES cells can form TE *in vitro* *(8)* and probably give rise to germ cells, although the latter awaits formal proof.

In our current molecular understanding of pluripotency, which is mainly based on studies of mouse and human ES cells, the transcription factors Oct4, Sox2, and Nanog are crucial to keep cells self-renewing and preserve their pluripotency *(9)*. These functions are implemented via a network of downstream transcriptional activators and repressors *(5, 10–12)*. Studying how the network is regulated is complicated since responses are combinatorial *(13)* and depend on the protein level of more than a few factors. The case of Oct4 may be regarded paradigmatic for a dose-dependent effect on pluripotency, but it is not exclusive for Oct4. Mouse ES cells heterozygous for a mutation of Nanog are characterized by unstable pluripotency, as seen by *in vitro* cell differentiation in the absence of feeder cells *(14)*. When Nanog is overexpressed, human ES cells acquire the ability to grow feeder-free, while features of the primitive ectoderm are induced *(15)*, whereas hematopoietic stem cells suffer a disorder upon forced expression *(16)*.

In this chapter we discuss the induction of pluripotency in somatic nuclei by means of the nuclear transfer technology that uses the ooplasm as a kind of bioreactor. Establishment of full pluripotency is the most defining aspect of the reprogramming process of somatic nuclei. Although complete reprogramming requires also the silencing of tissue-specific genes, we interchangeably use the terms "reprogramming" and "pluripotency induction." Here we focus on Oct4 as the best (albeit not the only) known transcription factor associated with pluripotency in a dose-dependent manner. Because there can be no educated manipulation nor application without prior description of any biologic phenomena, first we summarize our observations from development of cloned and fertilized mouse embryos that hold different amounts of Oct4. In general, cloned embryos have lower levels of Oct4 than fertilized counterparts. Therefore, we hypothesize that the abnormal development of cloned mouse embryos is contributed, at least in part, by an insufficient level of Oct4. Based on these observations, it becomes a sensible deed to manipulate the level of Oct4 in cloned embryos without this necessarily making sense for fertilized embryos. In fact, the establishment of natural and induced pluripotency might follow different pathways, similar to the case of ES cells derived from blastocyst as opposed to induced pluripotent stem (iPS) cells formed directly from fibroblasts without an intermediate embryo *(17)*. Finally, we anticipate an experimental approach to support nuclear reprogramming and pluripotency in mouse clones via manipulation of the cellular level of Oct4.

Oct4 as the Prototypic Determinant of Pluripotency

Oct4 is historically *the* pluripotency-associated factor *(18–20)*. Somatic stem cells do not depend on Oct4 for self-renewal *(20)*, and since they are also not pluripotent, this confirms the role of Oct4 in pluripotency rather than multi- or oligopotency. Yet we should resist the temptation to adopt an Oct4-centered view when dealing with pluripotency. Forced expression of Sox2 in the absence of Oct4 also keeps mouse ES cells pluripotent *(21)*. In this chapter, we choose to confine our study to Oct4 in mouse development, being aware that there is a lot more to pluripotency than just Oct4.

Homozygous mouse embryos lacking the *Oct4* locus form morphologically normal blastocysts but fail to form a pluripotent ICM *(22)*. During embryogenesis, Oct4 knock-down in the ICM by RNA interference impairs cardiogenesis *(23)*, whereas ubiquitous Oct4 expression driven by the chicken β-actin promoter and the cytomegalovirus enhancer affects midhindbrain patterning *(24)*. In the adult mouse, epithelial cells respond to a bulk increase in Oct4 level and become neoplastic *(25)*. In mouse ES cells, experimental manipulation of Oct4 protein levels below 50% or above 150% of the physiologic amount leads to differentiation along different lineages *in vitro (6)*. Similarly, a bias in differentiation *in vitro* is observed in human ES cells upon manipulation of Oct4 level *(26)*. *In vivo*, overproduction of Oct4 increases the malignant potential of mouse ES cell-derived tumors, while reduction of Oct4 induces tumor regression *(27)*. In embryoid bodies, Oct4 overproduction inhibits hematopoietic differentiation in a dose-dependent manner *(28)*. Maintenance of a certain protein level may be under post-translational control *(29)*.

Induced Pluripotency

Pluripotency can be induced in somatic cells by nuclear transfer in the ooplasm, by cell–cell fusion where the partner cell is already pluripotent, by exposure to pluripotent cell extracts or specific molecules, and by retroviral delivery of genes encoding pluripotency factors. Induction of pluripotency becomes apparent from several landmarks, which include but are not limited to:

1. Reactivation of somatic cell-encoded *Oct4* and *Nanog (14, 30)* with demethylation of their relevant control genomic sequences
2. Silencing of tissue-specific genes *(31)*
3. Silencing of *Xist* and reactivation of the inactivated X-chromosome in female somatic cell *(32, 33)*

4. Teratoma formation by injection of pluripotent cells into severe combined immunodeficiency (SCID) mice
5. Formation of endodermal, mesodermal, and ectodermal tissues in embryoid bodies or contribution of such tissues in chimeric embryos

Historically, the chief method for pluripotency induction, known since 1928 from Hans Spemann and practiced since the 1950s by Briggs, King, and Gurdon, is to transplant the nucleus of a differentiated cell in an ooplasm. More than half a century later, transferring metaphase or interphase somatic genomes into metaphase zygotes or unfertilized oocytes, respectively, is still used to induce pluripotency *(34)*. However, in humans this method is widely regarded as ethically objectionable. Cloned embryos are potentially implantable and may develop to term when transferred to a foster mother. In humans, in humans this potential of cloned embryos raises ethical concerns regarding the technique. Therefore, human reproductive cloning has not been permitted in any country to date. Another complication in oocyte-mediated nuclear reprogramming is the fact that special micromanipulation skills are needed, and freshly prepared unfertilized oocytes are required for making cloned blastocysts efficiently. In one day, only 100–200 nuclei may be successfully transplanted. The molecular and functional aspects of ooplasm-mediated nuclear reprogramming are dealt in the section "Ooplasm-Mediated Induction of Pluripotency."

Alternatively, pluripotency can be induced in somatic cells by fusion with ES cells *(30, 35)* or by delivering via retroviruses specific genes encoding pluripotency-associated factors. Cell fusion and direct gene transfer give rise to iPS cells without generating an intermediate embryo. However, these methods entail modification of the donor genome, i.e., tetraploidy after fusion or insertional mutagenesis after retroviral infection. Strategies are being developed to eliminate the ES cells' chromosomes after cell fusion *(36)* and to deliver the factors without insertional mutagenesis. So far, four factors have proven sufficient to induce pluripotency in mouse and human somatic cells after retrovirus-mediated delivery (OCT4, SOX2, C-MYC, KLF4 *(17, 37)*; OCT4, SOX2, NANOG, LIN28 *(38)*). Most recently the number of factors has been reduced to three (OCT4, SOX2, KLF4 *(39)*).

Multi- and pluripotency can also be induced by exposing somatic cells to pluripotent cell extracts *(40)* or synthetic molecules *(41)* and by transduction of defined factors as proteins *(42)*. When 293T cells are permeabilized and exposed to extracts of NCCIT human carcinoma cells, they form colonies that can be maintained for at least 23 passages *(40)*. In these colonies, markers of differentiation such as Lamin A are down-regulated, pluripotency associated genes *Oct4* and *Sox2* are up-regulated, the Oct4 protein is detected in the cell nucleus, and the *Oct4* promoter is demethylated (positions −1534 to −1773). Despite these remarkable observations, the nature of reprogramming induced via cell extracts is not yet firmly established. In particular, functional proof is lacking that cells reprogrammed by extracts can form unrelated tissues when placed in a proper proper developmental system, as is known for ES cells when injected into blastocysts. It is also unclear how the reprogramming factors reach their targets in the nucleus. This is an issue in protein transduction since a substantial proportion of the factor remains trapped inside cytoplasmic vesicles and thus is not (promptly) available *(43)*.

Ooplasm-Mediated Induction of Pluripotency

Although nuclear transfer in an ooplasm is ethically objectionable, it remains the most efficient method to induce pluripotency, as it allows up to 60% blastocyst formation and from these about 20% ES cell line derivation *(44)*. Full development of cloned embryos in the uterus provides the ultimate proof of effective reprogramming, as evidenced by the thousands of cloned animals produced worldwide *(45)*.

A central question in ooplasm-mediated induction of pluripotency is why gene expression of resultant cloned embryos does not resemble that of fertilized embryos. The obvious answer is that a somatic genome does not have the epigenetic makeup of a gametic genome. The ooplasm machinery modifies the somatic chromatin in a way the chromatin was not prepared for *(46)*; hence, gene expression tends to be abnormal. Another possibility is that nuclear reprogramming is random or stochastic and thereby generates gene expression patterns that are mostly abnormal, no matter which type of input. Bortvin et al. *(47)* analyzed the expression of *Oct4* and 10 *Oct4*-related genes in individual cloned mouse blastocysts derived from cumulus cells. They found that 62% of these correctly expressed all tested genes. In contrast, ES cell-cloned embryos expressed these genes quite normally, although later studies exposed that levels of *Oct4* mRNA in ES cell clones are actually lower than in somatic cell clones *(48, 49)*. Additionally, ES cells derived from cumulus cell-cloned mouse embryos accumulate chromosomal aneuploidies *(50)*. Although these views imply that nuclear reprogramming is bound to go wrong, they do not mean that it cannot be normalized, rescued, or prevented from getting worse. Recent observations suggest that cell fate in mouse embryos may be manipulated by changing the level of defined factors.

In a recent paper, Torres-Padilla et al. *(51)* showed that microinjection of mRNA encoding CARM1 (coactivator arginine methyltransferase 1) in one mouse blastomere at the 2-cell stage causes a developmental bias. At the blastocyst stage, the cell progeny of the injected blastomere is found enriched, albeit not exclusively localized, in the ICM. Notably, deletion of the CARM1 gene allows mouse development to E18.5 *(52)*. Because depletion of CARM1 does not impair development to near term, increase, rather than reduction, of this factor appears to be more consequential for pluripotency. As CARM1 is normally present in the embryo, it would be even more interesting to introduce factors that are lacking in cloned embryos, such as Oct4.

In our so far unpublished studies, we have found that the level of Oct4 protein is 40% lower in cloned than in fertilized mouse embryos produced by intracytoplasmic sperm injection (ICSI) (Fig. 1). This is in line with the report that cloned mouse embryos exhibit correct timing but the wrong level of gene expression *(53)*. The immunoblotting method used to measure the amount of Oct4 in embryos is obviously

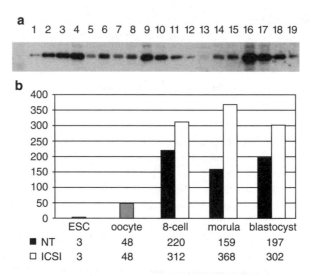

Fig. 1 Quantitation of Oct4 in nuclear transfer (NT) and intracytoplasmic sperm injection (ICSI) mouse embryos by immunoblotting. Calibrator: mouse embryonic stem (ES) cells. Antibody anti-Oct4 from Santa Cruz (sc-5279). (**a**) Western blotting after SDS-PAGE (**b**) histogram of the western blot in (**a**) after densitometry, presenting the amount of Oct4 per developmental unit (one embryonic stem cell [ESC] or one oocyte or one embryo). Lanes of the gel: 1: 5,000 ESC; 2: 10,000 ESC; 3: 25,000 ESC; 4: 50,000 ESC; 5: 120 NT morulae; 6: 120 ICSI morulae; 7: 120 NT blastocysts; 8: 120 ICSI blastocysts; 9: 50,000 ESC; 10: 25,000 ESC; 11: 10,000 ESC; 12: 5,000 ESC; 13: 140 oocytes; 14: 140 8-cell NT embryos; 15: 140 8-cell ICSI embryos; 16: 50,000 ESC; 17: 25,000 ESC; 18: 10,000 ESC; 19: 5,000 ESC

destructive, thereby preventing test hypotheses on subsequent development. Therefore, we sought to measure embryonic pluripotency without consuming the embryos for the assay. This requires a suitable reporter.

Visualization of Mouse Embryo Pluripotency Using the Oct4-GFP Transgene

Transgenic strains of mice with the green fluorescent protein (GFP) as a reporter allow tracking of embryonic blastomeres and discrimination between the G1 and G2 phase of their cell cycle (H2B-GFP *(54)*), monitoring of X-chromosome activity (X-linked GFP *(55)*), or monitoring of organogenesis (Hox-GFP *(56)*). As of May 2007, as many as 53 GFP mouse strains were available from the Jackson Laboratory (Bar Harbor, Maine, USA). Among them is one strain that carries GFP under control of the *Oct4* promoter.

An 8.5 kb DNA region upstream the start codon of mouse *Oct4* gene was found to be capable to drive expression of β-galactosidase (LacZ) indistinguishable from endogenous *Oct4 (57)*. This region contains regulatory sequences, chiefly the proximal

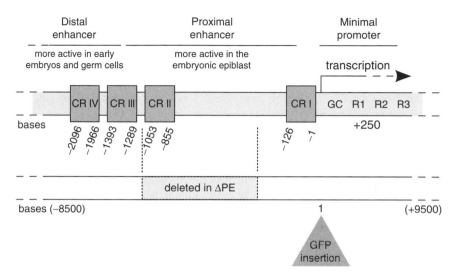

Fig. 2 Schematic of the structure of the Oct4-GFP (green fluorescent protein) transgene

enhancer (PE) which is relevant to expression in the epiblast, and the distal enhancer (DE), which is relevant to expression in preimplantation stages and in germ cells. To allow for live cell imaging, the LacZ-encoding sequence was replaced with enhanced GFP (EGFP). The construct was used to generate a transgenic mouse strain known as GOF18 *(58)*. In a subsequent variant, the 8.5 kb DNA control region was shortened by excision of the PE *(59)*. Linked to EGFP, the shortened region was used to generate a transgenic mouse strain known as OG2 (GOF18-ΔPE-EGFP; deposited at the Jackson Laboratory as B6;CBA-Tg(Pou5f1-EGFP)2Mnn/J *(60)*) (Fig. 2). By further reducing the size of the Oct4 control region in the transgene, Hübner et al. *(7)* obtained a marker solely for germ cells, termed gc-Oct4, which provides another tool to track primordial germ cells *(61, 62)*.

It should be emphasized that both the GOF18 and OG2 mouse strains are transgenic models that have both advantages and disadvantages. Although knock-in reporters obtained by homologous recombination allow faithful regulation of the reporter by the full set of genomic control sequences, this usually impairs one of the two alleles and thus introduces possible dosage effects. Some genes are indeed haplo-insufficient (e.g., *Berf-1 (63)*). Transgenes instead may not always be faithful mirrors of the endogenous gene expression, but they also do not directly interfere with its function. The *Oct4* promoter is silent in sperm and somatic cells, and its activation provides a direct measure of the reprogramming process. Work from our laboratory until recently *(64)* has shown that the quantitative expression of *GOF18-ΔPE-EGFP* transgene is proportional to that of the endogenous *Oct4*. This, together with the fact that OG2 probably has only two insertions in the genome (as detected by Southern blotting of EcoR1 digest with GFP probe; Konstantinos Anastassiadis, personal communication), portrays GOF18-ΔPE-EGFP as a valid tool to monitor nuclear reprogramming.

Live Cell Confocal Imaging of Cloned Mouse Embryos Expressing Oct4-GFP (GOF18-ΔPE-EGFP)

To study embryonic potential in relation to Oct4 level, we produced mouse embryos that carry the *GOF18-ΔPE-EGFP* transgene from either sperm of homozygous (t/t) donors and wild-type oocytes (ICSI) or cumulus cell of hemizygous (t/+) donors (nuclear transfer; NT) so as to have the same number of copies of *GOF18-ΔPE-EGFP* in both cloned and fertilized embryos.

In previous work, we used wide-field fluorescence microscopy and qualitative criteria to score the pattern of GOF18-ΔPE-EGFP in live mouse blastocysts (96 hpa) cloned from OG2 cumulus cells *(65)*. Because our checkpoint was at least 24 h past the onset of *Oct4* expression at the morula stage as we scored the blastocysts, we might have missed information about Oct4 induction. One problem with early assessment is that early stages of development are more sensitive to manipulation including imaging and photo damage. Confocal microscopy might serve our purpose as it has been shown to preserve the full developmental potential of preimplantation embryos *(66)*.

Using the Perkin-Elmer UltraView RS3 system, we first detected GOF18-ΔPE-EGFP fluorescence at the 4-cell stage and were able to score the pattern of GOF18-ΔPE-EGFP in live cloned morulae (78 hpa) in a quantitative manner (Fig. 3a–c). Thanks to a spinning (>1,800 rpm) disk with 20,000 pinholes that let only 1–4% of the incident light pass through, the energy delivered to the embryo is very low (below the saturation threshold of the fluorophore), yet an image is produced due to the high quantum efficiency (up to 75%) of a charge-coupled device (CCD) light detector. A Nikon CFI Plan Apochromat VC 60× WI objective lens (N.A. 1.20) was used to convey 488-nm laser excitation to the GFP-expressing morula. The source was a three-line (488 nm, 568 nm, 647 nm) Argon/Krypton

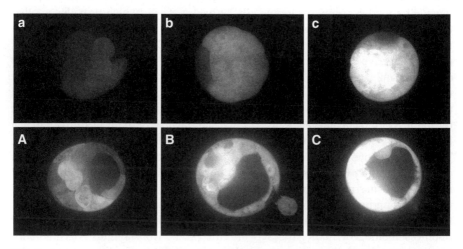

Fig. 3 Representative images of three mouse morulae (**a–c**) expressing GOF18-ΔPE-EGFP and their transition to blastocyst after imaging with UltraView RS3 (**a→A**; **b→B**; **c→C**)

laser (Melles Griot). No auxiliary lens (e.g., optovar) or optical device (e.g., filter) was present in the optical path except for the UltraView's own, and the laser was not attenuated (100% AOTF). Optical sections were captured using a 1.3 megapixel Hamamatsu ORCA ER digital camera with standardized settings (999 ms camera exposure, 2×2 binning hence 672×512 image pixels, no electronic gain).

A total of 794 morulae cloned from OG2 t/+ cumulus cells and 546 morulae obtained from fertilization with sperm of OG2 t/t males were imaged. Morulae were placed individually in 1.0 μL drops of α-MEM arrayed 5×5 on a 50-mm thin-bottom plastic dish (Greiner Bio-One, Lumox dish, catalog 96077303) over-laid with mineral oil (Sigma catalog M5310). A Tokai-Hit environmental mini-chamber maintained the dish in a gas phase of 5% CO_2 at the temperature of 37°C during the time of imaging (about 20 min. per dish).

Photodamage to the embryo would be readily apparent from the inability of the morula to form a blastocyst (cavitation) within 10 h (Fig. 3a–c). To test this, we compared imaged to non-imaged morulae. Not only the rates of morula-to-blastocyst transition were very similar in imaged versus nonimaged morulae, but cloned morulae exposed to GFP excitation formed fetuses at a similar rate to unexposed controls *(64)*.

GFP Image Analysis

For each morula, five fluorescence confocal sections were captured 5 μm apart from each other in the equatorial region. In order to match the pattern of GFP in morulae with the ability to form blastocyst, combined (maximum projection) images were analyzed with the software ImageJ *(67)*. It was obvious that images had complex pat-terns (Fig. 3). Our group has shown that "patches" of higher or lower GFP intensity in the morula are not contributed by mosaic aneuploidy *(50)*, thereby warranting further analysis with ImageJ. A region of interest (ROI) was drawn by hand using the polygon selection tool of ImageJ including only pixels belonging to the embryo. For each maximum projection, we measured 15 image parameters extracted by ImageJ and we performed a principal component analysis (PCA) to check whether the images of embryos contain information about the probability of embryo survival. The proportion of variability explained by the first three principal components was 84.6% for the ICSI morulae and 80.6% for the NT morulae. We plotted the components in three-dimensional projections (Fig. 4a, a′) and observed that the nonsurvival cases (in black) tend to have lower values of their first principal component than the survival cases (in white). To better expose the degree to which the first three principal compo-nents of the survival population overlap with the components of the nonsurvival population, we performed a Delaunay triangulation. Although the tetrahedrons of the non-survival (in black) and survival cases (in white) overlap in a central region, they cover distinguishable nonsurvival and survival regions (Fig. 4b, b′).

Given the results of the PCA, we focused on searching the more informative image parameter and found that this was GFP intensity. Distributions of GFP intensity

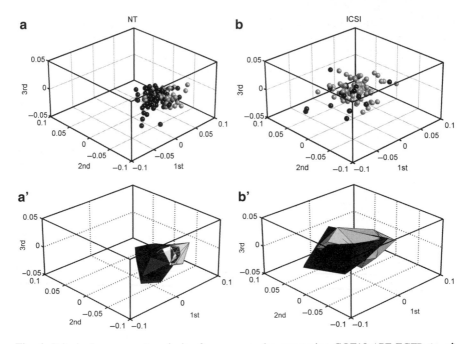

Fig. 4 Principal component analysis of mouse morulae expressing GOF18-ΔPE-EGFP. (**a, a′**) nuclear transfer (NT) morulae. (**b, b′**) intracytoplasmic sperm injection (ICSI) morulae. The *spheres* in (**a, b**) correspond to the three-dimensional projection of the principal components, and the *tetrahedrons* in (**a′, b′**) represent their connection using the Delaunay triangulation. *White colors* represent survival cases; *black colors*, nonsurvival cases

were noticeably different in clones and ICSI morulae, and the average GFP intensity was lower in clones than ICSI (Fig. 5). This was confirmed by immunoblotting of the endogenous Oct4 protein. A total of 794 cloned and 546 ICSI morulae were allowed to develop to blastocyst, while the corresponding images were processed with ImageJ.

Intervals of GOF18-ΔPE-EGFP Intensity Define Morulae that Have Distinct Blastocyst Potentials (Subsets)

In the following experiments, we used only GFP intensity as the image component that gives the most information about embryonic potential. Absolute intensity values of GFP depend on biological as well as non-biological factors, such as type and age of the light source, objective lens, and camera, which are difficult to reproduce exactly in different laboratories. For better reproducibility, we ranked the morulae by GFP intensity (Fig. 5) and allocated them into quartile intervals:

1. 0–25th percentile, low GFP, subset 1
2. 26th–50th percentile, low-medium GFP, subset 2

3. 51th–75th percentile, medium-high GFP, subset 3
4. 76th–100th percentile, high GFP, subset 4

On the day following imaging, the records of image analysis and blastocyst formation were matched. Table 1 shows the data from 176 ICSI and 316 NT morulae that

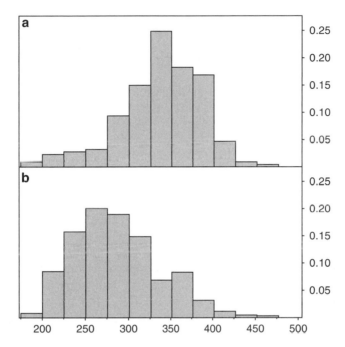

Fig. 5 Distribution of green fluorescent protein (GFP) intensity among mouse morulae expressing GOF18-ΔPE-EGFP. (**a**) intracytoplasmic sperm injection (ICSI) morulae and (**b**) nuclear transfer (NT) morulae

Table 1 Unequal allocation of developmental potential to subsets of cloned morulae defined by GFP intensity

	N morulae	N blastocysts (%)	Mean GFP intensity[a]	SD	Skewness	Kurtosis
ICSI subset						
1	44	13 (18.6%)	263.5	18.7	−0.16	−0.28
2	44	20 (28.6%)	317.7	31.1	−0.42	−0.17
3	44	18 (25.7%)	344.8	37.3	−0.32	−0.45
4	44	19 (27.1%)	383.5	41.5	−0.39	−0.23
NT subset						
1	79	7 (7.6%)	235.0	13.1	+0.04	−0.46
2	79	14 (15.2%)	268.6	22.4	−0.19	−0.56
3	79	33 (35.9%)	300.5	28.6	−0.20	−0.51
4	79	38 (41.3%)	353.5	40.2	−0.24	−0.36

Abbreviations: ICSI intracytoplasmic sperm injection; *GFP* green fluorescent protein; *NT* nuclear transfer; *SD* standard deviation
[a] Arbitrary Unit of intensity of 16-bit grayscale image analyzed using ImageJ.

were analyzed by quartile allocation, until practical reasons emerged and called for the introduction of tertiles (see the section "Success in Deriving ES Cell Lines Increases with Increased GOF18-ΔPE-EGFP Intensity of Morulae"). Blastocyst formation was associated with GFP intensity in NT morulae ($r^2 = 0.42$) more than in ICSI morulae ($r^2 = 0.26$) (Table 1).

Subsets of Morulae Defined by GOF18-ΔPE-EGFP Contain Specific Levels of Oct4, Nanog, Sox2, and Cdx2 Transcripts

An important question in oocyte-mediated nuclear reprogramming is whether genes are reprogrammed independently of one another. Thus we analyzed the GFP subsets for expression of the endogenous genes *Oct4, Nanog,* and *Sox2* (pluripotency- and ICM-associated), and *Cdx2* (TE-associated) *(64)*. Levels of mRNA (cDNA) were measured using TaqMan real-time PCR and normalized to the endogenous control *Hprt* (Table 2).

In general, correlation with GFP intensity was substantial for all genes analyzed. Except for *Cdx2*, cloned morulae had lower mRNA values than those of ICSI morulae, with *Sox2* mRNA at bare levels ($p < 0.01$). Differences of mRNA levels between cloned and ICSI blastocysts were less prominent; only for *Sox2* was the difference significant. This led us to speculate *(64)* that embryonic potential might be limited by the factor present in the least amount, reminiscent of the Liebig's Law of the Minimum, a principle developed in agricultural science whereby plant growth is limited by the scarcest resource available. It would therefore be interesting to increase the level of Sox2 in the embryo.

Table 2 Correlates of gene expression (mRNA) and GFP intensity (subsets 1–4) in NT and ICSI embryos allocated to subsets

	ICSI	ICSI	ICSI	ICSI	NT	NT	NT	NT
	1	2	3	4	1	2	3	4
Morulae								
Oct4	0.412	0.540	0.610	0.671	0.370	0.424	0.543	0.527
Nanog	0.083	0.174	0.129	0.162	0.063	0.074	0.109	0.122
Sox2	0.002	0.034	0.050	0.052	0.004	0.003	0.002	0.006
Cdx2	0.082	0.162	0.159	0.179	0.110	0.172	0.205	0.197
Hprt	1.000	1.000	1.000	1.000	1.000	1.000	1.000	1.000
Blastocysts								
Oct4	0.530	0.879	0.774	0.715	0.429	0.421	0.360	0.465
Nanog	0.121	0.125	0.125	0.127	0.182	0.133	0.214	0.194
Sox2	0.015	0.139	0.228	0.195	0.037	0.044	0.086	0.150
Cdx2	0.232	0.384	0.565	0.485	0.525	0.443	0.483	0.472
Hprt	1.000	1.000	1.000	1.000	1.000	1.000	1.000	1.000

Abbreviations: ICSI intracytoplasmic sperm injection; *GFP* green fluorescent protein; *NT* nuclear transfer

Transcript levels normalized to *Hprt*. *Hprt* internal control and normalizer, therefore set to value 1

Success in Deriving ES Cell Lines Increases with Increased GOF18-ΔPE-EGFP Intensity of Morulae

Since the blastocyst rate increases with higher intensity of GOF18-ΔPE-EGFP, we hypothesized that the level of Oct4 in the embryo could have an effect on the pluripotent founder cells in the ICM and ES cell derivation. Indeed, a statistically not significant increase of the total cell number was observed in cloned blastocysts that had Oct4 riboprobe signal localized more prominently to the putative ICM. We set out to follow this up by deriving ES cells.

ES cell derivation was conducted according to a standard protocol *(44, 64)*. Since derivation is labor-intensive and a substantial number of lines is required for each subset, we reduced the number of subsets from four to three. Practically, the morulae were scored for GFP as described, and allocated to tertile instead of quartile intervals of GFP intensity (0–33%, subset 1; 34–67%, subset 2; 68–100%, subset 3).

Overall, 104 and 21 ES cell lines were derived from NT and ICSI embryos, respectively. NT-ES cell derivation was more efficient (χ^2 test, $p = 0.000012$) for clones of GFP subsets 3 and 2 (36%, 27%) than for subset 1 (9% of blastocysts form ES cells). ES cell derivability was almost equal for ICSI embryos across the three groups (χ^2 test, $p = 0.49$). All ES cell lines displayed characteristic morphology of undifferentiated ES cells. The pluripotent status was confirmed by immunohisto-chemical analysis of SSEA-1, alkaline phosphatase (AP), and Oct4 proteins.

ES cells derived from cloned or fertilized embryos are held to be transcriptionally and functionally indistinguishable *(68, 69)*. However, Gidekel et al. *(27)* reported that ES cells transplanted *in vivo* behave differently depending on the residual amount of Oct4. To probe if the different rates of ES cell derivation from embryos that had different amounts of Oct4 corresponded to qualitative differences of the ES cells, we allowed them to differentiate *in vivo* by injecting them into recipient blastocysts followed by embryo transfer.

Six lines per subset were selected for prevalent normal karyotype (4 of 6 lines, 3 of 6 lines, and 4 of 6 lines of GFP intervals 1, 2, 3, respectively), each presenting greater than 50% euploid chromosome sets. Approximately 15 cells were injected into fertilization-derived blastocysts: 286, 347, and 334 blastocysts were injected with ES cells of GFP subset 1, 2, and 3, respectively. The injected blastocysts were transferred into uteri of pseudopregnant mice. On average, 17% of the injected blastocysts had formed proper fetuses at 14.5 dpc (166 of 967) and 28% fetuses had GFP-positive gonads (46 of 166). Distribution of the chimeric fetuses according to the subset of origin of the ES cells was unbiased ($n = 54, 61, 51$). Germline contribution was observed regardless of the GFP subset (subset 1, 28%; subset 2, 38%; subset 3, 16%; defined as the proportion of fetuses with GFP+ gonads).

Although our data corroborate the view that ES cells derived from cloned or fertilized embryos are indeed indistinguishable from each other, we maintain that further investigation is necessary. In two recent studies, NT-ES cells were found to have higher aneuploidy rates than ES cells from fertilized embryos *(50)*, and the nuclear transfer procedure was found to leave a specific mark on NT-ES cells *(70)*.

Manipulating Pluripotency in Cloned Mouse Embryos

Oct4 is deficient in cloned mouse embryos after ooplasm-mediated induction of pluripotency (broad literature, e.g., Sebastiano et al. *(53)*) and clones with lower or higher Oct4 level attain lower or higher developmental rates, respectively *(64)*. Oct4 exhibits a dose–effect response in pluripotent ES cells *(6)*, and it is among the three or four factors that converted mouse and human fibroblasts into ES-like cells after retrovirus-mediated delivery to the nucleus *(17, 37–39)*. For these reasons, we became interested whether the level of Oct4 is a determinant of blastomere fate in normal and cloned mouse embryos.

It is possible that subsets of cloned mouse embryos with low, intermediate, and higher levels of GOF18-ΔPE-EGFP represent discrete stages in the induction of pluripotency. Microinjection of the arginine methyltransferase CARM1 mRNA into blastomeres has recently been shown to be feasible and resulted in alteration of their fate *(51)*. Therefore, using a similar approach we decided to examine the impact of changing the Oct4 level on ooplasm-induced pluripotency. To avoid any genetic manipulation, we did not use retrovirus-mediated delivery of Oct4 cDNA, which could result in insertional mutagenesis. Instead, we implemented a dual strategy based on microinjection of known amounts of Oct4 mRNA or recombinant protein into the ooplasm prior to somatic nuclear transfer or into a blastomere of a 2-cell stage embryo. This approach allows us to monitor the effects of changing Oct4 levels without modifying the genome of the embryo.

For recombinant Oct4 (rOct4) production, we expressed a GST-Oct4-His construct in the *Escherichia coli* BL21 strain and purify the recombinant protein in two-steps, using the Äkta Purifier chromatographic system (GE Healthcare). First, GST-Oct4-His was captured by Ni-affinity chromatography, eluted by high concentrations of imidazole and the GST tag cleaved by thrombin. In the second step, the resulting Oct4-His protein was purified from thrombin, imidazole, and other contaminants by size-exclusion chromatography. To verify the activity of the rOct4 obtained in this way, we made use of 2-cell stage embryos derived from the GOF18-(ΔPE)-EGFP transgenic mouse described above. If the rOct4 injected is functional, it can bind to the Oct4 promoter and induce higher levels of expression of EGFP as compared to embryos injected with recombinant GST as a control.

In a pilot experiment, one blastomere of the 2-cell stage mouse embryo was injected with rOct4 protein, along with a tracer (Alexa 488 dextran beads 40 kDa) (Fig. 6). Blastocyst formation was not affected by the procedure. The injected blastomere was tracked to see its contribution to the ICM or TE. We examined six blastocysts and observed 4 instances of blastocysts with the green fluorescent tracer in the ICM and 2 instences in the TE *(2)*. Torres-Padilla et al. *(51)* showed that the pattern of the second cell division (meridian, equatorial) determines the shape of the 4-cell embryo, and according to this shape, the allocation of blastomere in the blastocyst is biased. Therefore, it will be very interesting to track the cleavage timing and geometry of the injected blastomere.

In parallel to this approach, we have also implemented a second strategy, based on microinjection of *in vitro* transcribed mRNA encoding for Oct4. We have cloned

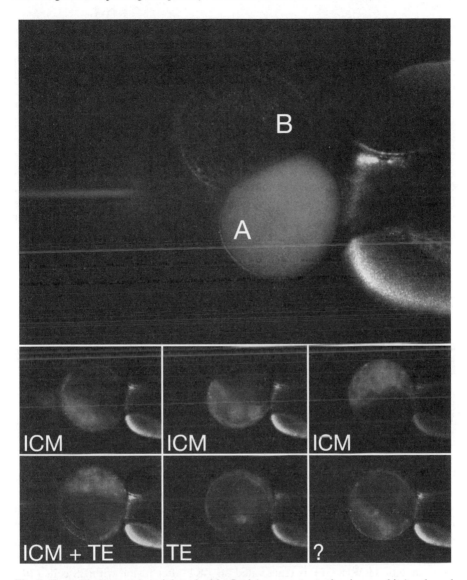

Fig. 6 Two-cell mouse embryo injected with rOct4 (**a**, green tracer showing as white) and mock (**b**, no tracer) developed to the blastocyst stage and the progeny of the injected blastomere localized in the inner cell mass (ICM) or trophectoderm (TE) of the blastocyst

the Oct4 cDNA in the pcDNA3.1-FLAG vector, which allows its *in vitro* transcription by the bacteriophage T7 polymerase. The resulting mRNA is then *in vitro* polyadenylated at the 3′ and capped at the 5′ terminus and finally injected in 1- or 2-cell stage embryos. The ability of the produced mRNAs to be properly translated will be tested using the reticulocyte lysate system (Promega) and 1- or 2-cell stage embryos derived from *GOF18-(ΔPE)-EGFP* transgenic mice, as for rOct4.

Compared with the previous strategy, this method offers the advantage of being able to deliver mRNAs encoding not only for wild-type Oct4, but potentially for any mutant of interest. In addition, this method offers a higher precision in the amount of mRNA delivered (measured by a Nanodrop® system) as compared to the injection of rOct4 protein, whose amount we roughly estimate using the Bradford method. Lastly, the injection of both rOct4 and of mRNA appears to be advantageous over the use of retro- or lentiviruses because there are no potential issues of gene reactivation at later stages and because the time window in which the exogenous protein will be present in the embryo is short and defined. Although we do not know the half-lives of our rOct4 or Oct4 mRNA in the embryo, experiments can be designed in which various amounts of proteins and mRNA are injected. In any case, the effects of our method are likely to be transient and limited to the very early stages of development, thus allowing us to understand the impact of transiently modulating the levels of Oct4 in the zygote or in the 2-cell stage embryo.

The combination of the above described two methods will hopefully provide exciting insights in the effects of engineering Oct4 levels in the early embryo without affecting its genome. A major question that still needs to be addressed is what the consequences of transiently changing Oct4 levels in a blastomere of a 2-cell stage embryo would be. Will the subsequent cleavage pattern be influenced? Will the fate of the injected blastomere be biased? Our strategy will possibly also allow us to map the region(s) in Oct4 responsible for a given effect by injecting mRNAs encoding for certain point mutants or deletions of Oct4.

What GOF18-ΔPE-EGFP May and May Not Tell Us About Nuclear Reprogramming

We used the *GOF18-ΔPE-EGFP* transgenic tool to investigate why gene expression of cloned embryos does not resemble that of fertilized embryos. In some cloned embryos, somatic cell-encoded GFP under control of the GOF18-ΔPE promoter is visible at the late 4-cell stage, i.e., 48 h after nuclear transfer. This is rather unexpected based on a previous report where GOF18-GFP and GOF18-ΔPE-EGFP were detected at the 8-cell stage after fertilization *(58)*. In case of reprogramming by fusion with ES cells, somatic GOF18-GFP is reactivated 36–48 h after fusion with ES cells *(30)*. In both cases, nuclear transfer and cell fusion data imply that two to four cell cycles are sufficient to enable significant induction of pluripotency in the somatic genome. However, the level of expression, whether mRNA or protein, is lower than normal while the timing is correct *(53)*. This suggests that erasure of somatic, repressive mechanisms is partial, and only some of the gene regulatory elements can be used by the cellular machinery of the ooplasm. It is well possible that repeated exposure of the somatic genome to reprogramming factors is a key event in the erasure of somatic cell memory. Therefore, it will be interesting to test if preventing cell cycle progression after nuclear transfer allows for the expression of pluripotency-associated genes.

Expression of these genes when the cell cycle is blocked would be a strong indication that the ooplasm can actively reprogram the nucleus transplant.

The underlying mechanism to reprogramming is not known, but there are clues to it. Pronuclear mouse eggs microinjected with unmethylated GOF18-ΔPE-EGFP plasmid show GFP expression already at the 2-cell stage *(71)*, therefore, the mechanism underlying reprogramming may involve DNA demethylation. This is in accord with our observation that the Oct4 DE is more methylated in cloned morulae when GOF18-ΔPE-EGFP levels are lower *(64)*. Thus it would be sensible to reduce the DNA methylation of the somatic genome prior or immediately after nuclear transfer. Regardless of the agent used, genomic imprints relying on CpG methylation should not be modified, especially if applications in tissue regeneration are envisioned. In fact, Tada et al. *(30)* showed that when using embryonic germ (EG) cells for fusion, imprints are erased, while they are retained when ES cells are used.

In spite of lower expression levels, Oct4 and GOF18-ΔPE-EGFP are informative of the developmental potential of cloned morulae. GOF18-ΔPE-EGFP level predicts development to blastocyst, and the correlation is higher for cloned than for fertilized embryos *(64)*. Differences in GOF18-ΔPE-EGFP expression are mirrored by the expression of the pluripotency-related transcripts Oct4, Nanog, Sox2, and Cdx2. There are clones that appear indistinguishable from fertilized embryos, but they are far less competent to develop to later stages *in vivo* than fertilized counterparts. In contrast, ES cell formation is less clear-cut. Although the rate of NT-ES cell derivation was proportional to GFP intensity, no differences could be detected subsequently. Established ES cells might always contribute to organogenesis *in vivo* after prior selection for *in vitro* proliferation during derivation. Indeed ES cells derived from cloned blastocysts are reported to be equivalent to those derived from fertilized blastocysts *(68, 69)*. It would be interesting to know, before selection occurs, if precursor ICM cells give rise to NT-ES cells in a similar or different way than ICM cells of normal embryos. The different responses of clones and ICSI embryos to Oct4 level suggest that the dose–effect response does not only apply to ES cells but also to embryo development, albeit in different ways. Although some aspects are conserved between natural and induced pluripotency, other aspects are distinct.

In conclusion, flawed resetting of epigenetic marks may explain why gene expression levels in clones are abnormal. Given the 30,000 genes that make up the mouse genome and the long list of genes that are dysregulated besides Oct4, it seems unlikely that reprogramming is only random and subject to selection, otherwise no embryo would survive. The "antichaos" model proposed by Stuart Kauffman *(72)* explains how a complex system that is disordered at the beginning spontaneously organizes into a high degree of order. We propose that all embryos are inherently dysregulated at the outset and self-organize out of such chaos. Self-organization needs a certain amount of time, and a somatic nucleus transplant commences this process from a more distant starting point than a gamete. Most cloned embryos form blastocysts *in vitro* but are doomed *in utero*, while ES cells derived from those blastocysts appear normal. This suggests that the *in vivo* environment allows minimal flexibility, whereas the *in vitro* environment allows substantial flexibility.

Acknowledgments We are grateful to Dr. Konstantinos Anastassiadis (Technical University Dresden) for sharing unpublished data on GOF18-ΔPE-EGFP; to Dr. Yong-Mahn Han (KAIST, Korea) for contributing to the article of Cavaleri et al., 2008 *(64)* which provides the groundwork for this chapter; and to Prof. Ivan Dikic (Goethe University, Frankfurt am Main) for supporting the quantitative analysis of Oct4 protein in preimplantation mouse embryos and the synthesis of recombinant Oct4 for microinjection.

References

1. Tarkowski AK, Ozdzeński W, Czołowska R. How many blastomeres of the 4-cell embryo contribute cells to the mouse body? Int J Dev Biol 2001;45:811–6.
2. Tarkowski AK, Ozdzenski W, Czolowska R. Identical triplets and twins developed from isolated blastomeres of 8- and 16-cell mouse embryos supported with tetraploid blastomeres. Int J Dev Biol 2005;49:825–32.
3. Chung Y, Klimanskaya I, Becker S, et al. Embryonic and extraembryonic stem cell lines derived from single mouse blastomeres. Nature 2006;439:216–9.
4. Rossant J. Postimplantation development of blastomeres isolated from 4- and 8-cell mouse eggs. J Embryol Exp Morphol 1976;36:283–90.
5. Boiani M, Schöler HR. Regulatory networks in embryo-derived pluripotent stem cells. Nat Rev Mol Cell Biol 2005;6:872–84.
6. Niwa H, Miyazaki J, Smith AG. Quantitative expression of Oct-3/4 defines differentiation, dedifferentiation or self-renewal of ES cells. Nat Genet 2000;24:372–6.
7. Hübner K, Fuhrmann G, Christenson LK, et al. Derivation of oocytes from mouse embryonic stem cells. Science 2003;300:1251–6.
8. Seguin C, Draper JS. Chapter 11. Extraembryonic differentiation of human ES cell. In: Sullivan S, Cowan CA, Eggan K, editors. Human Embryonic Stem Cells: The Practical Handbook. Wiley, West Sussex, England, 2007:404.
9. Masui S, Nakatake Y, Toyooka Y, et al. Pluripotency governed by Sox2 via regulation of Oct3/4 expression in mouse embryonic stem cells. Nat Cell Biol 2007;9:625–35.
10. Boyer LA, Lee TI, Cole MF, et al. Core transcriptional regulatory circuitry in human embryonic stem cells. Cell 2005;122:947 56.
11. Babaie Y, Herwig R, Greber B, et al. Analysis of OCT4 dependent transcriptional networks regulating self renewal and pluripotency in human embryonic stem cells. Stem Cells 2006;25:500–10.
12. Greber B, Lehrach H and Adjaye J. FGF2 modulates TGFβ signaling in MEFs and human ES cells to support hESC self-renewal. Stem Cells 2006;25:455–64.
13. Reményi A, Schöler HR, Wilmanns M. Combinatorial control of gene expression. Nat Struct Mol Biol 2004;11:812–5.
14. Hatano SY, Tada M, Kimura H, et al. Pluripotential competence of cells associated with Nanog activity. Mech Dev 2005;122:67–79.
15. Darr H, Mayshar Y, Benvenisty N. Overexpression of NANOG in human ES cells enables feeder-free growth while inducing primitive ectoderm features. Development 2006;133: 1193–201.
16. Tanaka Y, Era T, Nishikawa S, Kawamata S. Forced expression of Nanog in hematopoietic stem cells results in a gamma delta T-cell disorder. Blood 2007;110:107–15.
17. Takahashi K, Yamanaka S. Induction of pluripotent stem cells from mouse embryonic and adult fibroblast cultures by defined factors. Cell 2006;126:663–76.
18. Schöler HR, Hatzopoulos AK, Balling R, Suzuki N, Gruss P. A family of octamer-specific proteins present during mouse embryogenesis: evidence for germline-specific expression of an Oct factor. EMBO J 1989;8:2543–50.
19. Okamoto K, Okazawa H, Okuda A, Sakai M, Muramatsu M, Hamada H. A novel octamer binding transcription factor is differentially expressed in mouse embryonic cells. Cell 1990;60:461–72.

20. Rosner MH, Vigano MA, Ozato K, et al. A POU-domain transcription factor in early stem cells and germ cells of the mammalian embryo. Nature 1990;345:686–92.

21. Lengner CJ, Camargo FD, Hochedlinger K, et al. Oct4 expression is not required for mouse somatic stem cell self-renewal. Cell Stem Cell 2007;1:403–15.

22. Nichols J, Zevnik B, Anastassiadis K, et al. Formation of pluripotent stem cells in the mammalian embryo depends on the POU transcription factor Oct4. Cell 1998;95:379–91.

23. Zeineddine D, Papadimou E, Chebli K, et al. Oct-3/4 dose dependently regulates specification of embryonic stem cells toward a cardiac lineage and early heart development. Dev Cell 2006;11:535–46.

24. Ramos-Mejía V, Escalante-Alcalde D, Kunath T, et al. Phenotypic analyses of mouse embryos with ubiquitous expression of Oct4: effects on mid-hindbrain patterning and gene expression. Dev Dyn 2005;232:180–90.

25. Hochedlinger K, Yamada Y, Beard C, Jaenisch R. Ectopic expression of Oct-4 blocks progenitor-cell differentiation and causes dysplasia in epithelial tissues. Cell 2005;121:465–77.

26. Rodriguez RT, Velkey JM, Lutzko C,et al. Manipulation of OCT4 levels in human embryonic stem cells results in induction of differential cell types. Exp Biol Med 2007;232:1368–80.

27. Gidekel S, Pizov G, Bergman Y, Pikarsky E. Oct 3/4 is a dose-dependent oncogenic fate determinant. Cancer Cell 2003;4:361–70.

28. Camara-Clayette V, Le Pesteur F, Vainchenker W, Sainteny F. Quantitative Oct4 overproduction in mouse embryonic stem cells results in prolonged mesoderm commitment during hematopoietic differentiation in vitro. Stem Cells 2006;24:1937–45.

29. Wei F, Schöler HR, Atchison ML. Sumoylation of Oct4 enhances its stability, DNA binding, and transactivation. J Biol Chem 2007;282:21551–60.

30. Tada M, Takahama Y, Abe K, Nakatsuji N, Tada T. Nuclear reprogramming of somatic cells by in vitro hybridization with ES cells. Curr Biol 2001;11:1553–8.

31. Ng RK, Gurdon JB. Epigenetic memory of active gene transcription is inherited through somatic cell nuclear transfer. Proc Natl Acad Sci U S A 2005;102:1957–62.

32. Eggan K, Akutsu H, Hochedlinger K, Rideout W3rd, Yanagimachi R, Jaenisch R. X-chromosome inactivation in cloned mouse embryos. Science 2000;290:1578–81.

33. Kimura H, Tada M, Hatano S, Yamazaki M, Nakatsuji N, Tada T. Chromatin reprogramming of male somatic cell-derived XIST and TSIX in ES hybrid cells. Cytogenet Genome Res 2002;99:106–14.

34. Egli D, Rosains J, Birkhoff G, Eggan K. Developmental reprogramming after chromosome transfer into mitotic mouse zygotes. Nature 2007;447:679–85.

35. Do JT, Schöler HR. Cell–cell fusion as a means to establish pluripotency. Ernst Schering Res Found Workshop 2006;60:35–45.

36. Matsumura H, Tada M, Otsuji T, et al. Targeted chromosome elimination from ES-somatic hybrid cells. Nat Methods 2007;4:23–5.

37. Takahashi K, Tanabe K, Ohnuki M, et al. Induction of pluripotent stem cells from adult human fibroblasts by defined factors. Cell 2007;31:861–72.

38. Yu J, Vodyanik MA, Smuga-Otto K, et al. Induced pluripotent stem cell lines derived from human somatic cells. Science 2007;318:1917–20.

39. Nakagawa M, Koyanagi M, Tanabe K, et al. Generation of induced pluripotent stem cells without Myc from mouse and human fibroblasts. Nat Biotechnol 2008;26:101–6.

40. Taranger CK, Noer A, Sørensen AL, Håkelien AM, Boquest AC, Collas P. Induction of dedifferentiation, genome wide transcriptional programming, and epigenetic reprogramming by extracts of carcinoma and embryonic stem cells. Mol Biol Cell 2005;16:5719–35.

41. Ding S, Schultz PG. Small molecules and future regenerative medicine. Curr Top Med Chem 2005;5:383–95.

42. Patsch C, Edenhofer F. Conditional mutagenesis by cell-permeable proteins: potential, limitations and prospects. Handbook Exp Pharmacol 2007;178:203–32.

43. Fittipaldi A, Giacca M. Transcellular protein transduction using the Tat protein of HIV-1. Adv Drug Deliv Rev 2005;57:597–608.

44. Cavaleri F, Gentile L, Schöler HR, Boiani M. Recombinant human albumin supports development of somatic cell nuclear transfer embryos in mice: toward the establishment of a chemically defined cloning protocol. Cloning Stem Cells 2006;8:24–40.

45. Kues WA, Niemann H. The contribution of farm animals to human health. Trends Biotechnol 2004;22:286–94.
46. Fulka JJr, Miyashita N, Nagai T, Ogura A. Do cloned mammals skip a reprogramming step. Nat Biotechnol 2004;22:25–6.
47. Bortvin A, Eggan K, Skaletsky H, et al. Incomplete reactivation of Oct4-related genes in mouse embryos cloned from somatic nuclei. Development 2003;130:1673–80.
48. Li X, Kato Y, Tsunoda Y. Comparative analysis of development-related gene expression in mouse preimplantation embryos with different developmental potential. Mol Reprod Dev 2005;72:152–60.
49. Li X, Amarnath D, Kato Y, Tsunoda Y. Analysis of development-related gene expression in cloned bovine blastocysts with different developmental potential. Cloning Stem Cells 2006;8:41–50.
50. Balbach ST, Jauch A, Böhm-Steuer B, Cavaleri FM, Han YM, Boiani M. Chromosome stability differs in cloned mouse embryos and derivative ES cells. Dev Biol 2007;308:309–21.
51. Torres-Padilla ME, Parfitt DE, Kouzarides T, Zernicka-Goetz M. Histone arginine methylation regulates pluripotency in the early mouse embryo. Nature 2007;445:214–8.
52. Yadav J, Lee J, Kim J, et al. Bedford. Specific protein methylation defects and gene expression perturbations in coactivator-associated arginine methyltransferase 1-deficient mice, Proc Natl Acad Sci U S A 2003;100:6464–8.
53. Sebastiano V, Gentile L, Garagna S, Redi CA, Zuccotti M. Cloned pre-implantation mouse embryos show correct timing but altered levels of gene expression. Mol Reprod Dev 2005;70:146–54.
54. Hadjantonakis AK, Papaioannou VE. Dynamic in vivo imaging and cell tracking using a histone fluorescent protein fusion in mice. BMC Biotechnol 2004;4:33.
55. Hadjantonakis AK, Gertsenstein M, Ikawa M, Okabe M, Nagy A. Non-invasive sexing of preimplantation stage mammalian embryos. Nat Genet 1998;19:220–2.
56. Godwin AR, Stadler HS, Nakamura K, Capecchi MR. Detection of targeted GFP-Hox gene fusions during mouse embryogenesis. Proc Natl Acad Sci U S A 1998;95:13042–7.
57. Yeom YI, Fuhrmann G, Ovitt CE, et al. Germline regulatory element of Oct-4 specific for the totipotent cycle of embryonal cells. Development 1996;122:881–94.
58. Yoshimizu T, Sugiyama N, De Felice M, et al. Germline-specific expression of the Oct-4/green fluorescent protein (GFP) transgene in mice. Dev Growth Differ 1999;41:675–84.
59. Nordhoff V, Hübner K, Bauer A, Orlova I, Malapetsa A, Schöler HR. Comparative analysis of human, bovine, and murine Oct-4 upstream promoter sequences. Mamm Genome 2001;12:309–17.
60. Szabo PE, Hübner K, Schöler H, Mann JR. Allele-specific expression of imprinted genes in mouse migratory primordial germ cells. Mech Dev 2002;115:157–60.
61. Anderson R, Fässler R, Georges-Labouesse E, et al. Mouse primordial germ cells lacking β1 integrins enter the germline but fail to migrate normally to the gonads. Development 1999;126:1655–64.
62. Molyneaux KA, Stallock J, Schaible K, Wylie C. Time-lapse analysis of living mouse germ cell migration. Dev Biol 2001;240:488–98.
63. Takeuchi A, Mishina Y, Miyaishi O, Kojima E, Hasegawa T, Isobe K. Heterozygosity with respect to Zfp148 causes complete loss of fetal germ cells during mouse embryogenesis. Nat Genet 2003;33:172–6.
64. Cavaleri F, Balbach ST, Gentile L, et al. Subsets of cloned mouse embryos and their non-random relationship to development and nuclear reprogramming. Mech Dev 2008;25:153.
65. Boiani M, Eckardt S, Schöler HR, McLaughlin KJ. Oct4 distribution and level in mouse clones: consequences for pluripotency. Genes Dev 2002;16:1209–19.
66. Ross PJ, Perez GI, Ko T, Yoo MS, Cibelli JB. Full developmental potential of mammalian preimplantation embryos is maintained after imaging using a spinning-disk confocal microscope. Biotechniques 2006;41:741–50.
67. Abramoff MD, Magelhaes PJ, Ram SJ. Image processing with ImageJ. Biophoton Int 2004;11:36–42.
68. Brambrink T, Hochedlinger K, Bell G, Jaenisch R. ES cells derived from cloned and fertilized blastocysts are transcriptionally and functionally indistinguishable. Proc Natl Acad Sci U S A 2006;103:933–8.

69. Wakayama S, Jakt ML, Suzuki M, et al. Equivalency of nuclear transfer-derived embryonic stem cells to those derived from fertilized mouse blastocysts. Stem Cells 2006;24:2023–33.
70. Hikichi T, Wakayama S, Mizutani E, et al. Differentiation potential of parthenogenetic embryonic stem cells is improved by nuclear transfer. Stem Cells 2007;25:46–53.
71. Kirchhof N, Carnwath JW, Lemme E, Anastassiadis K, Schöler H, Niemann H. Expression pattern of Oct-4 in preimplantation embryos of different species. Biol Reprod 2000;63:1698–705.
72. Kauffman SA. Antichaos and adaptation. Sci Am 1991;265:78–84.

Current Developments in Genetically Manipulated Mice

Klaus I. Matthaei

Abstract The use of genetically manipulated mice in medical research is one of the premier tools for the study of genetic diseases. I describe here our routine methods to produce these animals that have proven to be highly reliable as well as give exceptionally high rates of germline transmission with all strains of embryonic stem cells that we have used.

Keywords Embryonic stem cells • Homologous recombination • Germline transmission • Knockout mice

Introduction

The ability to understand gene function in vivo in an entire mammal has been enormously improved in the past 25 years. This has come about first by the ability to introduce new genes into the germline of mice by injecting DNA constructs directly into the pronucleus of a fertilised mouse egg where it becomes integrated, resulting in a "transgenic" mouse in which a new protein is expressed. In the first "proof of principle" example by Palmiter et al. *(1)* in 1982 the gene for growth hormone was overexpressed, resulting in giant mice. The second major breakthrough resulted from the ability to isolate the inner cell mass cells from a blastocyst stage embryo (embryonic stem [ES] cells), which when reinjected into a new blastocyst resulted in chimaeric mice derived from both the host cells and the injected ES cells, thereby re-creating a live animal that contained tissue culture cells capable of contributing to the germline and produce ES cell derived offspring *(2)*. Thereafter the ability to alter the genetic composition of the ES cells while they were in tissue culture was

K.I. Matthaei (✉)
Stem cell and Gene Targeting Laboratory, The Division of Molecular Bioscience,
The John Curtin School of Medical Research, The Australian National University,
Canberra, ACT 0200, Australia
e-mail: Klaus.Matthaei@anu.edu.au

H. Baharvand (ed.), *Trends in Stem Cell Biology and Technology*,
DOI 10.1007/ 978-1-60327-905-5_8,
© Humana Press, a Part of Springer Science+Business Media, LLC 2009

developed *(3, 4)*, and it was then possible to make predictable modifications to any gene in the mouse genome, leading to literally thousands of genetically altered mice. However, the process of generating each mouse is labour intensive and usually takes more than a year to generate a new strain; from cloning of the gene to obtaining the targeted mouse. The methods therefore develop slowly as each change must be carefully assessed and analysed to ensure optimal conditions. Indeed most laboratories do not experiment with methods since they are only involved with a small number of genes to target. My laboratory has been involved in generating these animals since 1991 and we have made a number of often minor but considerably important changes. I present here our experience over the past 17 years that has proven to be highly reliable as well as give exceptionally high rates of germline transmission with all strains of ES cells that we have used.

ES Cells and Culture

The culture of ES cells and their subsequent ability to contribute to the germline of mice has become routine. Most of the reagents are commercially available and they are sufficiently robust to allow the isolation of new ES lines *(5)*. However, there are a number of different minor variations that we have found that can influence germline transmission. Indeed our methods have allowed rates of germline transmission that approach 100% (almost all clones of ES cells [129SV, C57BL/6, or BALB/c] when injected into host blastocysts result in chimaera that produce ES cell derived offspring). Although it is possible to culture ES cells in the absence of feeder layers, we routinely use mitomycin C–treated embryonic feeder (mEF) cells (less than passage 5) (for method of preparation see Abbandanzo et al. *(6)*) as well as 1,000 U/mL leukemia inhibitory factor (LIF) (ESGRO ESG1106, Millipore) in Dulbecco's modified Eagle's medium (DMEM) (Gibco ES cell qualified 1829, Invitrogen) supplemented with 15% ES grade fetal calf serum (FCS) (Gibco ES cell qualified, Invitrogen) and 2 mM glutamine (Sigma-Aldrich, St. Louis, USA), 0.1 mM MEM nonessential amino acids and 0.1 mM 2-mercaptoethanol, 50 U penicillin, and 50 U streptomycin (GIBCO 15140–148, Invitrogen), called complete ES medium. It is not necessary to heat inactivate the FCS (traditionally performed for 30 min at 56°C) for ES cell culture since untreated FCS supports superior ES cell growth. The medium must be replaced daily and it is important that once glutamine is added, the medium must be used within 1 week even if stored at 4°C. Moreover, if older than 1 week, fresh glutamine cannot just be re-added since the breakdown products of the old glutamine (ammonia) are toxic to ES cells and decrease germline transmission. The cells are cultured in 25 cm (TTP, Techno Plastic Products, Switzerland) culture flasks at 37°C, 10% CO_2 and are passaged every 2 days at a ratio of about 1:3. Prior to the addition of 0.05% trypsin-ethylenediaminetetraacetic acid (EDTA) (Gibco 25300, Invitrogen) we briefly wash the adhered cells twice with $Ca^{2+}Mg^{2+}$ free phosphate-buffered saline (PBS) (Gibco 14190, Invitrogen) and then incubate them at room temperature for 2 min with PBS containing 0.5 mM ethylene glycol tetraacetic acid (EGTA) (Sigma-Aldrich E3889, St. Louis, USA). This treatment chelates calcium and disrupts the tight intercellular bonds between the ES cells, and this can be easily observed under a phase microscope as the colonies lose their tightly packed appearance. Importantly, this allows easy access for the trypsin to rapidly dissociate the cells within 2 min. Single cell suspensions are obtained by rapid repeated passage of the cells while in trypsin (usually 1 mL for a 25-cm flask) through a small-bore pipette (2 mL TPP). The trypsin is inactivated with the addition of 5 mL of complete ES medium and the cells (2 mL) are plated immediately into three fresh flasks containing feeder cells and 5 mL of fresh equilibrated ES medium. It is not necessary to remove the trypsin-containing medium by the usual practice of centrifugation of the cells and aspiration followed by resuspension in fresh medium.

Cryopreservation

The ES cells are harvested as described above, pelleted by centrifugation ($600 \times g$) and resuspended as a single cell suspension in 40% complete ES medium containing LIF, 50% ES cell qualified FCS, and 10% dimethylsulfoxide (DMSO) (SIGMA-Aldrich D2650 Hybrimax, St. Louis, USA) prepared on ice. The 1-mL aliquots are rapidly distributed into cryovials (NUNC, Denmark) and these are placed into a precooled (4°C) freezer box (StrataCooler Stratagene, Fisher Scientific, USA), which is then stored overnight at −70°C. The frozen vials are then plunged directly into liquid nitrogen for long-term storage. The cells are recovered by rapid thawing to just room temperature by immersion in a 37°C waterbath and dilution with 7 mL of room temperature complete ES medium. The cells are plated immediately onto feeder cells. It is not necessary to remove the DMSO by centrifugation and aspiration since this is sufficiently diluted by the fresh medium.

Transfection of Constructs into ES Cells

DNA constructs are introduced into the ES cells by electroporation. We use approximately 5×10^7 ES cells (3×150 mM T flasks) for homologous recombination (HR) and about 10^6 ES cells (1×25 mM T flask) for transgenic (Tg) transformation. The cells are harvested as above, washed twice in complete medium by centrifugation and aspiration, then resuspended in complete medium with FCS and LIF in an 800 µL volume. The DNA in a maximum volume of 50 µL of $T_{10, E0.1}$ (10 mM Tris, 0.1 mM EDTA pH 8.0) is added to the cells, mixed, and incubated at room temperature for 10 min. The cells are transfected at room temperature by electroporation at 0.25 kV, 500 µF in a 0.4-mm cuvette using a Gene Pulser electroporator fitted with a capacitance extender (Biorad, Hercules, CA, USA) and allowed to rest at room temperature for 10 min. We routinely use 40 µg of linearised construct for HR and 5–10 µg of linearised Tg construct (all vector sequences must be removed), plus 5 µg of circular neomycin-resistance plasmid for G418 selection (pMC1Ne-oPolyA, Stratagene). The cells are plated onto neomycin resistant mEF layers in 100-mm tissue culture dishes (Falcon, Becton Dickinson, USA) (ten dishes for HR and one for Tg) and incubated overnight at 37°C, 10% CO_2. Neomycin resistant ES cells are selected by the addition of 175 µg/mL (active) Geneticin (G418, GibcoBRL) for 10 days. The medium is replaced daily. Ganciclovir (Cymevene, Syntex Corp, USA) at 2 µM is added for negative selection of random integrants if the thymidine kinase gene is used. More recently random integrants can be negatively selected by the use of the diphtheria toxin gene (see for example Araki et al. (7)). G418 resistant clones are picked with a Gilson pipettor set at 20 µL. The plates are aspirated, washed with 10 mL of PBS, and then 10 mL of PBS is readded. Each clone is individually circumscribed with the pipette tip to loosen the feeder cells and the clone is sucked up into the pipette. The clone is then transferred into a 96-well plate containing 180 µL of PBS. The wells are set up in a grid pattern (Fig. 1) to allow direct transfer of the ES cells into 48-well plates containing feeders using a multichannel pipettor (Finnpipette, Labsystems). A single cell suspension of each colony is made by rapid pipetting (40 strokes) using the multichannel pipettor. One-third of the cells are then added to a 48-well plate containing feeder cells for continued growth and subsequent cryopreservation (growth plate), while the remaining two-thirds are plated into a replica 48-well plate for growth and DNA isolation (DNA plate). The HR plates are incubated for 4 days, refreshing the medium containing G418 daily. Due to the transient transfection of the neomycin resistance gene in the Tg colonies, G418 is removed from the culture medium 1 day

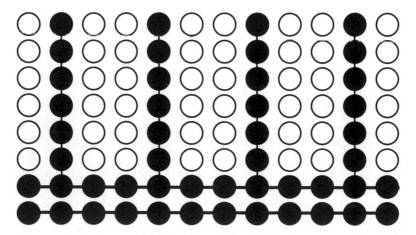

Fig. 1 A 96-well template for colony collection, single cell suspension, and transfer to 48-well plate for culture. The *blacked* out wells are not used for culture. The multichannel pipette does not have tips on rows 2, 5, 8, or 11, which allows transfer of cells directly from the 96-well plate to a 48-well plate

after transfer to 48-well plates. DNA is isolated from wells with robust growth in the DNA plate using a kit (Qiamp DNA Blood kit, Qiagen GmbH, Hilden Germany) and homologous recombinants or Tg constructs detected by PCR (but see also Kontgen et al. *(8))*. Homologous recombination is usually detected by a unique PCR fragment where one primer lies within the neomycin gene and one outside the targeting construct (for high sensitivity PCR see Nitschke et al. *(9))*. Tg constructs are also detected by unique PCR fragment, where one primer lies within the promoter and another lies in the expressed gene. Positive clones are then identified, expanded to a 25-cm T flask, and cryopreserved as indicated above. Larger amounts of DNA are also isolated from specific clones for Southern blotting and further PCR verification.

Embryonic Feeder Cells

Embryonic feeder cells are a mixture of cell types with a largely fibroblastic morphology and are isolated from whole day 13 postcoitum embryos after removal of the head and liver as described *(6)*. It is important to use drug-resistant mEF to enable their survival during the selection process. For selection with G418 we use cells from the M-TKnco mice *(10)*, but any neomycin-resistant animal would suffice that is not detective in mEF supplementation activity (we have used the interleukin-5 knockout, for example *(11)*). A very useful multipurpose mouse for mEF cells was developed that is resistant to four drug-selectable markers, neomycin, puromycin, hygromycin, as well as a mutation in the *Hprt* gene *(12)*, and is available from the Jackson Laboratory (http://www.jax.org/). The cells are grown at 37°C, 10% CO_2 in DMEM (Gibco 10313, Invitrogen) with 2 mM glutamine and 15% FCS (complete feeder medium) to complete confluence and passaged 1 in 5 every 4–5 days. They are not used after 5 passages and are best used at passage 3. We mitotically arrest the cells prior to use with ES cells by treatment with 10 µg/mL mitomycin C (Sigma-Aldrich M0503, St. Louis, USA) that is added directly into the culture medium for 1.5 h. We do not count the cells but rather take a confluent stock flask (say a 150 cm large T), treat with mitomycin C, wash twice with 10 mL PBS, obtain single cells by treatment with 0.05% trypsin-EDTA for 5 min and disruption by pipetting, and then plate these in complete EF medium into 6 × 25 cm small flasks, or 3 × 1 cm 48-well plates, or 3 × 9 cm 6-well plates (all = 150 cm equivalent). The mEF are allowed to adhere for at least 4 h in mEF DMEM (preferably overnight to allow for complete leaching of mitomycin C from the cells). The feeder culture medium is then aspirated and replaced with ES DMEM just prior to the addition of the ES cells. It is noteworthy that in our hands, for reasons that are unclear, the mEF after mitomycin C treatment does not adhere well if plated in ES DMEM, so it must be played in feeder DMEM. We do not use treated mEF after 1 week. Notably we always prepare fresh mEF the day prior to electroporation for the 100-mm selection plates since this will require culture for at least 10 days.

Additional Points on ES Cell Culture

We do not routinely count the ES cells prior to harvest for electroporation or for subculture. This procedure requires that the cells be left in the absence of feeder cells for periods of time and, although cell counting should only require a few minutes and ES cells can be grown in the complete absence of feeder cells, we believe that the rate of germline transmission can be negatively affected. We also think that it is detrimental to have the cells kept in the ES medium in the absence of proper 10% CO_2 equilibration, allowing possible changes in pH. We therefore estimate cell numbers by eye using an inverted phase microscope. Importantly we do not allow the individual colonies to grow to large sizes, and we routinely passage every 2 days due to the rapid replication times of these cells under ideal conditions. It is also important to passage cells into to preequilibrated ES medium to ensure the correct pH. This can be achieved by placing the culture flask for a period of time into the 10% CO_2-in-air incubator. We also regas the culture flask with sterile 5% CO_2-in-air once the flask has been opened in the tissue culture hood prior to replacement into the incubator to ensure rapid equilibration of the correct pH of the ES medium. Gassing with 5% CO_2 is particularly important when the ES cells are thawed after cryopreservation, they then require plating directly into room temperature ES medium and cannot be preequilibrated in the incubator.

We do not routinely count chromosomes in metaphase spreads of ES cells since we have found that in our hands the karyotype is very stable. However, on one occasion we had a female chimaera transmit a genetic manipulation to the germline indicating the loss of the Y-chromosome from the male ES cells. Karyotyping of the parent ES cells indicated a normal number of chromosomes including the Y, indicating that the loss of the Y-chromosome was most likely a spontaneous event in the selected colony bearing the modification. We would recommend to karyotype any new ES cells obtained from other sources.

Sources of Embryonic Stem Cells

129

The first stem cells to be used for gene targeting were obtained from the 129 strain that has an agouti coat colour. When injected into C57BL/6 (B6) blastocysts they produced chimaera with coat colours that were both black (from the host embryo) and agouti (from the stem cells). The degree of chimaerism could be easily seen by the amount of agouti coat colour that was present. When the male chimaera were bred with B6 females and if the 129 cells had contributed to the germline (produced 129 derived sperm), some of the resultant offspring were also agouti since this is dominant over the B6 genotype. Nongermline offspring produced from the resident B6 sperm were easily identified by their black coat colour. Notably the agouti offspring

Table 1 Embryonic stem cell host embryo combinations used for generating germline transmission

Embryonic stem cell	Coat colour	Host embryo	Chimaera colour	Breeding partner	Germline colour	Genetic background
129	Agouti	B6 black	Agouti/black	B6 black	Agouti	129/B6 F2
129[a]	Agouti	129 albino	Agouti/white	129 albino	Agouti	129[a]
C57BL/6[b]	Black	BALB/c albino	Black/white	B6 black	Black	B6
C57BL/6	Black	B6 black	Black	B6 black	Black	B6
C57BL/6	Black	WB6 albino	Black/white	WB6 albino	Black	B6
BALB/c	Albino	B6 black	Black/white	BALB/c albino	Albino	BALB/c

[a] There are a number of 129 substrains that are sufficiently genetically different that they will not accept skin grafts between different substrains *(15)*. Caution should therefore be taken when choosing a 129 strain
[b] It is also becoming apparent that there are now substrains of the C57BL/6 strain since the mice held at Jackson Laboratory and those at National Institutes of Health have now sufficiently diverged to be considered genetically different (16, 17) (although not as widely diverse as those of the 129 strain). The substrain of the C57BL/6 ES cells is therefore now also of importance

were an equal mix of both the 129 and B6 genotypes. Interbreeding of these mice therefore produces offspring that have a 129:B6 F2 genotype, which may be problematical in the analysis of the phenotype *(13, 14)*. Germline chimaera can also then be bred with 129 females to produce complete 129 strain offspring. Transgenic offspring can then be identified by genotyping, resulting in a strain that is all 129 derived (or by coat colour if albino 129 mice are used (Table 1). Notably the 129 strain does not reproduce well, and most targeted strains have been kept in the 129:B6 F2 or crossed for ten generations to the B6 strain. Importantly there is a large genetic variation in the 129 strain, resulting in a large cohort of 129 substrains *(15)*, further complicating the use of these ES cells. However, with the extensive single nucleotide polymorphisms (SNPs) now becoming available to accurately genotype these strains in specific regions *(18)*, these strains of 129 ES cells may regain their usefulness in generating valuable congenic strains.

C57BL/6

The first B6 ES cells were isolated by Ledermann and Bürki *(19)* and Kontgen et al. *(20)*. These cells were injected into blastocysts derived from the albino (white) BALB/c strain, giving rise to white/black chimaera (Table 1). Germline transmission was easily detected in these males by breeding with B6 females, giving rise to black offspring, while nongermline offspring had an agouti coat colour. Notably this germline transmission produces mice entirely derived from the inbred B6 strain, thus avoiding the problems with mixed genetic backgrounds *(14)*. However, the BALB/c strain has very poorly timed embryos where at day 3.5 postconception there are very few expanded blastocysts, making injection of ES cells into the embryo difficult. Although an overnight culture method has been developed to help synchronise these embryos (21), the numbers of injectable embryos still do not compare favourably with what is possible with other strains. Some laboratories have overcome this problem by injecting the B6 ES cells into B6 blastocysts. This produces chimaera but these cannot be identified by coat colour, and all offspring must therefore be genotyped to detect the transgene. We have therefore developed albino B6 mice (white B6, WB6). These mice produce high ratios of expanded blastocysts that can be easily injected with B6 ES cells to produce excellent numbers of high-grade chimaera that have high rates of germline transmission. However, the chimaera must be bred with WB6 females since B6 and WB6 males when mated with B6 females produce black offspring.

BALB/c

The first BALB/c ES cells were derived by Noben-Trauth et al. (22) and more recently by others (5). These ES cells, if cultured under our conditions described above, when injected into B6 blastocysts give very high rates of germline transmission. Immunologically these cells are of particular interest since the BALB/c strain responds to certain stimuli such as Leishmania parasites in a T helper 2 (TH2) manner expressing interleukin-4 and -13. The B6 strain, on the other hand, responds to Leishmania infection in a TH1 manner, expressing interferon-γ. Comparison of mice derived in the B6 as well as the BALB/c genetic background is of considerable immunological significance. Moreover, it is not possible to perform pronuclear injections in BALB/c embryos since the nucleus is not visible in this strain at the zygote stage. It is impossible therefore to generate transgenic mice in the BALB/c strain by zygote injection. However, it is possible to generate transgenic mice using BALB/c ES cells (23), making these ES cells of further importance.

Choice of ES Cells

To avoid problems with genetic backgrounds and the "hitch-hiker effect" (14) caused by genetic linkage, which cannot be avoided during backcrossing to a new strain, it is advisable to generate your targeted or transgenic mouse in the appropriate ES cells in the first instance. Some of these cells are available (see for example Ledermann (24) for DBA/1 and MRL ES cells). However, it is difficult to isolate ES cells with good rates of germline transmission from a number of other strains, hence these ES cells are not freely available. There are several ways of overcoming this problem. It is well known that all of the standard inbred mouse strains that are commonly used (C57BL/6, BALB/c, NOD, NZW, DBA) were derived from parental stocks that were essentially the same (25). As a consequence there are regions on chromosomes of different strains as determined by SNP analysis that have identical or almost identical sequences in regions called "identical by descent" (IBD) *(18)*. It is possible therefore by characterising the chromosomal location of the gene of interest to identify a mouse strain in which the IBD is identical or most similar to the strain to which you will finally need to cross. One example of this was the transfer of the interleukin-2 gene from the 129 genetic background to the NOD background *(26)* since the mutated region in the 129 strain was in an IBD of the NOD strain *(18)*, hence the two regions were identical. Another approach was to generate a hybrid ES cell line with excellent rates of germline transmission between 129 and NOD mice *(27)*. These ES cells were derived from a (NOD×129 F1) × 129 backcross strain that had been intercrossed and selected for homozygosity of particular regions of the NOD genome. Targeting of these regions in the ES cells then allowed the generation of true NOD congenic mice after several backcrosses of the subsequent mice to the NOD background, resulting in the removal of the 129 DNA and thereby avoiding the hitch-hiker effect *(27)*.

Generating Transgenic Mice

Standard "Minimal" Promoter

The first transgenic mouse generated by the direct injection of a transgene construct into the male pronucleus of a single cell fertilised embryo occurred in 1982 *(1)*, and there are many publications that give the methods in detail *(28)*. I will therefore not provide methods here except to say that these mice usually use a small size promoter driving a specific gene that must be freed from any plasmid DNA, and the construct is directly injected into the male pronucleus of a single cell fertilised embryo under a microscope. Moreover, recently it has become evident that there are a number of genetic differences between the different strains. In particular previous methods used embryos from the FVB/n strain. These mice naturally produced very large litters (15–20 pups as opposed to 5–6 pups from the B6 strain) and more importantly the FVB/n zygotes had easily visible pronuclei under the light microscope, making pronuclear injection simpler. Since most research is commonly performed in the B6 and BALB/c strains, the transgene was subsequently crossed to these strains from the FVB/n for ten generations. However, it is now known that by crossing to the new strain the local DNA surrounding the transgene from the original strain will continue to be present in the backcrossed strain (the hitch-hiker effect) *(14)*. Strain differences between the transgenic mouse and its wild-type counterpart may still therefore be present. More recently, therefore, the choice of embryo has been from the B6 (or for us the BALB/c, see above). Of course this modification does not remove the problems with random integration, causing the "position effect" where gene expression can be modified by its specific genetic location *(13)*.

Transgenes Using Large Promoters

Attempts have been made to overcome the position effect by increasing the size of the promoter from the usual 1–5 kb to fragments as large as 300 kb using bacterial artificial chromosomes (BAC). These transgenic mice have been shown to produce excellent tissue specificity with reporter genes *(29)*. However, caution should be taken when using this approach. We, for example, have used a cosmid transgene to rescue the embryonic lethality of the deletion of FliI *(30)*. However, since the cosmid also contains other genes, we could not be certain that it was the FliI transgene alone that was responsible for the rescue until we could show that a minimal promoter containing only 17.6 kb of the promoter was also successful (unpublished). Moreover, as was pointed out earlier *(13)*, the large cosmid transgenes will likely produce excellent tissue specificity as a reporter, but caution should be exercised when a transgene such as a cytokine is expressed. An interleukin-5 containing cosmid will also contain a number of other cytokines that are located nearby, such as interleukin-3, -4, -13, and granulocyte-macrophage colony-stimulating factor, making interpretation of the phenotype of such a mouse problematical if not impossible. Moreover, in some instances even bacterial artificial chromosome-based transgene expression may not show true tissue specific fidelity since in the BAC CD21-cre3A transgenic mouse *(31)*. Cre was also expressed in unexpected tissues *(32)*. Careful characterisation of all transgenic mice is therefore still of utmost importance even with BAC transgenic mice. Of most importance is to analyse several founders for all transgenic constructs, whether using small or large promoters, to ensure that the phenotype observed is due to the transgene expression and not from some random integration event.

Knockin Transgenic Mice

Attempts have also recently been made to overcome the position effect by introducing the gene of interest into the genome by homologous recombination so that the transgene is expressed from a specific promoter in its natural location. Again caution should be exercised. Deletion of one copy of a number of genes has been shown to reduce the amount of the gene expressed (called haploinsufficiency), resulting in a change in phenotype *(33)*. It is recommended, therefore, that the transgene be expressed using an internal ribosome entry site (IRES) sequence to allow normal expression of the "host" gene as well as the transgene, thereby avoiding change of function.

Generating a Knockout Mouse

In the paragraphs above I have given our proven methods for introducing targeting and transgenic constructs into ES cells. It is advisable at the outset that care is taken in designing the appropriate construct and then introducing it into the best genetic background available. It is also recommended that for targeting and the most flexibility an inducible system like the Cre/LoxP is considered (but see also below).

In the generation of gene targeting constructs there are now excellent Web sites that describe gene targeting for all levels of expertise (http://www.cellmigration.org/ resource/komouse/komouse_approaches.shtml and for methods komouse_resources. shtml#protocols, or at the same site targeting4beginners.shtml#1), so I will not describe these here. Using these sites any laboratory competent in molecular biology techniques should be able to generate such a construct. However, generating and characterising a knockout or transgenic mouse requires a long period of extensive effort, and collaboration with any other laboratory that may have similar interests in generating these animals is recommended so that the workload can be shared.

An alternative should also be considered. By carefully screening what is available on the Internet it may be possible to find a collaborator. There are many consortia that aim to target specific classes of genes, e.g., for genes involved in cell migration see the Web site for Cell Migration Gateway (http:// www.cellmigration.org/index.shtml). This site has a searchable database where all current targeted and transgenic (reporter, RNAi knockdown) mice are listed as well as mice that are in progress, allowing rapid determination whether your specific gene is or will soon be available. Similarly there are more generalised consortia, e.g., the International gene Trap Consortium ITGC (for an overview see http://www.genetrap.org/tutorials/overview.html#intro). At these sites it is again possible to search for specifically targeted ES cells (http://www.sanger. ac.uk/cgi-bin/PostGenomics/genetrap/browser); or fully completed mice such as at the Mutant mouse Regional Resource Center (http://www.mmrrc.org/cata- log/StrainCatalogSearchForm.jsp?SourceCollection = SIGTR&pageSize = 50&jboEvent = Search). Alternatively if no mice or targeted ES cells are found, a construct to make a conditional knockout can be designed by contacting the European Conditional Mouse Mutagenesis program (http://www.sanger.ac.uk/ htgt/welcome?style%20=%20EUCOMM), which is part of the International Knockout Mouse Consortium (IKMC) also consisting of KOMP (Knock Out Mouse Project, USA), NorCOMM (North American Conditional Mouse Mutagenesis program, Canada), and TIGM (Texas Institute for Genomic Medicine, USA). All of these sites deserve scrutiny. The need to generate tar- geting constructs in-house, or even to generate targeted ES cells, may possibly become unnecessary in the future.

Conclusions

As with any scientific method there are always limitations and unsuspected problems. Genetically modified mice are no exception, and I have reviewed these caveats recently *(13, 14)*. However, conditional gene targeting based on the inclusion of *Lox*P sites to flank genes of interest and inactivating them using tissue specific Cre expression in cell-type specific or inducible way is still considered the method of choice. Notably even more problems with this system have recently been described *(32)*, relating to the toxicity of Cre recombinase when expressed at high levels *(34)*, presumably due to the cryptic *Lox*P sites *(35)*. Whether low levels of expression also have effects, or whether some cell types are more prone to toxicity than others, is at this stage not clear. Similarly, the tissue specificity of the Cre expression can still be a problem when it occurs in the wrong tissue type or at the wrong time, even in constructs using BACs *(31, 32)*. The efficiency of Cre excision in different cell types is at times also difficult to assess accurately. It is possible that limiting the level and extent of Cre expression may help due to the reduction of exposure to the Cre enzyme in the target tissue. One way to do this may be to only remove floxed neomycin cassettes in vitro in the ES cells using transient Cre expression rather than in vivo by crossing with a Cre transgenic mouse. Most importantly, since Cre toxicity has been unknown or ignored, it is now crucial that appropriate controls are employed that investigate the effect of Cre expression in the absence of the floxed target gene.

Moreover, since it is possible to have floxed DNA deleted between *Lox*P sites in animals not carrying a Cre transgene *(13)*, we will need to carefully reassess previous data using these animals as negative controls.

In conclusion, the more we know about possible problems the better we can design our experiments. Although a number of the problems have either been unknown or ignored in the past, the improved knowledge now gives us the ability to include a number of more appropriate controls. Clearly it is now much more possible to avoid problems with strain variation by using the information about IBD regions in different strains. Undoubtedly other new methodologies will be developed to allow a more complete understanding of gene function both in health and disease.

Acknowledgements I am indebted to my many collaborators with whom I have discussed ES cell culture. In my laboratory I am particularly indebted to my long-term technicians and valued colleagues Vane (Wayne) Damcevski and Helen Taylor without whom many of the mice would not have been so successfully generated.

References

1. Palmiter RD, Brinster RL, Hammer RE, et al. Dramatic growth of mice that develop from eggs microinjected with metallothionein-growth hormone fusion genes. Nature 1982;300:611–5.
2. Evans MJ, Kaufman MH. Establishment in culture of pluripotential cells from mouse embryos. Nature 1981;292:154–6.
3. Doetschman T, Gregg RG, Maeda N, et al. Targeted correction of a mutant HPRT gene in mouse embryonic stem cells. Nature 1987;330:576–8.
4. Thomas KR, Capecchi MR. Site-directed mutagenesis by gene targeting in mouse embryo-derived stem cells. Cell 1987;51:503–12.
5. Baharvand H, Matthaei KI. Culture condition difference for establishment of new embryonic stem cell lines from the C57BL/6 and BALB/c mouse strains. In Vitro Cell Dev Biol Anim 2004;40:76–81.
6. Abbondanzo SJ, Gadi I, Stewart CL. Derivation of embryonic stem cell lines. Methods Enzymol 1993;225:803–23.

7. Araki K, Araki M, Yamamura K. Negative selection with the diphtheria toxin A fragment gene improves frequency of Cre-mediated cassette exchange in ES cells. J Biochem 2006;140:793–8.

8. Kontgen F, Stewart CL. Simple screening procedure to detect gene targeting events in embryonic stem cells. Methods Enzymol 1993;225:878–90.

9. Nitschke L, Kopf M, Lamers MC. Quick nested PCR screening of ES cell clones for gene targeting events. BioTech 1993;14:914–6.

10. Stewart CL, Schuetze S, Vanek M, Wagner EF. Expression of retroviral vectors in transgenic mice obtained by embryo infection. EMBO J 1987;6:383–8.

11. Kopf M, Brombacher F, Hodgkin PD, et al. IL-5-deficient mice have a developmental defect in CD5+ B-1 cells and lack eosinophilia but have normal antibody and cytotoxic T cell responses. Immunity 1996;4:15–24.

12. Tucker KL, Wang Y, Dausman J, Jaenisch R. A transgenic mouse strain expressing four drug-selectable marker genes. Nucleic Acids Res 1997;25:3745–6.

13. Matthaei KI. Genetically manipulated mice: a powerful tool with unsuspected caveats. J Physiol 2007;582:481–8.

14. Matthaei KI. Caveats of gene targeted and transgenic mice. In: Lanza R, Gearhart J, Hogan B, et al., editors. Handbook of Stem Cells, Elsevier, Amsterdam, 2004:589–98.

15. Simpson EM, Linder CC, Sargent EE, Davisson MT, Mobraaten LE, Sharp JJ. Genetic variation among 129 substrains and its importance for targeted mutagenesis in mice. Nat Genet 1997;16:19–27.

16. Moran JL, Bolton AD, Tran PV, et al. Utilization of a whole genome SNP panel for efficient genetic mapping in the mouse. Genome Res 2006;16:436–40.

17. Moran JL, Pollock JD, Fletcher CF, et al. Genetic variation among substrains of C57BL/6. Presented at Frontiers in Genome Engineering: Building a Better Mouse II 2007,1,47.

18. Ridgway WM, Healy B, Smink LJ, Rainbow D, Wicker LS. New tools for defining the 'genetic background' of inbred mouse strains. Nat Immunol 2007;8:669–73.

19. Ledermann B, Burki K. Establishment of a germ-line competent C57BL/6 embryonic stem cell line. Exp Cell Res 1991;197:254–8.

20. Kontgen F, Suss G, Stewart C, Steinmetz M, Bluethmann H. Targeted disruption of the MHC class II Aa gene in C57BL/6 mice. Int Immunol 1993;5:957–64.

21. Lemckert FA, Sedgwick JD, Korner H. Gene targeting in C57BL/6 ES cells. Successful germ line transmission using recipient BALB/c blastocysts developmentally matured in vitro. Nucleic Acids Res 1997;25:917–8.

22. Noben-Trauth N, Kohler G, Burki K, Ledermann B. Efficient targeting of the IL-4 gene in a BALB/c embryonic stem cell line. Transgenic Res 1996;5:487–91.

23. Dinkel A, Aicher WK, Warnatz K, Burki K, Eibel H, Ledermann B. Efficient generation of transgenic BALB/c mice using BALB/c embryonic stem cells. J Immunol Methods 1999;223:255–60.

24. Ledermann B. Embryonic stem cells and gene targeting. Exp Physiol 2000;85:603–13.

25. Witmer PD, Doheny KF, Adams MK, et al. The development of a highly informative mouse simple sequence length polymorphism (SSLP) marker set and construction of a mouse family tree using parsimony analysis. Genome Res 2003;13:485–91.

26. Yamanouchi J, Rainbow D, Serra P, et al. Interleukin-2 gene variation impairs regulatory T cell function and causes autoimmunity. Nat Genet 2007;39:329–37.

27. Brook FA, Evans EP, Lord CJ, et al. The derivation of highly germline-competent embryonic stem cells containing NOD-derived genome. Diabetes 2003;52:205–8.

28. Nagy A, Gertsenstein M, Vintersten K, Behringer R. Manipulating the Mouse Embryo, Third ed., Cold Spring Harbor Laboratory Press, Cold Spring Harbor, NY, 2003.

29. Moreira PN, Giraldo P, Cozar P, et al. Efficient generation of transgenic mice with intact yeast artificial chromosomes by intracytoplasmic sperm injection. Biol Reprod 2004;71:1943–7.

30. Campbell HD, Fountain S, McLennan IS, et al. FliI, a gelsolin-related cytoskeletal regulator essential for early mammalian embryonic development. Mol Cell Biol 2002;22:3518–26.

31. Kraus M, Alimzhanov MB, Rajewsky N, Rajewsky K. Survival of resting mature B lymphocytes depends on BCR signaling via the Ig alpha/beta heterodimer. Cell 2004;117:787–800.
32. Schmidt-Supprian M, Rajewsky K. Vagaries of conditional gene targeting. Nat Immunol 2007;8:665–8.
33. Cowin AJ, Adams DH, Strudwick XL, et al. Flightless I deficiency enhances wound repair by increasing cell migration and proliferation. J Pathol 2007;211:572–81.
34. Schmidt EE, Taylor DS, Prigge JR, Barnett S, Capecchi MR. Illegitimate Cre-dependent chromosome rearrangements in transgenic mouse spermatids. Proc Natl Acad Sci U S A 2000;97:13702–7.
35. Thyagarajan B, Guimaraes MJ, Groth AC, Calos MP. Mammalian genomes contain active recombinase recognition sites. Gene 2000;244:47–54.

Differentiating Gametes from Stem Cells

Ana Isabel Marqués-Marí, José Vicente Medrano, and Carlos Simón

Abstract Embryonic stem cell lines derived from the inner cell mass of the blastocyst are pluripotent (they can differentiate into all the different cell types) and have the ability to self-renewal in vitro, remaining undifferentiated.

It has been demonstrated that murine embryonic stem cells can give rise to structures very similar to sperm and oocytes *in vitro*. These differentiated cells are able to undergo meiosis, generating haploid gametes, which, in the case of oocytes, are able to form structures mimicking blastocysts. However, none of these blastocysts have survived to embryonic development. In the case of male gametes, successful progeny has been obtained after injection into normal oocytes, but the obtained progeny died prematurely.

Experimental studies have also demonstrated that it is possible to obtain germ cells from human embryonic stem cells, although their functionality to generate successful and healthy progeny has not been demonstrated to date. The problems of meiosis completion and acquisition of the proper epigenetic pattern remain to be surpassed.

Recent studies have reported the obtaining of germ cell-like cells from fetal and adult stem cells (from porcine skin and mice bone marrow, respectively). In addition, against present dogma, which supports that oocyte production in female mammals stops before birth, some recent studies have revealed that there is postnatal oogenesis in the adult mice ovaries. Nevertheless, there is much controversy regarding the results of these studies. Anyhow, a recent work has shown that meiosis, neo-oogenesis, and germ stem cells are unlikely to occur in normal adult human ovaries.

Keywords Stem cells • Differentiation • Primordial germ cells • Gametes • Meiosis

A.I. Marqués-Marí (✉), J.V. Medrano, and C. Simón
Valencia Stem Cell Bank, Prince Felipe Research Center (CIPF), Valencia, Spain
e-mail: amarques@cipf.es

H. Baharvand (ed.), Trends in Stem Cell Biology and Technology
DOI 10.1007/978-1-60327-905-5_9

Introduction

The discovery of the stem cells is the most revolutionary scientific breakthrough in the field of regenerative medicine. In general terms, these cells are classified, depending on their origin, as embryonic and nonembryonic stem cells.

The embryonic stem cell (ESC) lines are derived from the inner cell mass of blastocysts and are pluripotent *in vivo* as well as *in vitro*. These cells have the ability to self-renewal and remain undifferentiated in culture almost indefinitely *(1–3)*.

The nonembryonic stem cells are located in different extraembryonic tissues, including the umbilical cord, placenta, and amniotic fluid *(4–6)* and in specific niches in a wide range of adult tissues *(7–14)*.

Recently, it has been described that mouse ESC *in vitro* can give rise to structures similar to sperm atocytes and oocytes *(15–21)*. These differentiated cells are able to undergo meiosis, and in the case of oocytes, give rise to structures similar to blastocysts, although none of the blastocyst-like structures survived the development and birth. *In vitro* differentiated sperm have produced successful offspring, but unfortunately the pups died prematurely.

Experimental studies conducted to date have demonstrated that it is also possible to obtain germ cells from human ESC *(22–24)*. However, the effectiveness of these putative gametes to generate viable and healthy offspring has not yet been proven. On the other hand, contrary to the established dogma that states that eggs production in females of most mammal species ceases before birth, several studies published during the past few years described postnatal oogenesis in the mouse adult ovary *(25–27)*. The results reported in these studies are controversial, and, additionally a recent study developed with human shows that there is no meiosis, neo-oogenesis, or presence of germ stem cells in adult ovaries *(28)*. Thus, if this is confirmed in the mouse, this species could represent an exception to the general rule that there is no postnatal oogenesis in female mammals.

State of the Art: In Vitro Differentiation of Gametes from Embryonic Stem Cells

Evidence of *in vitro* generation of germ cells from mouse ESCs seems to be robust given that some groups have described obtaining cells that express markers related to the germ cell lineage *(15–19)*. The next step toward the terminal gametogenesis and creation of functionally mature gametes remains weak or anecdotal *(20)*. Developmental competence of the generated gametes in reproduction remains uncertain in terms of safety and effectiveness and needs further investigation. Establishment of the appropriate genetic imprinting is also a challenge yet to be solved.

Different strategies have been used for *in vitro* differentiation of gamete-like cells from ESC: spontaneous *in vitro* differentiation resulting in the formation of embryoid bodies (EBs) from ESC in a suspension culture *(22)*; addition of a specific cocktail of growth factors to the medium in a time specific manner *(17, 23)*; coculture with soluble factors from differentiated cells *(18)*; or genetic manipulation that may direct ESC to differentiate into a specific linage *(15, 16, 20)* (Table 1). Nevertheless, the efficiency of these protocols in target differentiation for somatic cells is limited and even more restricted to create germ line cells.

The work by Hubner et al. *(15)* provided the first evidence of the in vitro ability of ESCs to differentiate into germ cells. In this case, they use mouse ESCs transfected with green fluorescent protein (GFP) linked to Oct4 expression as an enrichment-selection system to further feeder-free culture with out leukemia inhibitory factor (LIF). These mouse ESCs spontaneously formed aggregates that gave rise to follicle-like structures that extruded oocyte-like cells, showing molecular markers of meiosis, which undergo parthenogenic activation forming pseudoblastocysts. However, the capacity of these generated oocyte-like structures to be fertilized or to be used as receptors in nuclear transfer techniques has not been analyzed or reported. The authors proposed that both XX and XY mouse ESCs develop into germ cells with the female phenotype because of the absence of appropriate Sry expression.

Toyooka et al. *(16)* reported in 2003 the differentiation of sperm-like cells from mouse ESCs. Using a similar strategy to that employed by the group of Hubner et al., they transfected mouse ESC in this case with the postmigratory germ cell marker Mvh (mouse VASA homologous). The culture conditions to stimulate differentiation consisted of LIF removal allowing EBs formation and addition of BMP4 to the culture medium. After detection of Mvh-positive cells, these cells were transplanted into host testis where they participated in spermatogenesis *in vivo*. However, no data about the fertilization capacity of these generated gametes were reported in this work.

The study of Geijsen et al. *(17)* also describes the differentiation of mouse ESCs into mature male gametes. They added retinoic acid (RA) to the culture media and analysed different markers to identify germ cells and their methylation status in several genes. The study demonstrates that male germ cell development and meiotic maturation are possible from EBs. The haploid gametes isolated from EBs were injected into oocytes, resulting in the formation of blastocysts. Although this work showed that postmeiotic male germ cells capable of fertilizing oocytes spontaneously arise from ESCs, further studies are required to determine its functional quality.

Lacham-Kaplan et al. *(18)* reported in 2006 the creation of oocyte-like cells derived from an XY mouse ESC line. In this study, authors induced differentiation to germ cells by adding testicular cell conditioned medium prepared from the testes of newborn male mice to the EBs culture. Since the testis is rich in numerous growth factors such as BMP4, stem cell factor (SCF), LIF, and GDF9, the authors suggested that these growth factors could be important in development of gametes from mouse ESC. Specific germ cell and oocyte markers were found, although the lack of zona pellucida and the small size of these structures suggest that putative oocytes are not completely formed. It seems likely that for oocyte differentiation, necessary requirements are also present in the male gonads, although this is not enough for the proper completion of the oocyte developmental process. Specific signals that lead to female gametes differentiation despite the chromosomal sex of the ESC line remain to be determined.

Table 1 Summary of the studies developed in germ cell differentiation from stem cells

Publication	Origin	Strategy for differentiation[a]	Cell type obtained	Offspring
Germ cell differentiation from embryonic stem cells				
Hubner et al. (*15*)	mESC: XX, XY	SD in adherent cultures	Oocyte-like cells	No, PB
Toyooka et al. (*16*)	mESC: XY	EBs formation + AC with growth factors	PGCs in vitro, sperm in vivo when transplanted	NT
Geijsen et al. (*17*)	mESC: XY	RA addition to EBs-derived cells	Spermatids	No, OF
Clark et al. (*22*)	hESC: XX, XY	SD through EBs formation	Oocyte-like cells (although TEKT1 expression was found)	NT
Lacham-Kaplan et al. (*18*)	mESC: XY	EBs with conditioned media from testicular cell culture	Immature oocyte-like cells	NT
Novak et al. (*19*)	mESC: XY	SD in adherent cultures and through EBs formation	Ovarian follicles	NT
Nayernia et al. (*20*)	mESC: XY	RA addition to SSC lines-derived EBs	Sperm	Yes
Kee et al. (*23*)	hESC: XX	SD through EBs + growth factors addition	Oocyte-like cells	NT
Qing et al. (*21*)	mESC: XY	Coculture of EBs with mouse granulosa cells	Oocyte-like cells	No, OA
Chen et al. (*24*)	hESC: XX	SD in adherent cultures and through EBs formation	Oocyte-like cells inside follicular-like structures	No, FD
Germ cell differentiation from nonembryonic stem cells				
Publication	Origin	Strategy/evaluation of differentiation	Cell type obtained	Offspring
Johnson et al. (*25*)	GSC in OSE	Histological analysis, Mvh and SCP3 IHC, BrdU-Mvh coexpression, ovarian fragments graft	Formation of ovarian follicles after graft	NT
Johnson et al. (*26*)	GSC in female BM and blood	Analysis of germ cells markers, BM and blood transplantations	Formation of ovarian follicles after graft	NT
Bukovsky et al. (*27*)	GSC in OSE	Culture medium with estrogenic stimuli Morphological and IHC analysis	Oocyte-like cells and granulosa-like cells	NT

Dyce et al. (34)	Fetal porcine skin	Medium with follicular fluid, addition of gonadotropins Oocytes markers and steroids production	Oocyte-like cells and Blastocyst-like structures	PB
Nayernia et al. (20)	Stem cells in mouse BM	BM cells in adherent culture, RA addition, transplantation into testis Analysis of PGC, SSC, and spermatogonia markers Analysis after graft	Spermatogonia-like cells	NT

Abbreviations: mESC mouse embryonic stem cells; *hESC* human embryonic stem cells; *PGCs* primordial germ cells; *GSC* germ stem cells; *IHC* immunohistochemistry; *SSC* spermatogonial stem cells; *OSE* ovarian surface epithelium; *BM* bone marrow; *SD* spontaneous differentiation; *EBs* embryoid bodies; *AC* aggregation cultures with cells producing growth factors; *RA* retinoic acid; *PB* parthenogenic blastocyst; *NT* not tested; *OF* oocyte fertilization; *OA* oocytes arrest; *FD* follicles degeneration

[a] Different strategies have been used to differentiate germ cells from embryonic stem cells, most of them based on spontaneous differentiation through EBs formation with or without combining addition of soluble factors to the culture media. Several studies have described the possibility of postnatal oocyte generation in females, not without controversy. The source of these cells (ovarian epithelium, bone marrow or fetal skin) still needs to be confirmed in most of them

The study published by Novak et al. *(19)* raises the difficulties related with meiosis accomplishment in the ESCs-derived oocytes and the related aneuploidies. They obtained follicular structures from mouse ESCs through EBs formation and analysed exhaustively these oocyte-like cells to detect evidences of meiosis. Despite the presence of the meiotic marker SCP3, they did not find expression of other important molecules implicated in meiosis such as SCP1, SCP2, REC8, STAG3, and SMC1-β. Moreover, chromosomal arrangement in these oocyte-like structures was different from synaptic disposition in oocytes *in vivo*.

The only study reported to date showing that haploid male germ cells derived from ESCs are capable to fertilize eggs and generate offspring was published by Nayernia et al. *(20)*. As a first step, they established spermatogonial stem cell lines from mouse ESCs, transfecting ESCs with two marker genes to detect and select cells (GFP and a red fluorescent protein linked to promoter regions of Stra8 and Prm1, respectively). To *in vitro* induce spermatogenesis, RA was added to the culture media and EBs were formed. They found that marked cells with motility were released into the medium, expressing meiotic and postmeiotic markers after RA addition. When transplanted into the testis of sterilized mice, they gave rise to haploid sperm with limited motility. After oocyte fertilization using intracytoplasmic sperm injection (ICSI) and embryo transfer techniques, live offspring were obtained, however, most pups died prematurely. The authors concluded that phenotypic alterations (both overgrowth and growth retardation) as well as short life of these mice were due to abnormal methylation patterns because of a failure in establishing germ line imprinting in ESCs-derived gametes.

The most recent study published concerning differentiation of oocyte-like cells from mouse ESCs has been reported by Qing et al. *(21)*. This group tried a two-step method for inducing oocyte-like cells from mouse ESC using mouse ovarian granulosa cells as feeder cells. First, they obtained primordial germ cells (PGCs) by spontaneous differentiation of mouse ESC within EBs. Second, EBs were plated in coculture with a monolayer of ovarian granulosa cells from fetal mouse ovaries. After 10 days of culture, expression of the germ cell markers Mvh and SCP3, as well as the oocyte-specific genes Fig α, GDF9, and ZP1, ZP2, and ZP3, was found in the formed colonies. Within these germ cell colonies, some cells were double-labeled for Mvh and GDF9. These oocyte-like cells were about 25 μm in size and larger than the accompanying GDF9-negative cells. These putative oocytes did not possessed zona pellucida and neither expressed ZP2, suggesting that these cells might be in an early developmental stage. Although the developmental stage of putative oocytes obtained by this group was immature, their results demonstrate that direct cell-to-cell contact may play an important role in inducing the differentiation of PGCs into oocyte-like cells.

The first study to demonstrate that human ESCs differentiate to germ cells *in vitro* was developed by Clark et al. *(22)*. Similar to previous studies, the strategy of this study relied on spontaneous differentiation of human embryonic stem cells (hESCs) to germ cells via EBs. Using gene expression sequence methods through spontaneous *in vitro* differentiation, the authors provided evidence that undifferentiated hESCs express some markers related to early development of germ cells such

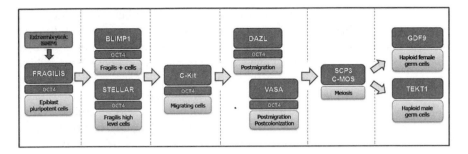

Fig. 1 During in vivo embryonic development, as a result of BMP4 stimulation, a subset of epiblast cells begins to express the Fragilis and Blimp1 genes and become primordial germ cells (PGCs). During migration into the embryo these cells express other germ cell–related markers, such as Stella and c-Kit, and initiate colonization of the gonadal ridge. At this time, PGCs express the premeiotic markers Dazl and VASA. The proteins SCP1, SCP2, and SCP3 are specific of meiotic germ cells, as well as the protooncogene c-Mos. The postmeiotic markers GDF9 and TEKT1 are specific of more mature haploid gametes

as c-Kit or DAZL, while VASA expression follows appearance of germ cell precursors *in vitro* (Fig. 1). In addition, genes involved in meiosis were also detected in putative germ cells, but haploid cells were not obtained. In contrast to the mouse model, markers of both female and male gametes such as GDF9 and TEKT1 were found in the differentiated cells regardless of the sex of the hESC line. A recent study by Kee et al. *(23)* showed that addition of BMP-4, BMP-7, and BMP-8 to the EBs culture medium resulted in an increase in the number of VASA-expressing cells, and thereby, an increase in the number of potential germ cells.

Chen et al. *(24)* demonstrated differentiation of germ cells from newly derived hESC lines either via EBs formation or not (it is via adherent culture). Structures mimicking ovarian follicles with c-Kit and VASA-positive cells appeared in both culture systems. Therefore, the environment provided by EBs may not be a strict necessity for female germ cell differentiation. However, as the number of VASA-positive cells identified by other studies was also low, a more efficient strategy to obtain germ stem cells is needed to improve germ cell differentiation.

State of the Art: Germ Cell Differentiation from Somatic Stem Cells

Several controversial reports have provided evidence that germ cells may be generated from somatic stem cells in adult gonads. For more than 50 years, we have assumed as a central dogma that mammalian females are born with a finite number of oocytes and this number decreases during their lifetime, whereas in adult males spermatogenesis is preserved throughout their complete adult lives. The study published by Johnson et al. *(25)* challenged this dogma, suggesting that a population of somatic stem cells exists in mice postnatal ovaries that contributes to oocyte formation through adult life. One year after, Johnson et al. *(26)* hypothesized that cells from bone marrow (BM) may be a source of germ cells capable of forming oocytes in adulthood when they enter the gonads. Furthermore, they propose the possibility that they could be present in peripheral blood, implying that bone marrow transplants and blood transfusions may collaborate in adult oogenesis.

This revolutionary hypothesis has sparked controversy, and several solid arguments have been raised against it *(29, 30)*. A recent work published by Liu et al. *(28)* showed that meiosis, neo-oogenesis, and germ stem cells are unlikely to occur in normal adult human ovaries. If postnatal oogenesis finally is confirmed in mice, then this species would represent an exception to the rule. Stronger evidence is needed to confirm this new theory that indicates that these somatic stem cell–derived oocytes enter meiosis or support the development of offspring in cases of patients with allogenic bone marrow transplant.

However, the authors of these controversial studies *(25, 26)* have come into discussion, refuting the arguments raised against postnatal oogenesis in adult human ovary arguing, among others, that these arguments derive from the inability to detect markers of germ cell mitosis and meiosis, and that an absence of evidence does not mean an absence of the possibility *(31)*. Curiously, a recent work published by Tilly et al. *(31)* presents a mouse model in which BM transplant helps to preserve or recover ovarian function of recipient females, but all offspring generated were derived from the host germline and not from BM donors *(32)*.

Nayernia et al. *(33)* have reported the derivation of male germ cells from mice BM cells. They isolated BM cells from a Stra8-enhanced green fluorescence protein (eGFP) transgenic mouse line and cultured them with RA. After RA addition they observed a small population of cells expressing eGFP as indication for differentiation of BM cells to male germ cells. Furthermore, after isolation of fluorescent cells by flow cytometry, it was observed that the BM stem cell–derived germ cells were positive for the germ cell markers *Fragilis*, *Stella*, Mvh, and Rnf17, as well as for markers of spermatogonial cells such as Rbm, c-Kit, Tex18, Piwil2, DAZL, Hsp90α, and α6 and β1-integrins. This work suggests that mesenchymal cells from BM can differentiate into male germ cells in culture, as well as in other cell types, depending on the culture conditions. These male gametes expressed PGCs markers, but the maturation process arrested in a certain moment without progress, even after transplantation into the testis.

Dyce et al. *(34)* differentiated oocyte-like cells from fetal porcine skin cells *in vitro* at rates of 0.1%. The skin stem cells were cultured with porcine follicular fluid. Addition of media containing gonadotrophins resulted in cells expressing oocyte markers followed by the appearance of follicle resembling aggregates containing oocyte-like cells. Some of these oocyte-like cells were spontaneously activated, forming parthenogenetic pseudoblastocysts similar to findings by Hubner et al. *(15)*. The authors suggest that nongonadal niches containing cells with stem cell–like characteristics exist, and that these cells are capable of giving rise to oocyte-like cells *in vitro*. As controversial as it may be, these results could support the work by Johnson et al. *(26)* in which oocytes were differentiated from BM derived stem cells.

Bukovsky et al. *(27)* reported that granulosa cells and oocytes may develop from cultured ovarian surface epithelial cells in adult human ovaries. But these affirmations are based solely in phase contrast microscopy observations and immunohistochemistry against several zona pellucida proteins. The authors argue that these *in vitro* derived "oocytes" underwent the first meiotic division, forming large secondary oocytes and becoming suitable for fertilization. In this culture system, the development and maturation of oocytes appear to be stimulated by estrogens provided by the phenol red present in the media components. In this case, more detailed studies with powerful techniques are needed to fully characterize the presumptive oocyte-like cells.

Future Prospects

Although in just a few years amazing progress has been made in germ cells differentiation *in vitro*, more efficient strategies are needed to achieve mature and functional gametes.

Currently, one of the most critical steps after germ cells differentiation is selection and isolation of differentiated germ cells. Since ESCs and PGCs share common markers, detection of postmigratory and meiotic markers is a useful method, as well as the use of transgenes with fluorescent reporter genes under control of specific promoters of male and female germ cells. These have been the most employed strategies in the studies of germ cell differentiation developed to date. The problem is that these methods are based on gene therapy, and retroviral vectors limit the use of gamete-like cells in future clinical treatments.

Future translational application of these ESC-derived gametes in assisted reproduction technology techniques when gametes are not available from patients still require an enormous amount of work in gamete differentiation to make it reproducible and efficient. The obstacle of meiotic completion remains to be surpassed to avoid harmful or incompatible with life chromosomic alterations. Furthermore, it is essential to strictly demonstrate that the obtained gametes possess the accurate epigenetic modifications and have acquired the correct imprinting status.

References

1. Evans MJ, Kaufman MH. Establishment in culture of pluripotential cells from mouse embryos. Nature 1981;292:154–6.
2. Martin GR. Isolation of a pluripotent cell line from early mouse embryos cultured in medium conditioned by teratocarcinoma stem cells. Proc Natl Acad Sci U S A 1981;78:7634–8.
3. Thomson JA, Itskovitz-Eldor J, Shapiro SS, et al. Embryonic stem cell lines derived from human blastocysts. Science 1998;282:1145–7.

4. Nakahata T, Ogawa M. Identification in culture of a class of hemopoietic colony-forming units with extensive capability to self-renew and generate multipotential hemopoietic colonies. Proc Natl Acad Sci U S A 1982;79:3843–7.

5. Miki T, Lehmann T, Cai H, et al. Stem cell characteristics of amniotic epithelial cells. Stem Cells 2005;23:1549–59.

6. De Coppi P, Bartsch G Jr., Siddiqui MM, et al. Isolation of amniotic stem cell lines with potential for therapy. Nat Biotechnol 2007;25:100–6.

7. Jackson KA, Mi T, Goodell MA. Hematopoietic potential of stem cells isolated from murine skeletal muscle. Proc Natl Acad Sci U S A 1999;96:14482–6.

8. Pittenger MF, Mackay AM, Beck SC, et al. Multilineage potential of adult human mesenchymal stem cells. Science 1999;284:143–7.

9. Gage FH. Mammalian neural stem cells. Science 2000;287:1433–8.

10. Bonner-Weir S, Sharma A. Pancreatic stem cells. J Pathol 2002;197:519–26.

11. Forbes S, Vig P, Poulsom R, et al. Hepatic stem cells. J Pathol 2002;197:510–8.

12. Zuk PA, Zhu M, Ashjian P, et al. Human adipose tissue is a source of multipotent stem cells. Mol Biol Cell 2002;13:4279–95.

13. Alonso L, Fuchs E. Stem cells of the skin epithelium. Proc Natl Acad Sci U S A 2003;100 Suppl 1:11830–5.

14. Zhao M, Momma S, Delfani K, et al. Evidence for neurogenesis in the adult mammalian substantia nigra. Proc Natl Acad Sci U S A 2003;100:7925–30.

15. Hubner K, Fuhrmann G, Christenson LK, et al. Derivation of oocytes from mouse embryonic stem cells. Science 2003;300:1251–6.

16. Toyooka Y, Tsunekawa N, Akasu R, Noce T. Embryonic stem cells can form germ cells in vitro. Proc Natl Acad Sci U S A 2003;100:11457–62.

17. Geijsen N, Horoschak M, Kim K, et al. Derivation of embryonic germ cells and male gametes from embryonic stem cells. Nature 2004;427:148–54.

18. Lacham-Kaplan O, Chy H, Trounson A. Testicular cell conditioned medium supports differentiation of embryonic stem cells into ovarian structures containing oocytes. Stem Cells 2006;24:266–73.

19. Novak I, Lightfoot DA, Wang H, et al. Mouse embryonic stem cells form follicle-like ovarian structures but do not progress through meiosis. Stem Cells 2006;24:1931–6.

20. Nayernia K, Nolte J, Michelmann HW, et al. In vitro-differentiated embryonic stem cells give rise to male gametes that can generate offspring mice. Dev Cell 2006;11:125–32.

21. Qing T, Shi Y, Qin H, et al. Induction of oocyte-like cells from mouse embryonic stem cells by co-culture with ovarian granulosa cells. Differentiation 2007;75:902–11.

22. Clark AT, Bodnar MS, Fox M, et al. Spontaneous differentiation of germ cells from human embryonic stem cells in vitro. Hum Mol Genet 2004;13:727–39.

23. Kee K, Gonsalves JM, Clark AT, Pera RA. Bone morphogenetic proteins induce germ cell differentiation from human embryonic stem cells. Stem Cells Dev 2006;15:831–7.

24. Chen HF, Kuo HC, Chien CL, et al. Derivation, characterization and differentiation of human embryonic stem cells: comparing serum-containing versus serum-free media and evidence of germ cell differentiation. Hum Reprod 2007;22:567–77.

25. Johnson J, Canning J, Kaneko T, et al. Germline stem cells and follicular renewal in the postnatal mammalian ovary. Nature 2004;428:145–50.

26. Johnson J, Bagley J, Skaznik-Wikiel M, et al. Oocyte generation in adult mammalian ovaries by putative germ cells in bone marrow and peripheral blood. Cell 2005;122:303–15.

27. Bukovsky A, Svetlikova M, Caudle MR. Oogenesis in cultures derived from adult human ovaries. Reprod Biol Endocrinol 2005;3:17.

28. Liu Y, Wu C, Lyu Q, et al. Germline stem cells and neo-oogenesis in the adult human ovary. Dev Biol 2007;306:112–20.

29. Byskov AG, Faddy MJ, Lemmen JG, Andersen CY. Eggs forever? Differentiation 2005;73:438–46.

30. Eggan K, Jurga S, Gosden R, et al. Ovulated oocytes in adult mice derive from non-circulating germ cells. Nature 2006;441:1109–14.

31. Tilly JL, Johnson J. Recent arguments against germ cell renewal in the adult human ovary: is an absence of marker gene expression really acceptable evidence of an absence of oogenesis. Cell Cycle 2007;6:879–83.
32. Lee HJ, Selesniemi K, Niikura Y, et al. Bone marrow transplantation generates immature oocytes and rescues long-term fertility in a preclinical mouse model of chemotherapy-induced premature ovarian failure. J Clin Oncol 2007;25:3198–204.
33. Nayernia K, Lee JH, Drusenheimer N, et al. Derivation of male germ cells from bone marrow stem cells. Lab Invest 2006;86:654–63.
34. Dyce PW, Wen L, Li J. In vitro germline potential of stem cells derived from fetal porcine skin. Nat Cell Biol 2006;8:384–90.

Spermatogonial Stem Cells

Dirk G. de Rooij

Abstract New developments in the field of spermatogonial stem cell (SSC) research have been reviewed. Novel techniques have rendered interesting results in studies on SSC kinetics in nonprimate mammals as well as in primates, and the classical views on the nature and the behavior of SSC are being challenged. However, no definite conclusions can yet be drawn. Many new proteins have been detected that function in the pathways that regulate SSC self-renewal and differentiation. Regretfully, no specific marker for SSCs has yet been detected. Furthermore, it has become clear that SSCs are located in specific niches that are related to the vasculature that surrounds the seminiferous tubules. Great progress has been made in the development of methods for culturing mouse and bovine SSCs. These cells can now be propagated in vitro for many months, while they retain their genomic integrity and capacity to colonize a recipient mouse testis. Finally, it has become abundantly clear that at least mouse SSCs can become multipotent embryonic stem–like cells again, capable of differentiation into many other cell lineages. Future study will determine whether the latter is also possible for human SSC.

Keyword Spermatogonial stem cells • Spermatogenesis • Multipotency • Testis • Stem cell culture

Introduction

In recent years, the stem cell field has become a major topic in biomedical research. It is important to understand how self-renewal and differentiation of these cells are regulated. This, because errors in the regulatory pathways involved, may lead either

D.G. de Rooij
Center for Reproductive Medicine, Academic Medical Center, Meibergdreef 9, 1105 AZ, Amsterdam, the Netherlands
e-mail: d.g.derooij@amc.uva.nl

H. Baharvand (ed.), *Trends in Stem Cell Biology and Technology*,
DOI 10.1007/ 978-1-60327-905-5_10,
© Humana Press, a Part of Springer Science + Business Media LLC 2009

to tumor formation or exhaustion of the stem cell pool of a tissue. Spermatogonial stem cells (SSCs) are important for male fertility and recently it has been shown that at least mouse SSC are able to transform into multipotent stem cells capable of differentiation into various other cell lineages. The recent developments in this field will be reviewed.

Spermatogenesis in Nonprimate Mammals

The process of spermatogenesis takes place in the seminiferous tubules and consists of a proliferative phase, a meiotic phase, and a transformation phase. The proliferative phase is carried out by way of a series of mitotic divisions of so-called spermatogonia. All spermatogonia are localized on the basal membrane of the tubules in between the somatic supporting cells named "Sertoli cells." At the beginning of the spermatogenic process there are SSCs that can either self-renew or differentiate. In the normal epithelium, to secure the steady state, about half of the number of stem cells will self-renew and the other half will differentiate.

Studying the topographical arrangement of early spermatogonia in whole mounts of rat seminiferous tubules, Huckins (1) observed that these spermatogonia consist of clones of one (A-single or A_s spermatogonia), two (A-paired or A_{pr} spermatogonia), or chains of 4, 8, or 16 cells (A-aligned or A_{al} spermatogonia). Huckins (1) and Oakberg (2) proposed that the A_s spermatogonia are the SSCs that can self-renew and form two new stem cells or form a pair of A_{pr} spermatogonia. The A_{pr} spermatogonia stay connected by an intercellular bridge and at all further divisions during spermatogenesis cytokinesis will be incomplete. Hence, in this scheme (Fig. 1), the formation of a pair is the first step along the differentiation pathway.

Many detailed morphological and cell kinetic studies, using ³H-thymidine, in several rodents and the ram followed (3–10), and it was concluded that the numbers of clones of various length and the proliferative activity, as measured in H³-thymidine incorporation studies, were completely compatible with the scheme of spermatogonial multiplication and stem cell renewal as first proposed by Huckins (1) and Oakberg (2). Cell counts revealed the presence of about 35,000 stem cells (A_s spermatogonia) in a C3H × 101 F1 mouse testis (10).

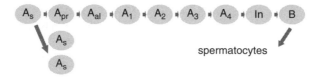

Fig. 1 Scheme of spermatogonial multiplication and stem cell renewal in rodents. The A_s (A-single) spermatogonia are the spermatogonial stem cells and can either self-renew or form a pair of A_{pr} (A-paired) spermatogonia that will differentiate and ultimately become spermatozoa. The subsequent mitotic divisions in the spermatogonial lineage are depicted. The last division in this series renders spermatocytes (spc) that carry out the meiotic divisions and form spermatids that will transform into spermatozoa

Recently, Nakagawa et al. *(11)* carried out a study in which they labeled A_s, A_{pr}, and A_{al} spermatogonia with enhance green fluorescent protein (EGFP). Having done this they were able to follow the labeled SSC in time. After 3 months only a few colonies were found and from this they concluded that there are very few actual SSC, much less than the 35,000 found in cell counts *(10)*. Together with the results of subsequent experiments Nakagawa et al. *(11)* concluded that A_{pr} and A_{al} spermatogonia can also have self-renewing capacity. However, these conclusions will need further study as their labeling technique probably preferentially labeled A_s spermatogonia that were about to differentiate. The construct with which the cells were transfected contained EGFP under the control of the promoter of Ngn3. Ngn3 is expressed by A_s, A_{pr}, and A_{al} spermatogonia, and has an unknown function in these cells, and its expression is inhibited by glial-cell line-derived, neurotrophic factor (GDNF). As described below, GDNF is the major growth factor that stimulates stem cell renewal. A low level of GDNF allows the SSC to differentiate and it also allows high levels of Ngn3 expression. Hence, the conclusions of Nakagawa et al. *(11)* will need to be confirmed in further studies.

Since 1994, it has also become possible to determine the presence of stem cells by way of a functional test, the SSC transplantation assay *(12–14)*. In this assay, cell suspensions containing SSCs are transplanted into the testes of recipient mice, the endogenous spermatogenesis of which has been removed by way of a cytostatic drug or irradiation *(12, 13, 15)*. Alternatively, W/W^v mice are used that have no endogenous spermatogenesis because mice that have this W allele of the c-Kit receptor lack SSCs. The transplanted SSCs are able to colonize the recipient mouse testis by forming repopulating colonies that can be quantified, the number of colonies being related to the number of transplanted stem cells. This method makes it possible to detect the presence of stem cells in a cell suspension and to compare stem cell numbers after various treatments or culture periods *(16, 17)*.

Regulation of Spermatogonial Stem Cell Renewal and Differentiation

In order to prevent exhaustion of the stem cell pool, stem cell renewal and differentiation must be regulated in such a way that there are always as many stem cells induced to self-renew as there are stem cells lost because of differentiation. In contrast, after cell loss recovery must take place by stimulation of self-renewal and/or inhibition of differentiation. Indeed, after irradiation during at least the first six divisions almost no stem cell differentiation takes place *(18)*. This implies that there are mechanisms that can alter the ratio between stem cell renewal and differentiation.

In recent years information has become available about the molecular mechanisms governing stem cell behavior (Fig. 2). First, an excess of GDNF in the mouse testis causes the formation of large clumps of SSCs in the seminiferous tubules, while in heterozygous GDNF knockout mice the seminiferous epithelium becomes depleted because of stem cell loss *(19–21)*. GDNF is secreted by Sertoli cells and is under the control of follicle-stimulating hormone (FSH) *(20)*, while SSCs express the

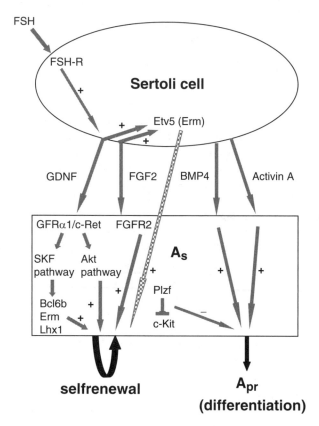

Fig. 2 Schematic representation of the present knowledge on the molecular regulation of spermatogonial stem cell renewal and differentiation in rodents. The pathway from Etv5 (Erm) expression
in Sertoli cells to stimulation of SSC renewal is presently unknown (*padded arrow*)

receptors for GDNF, c-Ret, and GFRα1 *(19)*. Recently, it was found that the effect
of GDNF on SSCs is mediated by the Akt pathway *(22)* as well as the Src family
kinase signaling pathway (SFK) *(23)*. The Akt pathway promotes self-renewing
divisions of SSCs *(22)*, while through the SFK signaling pathway GDNF up-regulates
the transcription factors Bcl6b, Erm, and Lhx1 *(23)*. The latter transcription factors are
important for the in vitro maintenance of SSCs *(23)*. Apparently, GDNF is important
for SSC self-renewal and survival through multiple pathways.

Besides GDNF, FGF2, another growth factor secreted by Sertoli cells, also plays
a role in the regulation of the behavior of SSC. Evidence for this comes from the
observation that the congenital malformation called Apert syndrome is mainly of
paternal origin in such a way that the chance of fathers to sire a child with this
malformation increases with age. This syndrome involves a gain of function mutation
in the fibroblast growth factor receptor 2 (FGFR2) gene. The mutation arises in an
SSC of the father and apparently this gives that SSC an advantage above other SSCs in
that it propagates itself by enhanced self-renewal, replacing more and more normal
SSCs. Eventually this leads to the production of more and more mutated sperm with

increasing age *(24)*. The FGFR2 was indeed found to be present in a SSC cell line *(24)*. This suggests that the molecular pathway downstream of the FGFR2 is also involved in regulating the balance between self-renewal and differentiation of SSC.

Mice that have the classical spontaneous mutation *luxoid* exhibit a progressive loss of SSCs. *Plzf* (promyelocitic leukemia zinc finger protein) was found to be the disrupted gene in the *luxoid* mutation *(25, 26)*. This finding indicates a role of this gene in the regulation of SSC renewal and differentiation. The Plzf protein is expressed by A_s, A_{pr}, and A_{al} spermatogonia. Although Plzf and Bcl6b are members of the same family of proteins and both act to prevent SSCs from differentiation, only Bcl6b is under the control of GDNF *(23, 27)*. Apparently, Plzf is involved in a different regulatory pathway. The molecular pathways downstream of Plzf and Bcl6b in SSCs have yet to be studied but, through the POZ domain, may involve the corepressors N-CoR and SMRT *(28)*. It was recently found that Plzf directly represses the transcription of the c-Kit receptor in early spermatogonia, and it was suggested that Plzf promotes SSC self-renewal by preventing the expression of c-Kit in these cells *(29)*. The c-Kit receptor is expressed in spermatogonia from late A_{al} spermatogonia onward, and after activation by its ligand SCF (stem cell factor) it is important in the transition of A_{al} spermatogonia into A1 spermatogonia *(30)*. It is possible that Plzf, by inhibiting c-Kit in earlier spermatogonial cell types, prevents differentiation into A1 spermatogonia.

Although the Sertoli cell secreted growth factors GDNF and FGF2 and the transcription factor Plzf promote self-renewal and inhibit differentiation of SSC, other factors promote differentiation. Activin A and bone morphogenic protein 4 (BMP4), when added to cultures of SSCs, reduce the maintenance of these stem cells, suggesting that these factors enhance stem cell differentiation *(31)*.

Taken together, Sertoli cells seem to control SSC behavior as these cells produce both factors that stimulate self-renewal as well as factors that promote differentiation of these cells. The next question is how Sertoli cells are directed to maintain the right balance between these factors with opposing effects. Recently, it was found that Ets-variant gene 5 (ETV5 or Erm), which is expressed in Sertoli cells, is necessary for SSC maintenance as ETV5 deficient mice quickly lose their SSCs *(32)*. Interestingly, in turn ETV5 expression in Sertoli cells is regulated by FGF2 and EGF *(33)*, and GDNF too was found to stimulate ETV5 expression *(23)*. Clearly, the regulation of SSC behavior is carried out by multiple factors via multiple pathways, all having their action within Sertoli cells. Although progress in this field is fast, no clear picture has emerged as yet about how all factors that play a role work together to ensure maintenance of the SSC population as well a constant production of differentiating cells destined to become spermatozoa.

Spermatogonial Stem Cell Niche

In several renewing tissues, stem cells reside in and depend on the presence of specific areas in the tissue called "niches." For example in the intestine, stem cells reside near the bottom of the crypts *(34)*, and stem cells in the bone marrow also occupy specific niches. In the testis, all mitotically active germ cells reside on the basal membrane of the seminiferous tubules, together with early spermatocytes and the Sertoli cells. Hence, only less than 1% of the cells on the tubule basal membrane are SSCs *(10)*. This implies that only some of the Sertoli cells could be part of such a SSC niche. When there would be a specific stem cell niche, one would have to find it in particular areas on the basal membrane. Studies on whole mounts of seminiferous tubules never suggested the presence of morphologically different Sertoli cells surrounding SSCs. Therefore, up until recently no clue existed for the existence of SSC niches.

However, detailed morphological studies in testis sections of mouse and rat have revealed that the population of A_s, A_{pr}, and A_{al} spermatogonia is preferentially located in those areas of the seminiferous tubules that border on interstitial tissue *(35, 36)*. Apparently, the vicinity of the interstitial tissue affects stem cell behavior in such a way that most early spermatogonia are formed near to it. Interestingly, high testosterone levels have been found to prevent spermatogonial differentiation *(37)*, and in normal epithelium testosterone levels will be highest in tubule areas bordering on interstitial tissue where Leydig cells reside.

Recently, the preferential position of the A_s, A_{pr}, and A_{al} spermatogonia with respect to the interstitial tissue has been superbly demonstrated by Yoshida et al. *(38)*. These authors succeeded in tagging A_s, A_{pr}, and A_{al} spermatogonia with EGFP and studied the position and behavior of these cells in vivo. A_s, A_{pr}, and A_{al} spermatogonia showed a biased localization to the vascular network, i.e., the interstitial tissue, and their differentiated descendants were shown to disperse to other regions of the seminiferous tubules *(38)*. It is noteworthy that the nature of the SSC niche in mammals is completely different from that in *Drosophila* and *Caenorhabditis elegans (38–40)*, although previously the situation in these animals has often been taken to somehow reflect that of higher animals.

Spermatogonial Stem Cell Markers

Extensive studies have been carried out to characterize mRNA and protein expression in SSCs in order to find specific markers for these cells. The first success in this direction was the finding that immunoselection using integrin-$\alpha6$ and integrin-$\beta1$, two receptors known to bind to laminin, resulted in enrichment of SSCs *(41)*. Furthermore, extracellular proteins, used for the isolation of stem cells of other tissues, were tested and it was found that rat and mouse SSCs can be enriched five- and sevenfold, respectively, by positive selection for CD9. CD9-positive cells were found on the basal membrane of the seminiferous tubules as well as in the interstitium *(42)*. Thy-1, a member of the immunoglobulin superfamily, was also shown to be highly expressed by neonatal rat SSCs *(43)*. Using immunohistochemistry, in the testis several proteins were found to be specific for early A spermatogonia: Plzf *(25, 26)*, Ngn3 *(44)*, Sox3 *(45)*, Gfrα-1 *(46, 47)*, Egr3 *(48)*, Notch1 *(46)*, and Oct3/4 *(49)*. Plzf(*Zfp145*) is expressed in A_s, A_{pr}, and A_{al} spermatogonia and Ngn3 is expressed in A_s, A_{pr}, and A_{al} spermatogonia that are c-Kit negative. However, faint NGN3 staining was also observed in some spermatocytes by another group *(45)*. Sox3 is expressed in A_s, A_{pr}, and A_{al} spermatogonia and probably plays a role in the progression of spermatogenesis *(45)*. Gfrα1 is one of the two receptors for GDNF, a growth factor that regulates the ratio of self-renewal and differentiation of SSCs, and Von Schönfeldt et al. *(46)* localized Gfrα1 to the single cell spermatogonia in the adult mouse testis. However, using Gfra-1 to isolate spermatogonia, about half of the Gfra-1-positive cells are c-Kit positive *(47)*, indicating that Gfrα-1 is not a completely specific SSC marker as SSCs lack c-Kit. Stra8 is expressed in premeiotic germ cells *(50)*. Later a 400-bp piece of the *stra8* promoter was used to direct the expression of luciferase to spermatogonia. Transplantation of germ cells selected for the expression of a marker under the control of this 400-bp piece of the *stra8* promoter led to a 700-fold enrichment of SSCs *(51)*.

Taken together, the extensive search for SSC markers has not yet led to the establishment of a really specific marker for this cell type. Generally, all genes investigated are also expressed in spermatogonia that have already taken one or more steps along the differentiation pathway, i.e., A_{pr} and A_{al} spermatogonia. Finding a specific marker for SSC would be an especially tremendous breakthrough in this area of research.

Culture of Spermatogonial Stem Cells

Already for quite some time people have been trying to develop a culture system to allow in vitro mammalian spermatogenesis *(50, 52)*. Although such a protocol has already been developed for the eel, where with the exception of the last step the whole process can take place in vitro *(53)*, for mammals these attempts have been unsuccessful. However, starting from differentiating spermatogenic cells these cells were sometimes able to proceed for one or more subsequent differentiation steps *(54–57)*. Fortunately, in recent years increasingly successful culture protocols have been developed for the culture of mouse and bovine SSC, which can be regarded as the first step of a future protocol to let the whole process of spermatogenesis take place in vitro *(21, 58–61)*. Generally, various growth factors are added to cultures of SSCs including GDNF, FGF2, epidermal growth factor (EGF), and leukemia inhibiting factor (LIF). As described above there is good reason to believe that GDNF and FGF2 enhance self-renewal of SSCs, but the possible roles of EGF and LIF in SSC cultures are still unclear. The cultures are carried out using a feeder layer of mouse embryonic fibroblasts (MEFs) *(58, 59)*, Sto cells *(62, 63)*, or the Sertoli cells, contaminating the isolated SSCs *(63, 64)*. However, it is also possible to culture the cells on laminin-coated dishes, provided serum is added to the cultures *(58)*. Finally, even anchorage-independent growth of mouse SSC has been achieved, although in that case the cells grew more slowly *(65)*.

For mouse SSC, long-term culture does not cause any genetic or epigenetic changes, and these cells remain capable of colonizing a recipient mouse testis after transplantation *(66)*. In addition, the spermatozoa arising from the SSC transplanted after long-term culture are still capable of producing normal offspring.

Some puzzling characteristics of the SSC culture systems remain. First, they have to be carried out at 37°C because SSC in culture do not survive well at 32–34°C, which is the normal testicular temperature. No data are available yet to explain this phenomenon. Second, in the successful culture system developed by Kanatsu-Shinohara et al. *(58, 59)* only a small percentage of the cultured cells are able to colonize a recipient mouse testis. This may mean that most of the cultured SSCs already have carried out a first differentiation step, making them unsuitable to give rise to colonies after transplantation. Alternatively, only SSC in a specific phase of the cell cycle are able to repopulate a recipient mouse testis. This will have to be studied in further detail, especially since this may imply that gene expression in the cultured cells is not completely representative of that in real SCCs.

Spermatogonial Stem Cell Lines

For a detailed characterization of proteins and mRNAs expressed in SSCs, it would be helpful to make use of cell lines of SSCs, especially when it cannot be determined whether all cells in SSC cultures are indeed stem cells or if one or more differentiation steps has already taken place in some of the cells. Unfortunately, it has proven difficult to generate cell lines from SSCs. Most cell lines described are formed by more differentiated cells *(67–69)*. One spermatogonial cell line was established using *telomerase* as an immortalizing gene *(70)*. Importantly, upon stimulation with SCF this cell line differentiates in vitro to spermatocytes and even spermatids. However, the cell line was not characterized for its ability to colonize a recipient testis after transplantation. Because it was positive for c-Kit, this cell line may actually be a differentiating type A spermatogonial cell line.

However, two rat cell lines are available that possess molecular SSC characteristics (GC-5spg and GC-6 spg) *(71)*. They are c-Kit negative and Oct-4 and Hsp90α positive. Furthermore, the cells are able to colonize a recipient mouse testis after transplantation. Unfortunately, so far these cell lines cannot be induced to differentiate in vitro or in vivo.

In conclusion, currently only two SSC lines are available, GC-5spg and GC-6spg *(71)*. Optimization of the culture conditions of these immortalized SSC lines might improve their capacity to study the regulation of the process of SSC self-renewal and differentiation in more detail.

Plasticity of Spermatogonial Stem Cells

In 2004 Kanatsu-Shinohara et al. *(72)* first indicated the possibility that SSCs could become pluripotent stem cells capable of forming teratomas with differentiation into mesodermal, endodermal, and neurectodermal cell lineages. This was done by culturing neonatal SSCs or SSCs from adult p53 knockout mice. During culture, besides colonies of normal SSCs, embryonic stem (ES) cell–like colonies were formed. This finding was a real breakthrough in the field and inspired a number of other studies. It was found that the multipotent stem cells derived from neonatal mouse testes are phenotypically similar to ES cells and have a similar potential to differentiate into cardiomyocytes and endothelial cells (73).

Guan et al. (74) showed ES-like cell formation in cultures of SSCs from normal 6-week-old mice, purified by selecting for Stra8 expressing germ cells. Again these ES-like cells are able to generate teratomas and to differentiate into cell lineages of all three embryonic germ layers. In subsequent work (75) this group reported that the multipotent stem cells derived in this way can produce functional cardiomyocytes both in vitro and in vivo. In the in vivo experiments no tumor formation was observed.

Comparable results were obtained by Seandel et al. (76) who purified male germ cells from adult mouse testes by using the spermatogonial surface marker Gpr125. These cells were cultured on a feeder layer in a culture medium comparable to that employed by Kanatsu-Shinohara et al. (72). After more than 3 months morphologically aberrant colonies were found that consisted of multipotent cells. Again these cells were able to form teratomas and to generate derivatives of all three germ layers. In experiments carried out by Hu et al. (77) SSCs were cultured in conditions promoting ES cells to become osteoblasts. The results suggested that the SSCs acquired characteristics of osteoblasts.

In a completely different approach Boulanger et al. (78) sorted testicular cells for the presence of the surface marker α6-integrin, which is expressed by SSCs. Together with dispersed mammary epithelial cells the cells were injected into epithelium-divested mammary fat pads. Testis cell–derived mammary epithelial cells were found, suggesting that the SSCs adapted to their new environment and gave rise to mammary epithelial cells.

Taken together, the conclusion can be drawn that indeed cultured mouse SSC can become multipotent ES-like cells that can generate various cell lineages. Intriguingly, the transition from SSC to multipotent cell can be achieved in a variety of culture conditions. An important focus of future research will be to develop a method to obtain multipotent stem cells from human SSCs. This will have great clinical implications. Interestingly, PrimeGen Biotech has announced it is able to derive human multipotent stem cells from human SSC, but this assertion has not yet been substantiated by a scientific publication (http://findarticles.com/p/articles/mi_m0EIN/is_2006_March_27/ai_n16113058).

Spermatogonial Stem Cells in Primates

In primates, as in nonprimate mammals, A and B spermatogonia are distinguished. However, the situation is more complicated in primates than in other mammals. First, the density of the A spermatogonia is too high to allow a proper study of their topographical arrangement in the normal testis as has been done for rodents.

Therefore, it is impossible to discern single spermatogonia or clones of two, four, or more cells. Second, similar to nonprimate mammals, the nuclei of the A spermatogonia do not show heterochromatin. However, some of these A spermatogonia have nuclei that are very dense, stain heavily with hematoxylin, and have a non-staining nuclear spot. The latter spermatogonia are called A_{dark} spermatogonia. The rest of the A spermatogonia stain less heavily and are called A_{pale}. Generally, in the adult there are equal numbers of pale and dark staining A spermatogonia *(79)*.

The A_{pale} spermatogonia divide once every epithelial cycle, which in the human means, once every 16 days *(80–82)*. Hence, spermatogonia in primates divide much less frequently than in other mammals. The A_{dark} spermatogonia normally do not divide and are apparently quiescent for long periods of time. The A_{dark} spermatogonia do not comprise a separate class of spermatogonia that renew themselves and give rise to differentiating type spermatogonia. In the monkey, after irradiation the A_{dark} spermatogonia transform into A_{pale} spermatogonia, probably in order to replenish the pool of A_{pale} spermatogonia that became depleted by the cell-killing effects of irradiation *(83)*. Subsequently, after transition into A_{pale} the cells start to proliferate again. During repopulation after irradiation, new A_{dark} spermatogonia are formed by A_{pale} spermatogonia *(84)*. A_{dark} spermatogonia seem to be cells set aside by the A_{pale} spermatogonia that can be recruited to become A_{pale} again when needed. Hence, the A_{dark} spermatogonia are a kind of reserve spermatogonia that are activated upon cell loss in the population of active spermatogonia. This phenomenon does not occur in the nonprimate testis.

Although in the normal epithelium the density of the A_{dark} and A_{pale} spermatogonia is such that in whole mounts it is not possible to distinguish the clones of these cells separately, after irradiation, when density is lower, it can be seen that both these types of A spermatogonia consist of single cells, pairs, and chains *(84)*. This suggests that spermatogonial multiplication and stem cell renewal in primates is in principle similar to that in other mammals, the single cells being the stem cells. In that case only the single A_{dark} and A_{pale} spermatogonia would have stem cell properties.

However, in recent years several alternative schemes of spermatogonial multiplication and stem cell renewal in primates have been published *(85–87)*. Bromodeoxyuridine (BrdU) labeling studies on monkey testes strongly suggest that in primates pairs and chains of A_{pale} spermatogonia after division can split in the middle, and in this way pairs and chains can carry out self-renewing divisions and are able to maintain a constant production of new clones of spermatogonia. In this way pairs and chains of spermatogonia can be the active stem cells in the primate seminiferous epithelium. Nevertheless, this group also found that the long-term BrdU label-retaining cells in the primate are single spermatogonia. In general, in adult tissues stem cells proliferate more slowly than their differentiating daughter cells, and because of that BrdU or H^3-thymidine incorporated in stem cells is diluted away more slowly. Therefore, long-term label-retaining cells are generally thought to be the tissue stem cells. Clearly, more studies will be needed to establish a comprehensive scheme of spermatogonial multiplication and stem cell renewal in primates.

Conclusions

Exciting progress has been made and new ideas have been developed in research on SSCs from many perspectives. New experimental possibilities have emerged to study SSC kinetics, and with respect both to nonprimate mammals and primates, the classical views on the scheme of spermatogonial multiplication and stem cell renewal are being challenged. Time will reveal which hypotheses will prevail. New genes involved in the pathways that regulate SSC behavior are detected regularly, although a marker really specific for SSCs alone has not yet been detected. Furthermore, it has become clear that SSCs, like stem cells in many other tissues, are located in specific niches, which are related to the vasculature that surrounds the seminiferous tubules. Great progress is being made with culturing mouse and bovine SSCs. These cells can now be propagated in vitro for many months and even a couple of years, while they retain their genomic integrity and capacity to colonize a recipient mouse testis. Finally, one of the most exciting developments is that it has become abundantly clear that at least mouse SSCs can become multipotent ES-like stem cells again, capable of differentiation into many other cell lineages. Future research will indicate whether this can be done with human SSC. If so, this would be of tremendous clinical importance.

References

1. Huckins C. The spermatogonial stem cell population in adult rats I. Their morphology, proliferation and maturation. Anat Rec 1971;169:533–57.
2. Oakberg EF. Spermatogonial stem-cell renewal in the mouse. Anat Rec 1971;169:515–31.
3. Huckins C. The spermatogonial stem cell population in adult rats. II. A radioautographic analysis of their cell cycle properties. Cell Tissue Kinet 1971;4:313–34.
4. Huckins C. Cell cycle properties of differentiating spermatogonia in adult Sprague–Dawley rats. Cell Tissue Kinet 1971;4:139–54.
5. Lok D, deRooij DG. Spermatogonial multiplication in the Chinese hamster I. Cell cycle properties and synchronization of differentiating spermatogonia. Cell Tissue Kinet 1983;16:7–18.
6. Lok D, de Rooij DG. Spermatogonial multiplication in the Chinese hamster. III. Labelling indices of undifferentiated spermatogonia throughout the cycle of the seminiferous epithelium. Cell Tissue Kinet 1983;16:31–40.
7. Lok D, Jansen MT, de Rooij DG. Spermatogonial multiplication in the Chinese hamster. II. Cell cycle properties of undifferentiated spermatogonia. Cell Tissue Kinet 1983;16:19–29.
8. Lok D, Jansen MT, de Rooij DG. Spermatogonial multiplication in the Chinese hamster. IV. Search for long cycling stem cells. Cell Tissue Kinet 1984;17:135–43.
9. Lok D, Weenk D, de Rooij DG. Morphology, proliferation, and differentiation of undifferentiated spermatogonia in the Chinese hamster and the ram. Anat Rec 1982;203:83–99.
10. Tegelenbosch RAJ, de Rooij DG. A quantitative study of spermatogonial multiplication and stem cell renewal in the C3H/101 F1 hybrid mouse. Mutat Res 1993;290:193–200.
11. Nakagawa T, Nabeshima Y, Yoshida S. Functional identification of the actual and potential stem cell compartments in mouse spermatogenesis. Dev Cell 2007;12:195–206.
12. Brinster RL, Avarbock MR. Germline transmission of donor haplotype following spermatogonial transplantation. Proc Natl Acad Sci U S A 1994;91:11303–7.

13. Brinster RL, Zimmermann JW. Spermatogenesis following male germ-cell transplantation. Proc Natl Acad Sci U S A 1994;91:11298–302.
14. Ogawa T, Arechaga JM, Avarbock MR, Brinster RL. Transplantation of testis germinal cells into mouse seminiferous tubules. Int J Dev Biol 1997;41:111–22.
15. Creemers LB, Meng X, Den Ouden K, et al. Transplantation of germ cells from glial cell line-derived neurotrophic factor-overexpressing mice to host testes depleted of endogenous spermatogenesis by fractionated irradiation. Biol Reprod 2002;66:1579–84.
16. Shinohara T, Orwig KE, Avarbock MR, Brinster RL. Spermatogonial stem cell enrichment by multiparameter selection of mouse testis cells. Proc Natl Acad Sci U S A 2000;97:8346–51.
17. Nagano MC. Homing efficiency and proliferation kinetics of male germ line stem cells following transplantation in mice. Biol Reprod 2003;69:701–7.
18. van Beek MEAB, Meistrich ML, de Rooij DG. Probability of self-renewing divisions of spermatogonial stem cells in colonies, formed after fission neutron irradiation. Cell Tissue Kinet 1990;23:1–16.
19. Meng X, Lindahl M, Hyvonen ME, et al. Regulation of cell fate decision of undifferentiated spermatogonia by GDNF. Science 2000;287:1489–93.
20. Tadokoro Y, Yomogida K, Ohta H, Tohda A, Nishimune Y. Homeostatic regulation of germinal stem cell proliferation by the GDNF/FSH pathway. Mech Dev 2002;113:29–39.
21. Kubota H, Avarbock MR, Brinster RL. Growth factors essential for self-renewal and expansion of mouse spermatogonial stem cells. Proc Natl Acad Sci U S A 2004;101:16489–94.
22. Lee J, Kanatsu-Shinohara M, Inoue K, et al. Akt mediates self-renewal division of mouse spermatogonial stem cells. Development 2007;134:1853–9.
23. Oatley JM, Avarbock MR, Brinster RL. Glial cell line-derived neurotrophic factor regulation of genes essential for self-renewal of mouse spermatogonial stem cells is dependent on Src family kinase signaling. J Biol Chem 2007;282:25842–51.
24. Goriely A, McVean GAT, van Pelt AMM, et al. Gain-of-function amino acid substitutions drive positive selection of FGFR2 mutations in human spermatogonia. Proc Natl Acad Sci U S A 2005;102:6051–6.
25. Buaas FW, Kirsh AL, Sharma M, et al. Plzf is required in adult male germ cells for stem cell self-renewal. Nat Genet 2004;36:647–52.
26. Costoya JA, Hobbs RM, Barna M, et al. Essential role of Plzf in maintenance of spermatogonial stem cells. Nat Genet 2004;36:653–9.
27. Oatley JM, Avarbock MR, Telaranta AI, Fearon DT, Brinster RL. Identifying genes important for spermatogonial stem cell self-renewal and survival. Proc Natl Acad Sci U S A 2006;103:9524–9.
28. Payne C, Braun RE. Glial cell line-derived neurotrophic factor maintains a POZ-itive influence on stem cells. Proc Natl Acad Sci U S A 2006;103:9751–2.
29. Filipponi D, Hobbs RM, Ottolenghi S, et al. Repression of kit expression by Plzf in germ cells. Mol Cell Biol 2007;27:6770–81.
30. Schrans-Stassen BHGJ, van de Kant HJG, de Rooij DG, van Pelt AMM. Differential expression of c-Kit in mouse undifferentiated and differentiating type A spermatogonia. Endocrinology 1999;140:5894–900.
31. Nagano M, Ryu BY, Brinster CJ, Avarbock MR, Brinster RL. Maintenance of mouse male germ line stem cells in vitro. Biol Reprod 2003;68:2207–14.
32. Morrow C, Hostetler C, Griswold M, et al. ETV5 is required for continuous spermatogenesis in adult mice and may mediate blood-testes barrier function and testicular immune privilege. Ann N Y Acad Sci 2007;1120:144–51.
33. Simon L, Ekman GC, Tyagi G, Hess RA, Murphy KM, Cooke PS. Common and distinct factors regulate expression of mRNA for ETV5 and GDNF, Sertoli cell proteins essential for spermatogonial stem cell maintenance. Exp Cell Res 2007;313:3090–9.
34. Barker N, van Es JH, Kuipers J, et al. Identification of stem cells in small intestine and colon by marker gene Lgr5. Nature 2007;449:1003–7.
35. Chiarini-Garcia H, Hornick JR, Griswold MD, Russell LD. Distribution of type A spermatogonia in the mouse is not random. Biol Reprod 2001;65:1179–85.
36. Chiarini-Garcia H, Raymer AM, Russell LD. Non-random distribution of spermatogonia in rats: evidence of niches in the seminiferous tubules. Reproduction 2003;126:669–80.

37. Meistrich ML, Shetty G. Inhibition of spermatogonial differentiation by testosterone. J Androl 2003;24:135–48.

38. Yoshida S, Sukeno M, Nabeshima Y. A vasculature-associated niche for undifferentiated spermatogonia in the mouse testis. Science 2007;317:1722–6.

39. Li L, Xie T. Stem cell niche: structure and function. Annu Rev Cell Dev Biol 2005;21:605–31.

40. Tulina N, Matunis E. Control of stem cell self-renewal in Drosophila spermatogenesis by JAK-STAT signaling. Science 2001;294:2546–9.

41. Shinohara T, Avarbock MR, Brinster RL. Beta(1)- and alpha(6)-integrin are surface markers on mouse spermatogonial stem cells. Proc Natl Acad Sci U S A 1999;96:5504–9.

42. Kanatsu-Shinohara M, Toyokuni S, Shinohara T. CD9 Is a surface marker on mouse and rat male germline stem cells. Biol Reprod 2004;70:70–5.

43. Ryu BY, Orwig KE, Kubota H, Avarbock MR, Brinster RL. Phenotypic and functional characteristics of spermatogonial stem cells in rats. Dev Biol 2004;274:158–70.

44. Yoshida S, Takakura A, Ohbo K, et al. Neurogenin3 delineates the earliest stages of spermatogenesis in the mouse testis. Dev Biol 2004;269:447–58.

45. Raverot G, Weiss J, Park SY, Hurley L, Jameson JL. Sox3 expression in undifferentiated spermatogonia is required for the progression of spermatogenesis. Dev Biol 2005;283:215–25.

46. von Schönfeldt V, Wistuba J, Schlatt S. Notch-1, c-Kit and GFRalpha-1 are developmentally regulated markers for premeiotic germ cells. Cytogenet Genome Res 2004;105:235–9.

47. Hofmann MC, Braydich Stolle L, Dym M. Isolation of male germ-line stem cells; influence of GDNF. Dev Biol 2005;279:114–24.

48. Hamra FK, Schultz N, Chapman KM, et al. Defining the spermatogonial stem cell. Dev Biol 2004;269:393–410.

49. Pesce M, Wang X, Wolgemuth DJ, Scholer H. Differential expression of the Oct-4 transcription factor during mouse germ cell differentiation. Mech Dev 1998;71:89–98.

50. Oulad Abdelghani M, Bouillet P, Decimo D, et al. Characterization of a premeiotic germ cell-specific cytoplasmic protein encoded by Stra8, a novel retinoic acid-responsive gene. J Cell Biol 1996;135:469–77.

51. Giuili G, Tomljenovic A, Labrecque N, Oulad-Abdelghani M, Rassoulzadegan M, Cuzin F. Murine spermatogonial stem cells: targeted transgene expression and purification in an active state. EMBO Rep 2002;3:753–9.

52. Aponte PM, van Bragt MPA, de Rooij DG, van Pelt AMM. Spermatogonial stem cells: characteristics and experimental possibilities. APMIS 2005;113:727–42.

53. Miura T, Ando N, Miura C, Yamauchi K. Comparative studies between in vivo and in vitro spermatogenesis of Japanese eel (*Anguilla japonica*). Zool Sci 2002;19:321–9.

54. Lee JH, Gye MC, Choi KW, et al. In vitro differentiation of germ cells from nonobstructive azoospermic patients using three-dimensional culture in a collagen gel matrix. Fertil Steril 2007;87:824–33.

55. Lee JH, Kim HJ, Kim H, Lee SJ, Gye MC. In vitro spermatogenesis by three-dimensional culture of rat testicular cells in collagen gel matrix. Biomaterials 2006;27:2845–53.

56. Movahedin M, Ajeen A, Ghorbanzadeh N, Tiraihi T, Valojerdi MR, Kazemnejad A. In vitro maturation of fresh and frozen–thawed mouse round spermatids. Andrologia 2004;36:269–76.

57. Tesarik J, Mendoza C, Greco E. In-vitro maturation of immature human male germ cells. Mol Cell Endocrinol 2000;166:45–50.

58. Kanatsu-Shinohara M, Miki H, Inoue K, et al. Long-term culture of mouse male germline stem cells under serum-or feeder-free conditions. Biol Reprod 2005;72:985–91.

59. Kanatsu-Shinohara M, Ogonuki N, Inoue K, et al. Long-term proliferation in culture and germline transmission of mouse male germline stem cells. Biol Reprod 2003;69:612–6.

60. Kubota H, Brinster RL. Technology insight: in vitro culture of spermatogonial stem cells and their potential therapeutic uses. Nat Clin Pract Endocrinol Metab 2006;2:99–108.

61. Izadyar F, Den Ouden K, Creemers LB, Posthuma G, Parvinen M, De Rooij DG. Proliferation and differentiation of bovine type a spermatogonia during long-term culture. Biol Reprod 2003;68:272–81.

62. Nagano M, Avarbock MR, Leonida EB, Brinster CJ, Brinster RL. Culture of mouse spermatogonial stem cells. Tissue Cell 1998;30:389–97.

63. Kubota H, Avarbock MR, Brinster RL. Culture conditions and single growth factors affect fate determination of mouse spermatogonial stem cells. Biol Reprod 2004;71:722–31.
64. Aponte PM, Soda T, van de Kant HJG, de Rooij DG. Basic features of bovine spermatogonial culture and effects of glial cell line-derived neurotrophic factor. Theriogenology 2006;65:1828–47.
65. Kanatsu-Shinohara M, Inoue K, Lee J, et al. Anchorage-independent growth of mouse male germline stem cells in vitro. Biol Reprod 2006;74:522–9.
66. Kanatsu-Shinohara M, Ogonuki N, Iwano T, et al. Genetic and epigenetic properties of mouse male germline stem cells during long-term culture. Development 2005;132:4155–63.
67. Hofmann MC, Narisawa S, Hess RA, Millan JL. Immortalization of germ cells and somatic testicular cells using the SV40 large T antigen. Exp Cell Res 1992;201:417–35.
68. Hofmann MC, Hess RA, Goldberg E, Millan JL. Immortalized germ cells undergo meiosis in vitro. Proc Natl Acad Sci U S A 1994;91:5533–7.
69. Tascou S, Nayernia K, Samani A, et al. Immortalization of murine male germ cells at a discrete stage of differentiation by a novel directed promoter-based selection strategy. Biol Reprod 2000;63:1555–61.
70. Feng LX, Chen Y, Dettin L, et al. Generation and in vitro differentiation of a spermatogonial cell line. Science 2002;297:392–5.
71. van Pelt AMM, Roepers-Gajadien HL, Gademan IS, Creemers LB, de Rooij DG, van Dissel-Emiliani FMF. Establishment of cell lines with rat spermatogonial stem cell characteristics. Endocrinology 2002;143:1845–50.
72. Kanatsu-Shinohara M, Inoue K, Lee J, et al. Generation of pluripotent stem cells from neonatal mouse testis. Cell 2004;119:1001–12.
73. Baba S, Heike T, Umeda K, et al. Generation of cardiac and endothelial cells from neonatal mouse testis-derived multipotent germline stem cells. Stem Cells 2007;25:1375–83.
74. Guan K, Nayernia K, Maier LS, et al. Pluripotency of spermatogonial stem cells from adult mouse testis. Nature 2006;440:1199–203.
75. Guan K, Wagner S, Unsold B, et al. Generation of functional cardiomyocytes from adult mouse spermatogonial stem cells. Circ Res 2007;100:1615–25.
76. Seandel M, James D, Shmelkov SV, et al. Generation of functional multipotent adult stem cells from GPR125+ germline progenitors. Nature 2007;449:346–50.
77. Hu H-M, Xu F-C, Li W, Wu S-H. Biological characteristics of spermatogonial stem cells cultured in conditions for osteoblasts. J Clin Rehab Tissue Eng Res 2007;11:6611–4.
78. Boulanger CA, Mack DL, Booth BW, Smith GH. Interaction with the mammary microenvironment redirects spermatogenic cell fate in vivo. Proc Natl Acad Sci U S A 2007;104:3871–6.
79. de Rooij DG, Russell LD. All you wanted to know about spermatogonia but were afraid to ask. J Androl 2000;21:776–98.
80. Clermont Y. The cycle of the seminiferous epithelium in man. Am J Anat 1963;112:35–51.
81. Clermont Y. Spermatogenesis in man. A study of the spermatogonial population. Fertil Steril 1966;17:705–21.
82. Clermont Y. Renewal of spermatogonia in man. Am J Anat 1966;118:509–24.
83. van Alphen MMA, van de Kant HJG, de Rooij DG. Depletion of the spermatogonia from the seminiferous epithelium of the rhesus monkey after X irradiation. Radiat Res 1988;113:473–86.
84. van Alphen MMA, van de Kant HJG, de Rooij DG. Repopulation of the seminiferous epithelium of the rhesus monkey after X irradiation. Radiat Res 1988;113:487–500.
85. Ehmcke J, Schlatt S. A revised model for spermatogonial expansion in man: lessons from non-human primates. Reproduction 2006;132:673–80.
86. Ehmcke J, Luetjens CM, Schlatt S. Clonal organization of proliferating spermatogonial stem cells in adult males of two species of non-human primates, *Macaca mulatta* and *Callithrix jacchus*. Biol Reprod 2005;72:293–300.
87. Ehmcke J, Simorangkir DR, Schlatt S. Identification of the starting point for spermatogenesis and characterization of the testicular stem cell in adult male rhesus monkeys. Hum Reprod 2005;20:1185–93.

Regulated Transcripts and Coregulated microRNAs in Male Spermatogonial Stem Cells

Virginie Olive, François Cuzin, and Minoo Rassoulzadegan

Abstract The spermatogonial stem cells (SSCs) are a key element in the biology of the germ line, critical for the individual as for the species. A precise balance between self-renewal and differentiation maintains the homeostasis of the testis. Identification of protein-coding transcripts and microRNAs (miRs) modulated in SSCs may lead to a better understanding of the molecular events critical for the maintenance of fertility, but this approach is hampered by the small size of the SSC population. We established a simple and efficient purification procedure starting from transgenic mice that express on the cell surface a neutral heterologous protein. Here we describe a gene expression profile of the adult SSC population, including both up- and down-regulated protein-coding transcripts and several differentially expressed miRs. We found 495 transcripts enriched in SSCs as compared with the bulk of differentiated germ cells and 133 decreased in abundance. Applying ontology criteria revealed candidate genes for a regulatory function in SSCs. A search in the available databases identified several miRs species, each one potentially interacting with a group of protein-coding transcripts either up- or down-regulated in the purified fraction. Among these candidates, quantitative reverse transcription-polymerase chain reaction (RT-PCR) assays confirmed the differential expression in the stem cells of miR-125, miR-141, and miR-181, each one either up- or down-regulated in the direction as its protein-coding targets.

Keywords Spermatogonial stem cell • microRNAs • Transcription

V. Olive, F. Cuzin, and M. Rassoulzadegan (✉)
Inserm U636, F-06108 Nice, France; Université de Nice–Sophia Antipolis, Laboratoire de Génétique du Développement Normal et Pathologique, F-06108 Nice, France; Equipe Labellisée Ligue Nationale Contre le Cancer, F-06108 Nice, France
e-mail: minoo@unice.fr

H. Baharvand (ed.), *Trends in Stem Cell Biology and Technology*,
DOI 10.1007/ 978-1-60327-905-5_11,
© Humana Press, a Part of Springer Science+Business Media LLC 2009

Introduction

Differentiation of the germ line includes both meiotic recombination and the generation of mature gametes. In the male as in the female, it is a stepwise process. In the testis, it is also a lifelong process, which starts with a strict periodicity from a pool of mitotically active diploid cells, the spermatogonia *(1)*. This pool itself is maintained by the entry of daughter cells issued from a small number of spermatogonial stem cells (SSCs) *(2)*. As with the hematopoietic stem cell reviewed in Charbord *(3)*, SSCs are defined by a rigorous experimental criterion and their ability to generate a functional tissue after transfer into a deficient, in this case, a sterile host *(4, 5)*. But while transfer and reconstitution experiments can be performed with crude or partially purified cell fractions, further analysis will in a number of instances require isolation of the stem cells. We reported the purification of a homogeneous SSC fraction fully active in the transplantation assay *(6)*. Unlike the promoter of the endogenous gene, expressed in the whole spermatogonial compartment, a fragment from the *Stra8* locus directed a transgene expression to a small number of cells whose morphology and localization in the tubule corresponded to the stem population. Purification close to homogeneity was achieved from transgenic mice in which this promoter directs expression of a neutral heterologous surface marker comprised of two domains of the human CD4 (huCD4) antigen linked to the transmembrane and cytoplasmic regions of the influenza virus hemagglutinin (CD4HAGlo). Expressing cells, which could then be purified by magnetic sorting, were homogeneous in shape and size and expressed the known markers of the stem population. They exhibited high efficiencies of colonization of recipient sterile testes, making them a unique material to investigate the distinctive properties of the SSC. In spite of their small number, we could prepare the necessary amounts of RNA at a sufficient degree of purity for a large-scale analysis on DNA array.

Expression profiles of the adult SSC population are obviously an important starting point, since identification of the transcripts either uniquely or preferentially present in this fraction may lead to that of the protein responsible for the "stemness" character. We also extended the analysis to the microRNAs (miRs). These small, 22-nt-long noncoding gene products were recently recognized as playing a role in the control of gene expression (reviewed in Alvarez-Garcia and Miska *(7)*). They regulate at the posttranscriptional level a variety of biological processes, including developmental timing, signal transduction, tissue differentiation and maintenance, diseases, and carcinogenesis. Emerging evidence demonstrates that miRs also play an essential role in stem cell self-renewal and differentiation. The limited amounts of RNA available combined with the small sizes of miRs made a direct screen for miR expression patterns unpractical. We attempted to gain relevant information by using computational analysis to evaluate the number of putative target genes for a given miR among the transcripts of protein-coding genes either enriched or repressed in SSCs. The miRs for which more targets were determined in the purified fraction than the numbers estimated from their frequency in the whole genome were, in three instances, found by quantitative reverse transcription-polymerase chain reaction (RT-PCR) to be differentially expressed in the stem fraction.

Materials and Methods

Transgenic Mice

Investigations were conducted in accordance with French and European rules for the care and use of laboratory animals. Founder transgenic animals bearing the pStra8 HAglo construct in C57BL/6 mice have been previously described (*6*). Progeny was genotyped by PCR analysis of tail DNA using primers hCD4F: GGACGGA ATTCCCATCGATCAAGGCCACAATGAACCG and hCD4R GGAGCTACCT CCGCTGGAGGCCTTCTGGA.

Immunomagnetic Sorting of SSCs

Total germ cells were prepared as described previously (*8*). Immunomagnetic isola tion of CD4-positive cells from total testicular cells of Stra8-CD4HAglo mice was performed using the CD4 Positive Isolation Kit (Dynal, Oslo, Norway) according to the manufacturer's instructions. For each preparation, 50 transgenic males were sacrificed in order to obtain enough stem cells for RNA preparation (20,000 stem cells per testis).

RNA Extraction

Total RNA was extracted using Trizol (Invitrogen, Carlsbad, CA) according to the manufacturer's instructions. Ten to 20 µg of RNA were obtained per SSC preparation. The quality of RNA was checked by electrophoresis by agarose gel electrophoresis and by the Agilent Bioanalyzer system (Agilent Technologies, Santa Clara, CA).

Quantitative Real-Time PCR

cDNA samples from SSC and total fractions were analyzed by real-time PCR using the qPCR MasterMix Plus for SYBR green (Eurogentec, Seraing, Belgium) and an ABI Prism 7000 (Applied Biosystem, Foster City, CA) sequence detector according to the manufacturer's protocol. For each gene of interest, each amplification was

Table 1 Oligonucleotides primers of eight genes considered as representative of the "enriched spermatogonial stem cell class"

Description and oligonucleotide sequences	Symbol	No. of °unigene
CD97 antigen	*Cd97*	334,648
F: CTCGGAGAAGGAATTGATCAC		
R: CTTCTGAGTCTCCCATCTGTG		
Stromal cell derived factor receptor 1 (neuroplastin)	*Sdfr1*	15,125
F: GAATTGTCACCAGTGAAGAGG		
R: CATCATAGCATCCTGCCCTTC		
FBJ osteosarcoma oncogene	*Fos*	246,513
F: CAGCCTTTCCTACTACCATTC		
R: CTGACACGGTCTTCACCATTC		
CD81 antigen	*Cd81*	806
F: GGGCATCTACATTCTCATTGC		
R: GTATGGTGGTAGTCAGTGTGG		
NIMA (never in mitosis gene a)-related	*Nek7*	143,817
F: GAGCATCCTGTCTCTTGGATG		
R: GACTCTTCGAGAATGCATGTG		
Cell division cycle 42 homolog	*Cdc42*	1,022
F: GGCAAGAGGATTATGACAGAC		
R: CAGCAGTCTCTGGAGTAATAG		
CD14 antigen	*Cd14*	3,460
F: GGAAGCCAGAGAACACCATCG		
R: GCAGGGCTCCGAATAGAATCC		
Yamaguchi sarcoma viral (v-yes-1) oncogene homolog	*Lyn*	317,331
F: GAGAGAAGGCTTCATCCCCAG		
R: GTCTTCACAGCCACCTTTGTG		

performed by applying the comparative C_t method, following the manufacturer's protocol. Primer sequences for Cd97, Sdfr1, c-fos, Cd81, Nek7, Cdc42, Cd14, and Lyn are listed in Table 1.

Microarray Analysis

Three independent RNA preparations were analyzed by in vitro amplification and hybridization to DNA microarrays (Affymetrix Mouse Genome 430A 2.0, Affymetrix, Santa Clara, CA) containing about 14,000 genes. Absolute and comparison analyses of the mouse oligonucleotide arrays were conducted using the statistics-based Affymetrix software MAS-5.0 (GeneChip Software MAS-5.0) with the default settings. The output of the GeneChip software for each microarray provided "Signal" values and "Detection" calls, "present" (P), "marginal" (M), or "absent" (A), for each probe set. The detection calls are based on statistical calculations of the difference in hybridization signals between PM (perfect match) and their control MM (mismatch) probe cells. The MAS-5.0 software provided P-values for the "detection calls" and defined a signal (arbitrary units of fluorescence) for genes/ESTs being expressed, as "present" (P) with $P < 0.05$, "marginal" (M) with P-value in the range 0.05–0.065, and "absent" (A) with $P > 0.065$. Before comparing data on microarrays, the average chip signal was normalized to an arbitrary value of 100. Affymetrix software uses two different and separate algorithms for evaluation of alterations in levels of gene expression between chips (e.g., SSCs and testicular differentiated cells) to calculate significant changes and change quantity metrics for every probe set. A change algorithm generates "Change calls" with the change P-values categorized by cutoff values called γ_1 ($\gamma_1 = 0.002$) and γ_2 ($\gamma_2 = 0.0027$). These cut-offs provide boundaries for the Change calls: "Increase," "Decrease," "Marginal Increase/Decrease," and "No Change," and a second algorithm produces a quantitative estimate of the change in gene expression in the form of "Signal Log Ratio" (Log fold change, see Gene-Chip_Software MAS-5.0 for details). Data were exported and analyzed in Excel (Microsoft).

Quantitative RT-PCR for microRNA Detection

RNA was polyadenylated with ATP by poly(A) polymerase (PAP) at 37°C for 1 h in a 20 µl reaction mixture following the manufacturer's directions for the Poly(A) Tailing kit (Ambion, Austin, TX). After phenol-chloroform extraction and ethanol precipitation, RNAs were dissolved in RNAse-free water and reverse transcribed with 200 U SuperScript II Reverse transcriptase (Invitrogen, Carlsbad, CA) and 0.5 µg poly(T) adapter (3' rapid amplification of complementary DNA ends [RACE] adapter in the FirstChoice RLM-RACE kit; Ambion, Austin, TX) according to the manufacturer's protocols (Invitrogen, Carlsbad, CA). Sequence of poly(T) adapter is GCGAGCACA GAATTAATACGACTCACTATAGG(T)12VN (V = A,G,C; N = A,T,G,C).

For the real-time PCR, the reverse primer (GCGAGCACAGAATTAATACGAC) was a 3' adapter primer (3' RACE outer primer in the FirstChoice RLM-RACE kit), and the forward primer was designed based on the entire miRNA sequence. Real-time PCR was performed using the qPCR MasterMix Plus for SYBR green and ABI Prism 7000 sequence detector: 15 s at 95°C, 15 s at a temperature 5°C below the primer's true T_m, and 20 s at 72°C for 45 cycles followed by the thermal denaturing step to generate the dissociation curves to verify amplification specificity. For each miR, amplification was performed by applying the comparative C_t method, as described by the manufacturer's protocol. 5.8S (ACGTCTGCCTGGGTGTCACAA) was used to normalized the results.

Results

DNA Array Analysis of Purified Fractions

For each RNA preparation, 50 adult Stra8HAGlo transgenic males (3–4 weeks old) were sacrificed. Total testicular cell suspensions were fractionated by immunomagnetic sorting to isolate the cells displaying the human CD4 epitope. Homogeneity of the huCD4-positive fraction had been previously established by a variety of markers *(6)* and was routinely checked (data not shown) by the combination of two criteria: in situ immunofluorescence detection in all the huCD4+ cells of the Stra8 protein, a marker of both the stem cells and the differentiated spermatogonia *(9)*, and RT-PCR analysis showing the absence of *Kit* RNA, a marker of spermatogonia absent in stem cells *(10)*. Yields of RNA were in the range of 10–20 μg per preparation. Three independent preparations were amplified in vitro and hybridized to DNA microarrays containing about 14,000 genes, in parallel with RNA isolated and amplified from the bulk of cells that had not been retained on the antiCD4-magnetic beads.

The number of transcripts responding to the criteria established for differential expression (see the section "Materials and Methods") was 495. A comparison with the published transcriptome of the embryonic testis showed that a large fraction (more than 55%) of these genes are expressed in the testis before birth *(11)*, in agreement with the notion that the prenatal gonocytes are the immediate precursors of SSCs. A smaller group was selected for further studies by applying ontology criteria with an emphasis on genes encoding known and potential transcription factors and elements of the signal transduction pathways (Table 2 and Fig. 1).

Table 2 Transcripts encoding candidate regulatory proteins up-regulated in spermatogonial stem cells

Description	Symbol	No. of unigene
Transcription factors		
Aryl-hydrocarbon receptor	*Ahr*	341,377
Catenin beta	*Ctnnb1*	291,928
Creg: cellular repressor of E1A-stimulated genes	*Creg1*	294,885
Delta sleep inducing peptide, immunoreactor	*Tsc22d3*	22,216
Early growth response 1	*Egr1*	181,959
ELK3, member of ETS oncogene family isoform b	*Elk3*	4,454
FBJ osteosarcoma oncogene	*Fos*	246,513
Forkhead box Q1	*Foxq1*	44,235
Interferon regulatory factor 7	*Irf7*	3,233
Jun oncogene	*Jun*	275,071
Myocyte enhancer factor 2A isoform 2	*Mef2a*	132,788
Myocyte enhancer factor 2C	*Mef2c*	24,001
Nuclear factor, erythroid derived 2, like 2	*Nfe2l2*	1,025
Reproductive homeobox 2	*Rhox2*	391,240
SFFV proviral integration 1	*Sfpi1*	1,302
Signal transducer and activator of transcription 6	*Stat6*	121,721
Thyroid hormone receptor alpha	*Thra*	265,917
Transcription elongation factor A (SII) 1	*Tceal*	207,263
Transcription factor 21	*Tcf21*	16,497
RNA binding proteins		
Aconitase 1	*Aco1*	331,547
DEAD (Asp-Glu-Ala-Asp) box polypeptide 5	*Ddx5*	220,038
Muscleblind-like 1	*Mbnl1*	255,723
Nucleolin	*Ncl*	154,378
Quaking protein	*Qk*	384,135

(continued)

Table 2 (continued)

Description	Symbol	No. of unigene
Ribosomal protein S4, X-linked	*Rps4x*	66
RNA binding motif protein 3	*Rbm3*	128,512
Cell cycle		
AXL receptor tyrosine kinase	*Axl*	4,128
Bridging integrator 1	*Bin1*	4,383
Colony stimulating factor 1 receptor	*Csf1r*	22,574
FBJ osteosarcoma oncogene	*Fos*	246,513
Heat shock protein 8	*Hspa8*	336,743
Integrin beta 1 (fibronectin receptor beta)	*Itgb1*	263,396
Jun oncogene	*Jun*	275,071
Microtubule-actin crosslinking factor 1	*Macf1*	3,350
Peripheral myelin protein	*Pmp22*	1,237
PYD and CARD domain containing	*Pycard*	24,163
Regulator of G-protein signaling 2	*Rgs2*	28,262
Ribosomal protein S4, X-linked	*Rps4x*	66
Stromal antigen 2	*Stag2*	290,422
Thyroid hormone receptor alpha	*Thra1*	265,917
Yamaguchi sarcoma viral (v-yes-1) oncogene homolog	*Lyn*	317,331
Intracellular signaling		
Cornichon homolog	*Cnih*	3,261
Phosphoprotein enriched in astrocytes 15	*Pea15*	544
Pleckstrin	*Plek*	98,232
Ras and Rab interactor 2	*Rin2*	228,799
RAS-related C3 botulinum substrate 1	*Rac1*	292,510
Signal transducer and activator of transcription 6	*Stat6*	121,721
Sorting nexin 5	*Snx5*	273,379
Sorting nexin 9	*Snx9*	89,515
Src-like adaptor	*Sla*	7,601
Yamaguchi sarcoma viral (v-yes-1) oncogene homolog	*Lyn*	317,331
Transcripts enriched in four distinct classes of stem cells (SSCs, ESCs, HSCs, NSCs)		
Acetyl-coenzyme A dehydrogenase	*Acadm*	10,530
Disintegrin and metalloprotease domain 9	*Adam9*	28,908
FK506 binding protein 9	*Fkbp9*	20,943
MyoD family inhibitor domain containing	*Mdfic*	1,314
Syntaxin binding protein 3A	*Stxbp3a*	316,894
Tyrosine 3-monooxygenase	*Ywhab*	34,319

The resulting short list comprises transcripts encoding 19 known or potential transcription control factors, 7 RNA binding proteins, 16 involved in cell cycle regulation, and 10 in intracellular signaling. On the other hand, 127 transcripts were either not represented in SSCs or detected at significantly lower levels than in the differentiated fraction. As a control of the procedure, we noted in the latter class the occurrence of genes known to be expressed either at later stages during spermatogenesis or in the somatic compartment of the testis. Neither the patterns of expression nor the possible functions of other genes listed as "down-regulated in SSCs" were investigated further at this stage.

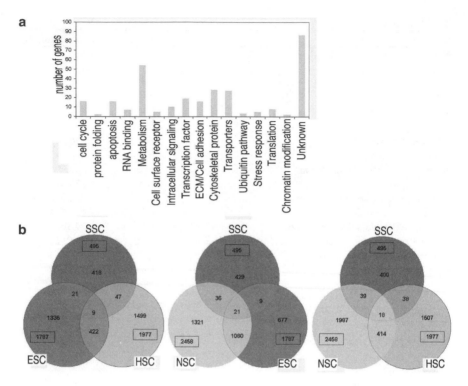

Fig. 1 Ontology of the genes enriched in the spermatogonial stem cell (SSC) fraction and overlap with genes previously reported as "stem cell-specific." (**a**) Ontology of the genes enriched in the SSC fraction according to National Center for Biotechnology Information (NCBI) classification (http://www.ncbi.nlm.nih.gov/locuslink). (**b**) The Venn diagrams show the number of gene products enriched in SSCs, neural stem cells (NSC), embryonic stem cells (ESC), and hematopoietic stem cells (HSC), and their overlaps results for ESCs, NSCs, and HSCs from (15). (**c**) Only six genes are found to be enriched in all the stem cells analyzed (Table 2)

For a series of eight genes considered as representative of the "enriched in SSC" class, oligonucleotide primers were designed and tested for detection of the corresponding RNAs (Table 1). Quantitative real-time PCR determinations confirmed in every instance their differential expression (Fig. 2).

Intersection of Embryonic Testis and SSC Transcriptomes

Previous analysis of the time course of gene expression during the development of the testis identified transcripts present in greater abundance at successive developmental stages (11). As shown in Table 3, a large fraction (55%) of the genes in the list of the 495 genes "up-regulated in SSCs" appears as being highly expressed in the embryonic testis, from the undifferentiated gonad (13.5 *dpc*) up to the gonocyte

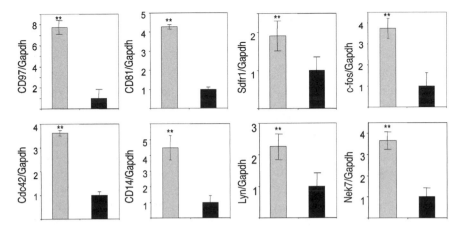

Fig. 2 Validation by quantitative reverse transcription-polymerase chain reaction (RT-PCR) of the differential expression of representative genes identified by array hybridization (*open bars* spermatogonial stem cells, *closed bars* differentiated testicular cells)

Table 3 Extensive overlap of the transcriptomes of spermatogonial stem cells and embryonic germ cells

Transcripts up-regulated in SSCs preferentially or	E11.5–14.5	22.2
uniquely expressed in the embryonic testis[a] (%)	E16.5 birth	33.3
	Total embryonic development	55.5
Transcripts up-regulated in SSCs and below detection level in the embryonic testis[a]		17.3

[a]According to *(11, 12)*

stage at birth. This observation confirms the close relationship of the adult stem cell to the embryonic germinal precursors. We note also that a sizable fraction of the 495 genes were not detectable on the Affymetrix arrays used in the previous analysis *(11)*, nor in a parallel analysis by the same approach of the total testis transcriptome *(12)*, in which the stem cell population, which could not be isolated, had become a quantitatively minor component.

Overlapping Patterns in Distinct Types of Stem Cells

Somewhat controversial studies have been published regarding the overlap of the transcriptomes of three types of stem cells—neural, hematopoietic, and embryonic—and the corresponding definition of "stemness genes" *(13–17)*. Venn diagrams shown in Fig. 1b show that only a small number of genes are common to SSCs and to any of the stem cells previously analyzed, and only six are common to all four (Fig. 1c and Table 2).

miR-125, miR-141, and miR-181 Are Differentially Expressed in the Germline

In view of the central role assumed for the noncoding class of miRs (reviewed in Alvarez-Garcia and Miska *(7)*), it was clearly of interest to determine whether a member or members of this large family is differentially expressed in the stem cell fraction. The limited amounts of RNA available from the small SSC population, combined with the small size of miRs, made, however, a direct approach unpractical. As a first stage to the evaluation of candidates for a role in SSC biology, we made use of the databases of target genes established by computer analysis (http://www. targetscan.org/) to evaluate the frequency of occurrence of the targets of a given miR in the short lists of up- and down-regulated transcripts. Given the large number of possible targets of any given miR, there is a general consensus that they could act to ensure the coordinated regulation of multiple genes. Accordingly, one would expect several genes in the same class to be targets of the same miR.

To test this hypothesis, we tabulated for each entry of the two short lists the miRs with whom the transcript would interact according to the databases. In the resulting extensive list of miRs, we checked whether some of them were more systematically appearing, in other words whether they would have multiple targets in either one of the two classes of transcripts, a number greater than expected in a random hypothesis based their overall target frequency in the genome. This first-degree approach led to possibly significant observations. Only a small number of miRs among those in the first list showed a nonrandom distribution of their targets (Table 4).

Quantitative RT-PCR after modification of the mature miR by polyadenylation generated variable results depending on the miR species. One extreme case is that of miR155, well expressed in SSCs, with ten possible targets in the up-regulated list and none among the down-regulated transcripts. Still, in this case, the mature miR was not found enriched in the stem population in comparison with the bulk of differentiated cells. We observed, on the other hand, that miR-125, miR-141, and miR-181 are expressed differentially (Fig. 3 and Table 4), but in a way that did not

Table 4 microRNAs with more putative targets than expected from their genomic averages among the genes differentially expressed in spermatogonial stem cells

miR	Estimated targets in genome[a]	Up-regulated transcripts (495)		Down-regulated transcripts (127)		Ratio of expression in SSCs to differentiated germ cells (PCR determination)[b]
		Targets[a]	Ratio to genome	Targets[a]	Ratio to genome	
miR-21	149	4	2.1	0	0	1
miR-125	414	9	1.7	1	0.8	1.95
miR-155	199	10	4.0	0	0	1
miR-181	571	14	1.9	1	0.6	2.5
miR-141	183	0	0	4	7.0	0.4

Abbreviations: miR microRNA; SSC spermatogonial stem cells; PCR polymerase chain reaction
[a]According to the Targetscan database (http://www.targetscan.org/).
[b]See Fig. 3.

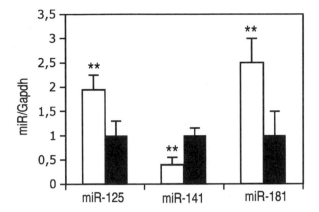

Fig. 3 MicroRNAs (miR)-125, miR-141, and miR-181 are expressed differentially during germ cells differentiation (*open bars* spermatogonial stem cells, *closed bars* differentiated testicular cells)

fit with our starting hypothesis. All three miRs showed variations in the SSC fraction parallel to those of their respective targets, miR-125 and miR-181 being enriched in SSCs, while miR-141 is preferentially expressed in differentiated germ cells. Since our current understanding is that miRs down-regulate gene expression, this result excludes the simple hypothesis of a miR-mediated regulation of these transcripts. It would rather suggest the existence of a complex regulatory network involving both positive and negative regulatory signals.

Discussion

This study was made possible by the availability of homogenous SSC preparations. Their characterization as SSCs, obviously critical to validate the results and provide grounds for further studies, was routinely controlled by expression of stage-specific genes and had been initially established by their high efficiency of colonization in transplantation assays (*6, 17*). The current results highlight a series of candidate transcriptional regulators, whose involvement in SSC function will eventually require targeted mutagenesis. In a separate series of experiments to be reported elsewhere (*17*), we started by conducting a more detailed analysis of the transcription factor PU.1 (*Sfpi1*), whose appearance in the short list of up-regulated transcripts was somewhat surprising. The protein is known as an important regulator of stem cell self-renewal and differentiation, but so far exclusively in hematopoietic lineages (*19, 20*). The results fully confirmed a critical function in germinal differentiation.

Transcriptome analysis also suggests a possible role in SSCs of three signalization pathways: JAK/STAT (Janus kinases/signal transducers and activators of transcription), TGF-β (transforming growth factor-β), and IGF (insulin-like growth factor).

The JAK/STAT pathway plays a role in the self-renewal of embryonic stem cells and is required for self-renewal of *Drosophila* germinal stem cells *(21)*. The TGF-β pathway is required for maintenance of quiescence of hematopoietic stem cells and for the development of mouse primordial germ cells *(22)* and of *Drosophila* germinal stem cells *(23)*. The JAK/STAT and the TGF-β pathways have been shown to interact and may characterize SSCs and others stem cell populations in which microarray analysis detected the corresponding transcripts *(14, 15)*. IGF-1, IGFBP (IGF binding protein)-4, and IGFBP-7 are also enriched in SSCs. Insulin-like growth factors have multiple functions in cellular growth, survival, and differentiation. IGFs have mitogenic and apoptotic functions and are modulated systemically and locally by the six high affinity IGF-binding proteins (IGFBP-1 to -6). IGF-1 and IGF-2 are locally produced in the testis, and previous results suggest that these factors have a paracrine or autocrine role in the regulation of spermatogonial proliferation and have been implicated in germ cell tumor progression.

It was of interest to compare the transcriptome established for adult SSCs and those previously published by the same techniques for the prenatal testis *(11)*. We noted (Table 3) that adult SSCs and the prenatal gonocytes share more than half of their specific transcriptomes. Schematically, one may then consider that expression of genes primarily involved in the maintenance of the stem cell state would be common to the two cell types, while the list of transcripts expressed in adult SSCs but not in gonocytes would rather correspond to genes involved in cell division and entry in the differentiation pathway. We searched for indications of the possible coregulation of groups of genes recognized by the same miRs. In three instances we observed a correlated expression of the miR and protein-coding transcripts in the same class of cells (Table 4 and Fig. 3). The functional significance of this coregulation remains to be established. It is evocative of a finely tuned regulation network, the levels of transcripts depending on a balance between one or several unknown positive factor(s) and the negative regulation exerted by the coregulated miRs. This model also remains to be tested experimentally, with the present approach only providing hypotheses to initiate a functional genetic analysis of the protein-coding genes and the regulatory mechanisms involved.

Acknowledgments We are indebted to Yan Fantei-Caujolles and Sandra Kanani for expert handling of databases, to Christelle Thibault for the generation of microarray data, and to Frank Paput and Frédérique Millot for the maintenance of the mouse facility. This work was made possible by grants from Région PACA-Inserm and Association pour la Recherche sur le Cancer to VO and from the Ligue Nationale Contre le Cancer and Ministère de la Recherche to MR.

References

1. Russel LD, EttlinRA, Sinha HikimAP, CleggED. Histological and histopathological evaluation of the testis. Cache River Press, Clearwater, 1990.
2. de Rooij DG, Grootegoed JA. Spermatogonial stem cells. Curr Opin Cell Biol 1998;10:694–701.
3. Charbord P. Hemopoietic stem cells: analysis of some parameters critical for engraftment. Stem Cells 1994;12:545–62.

4. Brinster RL. Germline stem cell transplantation and transgenesis. Science 2002;296:2174–6.
5. Brinster RL, Zimmermann JW. Spermatogenesis following male germ-cell transplantation. Proc Natl Acad Sci U S A 1994;91:11298–302.
6. Giuili G, Tomljenovic A, Labrecque N, Oulad-Abdelghani M, Rassoulzadegan M, Cuzin F. Murine spermatogonial stem cells: targeted transgene expression and purification in an active state. EMBO Rep 2002;3:753–9.
7. Alvarez-Garcia I, Miska EA. microRNA functions in animal development and human disease. Development 2005;132:4653–62.
8. Vidal F, Lopez P, Lopez-Fernandez LA, et al. Gene trap analysis of germ cell signaling to Sertoli cells: NGF-TrkA mediated induction of Fra1 and Fos by post-meiotic germ cells. J Cell Sci 2001;114:435–43.
9. Oulad-Abdelghani M, Bouillet P, Decimo D, et al. Characterization of a premeiotic germ cell-specific cytoplasmic protein encoded by Stra8, a novel retinoic acid-responsive gene. J Cell Biol 1996;135:469–77.
10. Shinohara T, Avarbock MR, Brinster RL. Beta1- and alpha6-integrin are surface markers on mouse spermatogonial stem cells. Proc Natl Acad Sci U S A 1999;96:5504–9.
11. Small CL, Shima JE, Uzumcu M, Skinner MK, Griswold MD. Profiling gene expression during the differentiation and development of the murine embryonic gonad. Biol Reprod 2005;72:492–501.
12. Shima JE, McLean DJ, McCarrey JR, Griswold MD. The murine testicular transcriptome: characterizing gene expression in the testis during the progression of spermatogenesis. Biol Reprod 2004;71:319–30.
13. Burns CE, Zon LI. Portrait of a stem cell. Dev Cell 2002;3:612–3.
14. Ivanova NB, Dimos JT, Schaniel C, Hackney JA, Moore KA, Lemischka IR. A stem cell molecular signature. Science 2002;298:601–4.
15. Ramalho-Santos M, Yoon S, Matsuzaki Y, Mulligan RC, Melton DA. "Stemness": transcriptional profiling of embryonic and adult stem cells. Science 2002;298:597–600.
16. Fortunel NO, Otu HH, Ng HH, et al. Comment on "Stemness": transcriptional profiling of embryonic and adult stem cells and a stem cell molecular signature. Science 2003;302:393.
17. Olive V, Wagner N, Chan S, Kastner P, Vannetti C, Cuzin F, Rassoulzadegan M. PU.1 (Sfpi1), a pleiotropic regulator expressed from the first embryonic stages with a crucial function in germinal progenitors. Development 2007; 134(21):3815–25.
18. Evsikov AV, Solter D. Comment on "Stemness": transcriptional profiling of embryonic and adult stem cells and a stem cell molecular signature. Science 2003;302:393.
19. Back J, Dierich A, Bronn C, Kastner P, Chan S. PU.1 determines the self-renewal capacity of erythroid progenitor cells. Blood 2004;103:3615–23.
20. Iwasaki H, Somoza C, Shigematsu H, et al. Distinctive and indispensable roles of PU.1 in maintenance of hematopoietic stem cells and their differentiation. Blood 2005;106:1590–600.
21. Tulina N, Matunis E. Control of stem cell self-renewal in Drosophila spermatogenesis by JAK-STAT signaling. Science 2001;294:2546–9.
22. Chuva de Sousa Lopes SM, van den Driesche S, Carvalho RL, et al. Altered primordial germ cell migration in the absence of transforming growth factor beta signaling via ALK5. Dev Biol 2005;284:194–203.
23. Shivdasani AA, Ingham PW. Regulation of stem cell maintenance and transit amplifying cell proliferation by tgf-beta signaling in Drosophila spermatogenesis. Curr Biol 2003;13:2065–72.

Human Mesenchymal Stem Cells: Basic Biology and Clinical Applications for Bone Tissue Regeneration

Basem M. Abdallah, Hamid Saeed, and Moustapha Kassem

Abstract Mesenchymal stem cells (MSCs) are a group of clonogenic cells present among the bone marrow stroma and capable of multilineage differentiation into mesoderm-type cells such as osteoblasts, adipocytes, and chondrocytes. Because of their ease of isolation and this wide differentiation potential, MSCs are being introduced into clinical medicine in a variety of applications. Here we discuss the characteristics of MSCs, their differentiation, as well as the challenges faced when they are used in cell therapy for bone regeneration.

Keywords Human bone marrow–derived mesenchymal stem cells • Mesenchymal stem cells • Bone regeneration • Cell therapy

Introduction

Human bone marrow–derived mesenchymal stem cells (hMSCs), also known as skeletal stem cells, bone marrow stromal cells, or, as recently suggested by the International Society for Cytotherapy, multipotent mesenchymal stromal cells (1), are a group of clonogenic cells present among the bone marrow stroma that are capable of multilineage differentiation into mesoderm-type cells such as osteoblast, adipocyte, chondrocyte (2) and possibly, but still controversial, other nonmeso-derm-type cells, e.g., neuronal cells or hepatocytes (3, 4). Moreover, hMSCs provide supportive stroma for growth and differentiation of hematopoietic stem cells (HSCs) and hematopoiesis (5).

The identification and characterization of MSCs have been initiated through the pioneering work of Friedenstein (6) in Russia and later Owen (7) in the United

B.M. Abdallah, H. Saeed, and M. Kassem (✉)

Endocrinology Research Laboratory (KMEB), Department of Endocrinology, Odense University Hospital, Kloevervaenget 6, 4th floor, DK-5000, Odense, Denmark

e-mail: mkassem@health.sdu.dk

H. Baharvand (ed.), *Trends in Stem Cell Biology and Technology*,
DOI 10.1007/ 978-1-60327-905-5_12,
© Humana Press, a Part of Springer Science+Business Media LLC 2009

Kingdom, where culture systems for expanding the cells and for studying their biological characteristics in vitro and in vivo have been established. Recently, there has been an increased interest toward understanding the biology of MSCs due to their potential use in therapy of a variety of diseases. The aim of this review is to provide an update related to the biology of hMSCs and the challenges facing their use in therapy.

Isolation and Characterization of MSCs

hMSCs are fusiform, fibroblastlike cells. During their initial growth in vitro, they form colonies (termed in analogy with HSCs: colony forming unit–fibroblasts) *(7–9)*. The cells are negative for hematopoietic surface markers CD34, CD45, and CD14 and positive for several surface markers, such as CD29, CD 73, CD90, CD105, CD166, and CD44 *(1, 10, 11)*. Traditionally, MSCs have been isolated based on their selective adherence to plastic surfaces, compared to hematopoietic cells *(9, 12, 13)*. One disadvantage of this method is the unavoidable hematopoietic cell contamination and the cellular heterogeneity of cultures. Clonal analysis of plastic-adhered bone marrow–derived hMSCs has demonstrated that around 30% of the clones are multipotent hMSCs *(14)*. Thus, one of the active areas of investigations is the identification of sensitive and specific markers for the multipotent hMSCs. In this context, Gronthos et al. *(15)* was the first to develop a monoclonal antibody, Stro-1, which has been used to isolate an enriched population of cells with MSC characteristics *(16)*. More recently, other surface markers have been utilized. For example, CD146, a marker known to be expressed by smooth muscle, endothelium, myofibroblasts, and Schwann cells, seems to identify populations of multipotent hMSCs in the bone marrow *(17)*. Also, coexpression of STRO-1bright/ CD146+ has been employed to enrich for clonogenic hMSCs *(18)*. Finally, two novel CD markers, CD200 and CD271, have been identified and employed to isolate bone marrow–derived hMSCs *(19, 20)*.

Other Bone Marrow–Derived MSC-Like Cells

Other investigators have reported the isolation of a more primitive MSC from the bone marrow using a variety of methods. For example, Reyes et al. *(21)* isolated a pluripotent MSC population (termed multipotent adult progenitor cells [MAPCs]) from CD45−/glycoprotein A− depleted bone marrow–derived mononuclear cell fraction that selectively adhered to laminin-coated plates under low serum conditions, and in a variety of publications the same group has demonstrated an extensive plasticity (and even pluripotency) of this cell type when derived from mice bone marrow *(22)*. So far, these findings have not been confirmed by other investigators. Murine side population (SP) cells isolated from bone marrow, based on their Hoechest dye exclusion, have been demonstrated to be able to develop into MSCs (and also HSCs), and thus it may represent an earlier, more primitive population of stem cells *(23)*. One study by Lodie et al. *(24)* described side-by-side comparison of the physical and functional similarities of adherent hMSCs isolated from a variety of sources. The authors reported that no major differences exist among these different cell populations. However, further studies are needed to standardize the isolation procedure for the cells and to provide a more detailed characterization of these different cells.

Other MSC Populations from Different Tissues

MSCs with similar biological characteristics to those derived from bone marrow have been isolated from other organs including peripheral blood *(25)*, umbilical cord blood *(26)*, synovial membrane *(27)*, adipose tissue *(28)*, lung *(29)*, fetal liver *(30)*, dental pulp *(31, 32)*, and deciduous teeth *(33)*. These data suggest the possible existence of an extensive network of MSCs in the body that is continuously replenished by MSCs from the bone marrow during time of need. However, the firm proof of this hypothesis is still lacking. Some recent studies comparing the "genetic signature" of these different cell populations have reported strong similarities and also differences related to the characteristics of the tissue of origin of the examined cells *(34–36)*.

The Use of MSCs in Regenerative Medicine

The emerging field of regenerative medicine holds the promise of treating a variety of degenerative and age-related diseases, where no specific or effective treatment is currently available, by transplanting biologically competent mature cells and tissues or through stimulation of the tissue-resident stem cells. Stem cells in general and MSCs in particular, with their versatile growth and differentiation potential, are ideal candidates for use in regenerative medicine protocols and are currently making their way into clinical trials. However, some biotechnological problems related to the biological characteristics of the cells need to be solved before we will be able to make use of the full potential of MSCs in therapy. Here we will discuss some of these issues and possible approaches for overcoming the limitations imposed by them on the clinical use thses cells.

Limited In Vitro Cell Growth and Replicative Senescence of hMSCs

The clinical use of hMSCs requires the availability of a large number of functionally competent cells with stable phenotype and genotype. However, in vitro expansion of hMSCs in long-term culture of these cells is limited. During long-term in vitro culture of hMSCs the cells exhibit reduced proliferation rate and finally enter a state of growth arrest, a phenomenon termed *in vitro replicative senescence(37–39)*. We have found that under current in vitro culture conditions, MSCs obtained from young donors can grow up to 40 population doublings (PD) *(39)*. Interestingly, we found no age-related changes in the number of hMSCs *(16)*, but hMSC obtained from elderly donors exhibited a decrease in their proliferative potential, with maximal cell proliferation of 25 PD *(39)*. In vitro replicative senescence is a general characteristic of cultured diploid cells, and it limits their ability for generating the large number of cells needed for therapy. Replicative senescence is caused by several mechanisms, including progressive telomere shortening during continuous subculture in vitro *(39, 40)* due to absence of telomerase activity *(41, 42)*. It is not clear whether the absence of telomerase activity in MSCs is secondary to nonoptimal culture conditions or an inherent property of the isolated MSCs. Interestingly, some investigators have isolated populations of MSCs that expressed low levels of telomerase activity that disappeared quickly after in vitro culture *(18)*. We have demonstrated that it is possible to overcome the senescence phenotype of cultured MSCs by overexpression of the human telomerase reverse transcriptase (hTERT) gene in MSCs to restore telomerase activity *(41)*. The telomerized cells exhibit an extended life span and maintain their "stemness" characteristics *(10)*. The combination of extensive proliferation and maintenance of functional activity of the cells make the telomerized cells attractive cells to be employed in a variety of regenerative medicine applications. Similar to our results, telomerized smooth muscle cells were able to form better-quality tissue-engineered arteries compared with nontelomerized counterparts *(43)*. Unfortunately, the extensive cell proliferation of telomerized cells in vitro led to genetic instability and resulted in MSC transformation after about 250 PD in culture *(44)*, thus limiting the clinical use of this approach. Conditional overexpression of the hTERT gene or intermittent chemical stimulation of its expression may be a more appropriate approach. In this respect, we have recently demonstrated that trichostatin A treatment, which affects DNA chromatin structure, resulted in transient expression of the hTERT gene *(45)*. These results demonstrate the feasibility of using chemicals (or small molecules) to stimulate hTERT gene expression and possibly obtaining controlled biological effects. However, more research needs to be performed in this direction.

Culturing MSCs in Xenofree Environment

The use of MSC in cell replacement therapy requires developing xenofree protocols for their culture. Although this is not currently a formal requirement from the regulatory authorities, such as the U.S. Food and Drug Administration, there is an increasing concern about the potential transmission of infection particles, e.g., viruses and prions, from animals to humans. Some previous reports have suggested that MSCs cannot be optimally cultured in human serum *(46)*. However, systemic studies in our lab on the effect of human serum on MSCs showed that MSCs can be cultured equally well in human serum compared to fetal calf serum with no observable effects on their proliferation or differentiation potential in vitro and in vivo *(47)*. Interestingly, sera obtained from elderly donors did not affect the proliferative potential of hMSCs in short-term cultures but caused a decrease in their osteoblastic differentiation *(47)*.

Directing Differentiation of MSCs into Specific Lineages

Although multipotentiality of MSCs is the basis for using these cells for generating different cells and tissues for cell replacement therapy, protocols that direct the differentiation of hMSCs into a particular lineage are still inefficient. Several approaches have been employed to direct the differentiation of MSCs to a specific differentiation lineage, e.g., osteoblasts by in vitro treatment with a mixture of growth factors, such as bone morphogenetic protein (BMP) or transforming growth factor β (TGF-β), that enhance osteoblast differentiation *(48, 49)*. Another possible strategy is to use the genetic approach by overexpression of genes that enhance bone cell differentiation *(50–53)*.

In order to identify novel targets for directing the differentiation of hMSCs, we have employed state-of-the-art proteomic approaches to study phosphor–tyrosine signaling pathways during hMSC differentiation into osteoblasts *(54)*. We compared the phosphor–tyrosine signaling pathway induced in hMSCs under either epidermal growth factor treatment, which leads to osteoblast differentiation, or platelet-derived growth factor (PDGF), which maintains hMSCs in undifferentiated state. Using this approach, we were able to identify PI3K as an important regulator of osteoblast differentiation, and by inhibiting PI3K in the context of PDGF treatment we were able to restore osteoblast-differentiation induction by PDGF *(54)*. This approach is an example of the power of the modern proteome technologies in identifying novel targets that can be employed to control stem cell differentiation.

Immunology of hMSCs

Although autologous hMSC transplantation is feasible and does not result in any immunological problems, allogenic cell transplantation is a more suitable approach for creating "off-shelf" allogenic cells for use in therapy. Interestingly, MSCs appear to be hypoimmunogenic, and thus allogenic mesenchymal stem cells transplantation is possible. MSCs express intermediate levels of HLA major histocompatibility complex (MHC) class I molecules, a less detectable level of HLA class II antigens, and no expression of costimulatory molecules like CD40, CD40L, CD80, or CD86 *(55, 56)*. Also, MSCs are reported to possess immunosuppressive properties in vitro by inhibiting T-cell alloreactivity induced in mixed lymphocyte cultures or by nonspecific mitogens *(57, 58)*. In addition, MSCs affect dendritic cell (DC) maturation, and their functional properties activate T cells by inhibiting the secretion of TNF-α and promoting interleukin 10 (IL-10) secretion and therefore directing the immune response toward more anti-inflammatory or tolerant phenotype *(59)*. In vivo, the immunosuppressive effect of MSCs was shown in their ability to prolong the histoincompatible skin graft survival when ex vivo expanded donor MSCs were administered to MHC-mismatched recipient baboons prior to placement of autologous, donor, and third-party skin grafts *(60)*. These immunoregulatory characteristics have been utilized to use hMSCs in the treatment of graft-versus-host (GVH) disease, which is one of the most serious side effects of allogenic hematopoietic stem cells transplantation and has resulted in spectacular results *(61–66)*.

Clinical Application of hMSCs in Bone Tissue Regeneration

MSCs are among the first stem cell types to be introduced in the clinic. Table 1 provides a list of the clinical trials where hMSCs have been employed *(67–76)*. Here we will discuss four areas for potential clinical use of MSCs in enhancing bone regeneration: the use of MSCs in tissue engineering protocols, local implantation of MSCs, combining stem cell therapy with gene therapy, and systemic transplantation.

Tissue Engineering

Tissue engineering may provide alternative ways for obtaining tissues and organs needed for transplantation due to a lack of sufficient number of organ donors and limitations attributable to immunological rejection and mismatch of physical dimensions. Tissue engineering utilizes patients own cells, seeding them on biodegradable scaffolds that allows formation of a particular tissue. These tissues can be employed to repair tissue defects due to disease or trauma. Also, tissue engineering can allow ex vivo engineering of tissue by means of three-dimensional bioscaffolds seeded with mature cell or stem cells and cultivated in bioreactors that lead to the formation of tissues or organs, e.g., liver, hearts, cartilage, or kidneys *(77)*.

MSCs are good candidates for use in tissue engineering protocols *(78)*. Several scaffolds are currently available and may be classified as either biologically derived polymers isolated from extracellular matrix, plants and seaweed, e.g., collagen type I or fibronectin; alginate from brown algae or synthetic, e.g., hydroxyapatite (HA), tricalcium phosphate (TCP) ceramics, polylactide, and polyglycolide; and a combination of these in the form of poly DL-lactic-co-glycolic acid (PLGA). There exists several animal experiments showing the success of using this approach, e.g., for treatment of large bone defects *(79)*. One of the most successful experiments in humans has been reported recently. A titanium mesh scaffold filled with bone mineral blocks infiltrated with BMP-7 and bone marrow mixture has successfully

Table 1 Clinical trials of using human stem cells in cartilage and bone regeneration

Author (ref)	Indication	MSCs source	No. of patients	Results
Goel et al. (*67*)	Tibial nonunion	Bone marrow	20	Clinical and radiological bone union following percutaneous injection in 15 of 20 patients
Herniguo et al. (*68*)	Nonunion sites	Autologous bone marrow	60	Percutaneous autologous bone marrow grafting is an effective and safe method for the treatment of an atrophic tibial diaphyseal nonunion. Bone unions were obtained in 53 patients
Kawate et al. (*69*)	Steroid-induced osteonecrosis of the femoral head	Autologous mesenchymal stem cells (MSCs) cultured with β-tricalcium phosphate	3	Osteonecrosis did not progress any further and early bone regeneration was observed. However, present procedure could not be used for cases with severe preoperative collapse
Kuroda et al. (*70*)	Cartilage defect in femoral condyle	Autologous human bone marrow stromal cells	1 (athlete)	Transplantation of autologous bone marrow stromal cells can promote the repair of large focal articular cartilage defects in young, active patients
Wakitani et al. (*71*)	Full-thickness articular cartilage defects in human patellae	Autologous bone marrow stromal cells	2	Autologous bone marrow stromal cell transplantation is an effective approach in promoting the repair of articular cartilage defects
Kitoh et al. (*72*)	Achondroplasia, congenital pseudarthrosis	Marrow derived mesenchymal stem cells	3	Transplantation of osteoblastlike cells and platelet rich plasma (PRP), which seemed to be a safe and minimally invasive cell therapy, could shorten the treatment period by acceleration of bone regeneration during distraction osteogenesis
Quarto et al. (*73*)	Segmental bone defect	Autologous bone marrow	—	Complete consolidation between the implant and bone at 6 months
Ohgushi et al. (*74*)	Prevention of loosening of total ankle arthroplasty	Bone marrow	3	Preliminary study: bone-prosthesis interface was established, no inflammatory reaction
Gangji et al. (*75*)	Femoral head osteonecrosis	Autologous bone marrow	13	Low rate of deterioration
Chang et al. (*76*)	Unicameral bone cysts	Bone marrow	79	Bone marrow can be used in the treatment of unicameral bone cysts but has a high failure rate; no difference compared with steroid injection

produced a large amount of bone tissue needed for reconstructing a mandibular defect in a human patient *(80)*. There is great expectation for more breakthroughs in creating osteogenic tissues, and probably other tissues based on these approaches, to be used in the clinic.

Local Implantation of MSCs

Several animal studies have also demonstrated the efficacy of using MSCs in treatment of bone defects *(81, 82)*. Also, some clinical case reports have demonstrated the success of locally injected ex vivo expanded autologous MSCs for treatment of large bone defects in patients with defective fracture healing *(73)*. Also, repair of cartilage defects has been attempted *(83)*.

Combining Stem Cell Therapy with Gene Therapy

The use of gene-modified stem cells in the context of gene therapy is an attractive option because of the theoretical advantage of using stem cells compared to somatic cells with respect to higher proliferative capacity and long-term survival. Genetically modified MSCs can deliver genes or proteins into organs or tissues with specific need for gene therapy. Some of these approaches have been attempted in animal models. MSCs expressing ectopic BMP-2 or insulinlike growth factor 1 (IGF-1) have been used successfully to repair bone defects and for bone regeneration in several animal models, and some of these studies are summarized in Table 2*(83–93)*.

Systemic Transplantation of MSCs

One human study has examined the effects of systemic transplantation of allogenic normal MSCs in children with severe osteogenesis imperfecta. Homing of MSCs in bone as well as the production of normal collagen by the transplanted MSCs have been demonstrated *(94)*. However, homing of hMSCs to bone is limited when tested in animal models *(95)*, and there is a need for developing approaches to enhance hMSC migration and homing.

Concluding Remarks and Future Perspectives

MSC population is one of the stem cell populations that have been introduced in the clinic for treatment of several disease conditions, and MSC-based therapy has demonstrated promising results in phase I clinical trials. We believe that combining

Table 2 Combining stem cell therapy with gene therapy for bone regeneration in animal models

Author (ref)	Gene and vector	Cell source	Model	Results
Lieberman et al. (*84*)	BMP-2, ADV	Bone marrow	SCID mouse and rat	Gene therapy with continuous delivery of osteoinductive factors to a specific anatomic site can enhance the formation and repair of bone
Gazit et al. (*85*)	BMP-2, none	C3H10T1/2	Mouse with radial segmental defect	The transfected mesenchymal cells enhanced segmental defect repair
Lee et al. (*86*)	BMP-2, AD	Mouse MDSCs	SCID mice	Healing of a critical-size skull defect in SCID mice
Laurencin et al. (*87*)	BMP-2, RV	Rat bone marrow	SCID hindlimb	In vivo, the scaffold successfully functioned as a delivery vehicle for bioactive BMP-2, as it induced heterotopic bone formation in a SCID mouse model
Moutsatsos et al. (*88*)	BMP-2	C3H10T1/2	C3H/HeN female mice	Regulated ectopic bone formation and regulated healing of critical-size radius defects in mice
Shen et al. (*89*)	IGF-1, RV	Mouse MSCs	Murine model	Mice receiving D1-IGF-1 cells demonstrated a greater percent of mineralized callus than controls at 2 weeks and average greater mineralized matrix at 4 and 6 weeks
Peterson et al. (*90*)	BMP-2, ADV	Human fat derived stem cells	Nude rat	Healing of critical femoral defects in nude rat
Hasharoni et al. (*91*)	BMP-2	C3H10T1/2 cells	Mouse	Spinal fusion in mouse model
Xu et al. (*92*)	BMP-2, ADV	Goat bone marrow stromal cells	Goat	Healing of critical size tibial defects in goat
Tobita et al. (*93*)	None	Adipose derived stem cell (rat)	Rat	ASCs can promote periodontal tissue regeneration in vivo

Abbreviations: ADV adenovirus vector; *SCID* severe combined immunodeficiency; *BMP* bone morphogenetic protein; *IGF-1* insulinlike growth factor 1; *RV* retroviral vector; *MSCs* mesenchymal stem cells; *ASCs* adipose derived stem cell; *MDSC* muscle derived stem cells

basic research studies identifying the mechanisms controlling MSC proliferation and differentiation with well-designed and controlled clinical trials will bring major advances in MSC-based therapy for several chronic and degenerative diseases

References

1. Dominici M, Le BK, Mueller I, et al. Minimal criteria for defining multipotent mesenchymal stromal cells. The International Society for Cellular Therapy position statement. Cytotherapy 2006;8:315–7.
2. Bianco P, Riminucci M, Gronthos S, Robey PG. Bone marrow stromal stem cells: nature, biology, and potential applications. Stem Cells 2001;19:180–92.
3. Dezawa M, Kanno H, Hoshino M, et al. Specific induction of neuronal cells from bone marrow stromal cells and application for autologous transplantation. J Clin Invest 2004;113:1701–10.
4. Luk JM, Wang PP, Lee CK, Wang JH, Fan ST. Hepatic potential of bone marrow stromal cells: development of in vitro co-culture and intra-portal transplantation models. J Immunol Methods 2005;305:39–47.
5. Dexter TM. Haemopoiesis in long-term bone marrow cultures. A review. Acta Haematol 1979;62:299–305.
6. Friedenstein AJ. Osteogenic stem cells in the bone marrow. Bone Miner 1991;7:243–72.
7. Owen M. Marrow stromal stem cells. J Cell Sci Suppl 1988;10:63–76.
8. Friedenstein AJ, Chailakhjan RK, Lalykina KS. The development of fibroblast colonies in monolayer cultures of guinea-pig bone marrow and spleen cells. Cell Tissue Kinet 1970;3:393–403.
9. Luria EA, Panasyuk AF, Friedenstein AY. Fibroblast colony formation from monolayer cultures of blood cells. Transfusion 1971;11:345–9.
10. Abdallah BM, Haack-Sorensen M, Burns JS, et al. Maintenance of differentiation potential of human bone marrow mesenchymal stem cells immortalized by human telomerase reverse transcriptase gene despite [corrected] extensive proliferation. Biochem Biophys Res Commun 2005;326:527–38.
11. Foster LJ, Zeemann PA, Li C, Mann M, Jensen ON, Kassem M. Differential expression profiling of membrane proteins by quantitative proteomics in a human mesenchymal stem cell line undergoing osteoblast differentiation. Stem Cells 2005;23:1367–77.
12. Kassem M, Mosekilde L, Eriksen EF. 1,25-Dihydroxyvitamin D3 potentiates fluoride-stimulated collagen type I production in cultures of human bone marrow stromal osteoblast-like cells. J Bone Miner Res 1993;8:1453–8.
13. Rickard DJ, Kassem M, Hefferan TE, Sarkar G, Spelsberg TC, Riggs BL. Isolation and characterization of osteoblast precursor cells from human bone marrow. J Bone Miner Res 1996;11:312–24.
14. Kuznetsov SA, Krebsbach PH, Satomura K, et al. Single-colony derived strains of human marrow stromal fibroblasts form bone after transplantation in vivo. J Bone Miner Res 1997;12:1335–47.
15. Gronthos S, Graves SE, Ohta S, Simmons PJ. The STRO-1+ fraction of adult human bone marrow contains the osteogenic precursors. Blood 1994;84:4164–73.
16. Stenderup K, Justesen J, Eriksen EF, Rattan SI, Kassem M. Number and proliferative capacity of osteogenic stem cells are maintained during aging and in patients with osteoporosis. J Bone Miner Res 2001;16:1120–9.
17. Sacchetti B, Funari A, Michienzi S, et al. Self-renewing osteoprogenitors in bone marrow sinusoids can organize a hematopoietic microenvironment. Cell 2007;131:324–36.
18. Gronthos S, Zannettino AC, Hay SJ, et al. Molecular and cellular characterisation of highly purified stromal stem cells derived from human bone marrow. J Cell Sci 2003;116(Pt 9):1827–35.
19. Delorme B, Ringe J, Gallay N, et al. Specific plasma membrane protein phenotype of culture-amplified and native human bone marrow mesenchymal stem cells. Blood 2008;111:2631–5.
20. Buhring HJ, Battula VL, Treml S, Schewe B, Kanz L, Vogel W. Novel markers for the prospective isolation of human MSC. Ann N Y Acad Sci 2007;1106:262–71.

21. Reyes M, Lund T, Lenvik T, Aguiar D, Koodie L, Verfaillie CM. Purification and ex vivo expansion of postnatal human marrow mesodermal progenitor cells. Blood 2001;98:2615–25.
22. Jiang Y, Jahagirdar BN, Reinhardt RL, et al. Pluripotency of mesenchymal stem cells derived from adult marrow. Nature 2002;418:41–9.
23. Olmsted-Davis EA, Gugala Z, Camargo F, et al. Primitive adult hematopoiletic stem cells can function as osteoblast precursors. Proc Natl Acad Sci U S A 2003;100:15877–82.
24. Lodie TA, Blickarz CE, Devarakonda TJ, et al. Systematic analysis of reportedly distinct populations of multipotent bone marrow-derived stem cells reveals a lack of distinction. Tissue Eng 2002;8:739–51.
25. Kuznetsov SA, Mankani MH, Gronthos S, Satomura K, Bianco P, Robey PG. Circulating skeletal stem cells. J Cell Biol 2001;153:1133–40.
26. Rosada C, Justesen J, Melsvik D, Ebbesen P, Kassem M. The human umbilical cord blood: a potential source for osteoblast progenitor cells. Calcif Tissue Int 2003;72:135–42.
27. De Bari C, Dell'Accio F, Tylzanowski P, Luyten FP. Multipotent mesenchymal stem cells from adult human synovial membrane. Arthritis Rheum 2001;44:1928–42.
28. Gronthos S, Franklin DM, Leddy HA, Robey PG, Storms RW, Gimble JM. Surface protein characterization of human adipose tissue-derived stromal cells. J Cell Physiol 2001;189:54–63.
29. in't Ankcr PS, Noort WA, Scherjon SA, et al. Mesenchymal stem cells in human second-trimester bone marrow, liver, lung, and spleen exhibit a similar immunophenotype but a heterogeneous multilineage differentiation potential. Haematologica 2003;88.845–52.
30. Campagnoli C, Roberts IA, Kumar S, Bennett PR, Bellantuono I, Fisk NM. Identification of mesenchymal stem/progenitor cells in human first-trimester fetal blood, liver, and bone marrow. Blood 2001;98:2396–402.
31. Gronthos S, Robey PG, Boyde A, Shi S. Human dental pulp stem cells (DPSCs): characterization and developmental potential. J Bone Miner Res 2001;16:S265.
32. Otaki S, Ueshima S, Shiraishi K, et al. Mesenchymal progenitor cells in adult human dental pulp and their ability to form bone when transplanted into immunocompromised mice. Cell Biol Int 2007;31:1191–7.
33. Miura M, Gronthos S, Zhao M, et al. SHED: stem cells from human exfoliated deciduous teeth. Proc Natl Acad Sci U S A 2003;100:5807–12.
34. Djouad F, Bony C, Haupl T, et al. Transcriptional profiles discriminate bone marrow-derived and synovium-derived mesenchymal stem cells. Arthritis Res Ther 2005;7:R1304–15.
35. Wagner W, Wein F, Seckinger A, et al. Comparative characteristics of mesenchymal stem cells from human bone marrow, adipose tissue, and umbilical cord blood. Exp Hematol 2005;33:1402–16.
36. Yamada Y, Fujimoto A, Ito A, Yoshimi R, Ueda M. Cluster analysis and gene expression profiles: a cDNA microarray system-based comparison between human dental pulp stem cells (hDPSCs) and human mesenchymal stem cells (hMSCs) for tissue engineering cell therapy. Biomaterials 2006;27:3766–81.
37. Kassem M, Ankersen L, Eriksen EF, Clark BF, Rattan SI. Demonstration of cellular aging and senescence in serially passaged long-term cultures of human trabecular osteoblasts. Osteoporos Int 1997;7:514–24.
38. DiGirolamo CM, Stokes D, Colter D, Phinney DG, Class R, Prockop DJ. Propagation and senescence of human marrow stromal cells in culture: a simple colony-forming assay identifies samples with the greatest potential to propagate and differentiate 6. Br J Haematol 1999;107:275–81.
39. Stenderup K, Justesen J, Clausen C, Kassem M. Aging is associated with decreased maximal life span and accelerated senescence of bone marrow stromal cells. Bone 2003;33:919–26.
40. Rattan SIS. Aging outside the body: usefulness of the Hayflick system. In: Kaul SC, Wadhwa R, editors. Aging of cells in and outside the body. Kluwer, London, 2003:1–8.
41. Simonsen JL, Rosada C, Serakinci N, et al. Telomerase expression extends the proliferative life-span and maintains the osteogenic potential of human bone marrow stromal cells. Nat Biotechnol 2002;20:592–6.
42. Zimmermann S, Voss M, Kaiser S, Kapp U, Waller CF, Martens UM. Lack of telomerase activity in human mesenchymal stem cells. Leukemia 2003;17:1146–9.

43. Mckee JA, Banik SSR, Boyer MJ, et al. Human arteries engineered in vitro. EMBO Rep 2003;4:633–8.
44. Serakinci N, Guldberg P, Burns JS, et al. Adult human mesenchymal stem cell as a target for neoplastic transformation. Oncogene 2004;23:5095–8.
45. Serakinci N, Hoare SF, Kassem M, Atkinson SP, Keith WN. Telomerase promoter reprogramming and interaction with general transcription factors in the human mesenchymal stem cell. Regen Med 2006;1:125–31.
46. Krebsbach PH, Kuznetsov SA, Satomura K, Emmons RV, Rowe DW, Robey PG. Bone formation in vivo: comparison of osteogenesis by transplanted mouse and human marrow stromal fibroblasts. Transplantation 1997;63:1059–69.
47. Abdallah BM, Haack-Sorensen M, Fink T, Kassem M. Inhibition of osteoblast differentiation but not adipocyte differentiation of mesenchymal stem cells by sera obtained from aged females. Bone 2006;39:181–8.
48. Luu HH, Song WX, Luo X, et al. Distinct roles of bone morphogenetic proteins in osteogenic differentiation of mesenchymal stem cells. J Orthop Res 2007;25:665–77.
49. Noel D, Gazit D, Bouquet C, et al. Short-term BMP-2 expression is sufficient for in vivo osteochondral differentiation of mesenchymal stem cells. Stem Cells 2004;22:74–85.
50. Ichida F, Nishimura R, Hata K, et al. Reciprocal roles of MSX2 in regulation of osteoblast and adipocyte differentiation. J Biol Chem 2004;279:34015–22.
51. Kojima H, Uemura T. Strong and rapid induction of osteoblast differentiation by Cbfa1/Til-1 overexpression for bone regeneration. J Biol Chem 2005;280:2944–53.
52. Tang Z, Sahu SN, Khadeer MA, Bai G, Franklin RB, Gupta A. Overexpression of the ZIP1 zinc transporter induces an osteogenic phenotype in mesenchymal stem cells. Bone 2006;38:181–98.
53. Wu L, Wu Y, Lin Y, et al. Osteogenic differentiation of adipose derived stem cells promoted by overexpression of osterix. Mol Cell Biochem 2007;301:83–92.
54. Kratchmarova I, Blagoev B, Haack-Sorensen M, Kassem M, Mann M. Mechanism of divergent growth factor effects in mesenchymal stem cell differentiation. Science 2005;308:1472–7.
55. Le BK, Tammik C, Rosendahl K, Zetterberg E, Ringden O. HLA expression and immunologic properties of differentiated and undifferentiated mesenchymal stem cells. Exp Hematol 2003;31:890–6.
56. Pittenger MF, Mackay AM, Beck SC, et al. Multilineage potential of adult human mesenchymal stem cells 6. Science 1999;284:143–7.
57. Di NM, Carlo-Stella C, Magni M, et al. Human bone marrow stromal cells suppress T-lymphocyte proliferation induced by cellular or nonspecific mitogenic stimuli. Blood 2002;99:3838–43.
58. Krampera M, Glennie S, Dyson J, et al. Bone marrow mesenchymal stem cells inhibit the response of naive and memory antigen-specific T cells to their cognate peptide. Blood 2003;101:3722–9.
59. Aggarwal S, Pittenger MF. Human mesenchymal stem cells modulate allogeneic immune cell responses. Blood 2005;105:1815–22.
60. Bartholomew A, Sturgeon C, Siatskas M, et al. Mesenchymal stem cells suppress lymphocyte proliferation in vitro and prolong skin graft survival in vivo. Exp Hematol 2002;30:42–8.
61. Le BK, Rasmusson I, Sundberg B, et al. Treatment of severe acute graft-versus-host disease with third party haploidentical mesenchymal stem cells. Lancet 2004;363:1439–41.
62. Inoue S, Popp FC, Koehl GE, et al. Immunomodulatory effects of mesenchymal stem cells in a rat organ transplant model. Transplantation 2006;81:1589–95.
63. Lee ST, Jang JH, Cheong JW, et al. Treatment of high-risk acute myelogenous leukaemia by myeloablative chemoradiotherapy followed by co-infusion of T cell-depleted haematopoietic stem cells and culture-expanded marrow mesenchymal stem cells from a related donor with one fully mismatched human leucocyte antigen haplotype. Br J Haematol 2002;118:1128–31.
64. Maitra B, Szekely E, Gjini K, et al. Human mesenchymal stem cells support unrelated donor hematopoietic stem cells and suppress T-cell activation. Bone Marrow Transplant 2004;33:597–604.
65. Ringden O, Uzunel M, Rasmusson I, et al. Mesenchymal stem cells for treatment of therapy-resistant graft-versus-host disease. Transplantation 2006;81:1390–7.
66. Dean RM, Bishop MR. Graft-versus-host disease: emerging concepts in prevention and therapy. Curr Hematol Rep 2003;2:287–94.

67. Goel A, Sangwan SS, Siwach RC, Ali AM. Percutaneous bone marrow grafting for the treatment of tibial non-union. Injury 2005;36:203–6.
68. Hernigou P, Poignard A, Beaujean F, Rouard H. Percutaneous autologous bone-marrow grafting for nonunions. Influence of the number and concentration of progenitor cells. J Bone Joint Surg Am 2005;87:1430–7.
69. Kawate K, Yajima H, Ohgushi H, et al. Tissue-engineered approach for the treatment of steroid-induced osteonecrosis of the femoral head: transplantation of autologous mesenchymal stem cells cultured with beta-tricalcium phosphate ceramics and free vascularized fibula. Artif Organs 2006;30:960–2.
70. Kuroda R, Ishida K, Matsumoto T, et al. Treatment of a full-thickness articular cartilage defect in the femoral condyle of an athlete with autologous bone-marrow stromal cells. Osteoarthritis Cartilage 2007;15:226–31.
71. Wakitani S, Mitsuoka T, Nakamura N, Toritsuka Y, Nakamura Y, Horibe S. Autologous bone marrow stromal cell transplantation for repair of full-thickness articular cartilage defects in human patellae: two case reports. Cell Transplant 2004;13:595–600.
72. Kitoh H, Kitakoji T, Tsuchiya H, et al. Transplantation of marrow-derived mesenchymal stem cells and platelet-rich plasma during distraction osteogenesis—a preliminary result of three cases. Bone 2004;35:892–8.
73. Quarto R, Mastrogiacomo M, Cancedda R, et al. Repair of large bone defects with the use of autologous bone marrow stromal cells. N Engl J Med 2001;344.385–6.
74. Ohgushi H, Kotobuki N, Funaoka H, et al. Tissue engineered ceramic artificial joint—ex vivo osteogenic differentiation of patient mesenchymal cells on total ankle joints for treatment of osteoarthritis. Biomaterials 2005;26:4654–61.
75. Gangji V, Hauzeur JP. Treatment of osteonecrosis of the femoral head with implantation of autologous bone-marrow cells. Surgical technique. J Bone Joint Surg Am 2005;87 Suppl 1(Pt 1).106–12.
76. Chang CH, Stanton RP, Glutting J. Unicameral bone cysts treated by injection of bone marrow or methylprednisolone. J Bone Joint Surg Br 2002;84:407–12.
77. Stock UA, Vacanti JP. Tissue engineering: Current state and prospects. Annu Rev Med 2001;52:443–51.
78. Bianco P, Robey PG. Stem cells in tissue engineering. Nature 2001;414:118–21.
79. Kon E, Muraglia A, Corsi A, et al. Autologous bone marrow stromal cells loaded onto porous hydroxyapatite ceramic accelerate bone repair in critical-size defects of sheep long bones 10. J Biomed Mater Res 2000;49:328–37.
80. Warnke PH, Springer IN, Wiltfang J, et al. Growth and transplantation of a custom vascularised bone graft in a man. Lancet 2004;364:766–70.
81. Ohgushi H, Goldberg VM, Caplan AI. Repair of bone defects with marrow cells and porous ceramic. Experiments in rats 116. Acta Orthop Scand 1989;60:334–9.
82. Bruder SP, Fink DJ, Caplan AI. Mesenchymal stem cells in bone development, bone repair, and skeletal regeneration therapy 172. J Cell Biochem 1994;56:283–94.
83. Diduch DR, Jordan LC, Mierisch CM, Balian G. Marrow stromal cells embedded in alginate for repair of osteochondral defects. Arthroscopy 2000;16:571–7.
84. Lieberman JR, Le LQ, Wu L, et al. Regional gene therapy with a BMP-2-producing murine stromal cell line induces heterotopic and orthotopic bone formation in rodents. J Orthop Res 1998;16:330–9.
85. Gazit D, Turgeman G, Kelley P, et al. Engineered pluripotent mesenchymal cells integrate and differentiate in regenerating bone: a novel cell-mediated gene therapy. J Gene Med 1999;1:121–33.
86. Lee JY, Musgrave D, Pelinkovic D, et al. Effect of bone morphogenetic protein-2-expressing muscle-derived cells on healing of critical-sized bone defects in mice. J Bone Joint Surg Am 2001;83-A(7):1032–9.
87. Laurencin CT, Attawia MA, Lu LQ, et al. Poly(lactide-co-glycolide)/hydroxyapatite delivery of BMP-2-producing cells: a regional gene therapy approach to bone regeneration. Biomaterials 2001;22:1271–7.

88. Moutsatsos IK, Turgeman G, Zhou S, et al. Exogenously regulated stem cell-mediated gene therapy for bone regeneration. Mol Ther 2001;3:449–61.
89. Shen FH, Visger JM, Balian G, Hurwitz SR, Diduch DR. Systemically administered mesenchymal stromal cells transduced with insulin-like growth factor-I localize to a fracture site and potentiate healing. J Orthop Trauma 2002;16:651–9.
90. Peterson B, Zhang J, Iglesias R, et al. Healing of critically sized femoral defects, using genetically modified mesenchymal stem cells from human adipose tissue. Tissue Eng 2005;11:120–9.
91. Hasharoni A, Zilberman Y, Turgeman G, Helm GA, Liebergall M, Gazit D. Murine spinal fusion induced by engineered mesenchymal stem cells that conditionally express bone morpho-genetic protein-2. J Neurosurg Spine 2005;3:47–52.
92. Xu XL, Tang T, Dai K, et al. Immune response and effect of adenovirus-mediated human BMP-2 gene transfer on the repair of segmental tibial bone defects in goats. Acta Orthop 2005;76:637–46.
93. Tobita M, Uysal AC, Ogawa R, Hyakusoku H, Mizuno H. Periodontal tissue regeneration with adipose-derived stem cells. Tissue Eng Part A 2008;14(6):945–53.
94. Horwitz EM, Gordon PL, Koo WKK, et al. Isolated allogeneic bone marrow-derived mesenchymal cells engraft and stimulate growth in children with osteogenesis imperfecta: implications for cell therapy of bone. Proc Natl Acad Sci U S A 2002;99:8932–7.
95. Bentzon JF, Stenderup K, Hansen FD, et al. Tissue distribution and engraftment of human mesenchymal stem cells immortalized by human telomerase reverse transcriptase gene. Biochem Biophys Res Commun 2005;330:633–40.

Clinical Cell Therapy for Heart Disease

Christof Stamm, Boris Nasseri, and Roland Hetzer

Abstract Despite improvements in medical and surgical therapy, myocardial infarction or nonischemic heart disease is often the beginning of a downward spiral leading to congestive heart failure. Other than heart transplantation or implantation of a ventricular assist device "artificial heart," current therapies merely help the organism to survive with a heart that is working at a fraction of its original capacity. It is therefore no surprise that cardiac stem cell therapy, with its promise to regenerate or rejuvenate the heart, has raised many hopes.

Although extensive experimental data support the concept of cardiac cell therapy, neither the ideal source nor the type of cell or the critical quantity and mode of application in the clinical setting has been defined so far. In patients with acute myocardial infarction, several cell-based approaches, such as intracoronary delivery of mononuclear bone marrow cells or enriched hematopoietic progenitor cell products; systemic cytokine stimulation with release of bone marrow progenitor cells into the systemic circulation; and both intravenous or intracoronary delivery of allogenic mesenchymal stem cells, are currently being tested. There are encouraging data for each of these strategies based on small cohorts, but the results regarding recovery of function are ambiguous. For treatment of patients with chronic heart failure, other approaches, such as catheter-based intramyocardial delivery of various cell types and surgical implantation of bone marrow or blood-derived cells in conjunction with bypass surgery or as surgical stand-alone procedures, are being evaluated. Moreover, skeletal muscle-derived myoblasts have been used with various delivery techniques. Again, a number of controlled trials have produced conflicting results, and multicenter studies are currently being conducted. Currently, more advanced forms of cardiac cell therapy are attracting a lot of attention, including genetic cell engineering, coimplantation of specific extracellular matrix components, preclinical testing of embryonic stem cell–derived cells, as well as imaging techniques for in situ cell tracking and imaging of cell function. Prior to clinical

C. Stamm (✉), B. Nasseri and R. Hetzer
Deutsches Herzzentrum Berlin, Cardiothoracic Surgery, Augustenburger Platz 1, 13353, Berlin, Germany

H. Baharvand (ed.) *Trends in Stem Cell Biology and Technology,*
DOI 10.1007/ 978-1-60327-905-5_13,
© Humana Press, a Part of Springer Science+Business Media, LLC 2009

routine use, however, numerous technical, medical, and regulatory obstacles will have to be overcome. Here we will briefly summarize the basic research background of cardiac regenerative medicine and will attempt a critical appraisal of the current efforts to translate the experimental approaches into the clinical setting.

Keywords Cell delivery • Cell therapy • Heart disease • Stem cells

Introduction

In the public perception, medicine has traditionally progressed step-wise rather than continuously. Decades of tedious work and incremental advancements in the laboratory are often overlooked, when communication-active scientists and the media proclaim that a "new era" has begun. For instance, the last decades of the past century were dominated by advances in biomedical engineering, producing mechanical replacements for various organs and organ system functions that range from renal replacement therapy via orthopedic prostheses to cochlear implants. In cardiovascular medicine, electrical pacemakers, artificial heart valves, extracorporeal circulation technology, and implantable cardiac assist devices have undoubtedly saved many lives. After some time, however, it became clear that the human body tolerates such artificial replacement parts only up to a point. Hip prostheses often do not really heal in, pacemaker batteries expire, and many implants with blood contact are constant sources of thromboembolic problems and require lifelong treatment with anticoagulation drugs. There is a permanent risk of infection, especially when parts of the implant cross the skin barrier (i.e., drivelines of artificial hearts). Other organs, such as the immune system, simply cannot be supported by mechanical means, with the notable exception of eliminating unwanted immune system components by apheresis techniques. Solid organ transplantation, although invaluable for selected patients, also proved to be a limited and temporary solution because of the lingering shortage of donor organs and the still unresolved immunologic problems. Physicians and scientists therefore have come to realize that one should aim at a reconstitution of organ function rather than artificially replacing it. Today those efforts are summarized as "regenerative medicine," a ubiquitous term that describes essentially all strategies to improve health by stimulating the body's own capacity for healing. Many of these approaches have been around for a long time (such as cell transplantation for bone marrow reconstitution, tissue preparations for wound healing, or the dubious "organotherapy or fresh cell therapy") and today enjoy renewed attention.

Other components of today's regenerative medicine, however, are indeed fundamentally new developments, and one of those is the concept of increasing the number of contractile cells in the heart to cure heart failure, either by stimulating intrinsic regeneration processes or by adding exogenous cells. A summary of the overall perception of basic myocardial biology as predominant until the late 1990s

Table 1 A summary of the overall perception of basic myocardial biology as predominant until the late 1990s and in the year 2008

1990s	The heart does not regenerate at all. After birth, there is no hyperplasia, only hypertrophy of cardiomyocytes in response to injury or increased load. Collateral vessel growth in ischemic hearts is merely an enlargement of preexisting microvessels
2008	Cells that express myogenic stem cell markers reside in the normal heart and may participate in regeneration processes. By genetic cell cycle interference, myocytes can be forced to proliferate in rodent models in vivo. There is a constant turnover of blood vessel cells, and possibly myocytes as well, in the heart. New blood vessels form in response to ischemia, and their development can be enhanced by delivery of exogenous bone marrow-derived cells

and in the year 2008 is presented in Table 1. This chapter will briefly summarize the basic research background of cardiac regenerative medicine and will present a critical appraisal of the current efforts to translate the experimental approaches into the clinical setting. Since some potential readers are not cardiologists or cardiac surgeons, a brief overview of the underlying disease and treatment options will also be provided.

The Biologic Basis of Heart Failure

Heart failure is not a uniform disease but has a variety of causes. Quantitatively, the most important is clearly ischemic heart disease, i.e., coronary artery disease. Although the risk factors for coronary arteriosclerosis have been well established, it remains unclear why a particular individual develops coronary artery disease while another with a similar combination of risk factor does not. Other mysterious phenomena are the extent and the progression rate of the disease. In some patients, an obstructive coronary artery lesion remains unchanged for many years, while it can progress to subtotal occlusion of the vessel in a matter of several months in others. Similarly, many patients present with isolated proximal stenoses of one or more coronary arteries, while the more distal part of the vessel appears completely normal, whereas other patients develop diffuse disease of the entire coronary vascular tree. The other causes of heart failure are usually grouped together as "nonischemic" and include genetic predisposition, inflammatory heart disease (viral myocarditis, Chagas disease), toxic myocardial damage (doxorubicin, alcohol, cocaine), and structural defects such as valvar disease or congenital abnormalities. Often, however, the underlying cause cannot be clearly established and the term *idiopathic dilated cardiomyopathy* is used. Although in the majority of patients heart failure is predominantly systolic, i.e., the heart fails to eject a normal amount of blood at normal blood pressure, there are some patients in whom the heart cannot properly fill during diastole because the relaxation and expansion of the cardiac chambers are restricted. If the onset of heart failure is sudden (acute heart failure) and neither the heart itself nor the other organ systems have had time to adjust to the reduced cardiac output, cardiogenic shock results. The symptoms of this life-threatening situation are mainly the result of a reduced blood flow *from* the heart (termed *forward failure*). If heart failure develops more gradually and allows the organism to exert compensatory mechanisms, symptoms are often those of a reduced blood flow *to* the heart (termed *congestive heart failure*).

Myocardial Infarction

The mechanisms leading to destruction of cardiomyocytes in nonischemic heart failure are still poorly understood. In ischemic heart failure, however, the loss of contractile tissue is readily explainable. Even at rest, the heart must perform a tremendous amount of work to maintain the circulation. Therefore, oxygen consumption is very high as compared with other organs, and the difference in coronary arterial and venous oxygen content ($AVDO_2$) is large even at rest. During exercise, cardiac output and hence work load can rapidly increase severalfold, and the myocardial demand for oxygen rises accordingly. Because $AVDO_2$ is already very high at baseline, the increased demand can only be met by an increase in coronary blood flow. This explains why the heart has such a low tolerance to coronary perfusion problems. A complex network of intrinsic and extrinsic modulators serves to adjust coronary flow when work load changes, but this regulation is effectively shut down when a relevant obstruction to coronary flow is present. Once the narrowing of a major coronary artery exceeds 50% of its cross-sectional area, blood flow to the downstream myocardium becomes insufficient when demand increases. A narrowing greater than 90% is often associated with angina at mild exertion. Acute myocardial infarction occurs when the endothelial surface of an atheromatous plaque ruptures, exposing highly thrombogenic subendothelial matrix components and thus inducing platelet aggregation and thrombus formation. The resulting complete interruption of coronary blood flow leads to an immediate cessation of contractility in the downstream myocardium. Within minutes, oxygen tension falls to almost zero, and the highly energy-dependent cardiomyocytes react by switching to anaerobic energy production. This compensation attempt is not sufficient, and the concentration of energy-rich phosphates inevitably declines, while protons, adenosine diphosphate, phosphate, potassium, and calcium accumulate and contribute to the complete breakdown of the fragile cellular energy balance. Unless the myocardium is reperfused within a few hours, a substantial number of cardiomyocytes will succumb to necrosis and apoptosis. The release of intracellular contents from necrotic cardiomyocytes attracts phagocytic leukocytes, leading to inflammatory infiltration of the ischemic tissue. For some time, lytic processes dominate and render the ischemic myocardium very fragile and prone to sudden rupture. Within several weeks, however, collagenous scar tissue develops and reinforces the necrotic myocardium.

The Heart's Response to Disease

Once a substantial amount of contractile tissue is lost and systolic function is impaired, the heart chambers progressively enlarge. Within certain limits, the increase in size helps to maintain cardiac output by preloading the left ventricle with a higher blood volume and by exploiting the preload-dependent regulation of cardiomyocyte contractility. When the wall tension of the enlarged ventricle exceeds a certain limit, however, those compensatory mechanisms give out and heart failure decompensates (Fig. 1). Complex changes in response to injury also occur on the cellular level. There are numerous adaptive and maladaptive alterations of protein expression that affect membrane receptor composition, ion homeostasis, oxidative metabolism, and contractile protein function; a detailed description of those is beyond the aims of this chapter. Particularly in response to increased pressure load, the most obvious response of the myocardium is hypertrophy. Here, myocyte diameter increases, as does the cellular content of contractile protein units. The traditional paradigm is that hypertrophy is not associated with an increase in cardiomyocyte number in the heart. Such hyperplasia occurs in many other organs, i.e., the liver, but cardiomyocytes have been said to permanently rest in the G_1/G_0 phase of the cell cycle, precluding mitotic cell division. Indeed, mitotic nuclei are extremely difficult to detect in the heart by light microscopy. The concept of the heart as a completely postmitotic organ is also based on the lack of macroscopically observable myocardial regeneration after myocardial infarction. On the other hand, it has been difficult to reconcile with the large body of quantitative data on cardiomyocyte apoptosis in various diseases. Even when the most cautious number of cell undergoing apoptotic cell death at a given time point is used for computation, every estimate indicates that the entire mass of cardiomyocytes would have disappeared

LVEF	NYHA	Relevance
<25%	III – IV	Symptoms at rest, survival a few years
25-35%	III	Symptoms at mild exercise, survival several years
35-45%	II - III	Symptoms at strong exercise, survival many years
>45%	I - II	No symptoms, survival normal

Fig. 1 Makeshift classification of left ventricular ejection fraction, the key end point of virtually all clinical cell therapy trials, grouped according to the usual clinical symptoms. Obviously, there are numerous exceptions, i.e., a patient with very low left ventricular ejection fraction (LVEF) can have very few symptoms, whereas as patient with near-normal LVEF can suffer from congestive heart failure. The intention is to show that, on average, one needs to achieve a greater than 10% increase in LVEF so as to clearly improve the patient's general state of health (*see Color Plates*)

after a rather brief period of time. Unless, of course, myocytes are constantly replaced *(1)*. The crucial questions here is whether myocytes are replaced (a) by mitotic division of preexisting cardiomyocytes, (b) by proliferation and differentiation of endogenous myocyte progenitor cells (see below), or (c) by cells originating from the bone marrow—or elsewhere—that have migrated to the heart. The most convincing evidence of a permanent cellular turnover in the adult human myocardium, supporting the notion that cells, including cardiomyocytes, are indeed being generated in the adult heart, comes from allogenic, gender mismatched human organ transplantation. After a female has received a bone marrow transplant donated by a male, male cells can be detected in the heart many years after the transplant, predominately in the coronary vasculature, but also in the interstitium and the myocardium *(2–4)*. Similar observations have been made in donor hearts from females that were transplanted into a male recipient *(5–10)*. Again, one can detect numerous male cells in the originally female heart. Even in mothers who gave birth to boys, male cells that have entered the maternal bloodstream before or during birth have been detected in the heart. The same phenomenon should also occur with female cells, but male cells in female hearts are the easiest to detect.

The Original Concept of Cardiac Cell Therapy

Replacing Muscle Cells

The primary goal of cardiac cell therapy is to increase the number of contractile cells in the ventricular myocardium so that systolic heart function can return to normal. Although this sounds self-evident, this concept has recently faded from the spotlight in favor of surrogate theories that might help explain a functional benefit of cardiac cell therapy when clearly no or very few contractile neocells are produced. Those include poorly understood paracrine effects that may support angiogenesis, endothelial differentiation as another means to stimulate blood vessel growth, modulation of extracellular matrix components either strengthening the fibrous scaffold or the heart or preventing unwanted scarring, direct supportive effect on cardiomyocytes suffering from ischemic stress, and even stimulating interactions with resident cardiac progenitor cells. Unfortunately, the initially straightforward approach (i.e., replacing diseased cardiomyocytes with transplanted contractile cells) has turned into a virtually inscrutable, dazzling array of different working hypotheses.

In the 1990s, a few investigators had the idea to apply the then very new concept of cell transplantation for solid organ failure to heart disease *(11–13)*. Initially, readily available myocyte cultures and neonatal rat cardiomyocytes were used for transplantation in the heart *(14–17)*, and the mere notion that transplanted contractile cells may be able to be incorporated in postnatal myocardium was revolutionary. Once the survival of transplanted cells in the heart had been demonstrated and the huge potential of this finding was slowly comprehended, those cells were applied to experimental models of myocardial infarction or nonischemic cryolesions, and it was shown that there may indeed be a benefit in terms of contractile function. At the same time it became clear that transplanted cardiomyocytes, being just as vulnerable as native cardiomyocytes, will have problems surviving in terminally ischemic infarcted tissue (18). Before long a very elegant solution for this problem emerged, that is to use skeletal muscle progenitor cells derived from ubiquitous satellite cells for heart repair (12, 19). Those cells, as well as their progeny, skeletal myofibers, have a very high tolerance to ischemia and the capacity to maintain contractile work even through prolonged periods of anaerobic metabolism. Following the first ground-breaking report, research activities on skeletal myoblast transplantation in the heart rapidly escalated. Several groups showed that myoblasts form contractile neotissue even in scar tissue after myocardial infarction, and the data were very consistent for many different experimental models. Industry quickly sought to obtain intellectual property of the underlying technology, which proved challenging when dealing with an autologous cell product. However, the complex infrastructure necessary for clinical-grade manufacturing of myoblast products, as well as the technology for catheter-based delivery devices, provided a promising enough business model. Although the preclinical development of catheter-based treatment modalities took somewhat longer, myoblast transplantation as part of a surgical procedure was finally introduced into the clinical arena in 2001 (20). In patients who had suffered from myocardial infarction some time ago and who had to undergo coronary artery bypass grafting (CABG), autologous myoblasts were implanted directly into the postinfarct scar tissue by straightforward transepicardial injection. Initial feasibility studies were successful and laid the foundation for the avalanche of cell therapy studies that were to come later (21). Once a large number of patients had undergone CABG and myoblast transplantation, however, problems with ventricular arrhythmia were noted. Consequently, clinical studies were limited to patients who were already carrying an implantable defibrillator device, and the biologic basis of this pro-arrhythmic effect was studied in further experiments. It soon became clear that skeletal myoblasts lack the capacity to electrically couple with surrounding cardiomyocytes because they do not express the intercellular communication protein connexin 43 and thus do not form "connexon" ion channels that are part of the gap junction typical for cardiomyocytes (22). This is not surprising since individual skeletal myoblasts and their myocytes progeny fuse to form multinucleated myofibers, which are connected with one specific motoneuron and are electrically isolated from their neighbors. This is the prerequisite for the rapid and fine brain-controlled adjustment of skeletal muscle contractile force. Cardiomyocytes, in contrast, maintain their single-cell integrity but connect with their neighbors via gap junction connexons to form a functional syncytium that allows for the propagation of excitation throughout the entire myocardium. Therefore, skeletal myofibers in the heart, although they readily survive, remain isolated from the surrounding myocardium, disturb the excitation wave, and may act as arrhythmogenic foci.

Clinical trials using skeletal myoblasts and catheter-based delivery devices are still ongoing, and some investigators report that they have never encountered arrhythmia problems *(23–25)*. However, given that observed improvement in contractility is, at best, very mild, the majority of clinicians have abandoned skeletal myoblasts for treatment of heart failure.

Embryonic Stem Cell Technology

Although intramyocardial transplantation of somatic contractile cells or their immediate progenitors has lost a lot of its initial momentum, experimental work with embryonic stem cells for myocyte reproduction has steadily progressed *(26)*. There is evidence that the host myocardium can control the specific cardiomyocyte differentiation of a limited number of implanted embryonic stem cells, but once a certain threshold has been exceeded, uncontrolled proliferation and differentiation with teratoma formation occur. Researchers and industry have therefore focused on predifferentiated cardiomyocytes from embryonic stem cells, which can, theoretically, be produced in large quantities in vitro prior to implantation into the diseased heart *(27–30)*. Clinical translation of this technology, however, is still hampered by several fundamental biologic and biotechnological problems:

(a) Theoretically, even a single naive embryonic stem cell (ESC) can give rise to a teratoma in the heart. Therefore, ESC-derived myocytes or myocyte progenitor cell products must have 100% purity. Although this should theoretically be possible, it requires extremely complex and reliable cell processing techniques.
(b) The immunogenicity of RSCs and their in vitro progeny is incompletely understood, and it is unlikely that such cell products can be transplanted in allogenic fashion *(31)*.
(c) The debate on the ethics of ESC procurement from viable human embryos severely hampers further development of ESC technology in many countries.
(d) There is accumulating evidence that embryonic stem cell–like cells can be produced without the need to destroy an embryo, by therapeutic cloning, reprogramming of somatic cells, or from germ cell progenitors in adult organisms. However, nuclear transfer (therapeutic cloning, recently performed in primates *(32)*) results in a hybrid cell with nuclear DNA from one organism and mitochondrial DNA from another. The consequences are not known. Somatic cell reprogramming, i.e., de-differentiation (see below) requires virus-mediated transfection with several exogenous genes.
(e) Finally, ESC-like cell production from germ cell progenitors has so far only been successful in rodents *(33, 34)*.

Taken together, embryonic stem cell technology for myocardial cell therapy will probably not reach the clinical arena in the foreseeable future, at least not in the form of well-designed trials that are based on a substantial body of systematic preclinical data. Nevertheless, it must be assumed that individual researchers and clinicians push for "renegade" applications of embryonic stem cells or related cell products for treatment of heart disease, as has already been the case with other diseases.

The Pluripotent Bone Marrow Cell Letdown

Hematopoietic Stem Cells

As mentioned above, research on cardiac cell therapy focused initially on transplantation of contractile cells or their immediate progenitors. In parallel, the concept of circulating endothelial progenitor cells originating from the bone marrow evolved, and before long, marrow-derived cells of hematopoietic–pre-endothelial lineage were shown to be usable for induction of angiogenesis in the ischemic heart. Following the largely disappointing clinical results of growth factor protein or gene therapy for intractable myocardial ischemia, the prospect of using autologous cells for induction of blood vessel growth in the heart was intriguing. The potent pro-angiogenic capacity of marrow-derived cells was initially assessed in the standard mouse hindlimb ischemia model, and Kocher et al. *(35)* were among the first to successfully use human CD34+ cells in a rat model of myocardial infarction. As is often the case in rodent models, the increased growth of small blood vessels in the infarcted heart was associated with a marked improvement of contractility. In large animals, however, the impact of neoangiogenesis on contractility is less pronounced, while in humans it is often negligible. The most significant, apparent breakthrough, however, was reported in 2001 *(36)* c Kit+ lin-cells were isolated from the bone marrow of green fluorescent protein (GFP)–expressing transgenic mice (which, according to the usual classifications, are not strictly hematopoietic stem cells) and implanted in the infarcted myocardium of non-GFP expressing animals. The assumption was that cells and tissue arising from the transplanted cells will be GFP positive, appear green on fluorescence microscopy, and indeed both GFP positive blood vessels and contractile cells were visualized. The no less than revolutionary implication was that adult bone marrow stem cells can readily differentiate into both endothelial cells and cardiomyocytes, presumably driven by factors present in the surrounding infarcted host myocardium. This report found a tremendous echo and immediately led to clinical pilot studies all over the world, where mainly autologous bone marrow mononuclear cells were delivered to the heart of patients with myocardial infarction. Other investigators, however, doubted the surprising plasticity of unmodified adult bone marrow stem cells. In 2004 reports were published that, using state-of-the-art methods, failed to detect relevant cardiomyocyte differentiation of murine hematopoietic cells in vivo *(37)*. Controversy on this issue remains at this writing, and variations in the technical details of the various experiments are used to help explain the different outcomes. Although hematopoietic stem cells do seem to be helpful in the treatment of ischemic heart disease, it must be acknowledged that the evidence supporting myogenic differentiation of CD34+ cells and phenotypically similar cell populations is dwindling. Even if the occasional hematopoietic cell can be driven to express several myocyte-specific markers, the frequency of such events is too small to guarantee a significant clinical effect *(38)*.

Mesenchymal Stem Cells

Besides the dominant hematopoietic compartment, the bone marrow contains an extensive cell-rich stroma that was long believed to simply support the proliferation of hematopoietic cells and the surrounding bone *(39)*. In the 1990s, however, it became clear that many stroma cells have the capacity to self-renew and to differentiate into lineages that normally originate from the embryonic mesenchyme (connective tissues, blood vessels, blood-related organs) *(40–42)*. Using pharmacologic stimulation in standardized assay systems, bone marrow–derived mesenchymal stem cells (MSCs) can be easily induced to differentiate into bone, cartilage, and fat cells. Thus, MSC-based cell products have already found clinical applications for regeneration of cartilage and bone defects. Moreover, MSCs have been found to have strong immunosuppressive effects in vitro and in vivo *(43, 44)*. Intuitively, it appears logical that the marrow stroma consists of cells that can control the immunologic activity of billions of leukocytes before they are released into the circulation. MSC cultures are easily separated from other bone marrow cell populations based on their ability to adhere to plastic culture dish surfaces, and they usually proliferate well in vitro and obtain a fibroblast-like phenotype. There are numerous reports that describe the differentiation of MSCs in various nonconnective tissue phenotypes. Quantitative MSC differentiation into cells of cardiomyocyte morphology and function was first described in 1999 *(45)*. It is important to note, however, that here epigenetic modulation of the transcriptional profile by DNA-demethylation with 5-azacytidine was employed. Demethylation may reduce the stability of DNA silencing signals and thus confer nonspecific gene activation. Under the influence of 5-azacytidine, MSCs obtained a wide range of different phenotypes out of which spontaneously beating myocyte-like cells were isolated and selectively expanded. The resulting cell population showed many of the morphologic, proteomic, and functional characteristics of true cardiomyocytes *(46)*. Convincing strategies to achieve a similar result by small molecule drug stimulation of MSCs without direct epigenetic modification have not yet been described. Taken together, cardiomyocyte differentiation of mesenchymal stem cells appears to be possible. Whether physiologic in vivo signals are sufficient to drive naive MSCs into a myogenic lineage without significant exogenous influence, however, is still unclear *(47–49)*. On the other hand, the effects of chemically induced DNA demethylation on long-term cellular behavior are not known and may pose a significant limitation to cell product safety. Occasionally, subpopulations of MSCs have been described that seem to have a greater "stemness," such as multipotent adult progenitor cells (MAPCs) *(50, 51)*. Such cells are presumably immature progenitors of classic MSC, may have a higher proliferative capacity, and may be driven more easily to differentiate into various nonconnective tissue phenotypes. Isolation of such cell populations, however, is usually tedious and has been difficult to reproduce by other groups.

Autologous Marrow Cell Pitfalls

The clinical use of autologous marrow-derived cells for myocardial regeneration obviously avoids the immunologic problems of allo- or xenotransplants. On the other hand, it means that cells from a chronically ill and often elderly organism are expected to perform a tremendous amount of regeneration and "rejuvenation" work. Traditionally, the cardiovascular system on one hand and the bone marrow–blood system on the other have been considered as two completely separate entities. This view, however, can no longer be held. For instance, it has become clear that the human heart is populated by marrow-derived cells throughout adulthood, although it is still controversial whether this is true for cardiomyocytes as well as for blood vessel and interstitial cells *(52)*. In fact blood vessels of the entire vascular tree appear to be constantly repaired by marrow-derived cells *(4)*, and there is a close correlation between cardiovascular disease and bone marrow–derived progenitor cell function *(53–56)*. The cause–effect relationship, however, is still unclear, i.e., do impaired marrow cells cause cardiovascular disease or vice versa, or does a third pathomechanism affect both systems independently? Moreover, cardiovascular disease is predominantly a disease of aging patients, and advanced age alone has a strong impact on marrow cells. Along with osteoporosis and diminishing bone marrow volume, the percentage of $CD34^+$ and $CD133^+$ stem cells decreases in the elderly. The same is the case for bone marrow stroma cells and thus for mesenchymal stem cells. When MSCs from elderly patients are cultivated, they proliferate slower, and their colony-forming ability, differentiation capacity, and paracrine activity also decrease. Therefore, trying to regenerate the heart with autologous cells that have long suffered from the effects of aging and chronic disease is intrinsically problematic. Possible solutions other than the use of allogenic cell products would be "rejuvenating" preconditioning strategies of marrow-derived cells prior to their use for cellular therapy or the use of autologous cells that were harvested and stored early in life. The latter concept is actively advocated by the cord blood banking industry. Both proliferation rate and functional capacity are generally believed to be higher the younger the cell and the healthier its donor is. Whether all of this is true for cord blood cells cannot be said with certainty. Detailed comparative analyses regarding the nonhematopoietic regeneration capacity have just begun, and preliminary data vary greatly between different subsets cord blood cells *(57–60)*.

Clinical Translation Issues

Although the experimental basis of myocardial cell therapy is incomplete, numerous clinical trials, but also nonscientific applications, have already been initiated. When choosing a particular myocardial cell therapy approach in the clinical setting, attention typically focuses on the cell product to be used (Fig. 2). However, there are numerous other factors that need to be considered to maximize the likelihood of successful cell-based myocardial regeneration.

Patient Selection

The majority of patients who have undergone myocardial cell therapy suffer from ischemic heart disease. In most of those, treatment was performed within several

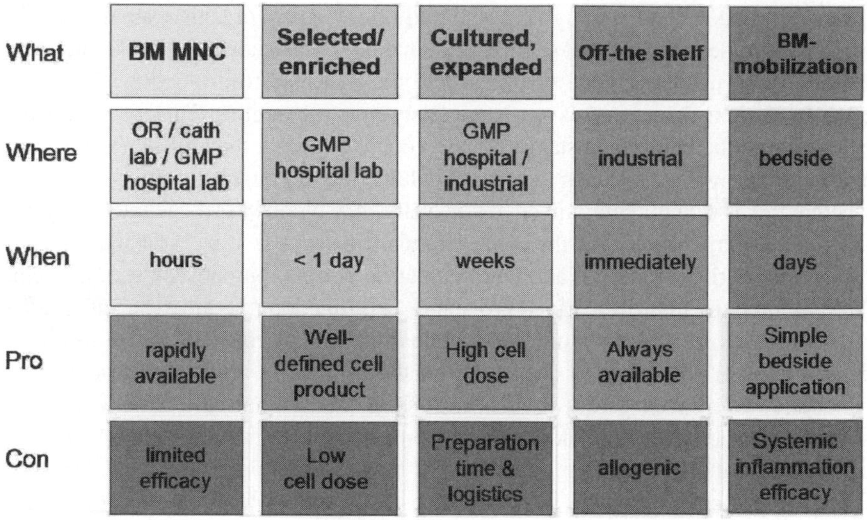

What	BM MNC	Selected/ enriched	Cultured, expanded	Off-the shelf	BM-mobilization
Where	OR / cath lab / GMP hospital lab	GMP hospital lab	GMP hospital / industrial	industrial	bedside
When	hours	< 1 day	weeks	immediately	days
Pro	rapidly available	Well-defined cell product	High cell dose	Always available	Simple bedside application
Con	limited efficacy	Low cell dose	Preparation time & logistics	allogenic	Systemic inflammation efficacy

Fig. 2 Clinically relevant principle features of the currently available cell produc ts (*see Color Plates*)

days after acute myocardial infarction. With few exceptions, such patients are treated by interventional cardiologists using catheter-based techniques, and cells are injected into the reopened coronary artery. Cell injection directly into the heart muscle is prohibitive in this situation, because the acutely ischemic myocardium is weakened and there is a high risk of mechanical injury. Intracoronary cell injection in acute myocardial infarction can be performed without major procedural and infrastructural problems, provided that the cell product is rapidly available. One major disadvantage, however, is that the patients cannot be selected according to their pretreatment heart function. Myocardial contractility is always somewhat compromised during the acute phase, but recovers spontaneously over a few weeks in many patients. Consequently, such patient cohorts cover a wide range of left ventricular contractility at baseline, rendering statistical analysis and interpretation of the data more difficult than in a uniform group of patients. A smaller number of patients with coronary artery disease have been treated for chronic myocardial ischemia, with varying degrees of resulting heart failure. Here it is easily possible to select patients according to their pretreatment heart function. At the same time, there are several options for cell delivery, including intracoronary injection (as in acute infarction patients), transepicardial injection in the myocardium involving surgical procedures, and transendocardial injection in the myocardium using a catheter-based system. Finally, patients with nonischemic heart disease can also be subject to cell therapy approaches, although the body of preclinical data is much smaller than for ischemic heart disease. A problem is the diversity of underlying etiologies for nonischemic heart failure, which include infectious and inflammatory processes as well as genetic diseases and a large group of idiopathic cases where no cause can be established. Ideally, a novel therapeutic approach is evaluated in a uniform cohort of patients with well-defined patient- and disease-related characteristics. With respect to cell therapy for heart disease, however, this is more difficult than it seems at first glance.

Concomitant Procedures

In acute myocardial infarction patients, cell therapy is usually performed several days after catheter-based revascularization of the occluded infarct vessel. In patients with chronic myocardial ischemia, intramyocardial cell delivery is most often done at the time of a coronary artery bypass operation that is indicated for triple vessel coronary artery disease. Some surgeons also perform intramyocardial cell injection as a stand-alone procedure through a small incision in the chest. There are also reports of studies in which intramyocardial cell therapy was performed with heart valve surgery, usually when the mitral valve is incompetent secondary to an enlargement of the failing left ventricle (61). Another option is the intramyocardial delivery of cells at the time of implantation of a ventricular assist device. Here end-stage heart failure patients with the most urgent need for innovative therapeutic strategies are addressed, and there is a chance to collect myocardial tissue samples before and after cell treatment in case the patient later undergoes explantation of the assist device or heart transplantation. A further possibility is the combination of myocardial cell therapy with transmural myocardial laser revascularization (TMLR), with or without additional bypass surgery (62). In TMLR, a laser is used to produce small channels of 1-mm diameter in the left ventricular wall. The initial concept was to induce transendocardial blood flow directly into the ischemic tissue. Today the working hypothesis is that the laser injury induces a local inflammatory response that may stimulate transplanted bone marrow cells and enhance their regenerative potential. However, little experimental evidence supports this hypothesis.

Cell Delivery Techniques

As mentioned above, the most frequently used way to delivery cells to the heart is catheter-based injection into the coronary arteries (Fig. 3). Normally, routine angiography catheters are being used, but there are also special cell injection catheter products that are said to minimize the loss of viable cells in the long catheter system. Often a balloon is inflated to occlude the coronary artery for a few minutes during cell injection, so that cells are not immediately flushed out. Whether this truly improves cell retention in the heart is unknown. Obviously, intravascular cell delivery requires migration of cells through the vascular wall into the myocardium, a phenomenon that is poorly understood. Using catheter-based systems, direct intramyocardial cell delivery is also possible. A catheter is guided into the left ventricular cavity, and an injection needle at the tip is pushed into the heart muscle.

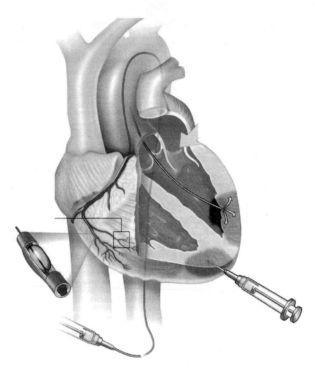

Fig. 3 Techniques to deliver cell products to the heart in the clinical setting. From *left* to *right*: Intracoronary injection with inflated percutaneous transluminal coronary angioplasty (PTCA) balloon; catheter-based transendocardial injection in the heart muscle; transepicardial injection requiring surgical access to the heart (*see Color Plates*)

This procedure is often combined with intracardiac electrical mapping of the left ventricle to identify the ischemic area of interest (i.e., NOGA® XP Cardiac Navigation System, Cordis Corp, Miami Lakes, FL). There is also a catheter system that is forwarded into the epicardial vein system and allows for injection into the heart muscle from the outer surface of the myocardium.

In contrast, cell delivery using surgical techniques is straightforward. Any commercially available syringe and needle system can be used to inject a cell suspension into the heart muscle under direct vision, but the industry has also developed special cell injection needles with side holes. The target area can be identified by visual inspection and consideration of preoperative imaging data. When TMLR is also planned, a device can be used that allows for one-step laser revascularization and cell injection.

In several clinical trials, cells (i.e., mesenchymal stem cell products) are being delivered by peripheral venous injection into patients with subacute myocardial infarction. It is believed that the myocardium produces homing factors, chemokines that are potent enough to attract stem cells from the periphery. A fraction of the intravenously injected cells indeed ends up in the heart, but the majority undergo first-pass trapping in the lung or are eliminated by the reticuloendothelial system in liver and spleen. An alternative systemic "delivery" approach is to induce the release of stem (and other) cells from the heart by cytokine stimulation. This can be achieved by treatment with granulocyte colony-stimulating factor (G-CSF) for several days, which leads to a significant increase in circulating $CD34^+$ and $CD133^+$ cells and is routinely done prior to stem cell harvest from peripheral blood. In experimental models it was demonstrated that such mobilized stem cells can migrate to the infarcted heart where they possibly participate in regeneration processes. A problem in the clinical setting is that stem cell mobilization is inevitable associated with the release of large numbers of mature leukocytes, mostly granulocytes, from the bone marrow. The resulting systemic inflammatory burden may be detrimental in very sick patients with acute infarction, patients undergoing surgery, and patients who have coronary artery stents implanted.

Timing

Acute Infarction

It is not known whether there is an ideal time point for cell therapy in patients with heart disease. As mentioned above, the majority of patients who are recruited for myocardial cell therapy are those with acute or subacute myocardial infarction. Here the onset of myocardial ischemia was usually between several hours and a few days ago. Typically a patient is admitted with acute chest pain and an electrocardiogram (ECG) shows signs of myocardial ischemia. Laboratory tests indicate the onset of myocardial necrosis and liberation of cardiomyocyte-specific intracellular proteins (i.e., creatine kinase MB, troponin) following loss of cell membrane integrity. If possible, the blocked coronary artery is immediately reopened by emergency catheterization, balloon dilation, and stent placement. The extent of myocardial necrosis and thus the impairment in contractility largely depend on the time that has passed until the infarct vessel was reopened. There is no way to predict the ultimate infarct size in a given patient. Ideally, cardiomyocyte necrosis is completely prevented because the coronary artery has been quickly reopened. Acutely ischemic cardiomyocytes can still be vital but have temporarily ceased contractile work (myocardial stunning). Tissue infiltration with inflammatory cells is beginning, but fibrous scarring has not yet occurred. The emergency treatment consists, whenever possible, of immediate cardiac catheterization with balloon dilation and stent placement in the infarct vessel. This is being done by the interventional cardiologist, who may also decide to perform intracoronary injection of a rapidly available cell product. It has been shown that the myocardium produces homing factors such as stromal cell–derived factor-1 (SDF-1) during the acute infarction phase, chemokines that may help attract stem cells from the bloodstream to the myocardium. When patients with acute infarction need emergency surgery, it is usually not feasible to arrange for concomitant cell therapy.

Chronic Ischemia

The situation in chronic ischemic heart disease is fundamentally different. The most recent myocardial infarction in those patients usually dates back several weeks, months, or even years. In the infarcted or chronically ischemic myocardium, a substantial net loss of contractile tissue mass has occurred, increased collagen deposition has led to a more or less pronounced diffuse or localized scar formation, and blood supply to the myocardium remains impaired, although there may have been collateral vessel growth. Usually the ischemic myocardium in those patients is not a complete transmural fibrous scar, which would eventually progress into a left ventricular aneurysm, but still vital cardiomyocytes are dispersed within the fibrous network. Theoretically, such "hibernating" cardiomyocytes can be rerecruited for contractile work once sufficient supply of oxygen and nutrients has been re-established, and stem or progenitor cell–induced growth of microvessels in the infarct border zone may thus translate into improved myocardial contractility. Nevertheless, it should be kept in mind that the chances to resuscitate hibernating cells in a functionally relevant fashion are likely to decrease with time. It is often said that the longer the interval between myocardial infarction and cell treatment is, the smaller is the chance to achieve a beneficial effect becomes. This notion, however, is largely intuitive and presently not supported by clear-cut data.

Outcome Measurement

To the clinician, the functional effects of myocardial cell therapy matter so much more than histologic evidence of neoangiogenesis or neomyogenesis. The most reliable of the many available surrogate indicators of heart function is ventricular ejection fraction, i.e., the percentage of the blood volume present in the filled ventricle that is ejected into the aorta during systole. Direct measurement of this blood volume involves elaborate invasive techniques, and an estimation of the ventricular cavity volume based on cardiac imaging techniques is used instead. Here a volume in three dimensions must be estimated based on two-dimensional images. Echocardiography is readily available but can be complicated by a poor image quality (termed *echo window*) in obese and in postoperative patients. Quantitative results are also said to be highly observer dependent. Cardiac magnetic resonance imaging (MRI) is considered the current gold standard, but it cannot be performed in patients who carry implants such as pacemakers or implantable defibrillators. The result of neoangiogenesis processes, improved myocardial perfusion, can be directly measured using scintigraphy scans with radioactive tracers. Supporting data on tissue viability can be obtained directly using positron emission tomography (PET) scan, where the uptake of radioisotopes in viable tissue is visualized. Myocardial viability can also be estimated indirectly, based on scintigraphic perfusion scans or MRI data. By combining several of these methods, a detailed picture of global and regional myocardial behavior can be obtained, but this does not necessarily correlate with the patient's overall state of health. Therefore, exercise tolerance measurements and quality-of-life questionnaires should always be performed in addition. The ultimate proof of efficacy in cardiovascular evidence-based medicine, however, is a reduction in major adverse cardiac events (MACEs), which include death, myocardial infarction, and the need for intervention or surgery. A meaningful analysis of such data, however, requires large patient cohorts and long follow-up periods.

Cell Imaging

The fate of cells in experimental models can be quite accurately assessed by histology, but in vivo imaging of cells in the human heart is much more challenging. Techniques that allow for visualization of biologic processes in particular cells in vivo are under development, but the clinical investigator is currently limited to physical cell labeling methods. For instance, radioisotopes or iron particles can be loaded into the cells or bound to the cell membrane prior to transplantation and later visualized by scintigraphy, MRI, or even echocardiography *(63–66)*. What all those methods have in common, however, is that the tracer material is visualized, not the cell itself. When the migration of cells to the heart over a brief period of time is in question, meaningful data can be obtained, but with respect to long-term persistence, viability, or even phenotype of cells in the heart, tracer-based imaging methods are of very limited use.

Regulatory Issues

The legal framework that governs the clinical application of myocardial cell therapy differs from country to country, and often even within a country. In most of the industrialized countries of the Western world, cardiac cell therapy is nowadays subject to medicinal product licensing legislation. The rationale is that in cardiovascular medicine, cell products are subject to nonhomologous use, that is, marrow cells normally involved in blood cell generation are asked to do an entirely different job in the heart. In contrast, when marrow cells are being used for bone marrow reconstitution or blood cells for blood replacement, their application is homologous and thus regulated under the transplantation or transfusion legislation. Therefore, every clinician involved in the collection of bone marrow, blood, or other cell sources, as well as those preparing the actual cell products, take part in the manufacturing process of a medicinal product, and all procedures involved must comply with good manufacturing practice (GMP) and good clinical practice (GCP). The legal issues become even more challenging when cells are being manipulated ex vivo or perhaps combined with biologic scaffolds or matrices. Such tissue engineering products are neither medical devices nor typical medicinal products, and the European Union has therefore recently published a novel regulation that governs the production and application of such "Advanced Therapies," including tissue engineering, cell therapy, and gene therapy. For most physicians and surgeons in clinical cardiovascular medicine, dealing with those complex regulations is new and often unnerving. Therefore, it is recommended to obtain counseling from the responsible government agencies in the early planning stages of a clinical cell therapy program.

Cell Survival

When cells are injected into diseased myocardium, it is likely that most of them will not survive. The cell preparation process is usually well tolerated. Skeletal myoblasts or other ex vivo expanded cells, as well as processed primary cells, normally have viability rates higher than 90%. It is often argued that injection of cell suspension through a needle or a catheter significantly compromises cell viability. In our experience, this is not the case. Even injection through a long cardiac catheterization device has very little effect on cell survival. Once the cells have entered the myocardial interstitium, however, many of them appear to succumb to necrotic or apoptotic death. Its magnitude is difficult to determine, but the suggested survival rate over days or weeks ranges between 0.1 and 10% (*18, 67*). The causality is probably multifactorial. First, the ischemic myocardium is obviously a hostile environment, due to local hypoxia, acidosis, lack of substrates, and accumulation of metabolites. Skeletal myoblasts are known to be quite resistant to ischemia, but little is known about the energy requirements of hematopoietic or mesenchymal stem cells. Second, necrotic myocardium is subject to infiltration with phagocytic cells that remove cell debris and initiate the scarring process. Although the stem or progenitor cells used clinically are autologous, many are probably lost in this clean-up process. Third, the mechanic forces that are present in the myocardium may play a role. Transmural pressure is high during systole, and there are shear forces between contracting myofibers and layers. Again, a skeletal myoblast might be able to tolerate this, but a marrow cell is certainly not well equipped to withstand such stress. Fortunately, the rate of cell death can be slowed by targeted manipulation of the cells to be injected. Transfection of marrow stromal cells with genes encoding for the antiapototic proteins Akt or Bcl-2 has been shown to greatly improve cell survival and regenerative capacity upon injection in infarcted myocardium (*68, 69*). Pretreatment of endothelial progenitor cells with endothelial NO synthase–enhancing substances also appears to have a beneficial conditioning effect (*70*), similar to the effects observed with statin drugs (*71*). Hypoxic preconditioning or heat shock prior to cell injection might also help, since it has been shown to activate both antiapoptotic and NO-related signaling pathways (*72*).

The Dosage Problem

The normal adult heart weighs between 250 and 350 g, and around 80% of the myocardial mass is comprised of approximately 1×10^9 cardiomyocytes. Following transmural myocardial infarction, a large percentage of those cells are lost to necrotic cell death. Assuming that in a given patient 20% of the cardiomyocytes are lost, one would ultimately need 20×10^8 surviving neomyocytes that form new contractile tissue weighing nearly 50 g to completely reconstitute the myocardium. Given the high rate of cell death upon transplantation into the heart and the presumably very low number of adult stem cells that indeed differentiate into myocytes, it becomes clear that, with currently available cell products, we are far from being able to truly replace all lost heart muscle tissue. Hypothetically one would either need to transplant a very large volume of cardiomyocyte suspension or progenitor cells with a high proliferative capacity in vivo to achieve restitution ad integrum (lat.) or complete restitution.

The Clinical Reality of Cardiac Cell Therapy

Skeletal Myoblasts in Chronic Heart Failure

As mentioned above, the first-ever clinical application of myocardial cell therapy was reported by Menasche et al. *(20)* in 2001. A patient with previous myocardial infarction underwent CABG surgery, during which skeletal myoblasts were injected into the infarct tissue. These so-called satellite cells reside in the periphery of skeletal muscle fibers and serve to regenerate injured skeletal muscle. Skeletal myoblasts can be isolated from a small muscle biopsy; in humans the thigh musculature is commonly used. The isolation process involves enzymatic digestion and mechanic destruction of myofibers, and the myoblasts are collected and enriched by filtration and plating. No specific surface marker–based cell selection is necessary. There is usually some fibroblast contamination, but the purity of the final product should exceed 80%. Myoblasts have a robust proliferation capacity and multiply in high-serum concentration for numerous passages without changes in phenotype. Once the serum concentration in the medium is lowered, they rapidly differentiate and fuse to form multinucleated myotubes. Besides their intrinsically preprogrammed differentiation in a myocyte phenotype, their most intriguing quality is resistance to hypoxia, and it has been shown that myoblasts survive in ischemic myocardium better and longer than cardiomyocytes. Following the pioneering work of Menasche et al., other groups initiated similar clinical trials, injecting skeletal myoblasts either surgically or via transendocardial catheter delivery *(21, 61)*. However, several reports have indicated that some of these patients developed ventricular arrhythmia, requiring antiarrhythmic medication and/or implantable cardioverter-defribrillator devices. The critical point is the integration of skeletal myoblasts or myotubes into the myocardial syncytium, i.e., the expression of cardiac-specific connexins and the formation of functioning gap junctions with surrounding viable cardiomyocytes. In principle, undifferentiated myoblasts can express connexin 43 (Cx43), the predominant gap junction protein in ventricular myocardium, and at least some of the myoblasts appear to be able to form functioning cell–cell communications with cocultured cardiomyocytes in vitro. Once they have differentiated and formed myotubes in conventional two-dimensional culture, the

Cx43 expression is down-regulated *(73)*. In vivo, skeletal myotubes are not integrated in the myocardial syncytium, instead they appear to form distinct islets in postinfarct tissue. Cx43 expression in transplanted myoblasts has been described in several animal models, but has not been detected in patients who underwent postmortem histology studies. In a careful study in mice, Rubart et al. *(22)* found that the majority of the intramyocardial myoblasts or myotubes are functionally isolated from the surrounding myocardium and suggested that the remaining cells connect with host cardiomyocytes as a result of cell fusion. The durations of calcium transients recorded from intramyocardial skeletal myoblasts were heterogeneous compared with those in neighboring host cardiomyocytes, which may interfere with the propagation of excitation across the ventricular myocardium and put the heart at risk of ventricular arrhythmia. Some clinical studies of catheter-based intramyocardial injection of skeletal myoblasts are still ongoing *(23–25)*, but a large international multicenter study on CABG surgery and concomitant myoblast transplantation has been terminated for lack of efficacy and safety problems. Overall, skeletal myoblasts for heart repair seem to have been largely abandoned.

Bone Marrow Mononuclear Cells in Acute Infarction

The proponents of bone marrow mononuclear cells (MNCs) for myocardial cell therapy often use the argument that by transplanting the entirety of nucleated marrow cells into the heart, no potentially critical cell population is missed, as may be the case when a particular subpopulation is isolated. On the other hand, a vast number of immediate leukocyte progenitor cells are being delivered in addition to the actual hematopoietic and nonhematopoietic stem cells, and some have argued that one induces more inflammation than regeneration in the myocardium. A clinically very relevant argument in favor of MNCs is their simple and speedy preparation. Traditionally, density gradient centrifugation is used to separate MNCs from other marrow components. This is usually performed in an open system and thus requires working in a class A or class B clean room environment. Alternatively, the total nucleated cell (TNC) fraction can be collected by several washing and centrifugation steps without the use of Ficoll (GE Healthcare) or similar density gradient media *(74, 75)*. Moreover, industry has developed several easy-to-use devices for one-step preparation of MNC or TNC products in closed systems, which can essentially be used at the patients bedside or in the procedure room. Bone marrow MNCs were the first cell products to be used in patients with acute myocardial infarction, where they are injected into the infarct vessel that had before been reopened by percutaneous balloon dilation and stent placement. MNC delivery requires a second cardiac catheterization procedure and is usually performed within a few days after the primary infarct treatment. This straightforward cell therapy approach has so far been performed in several hundred, if not thousands, patients worldwide. Following several small-scale pilot trails *(76, 77)*, the first randomized, placebo-controlled study comparing intracoronary MNC injection with standard treatment of acute myocardial infarction was the Hannover BOOST trial *(78–80)*. At 6 months' follow-up, cell-treated patients had a significantly higher left ventricular ejection fraction than control patients. Subsequently, a number of similar studies were conducted by other mainly European groups, including a multicenter study that enrolled 200 patients *(81–89)*. The results were ambiguous. Some of those trials clearly produced a negative result, i.e., there was no difference in outcome between cell-treated patients and control patients *(90, 91)*. In the multicenter trial coordinated by the Frankfurt group, left ventricular ejection fraction (LVEF) rose by 5.5% in cell-treated patients and by 3.0% in the control group *(92)*. The difference of an average of 2.5% proved statistically significant, but it remains controversial if such a small effect would translate into a relevant clinical benefit. Other reports focused on clinical exercise tolerance and quality-of-life data, and again there seemed to be a slight advantage for patients who had received cell therapy *(93)*. Accepting that intracoronary MNC injection in acute myocardial infarction produces a mild functional improvement in addition to the standard treatment, it now needs to be determined if there is a clear advantage in the long-term clinical course. Ultimately, a significant reduction of major cardiac adverse events will have to be proven so as to justify reimbursement of the substantial additional costs.

Bone Marrow Mononuclear Cells in Chronic Ischemia

Patients with chronic myocardial ischemia have also been treated with bone marrow MNC products in several clinical studies. Here direct intramyocardial injection during cardiac surgical dominates, but intracoronary cell delivery has also been performed. Moreover, there are some reports on catheter-based transendocardial intramuscular injection of bone marrow MNCs *(94)*. Again, some trials on catheter-based delivery of MNCs have shown a modest benefit, while others have produced an essentially negative result *(95, 96)*. The same must be said regarding surgical injection of MNCs in conjunction with bypass surgery. In initial pilot studies, an improvement of regional ventricular wall motion in cell-treated areas was observed, but this did not lead to a better global heart function as compared with routine bypass patients *(97)*. Our own experience was very similar. We treated 14 patients undergoing bypass surgery for chronic ischemic heart disease and compared their outcome with that of 10 patients who had had a standard CABG operation. Using a novel echocardiographic analysis tool (ventricular wall strain imaging), we were able to detect improved myocardial function in cell-treated segments. However, this did not result in better global ventricular function as assessed by left ventricular ejection fraction (unpublished data). In a recent elegant study, Galinanes et al. directly compared intracoronary and intramyocardial injection of MNCs in CABG patients by injecting intraoperatively either in the heart muscle or into the bypass graft (Galinanes et al., *American Heart Association Scientific Sessions*, 2007). Again, they found no relevant benefit of either delivery technique over placebo-treated patients. Overall it has become clear that by using unmodified, autologous bone marrow mononuclear cell products, little meaningful benefit in terms of contractile function can be achieved.

Enriched Progenitor Cell Products

Given that the functional and regenerative capacity of bone marrow mononuclear cells in terms of myocardial cell therapy is limited, it is warranted to test the hypothesis that enriched specific stem cells populations might be more efficacious. Aside from mesenchymal stem cell products, which will be discussed below, the options for preparation of specific human bone marrow stem cell products are few. Progenitor cell products can be prepared using clinical-grade immunomagnetic selection for either CD34 or CD133. In addition, negative selection for CD45 is also possible. The same applies when peripheral blood leukapheresis products are being used instead of bone marrow. An intermediate strategy is the in vitro cultivation and expansion of bone marrow mononuclear cells in open or commercially available closed systems, with or without addition of differentiation-inducing or -suppressing substances, which may give rise to less specific cell products with alleged pro-angiogenic potential. Attempts to selectively expand CD34 or CD133 positive cells without inducing lineage-specific leukocyte differentiation have proven disappointing in the clinical setting. The bottom line is that the clinician has to choose between a freshly isolated CD34- or CD133-enriched cell product with high purity, well-defined characteristics, but rather small cell dose that can be prepared within several hours, and a high-dose, ex vivo expanded cell product that contains a wide range of incompletely characterized cell types and takes several weeks to be prepared.

Few investigators have used purified hematopoietic stem cell products for treatment of acute or subacute myocardial infarction. In one such study, concerns were raised about a higher rate of stent occlusion following intracoronary injection of CD133[+] cells. Of note, Hofmann et al. (63) studied the cardiac retention of bone marrow cells after intracoronary injection using radioactively labeled cells. When mononuclear marrow cells were used, less than 1% of the activity was retained in the heart, while the vast majority of the cells accumulated in liver and spleen. However, when CD34 selected cells were used, cardiac cell retention was nearly 10%, and the activity was localized in the infarct region. In chronic ischemia, however, CD34 or CD133 enriched cell products have found more widespread application (98, 99). One of the first comparative trials was performed in Argentina. In conjunction with CABG surgery, intramyocardial injection of CD34[+] bone marrow cells resulted in nearly 10% higher left ventricular ejection fraction than CABG surgery alone (100). Our group has focused on CD133[+] cells given during CABG surgery because they are believed to contain a subpopulation of cells that are even more immature than CD34[+] cells. In 2001 we started a feasibility and safety study in ten patients, and no procedure-related adverse events were observed (101, 102). Subsequently, we conducted a controlled study in 40 patients. Here, CABG and

CD133+ cell injection led to a significantly higher left ventricular ejection fraction at 6 months' follow-up than CABG surgery alone *(103)*. Whether this benefit is maintained in the long term remains to be determined. A phase III, analogous, multicenter study has just been initiated and will hopefully clarify whether the marked benefit in terms of global left ventricular function at 6 months translates into a sustained clinical advantage. If so, autologous CD133+ bone marrow cell injection may well become a widespread adjunct therapy to CABG surgery in patients with coronary artery disease and impaired heart function.

A problem in the clinical setting, particularly in high-throughput surgical centers, is the rather complex procedure for CD34 or CD133+ cell isolation, requiring a clean room facility that complies with GMP standards. In a university hospital setting, the infrastructure will usually be provided by the hematology department. Some surgical groups have established a cell isolation facility in the immediate vicinity of the operating room. From the clinical standpoint, cell production may then be considered an integral part of the surgical procedure, bypassing the medicinal product legislation. It is unclear, however, whether this point of view is also shared by the regulatory bodies. By streamlining the cell preparation process, i.e., by skipping MNC preparation by Ficoll density centrifugation, cell preparation time can be cut to approximately 3 h. Thus, marrow harvest, cell preparation, and surgery can be performed within the same procedure. Other groups have isolated CD133 cells for surgical delivery from peripheral blood, following mobilization from the marrow with G-CSF. This procedure yields a substantially higher cell dose but requires several days for cell harvest. Moreover, it is unclear whether the massive inflammatory stimulus induced by G-CSF can negatively influence the outcome of the surgical procedure. Enriched bone marrow stem cell products (here: CD34+ cells) have also been injected intramuscularly using catheter-based systems. In a phase I or II pilot trial, this procedure has been shown to be safe, and preliminary data indicate an improvement of left ventricular function over placebo-treated patients.

Mesenchymal Stem Cells

Mesenchymal stem cells have reached the clinical cardiovascular arena later than their hematopoietic counterparts, partly because they require cultivation and expansion in vivo over several weeks when used in autologous fashion, partly because there were reports of microinfarction after intracoronary injection of MSC product in animal models, as well as a potential risk of bone, cartilage, and adipose tissue formation in the heart after direct intramuscular injection. Another drawback was the controversial data on cardiomyocyte differentiation requiring epigenetic modification by DNA-demethylating agents. The pro-angiogenic capacity of MSCs is undisputed but can also be achieved by using "simpler" cell products such as MNCs. An independent and unique characteristic of MSCs, however, is their immunomodulatory potency, which might also be exploited for treatment of heart disease. When transplanted in allogenic (and perhaps even xenogenic) fashion, MSCs may escape detection and elimination by the host immune system *(104, 105)*, because they often do not express MHC class II and only very low levels of MHC class I. Moreover, they can actively suppress immunologic processes and reduce inflammation by inhibiting T-cell proliferation in response to antigens and prevent the development of cytotoxic T cells. It is believed that MSCs can beneficially influence solid organ injury by shifting the local balance of pro- and anti-inflammatory cytokines.

To date there is only anecdotal information on the use of autologous, bone marrow–derived MSCs for cardiac repair. Systematic safety and/or efficacy studies have not yet been reported. Given the low immunogenicity of human MSCs, they may theoretically be used in allogenic fashion, in principle making an off-the-shelf cell product possible. A clinical pilot study testing such a product in patients with myocardial infarction by peripheral intravenous injection is currently being performed, and initial results are promising. A lot of preclinical work is currently being done on MSCs from nonmarrow sources, including MSCs from adipose tissue for autologous use, and from umbilical cord, cord blood, and placenta for possible allogenic application. The immunophenotype of MSC populations from different sources varies somewhat, as does their proliferation and differentiation capacity. Whether MSCs from autologous fat tissue are truly superior to those obtained from bone marrow remains to be determined *(106)*, as does the alleged greater "stemness" and even lower immunogenicity of MSCs from fetal or neonatal tissue. Given the high economic potential of those cell products and the related technology, clinical pilot trials will surely emerge very soon.

Cytokine-Induced Bone Marrow Cell Mobilization

Another fundamentally different approach aims at circumventing any invasive procedure for cell delivery, while minimizing the interval between the onset of myocardial infarction and cell therapy by mobilizing marrow cells using G-CSF. The idea is that stem or progenitor cells mobilized from marrow will be attracted to the ischemic heart and initiate regeneration events or at least modulate remodeling processes *(107)*. It has been well established that the number of circulating progenitor cells can be greatly enhanced by G-CSF stimulation *(108)*. However, the number of mature leukocytes also rises markedly, and this has raised concerns regarding the safety of G-CSF treatment. Because bone marrow cell mobilization did not result in a favorable outcome in a nonhuman primate model of myocardial infarction, this practicable pharmacological approach has been viewed with some restraint. Despite the safety of G-CSF mobilization with or without consecutive apheresis and favorable short-term results of intracoronary infusion of G-CSF mobilized peripheral blood cells in patients with myocardial infarction, an unexpectedly high rate of in-stent restenosis has been observed. In another study in 12 patients with intractable angina the administration of G-CSF was associated with two acute myocardial infarctions and one cardiac death *(109)*. Nevertheless, several controlled clinical efficacy studies have subsequently been performed, but the outcome data in terms of heart function improvement remain equivocal.

Nonischemic Heart Disease

For several reasons, very few investigators thus far have attempted to use cell therapy for treatment of nonischemic heart disease *(66)*. Much of the benefit of bone marrow–derived cells in the heart is based on their potent capacity to support angiogenesis processes in ischemic tissue. Lack of blood supply, however, is simply not a problem in patients with nonischemic heart disease. What would instead be necessary is an increase in cardiomyocyte number or contractile tissue mass, but this cannot be achieved using the currently available clinical cell products other than skeletal myoblasts. Hypothetically, the immunomodulatory action of MSC products may be helpful in inflammatory types of heart failure, but there are few preclinical data to support this idea. This, in turn, is partly because there is no good experimental model that mimics the biology of human inflammatory or idiopathic heart failure *(110)*. The most frequently used approach is the induction of heart failure in animal models by doxorubicin, which has a unique biology that leads to cardiomyocyte loss that occurs only in the occasional patient on doxorubicin antitumor medication. Last, but not least, nonischemic heart disease is much more rare than ischemic heart failure, rendering recruitment of patients for clinical trials very time consuming.

Our group has performed a clinical pilot study in patients with nonischemic heart disease who had to undergo implantation of a left ventricular mechanical assist device for terminal heart failure. Here autologous bone marrow mononuclear cells were implanted over the accessible surface of the left ventricle. In the first patient, a 14-year-old boy, this combination therapy proved very efficacious, his heart function returned to normal within 12 weeks, and the assist device was successfully removed (Fig. 4). In the other nine patients, who were treated in an identical fashion, however, heart function did not recover sufficiently, and they ultimately had to undergo heart transplantation or are still on mechanical circulatory support. As mentioned above, local immunomodulation by MSCs is a very intriguing concept for treatment of patients with inflammatory heart disease, and clinical trials are currently being planned. Reports on systematic safety and efficacy studies have so far not yet been published.

Combination Treatments

Numerous strategies to enhance the regenerative capacity of adult stem cells in the heart have been developed. Besides combining pro-angiogenic bone marrow cells with myogenic cells, primary muscle progenitor cells have been transplanted in

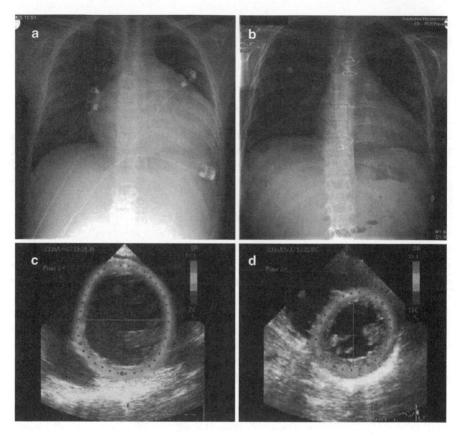

Fig. 4 Imaging data of a young patient undergoing left ventricular assist device (LVAD) implantation and intramyocardial injection of autologous bone marrow mononuclear cells. The combination treatment worked very well in this young patient, but unfortunately not in the other nine patients who were subsequently treated in similar fashion. (**a**) Chest X-ray prior to the operation. Note the massively enlarged heart. (**b**) Chest X-ray after the explantation of the LVAD. Heart size has returned to near normal. (**c**) Echocardiography with wall strain imaging. The LV cavity is extremely enlarged and the wall thinned. (**d**) Echocardiography after LVAD explantation. LV size and wall thickness have returned to normal values (*see Color Plates*)

conjunction with pro-angiogenic growth factors *(111)*. Very solid experimental evidence also exists for genetic manipulation of mesenchymal stem cells, which have been transfected to overexpress antiapoptotic proteins such as Akt or Bcl-2. Both approaches not only lead to a marked improvement of MSC survival following transplantation in the heart, but also seem to have beneficial effects on the surrounding host myocardium that comes in contact with secreted antiapoptotic factors. Similarly, conditioned media obtained from Akt or Bcl-2 transfected MSCs is also able to rescue ischemic cardiomyocytes *(112)*. However, due to patient safety concerns and the related regulatory restrictions, genetically modified adult stem cells have not yet been tested in the clinical setting. The situation is similar with respect to nitric oxide synthase modulation in marrow-derived cells. Clinically better applicable strategies to improve cell therapy in the heart are, for example, hypoxic preconditioning of bone marrow cells or heat shock treatment of myoblasts *(72)*, which increase both cellular resilience and paracrine activity, or pretreatment with pharmaceutics such as erythropoietin, which is said to increase stem cell proliferation and plasticity or parathyroid hormone. For both approaches, clinical trials have been initiated, but data are not yet available. The same is the case for the combination of cell therapy and shock wave treatment *(113)*. In experimental models, shock waves were shown to increase the functional capacity of bone marrow and endothelial progenitor cells *(114)*, and a clinical trial has been initiated to evaluate this combination treatment in patients with ischemic heart disease.

It is often said that transplanted cells survive and function better in their new environment when they are embedded in an adequate extracellular matrix *(115)*. Consequently, a clinical trial has been set up were bone marrow cells are mixed with collagen-rich semiliquid hydrogel prior to implantation in the heart. Again, definitive results are not yet available.

Finally, the combination of transmural laser revascularization and cell therapy deserves to be mentioned. As described above, TMLR is believed to provide a beneficial local inflammatory stimulus that supports marrow cells injected in the heart. Quite impressive heart function data in patients with ischemic cardiomyopathy have been reported from pilot studies, but a systematic controlled trial that would allow distinguishing the contribution of cell therapy and that of TMLR has not yet been reported.

The Future of Cardiac Cell Therapy

Endogenous Cardiac Stem Cells

The presence of myogenic progenitor cells in skeletal muscle (skeletal myoblasts, satellite cells) has been known for a long time, but the existence of similar progenitor cells in the postnatal myocardium of mammals was considered impossible. However, recent experimental data indicate the existence of several types of cardiac muscle stem cells that might be involved in physiologic regeneration attempts *(116–118)*. In rodents, putative cardiac stem cell populations have been described by various groups *(119–125)*. Identification is mainly based on the expression of c-Kit, Sca-1, and Isl-1, but also CD34, FLK-1, CD31, and GATA-4, in varying combinations with the presence or absence of other surface markers. As Torella et al. *(116, 126)* have pointed out, the seemingly different cardiac progenitor cells are probably phenotypic variations of a unique cell type, with the exception of Isl-1[+] cells in the right heart, which seem to be remnants from the cardiac primordia. Rodent cardiac stem cells have been expanded ex vivo and successfully been used for heart muscle regeneration in syngeneic and allogenic models of myocardial infarction. Extension of this concept to the human heart was reported by several groups, who isolated c-Kit⁺ cardiac stem cells from human endomyocardial biopsy or surgical samples, confirmed their cardiomyogenic differentiation potential in vitro, and applied them to xenogenic experimental models of myocardial infarction *(127–130)*. The concept of using autologous cardiac stem cells for clinical cardiac regeneration is clearly the most intriguing of all currently known cell therapy strategies. Nevertheless, there are several conceptual issues that still need to be addressed. The most obvious question is: why does the heart not regenerate in so many patients if it really contains substantial numbers of powerful tissue-specific stem cells? Is cardiac stem cell function impaired in patients with severe heart disease? If so, can autologous cardiac stem cells from chronically sick patients really be expected to do the trick? To answer these questions, many detailed basic research and preclinical studies will have to be carried out prior to serious clinical translation attempts.

Genetic Modification of Stem Cells

Preclinical attempts to enhance adult stem cell function by means of genetic manipulation have been described above, and the genetic reprogramming of somatic cells to regress into a pluripotent stem cell–like state can also be viewed in this context. Indeed, the hypothetic approaches to genetically manipulate cells for regenerative medicine application are limitless. Several fundamental problems, however, will have to be solved prior to serious clinical applications. One is that retroviral transfection is still the only way to efficiently channel exogenous DNA into mammalian cells. The clinical safety of this approach is still very controversial, while nonviral transfection techniques remain limited. Another issue is the possible activation of oncogenes in transduced cells, especially when genes are modified that regulate basic functions of cell cycle and cellular proliferation. Gene therapy itself has suffered a significant setback following the death of some of the first patients treated, and a lot of work will have to be done to make sure that similar problems do not occur when genetically modified cells are transplanted into the human heart.

Novel Somatic Stem Cell Types

Many researchers are actively seeking to detect novel pluripotent stem cell populations in the adult organism. In this context, the bone marrow has been pretty much exhausted, and a lot of the previously described pluripotent nonhematopoietic "miracle cells" have not held their promise for treatment of heart disease. The same must be said for other organs with high cellular turnover that contain tissue-specific progenitor cells, such as the hair follicle, intestine, and exocrine glands.

An exception may be the reproductive organs, at least in the male organism, where spermatogonial progenitor cells constantly produce offspring (sperm), which is obviously involved in generation of every tissue in the human body. Promising pluripotent stem cells have recently been produced from murine and human testicular tissue, but their cardiomyogenic differentiation capacity in vitro and in vivo remains to be determined (33, 34).

Embryonic Stem Cell Technology

The problems surrounding the clinical use of embryonic stem cells and their derivatives have been described above. Strategies are needed that replace embryo-consuming research, and the development of those progresses rapidly. Nuclear transfer of the somatic cell genome into a denucleated oocyte was recently performed in nonhuman primates. This therapeutic cloning technology, however, is controversial because of fears that it might eventually be used for reproductive cloning of humans and because nuclear transfer results in cells with nuclear DNA from one individual and mitochondrial DNA from another. Very recently, several milestone publications have reported the successful genetic reprogramming of skin fibroblasts, resulting in dedifferentiation of those into cells with embryonic stem cell characteristics (131–134). In principle, this technique solves all of the ethical problems surrounding ESC research and use, as well as the unclear situation regarding the immunogenicity of allogenic ESC-derived cell products. On the other hand, this method requires virus-mediated transfection of cells with several dedifferentiation inducing genes, and the in vivo consequences are completely unknown to date. Nevertheless, a door has been opened that leads to completely new options for obtaining pluripotent cells to be processed into autologous cardiomyocyte cell products, and this will surely stimulate a tremendous amount of research and development work.

Cardiomyocyte Cell Cycle Manipulation

Although there have been reports of mitotic cell division of postnatal human cardiomyocytes under physiologic conditions, there is not sufficient evidence to disprove the paradigm that adult cardiomyocytes can re-enter the cell cycle spontaneously or in response to tissue injury. It may, however, be possible to force cardiomyocytes to leave the G_0 resting phase and re-enter the S phase of the cell cycle, ultimately leading to cell replication. This has been achieved by inducing overexpression of cyclins and cyclin-dependent enzymes in rodent models; particularly cyclin D1, D2, and A2 have been targeted in the context of myocardial regeneration (135–140). Quite impressive histology data have been published that show how neomyocardium is formed in response to myocardial infarction. One problem is how to control cyclin-induced cell division so as to prevent overshooting myocyte growth and tumor formation, and it will be challenging to translate this approach from bench to bedside without being able to predict how the human myocardium will react to forced cell cycle reactivation in vivo. One might argue that nature must have had its reasons to have cardiomyocytes in higher mammals permanently rest in the G_0 phase. After all, there is at least no primary cardiomyocyte cancer known in man. In principle, however, this strategy may eventually prove superior to the concept of adding exogenous cells, and further research in this field is clearly warranted.

Highway or Dead End?

The prospect of curing heart failure by replacing diseased or dead cardiomyocytes with new ones has understandably caused tremendous excitement. It is not surprising that there have been, and will be more, obstacles and setbacks along the road to successful cardiac cell therapy (Table 2). It must not be forgotten that, after many decades, if not centuries, of research on palliative measures for treatment of patients with heart failure, only about 10–15 years have so far been spent trying to develop therapies based on cardiac regeneration. Although the early clinical use of bone marrow–derived cells for heart disease has been much criticized, it is understandable that physicians have begun by testing the clinical efficacy of simple cell products (such as bone marrow mononuclear cells) before moving on to do further research on more promising, but also more complex, cell products. It is still unclear whether a meaningful effect can be achieved by using any of the currently available cell products in patients with heart disease, but we will never know for sure unless we perform well-designed clinical studies. After all, we have learned that the comparability of cell therapy in experimental models and in clinical patients is much more limited than for conventional pharmaceutical compounds or medical devices. At the same time, clinicians doing these trials must proceed with great care, and if possible in close collaboration with the regulatory agencies. The latter will always have a lot more pull if controversy arises. In this respect, precocious enthusiastic reports based on anecdotal observations are clearly not helpful. There was, for instance, a recent report in a German medical journal on a single patient with acute myocardial infarction and cardiogenic shock who recovered after receiving standard intensive care treatment plus intracoronary injection of bone marrow mononuclear cells *(141)*. Publication of such a single observation, including a press release, is not only poor scientific conduct, but has also further damaged the already problematic public perception of stem cell research in this country. Moreover, it casts the shadow of "publicity at all costs" over all serious clinical investigators working on cardiac cell therapy.

However, ups and downs during the first stages of development of a new therapeutic concept are not exclusive to cell therapy, and the situation does not differ much from the early days of heart transplantation or interventional cardiology. What matters is that the fundamental concept of cardiac cell therapy is sound and will one day be turned into a valuable therapeutic option for many patients with severe heart disease. How much time this is going to take, however, cannot be foreseen. The time and effort we have to invest into cardiac cell therapy are inevitably longer

Color Plates

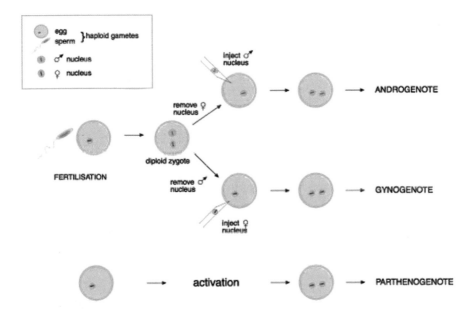

Chapter 3, Fig. 1 Construction of diploid embryos with genomes derived from a single sex. In a fertilized egg, the male and female pronuclei can be removed before they fuse, and a diploid egg re-created by injection of another pronucleus. The development of these diploid eggs can then be followed in vitro by culturing or in vivo by transplantation to pseudopregnant mice. Parthenogenetic embryos can be created by the chemical activation of an unfertilized egg *(12)*

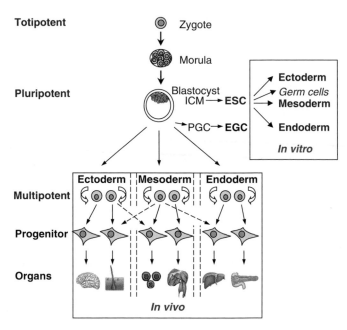

Chapter 3, Fig. 2 Stem cell hierarchy. Zygote and early cell division stages (blastomeres) to the morula stage are defined as totipotent because they can generate a complex organism. At the blastocyst stage, only the cells of the inner cell mass (ICM) retain the capacity to build up all three primary germ layers, the endoderm, mesoderm, and ectoderm as well as the primordial germ cells (PGC), the founder cells of male and female gametes. In adult tissues, multipotent stem and progenitor cells exist in tissues and organs to replace lost or injured cells. At present, it is not known to what extent adult stem cells may also develop (transdifferentiate) into cells of other lineages or what factors could enhance their differentiation capability (*dashed lines*). Embryonic stem (ES) cells, derived from the ICM, have the developmental capacity to differentiate in vitro into cells of all somatic cell lineages as well as into male and female germ cells (*14*)

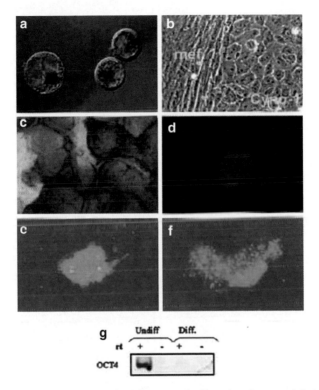

Chapter 3, Fig. 3 Characterization of parthenogenetic Cyno-1 embryos and derived cell lines. (**a**) Parthenogenetically activated eggs at day 8 of development before inner cell mass (ICM) isolation. (**b**) Phase contrast of Cyno-1 stem cells growing on top of mitotically inactivated mouse feeder layer (mef). (**c**) Alkaline phosphatase staining. (**d**) Stage-specific embryonic antigen 4. (**e**) Tumor rejection antigen 1–60. (**f**) Tumor rejection antigen 1–81 staining. (**g**) Reverse Transcription Polymerase Chain Rxn (RT-PCR) octamer-binding transcription factor 4 expression in undifferentiated Cyno-1 cells. (Scale bars = 50 μm in (**a**), 10 μm in (**b**) and (**d**–**f**), and 4 mm in (**c**).) (*16*)

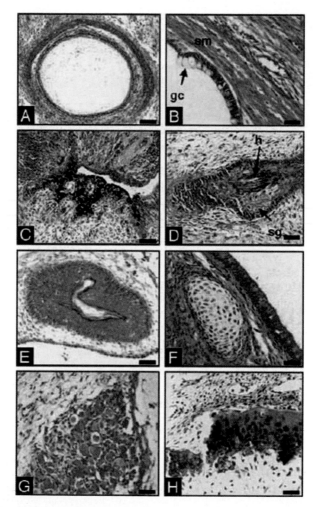

Chapter 3, Fig. 4 In vivo differentiation of Cyno-1 cells. Cells were injected intraperitoneal in severe combined immunodeficient mice. Eight and 15 weeks after injection, teratomas 12 and 30 mm in diameter, respectively, were isolated, fixed with 10% paraformaldehyde, and paraffin embedded. Sections were stained with hematoxylin-eosin. The following complex structures were observed: gut (**a**), intestinal epithelium with typical goblet cells (gc), and smooth muscle (sm) (**b**), neuronal tissue with melanocytes (**c**), hair follicle complex with evident hair (h), and sebaceous gland (sg) (**d**), skin (**e**), cartilage (**f**), ganglion cells (**g**), and bone (**h**). (Scale bars = 40 μm in (**a**), 10 μm in (**b**) and (**d–h**), and 20 μm in (**c**).) *(16)*

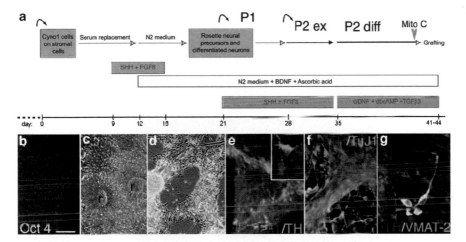

Chapter 3, Fig. 5 In vitro differentiation of primate Cyno-1 parthenogenetic embryonic stem (PGES) cells. (**a**) Schematic representation of the sequential steps designed to induce a dopamine neuronal phenotype from undifferentiated primate embryonic stem (ES) cells. ES cells are first grown in a coculture system on stromal feeders. At the rosette stage, these neuroepithelial structures are replated on coated dishes in a feeder-free system for differentiation. (**b–d**) Microscopic images illustrating the aspect of the colonies at each passage. Undifferentiated stage Cyno-1 cells grow in colonies and all cells (**b**) express the transcription factor Oct-4. (**c**) Soon after the first passage, cells are organized into typical neuroepithelial structures (rosettes, r). (**d**) At the differentiation stage neurons grow in clusters and extend out neurites and (**e**) express midbrain genes like Engrailed-1 (En1), which is colocalized in some neurons with TH. (**f**) TH neurons coexpressed the neuronal marker β-tubulin III (clone TuJ1) and (**g**) some coexpressed VMAT-2 (yellow coexpression). (Scale bars = 100 μm (**b–c**), 150 μm (**d**), 75 μm (**e**), 50 μm (**f**), 35 μm (**g**) and *inset* in (**e**).) *Abbreviations: BDNF brain-derived neurotropic factor; FGF8 fibroblast growth factor 8; GDNF glial-derived neurotropic factor; TH tyrosine hydroxylase; SHH sonic hedgehog; TGFβ3 transforming growth factor β3; TH tyrosine hydroxylase; VMAT vesicular monoamine transporter (17)*

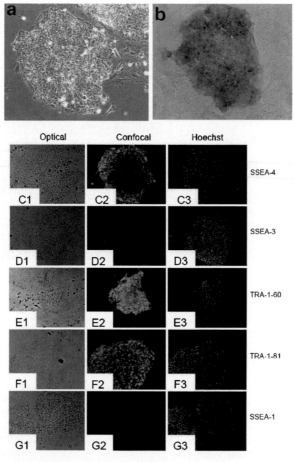

Chapter 3, Fig. 6 Morphology, alkaline phosphatase (AP), and immunostaining of stem cell markers for human parthenogenetic embryonic stem-1 (hPES-1) cell line. (**a**) Morphology of hPES-1 colony under invert microscope; (**b**) AP staining of hPES-1; (**c**) SSEA4; (**d**) SSEA3; (**e**) TRA-1-60; (**f**) TRA-1-81; (**g**) SSEA1. (**c–g**) Listed are optical, confocal images and the corresponding Hoechst staining for the hPES-1 cells (4)

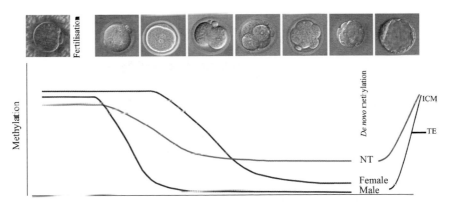

Chapter 4, Fig. 1 DNA methylation profile of mouse preimplantation development in embryos obtained after fertilisation or nuclear transfer [adapted from *(78)*]

LVEF	NYHA	Relevance
<25%	III – IV	Symptoms at rest, survival a few years
25-35%	III	Symptoms at mild exercise, survival several years
35-45%	II - III	Symptoms at strong exercise, survival many years
>45%	I - II	No symptoms, survival normal

Chapter 13, Fig. 1 Makeshift classification of left ventricular ejection fraction, the key end point of virtually all clinical cell therapy trials, grouped according to the usual clinical symptoms. Obviously, there are numerous exceptions, i.e., a patient with very low left ventricular ejection fraction (LVEF) can have very few symptoms, whereas as patient with near-normal LVEF can suffer from congestive heart failure. The intention is to show that, on average, one needs to achieve a greater than 10% increase in LVEF so as to clearly improve the patient's general state of health

What	BM MNC	Selected/ enriched	Cultured, expanded	Off-the shelf	BM mobilization
Where	OR / cath lab / GMP hospital lab	GMP hospital lab	GMP hospital / industrial	industrial	bedside
When	hours	< 1 day	weeks	immediately	days
Pro	rapidly available	Well-defined cell product	High cell dose	Always available	Simple bedside application
Con	limited efficacy	Low cell dose	Preparation time & logistics	allogenic	Systemic inflammation efficacy

Chapter 13, Fig. 2 Clinically relevant principle features of the currently available cell produc ts

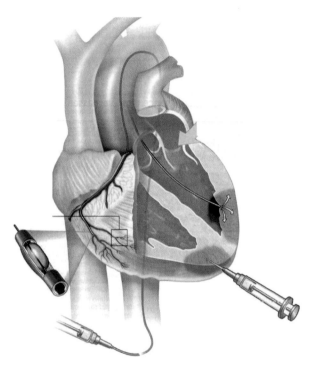

Chapter 13, Fig. 3 Techniques to deliver cell products to the heart in the clinical setting. From *left* to *right*: Intracoronary injection with inflated percutaneous transluminal coronary angioplasty (PTCA) balloon; catheter-based transendocardial injection in the heart muscle; transepicardial injection requiring surgical access to the heart

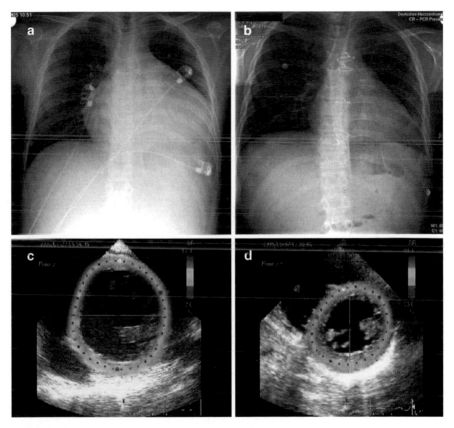

Chapter 13, Fig. 4 Imaging data of a young patient undergoing left ventricular assist device (LVAD) implantation and intramyocardial injection of autologous bone marrow mononuclear cells. The combination treatment worked very well in this young patient, but unfortunately not in the other nine patients who were subsequently treated in similar fashion. (**a**) Chest X-ray prior to the operation. Note the massively enlarged heart. (**b**) Chest X-ray after the explantation of the LVAD. Heart size has returned to near normal. (**c**) Echocardiography with wall strain imaging. The LV cavity is extremely enlarged and the wall thinned. (**d**) Echocardiography after LVAD explantation. LV size and wall thickness have returned to normal values

Table 2 Established evidence in favor of ("Pro") and against ("Contra") the validity of the concept of cardiac cell therapy; "Ambiguous" represents data that cannot be conclusively interpreted

Pro	Contra	Ambiguous
Cells expressing stem cell markers reside in the adult heart	The majority of transplanted cells die immediately upon transplantation in the heart	Transplanted marrow-derived cells fuse with cardiomyocytes. The relevance of this phenomenon is unknown
Cells of bone marrow origin migrate to the heart and persist there	Most surviving transplanted cells disappear within weeks after transplantation	
Bone marrow–derived cells readily differentiate in endothelial cells and support new blood vessel growth in the ischemic heart	Few, if any, marrow-derived cells obtain a myocyte phenotype in the heart	
Transplanted bone marrow derived cells secrete proteins that interfere with cardiac blood vessel cells, interstitial cells, and myocytes	Transplantation of autologous marrow-derived cells in the heart leads to little, if any, improvement of heart function in humans	
Marrow-derived and other somatic stem cells can be induced to express cardiomyocyte-specific markers in vitro	Noncardiac myogenic cells (i.e., skeletal myoblasts) remain electrically isolated from the surrounding myocardium	
Cell transplantation improves heart function in animal models of infarction		
In lower organisms (fish, amphibians), the myocardium regenerates completely and spontaneously in response to injury		

and greater than for any previous development, simply because the underlying biology and necessary technology are dramatically more complex. Unfortunately, the road to success of cardiac cell therapy is not a highway; nevertheless, it will eventually take us to our destination—slow and steady!

References

1. Rota M, Hosoda T, De Angelis A, et al. The young mouse heart is composed of myocytes heterogeneous in age and function. Circ Res 2007;101:387–99.
2. Bittmann I, Hentrich M, Bise K, Kolb HJ, Lohrs U. Endothelial cells but not epithelial cells or cardiomyocytes are partially replaced by donor cells after allogeneic bone marrow and stem cell transplantation. J Hematother Stem Cell Res 2003;12:359–66.

3. Deb A, Wang S, Skelding KA, Miller D, Simper D, Caplice NM. Bone marrow-derived cardiomyocytes are present in adult human heart: a study of gender-mismatched bone marrow transplantation patients. Circulation 2003;107:1247–9.

4. Jiang S, Walker L, Afentoulis M, et al. Transplanted human bone marrow contributes to vascular endothelium. Proc Natl Acad Sci U S A 2004;101:16891–6.

5. Fogt F, Beyser KH, Poremba C, Zimmerman RL, Ruschoff J. Evaluation of host stem cell-derived cardiac myocytes in consecutive biopsies in long-term cardiac transplant patients. J Heart Lung Transplant 2003;22:1314–7.

6. Bayes-Genis A, Salido M, Sole Ristol F, et al. Host cell-derived cardiomyocytes in sex-mismatch cardiac allografts. Cardiovasc Res 2002;56:404–10.

7. Hocht-Zeisberg E, Kahnert H, Guan K, et al. Cellular repopulation of myocardial infarction in patients with sex-mismatched heart transplantation. Eur Heart J 2004;25:749–58.

8. Thiele J, Varus E, Wickenhauser C, et al. Mixed chimerism of cardiomyocytes and vessels after allogeneic bone marrow and stem-cell transplantation in comparison with cardiac allografts. Transplantation 2004;77:1902–5.

9. Minami E, Laflamme MA, Saffitz JE, Murry CE. Extracardiac progenitor cells repopulate most major cell types in the transplanted human heart. Circulation 2005;112:2951–8.

10. Laflamme MA, Myerson D, Saffitz JE, Murry CE. Evidence for cardiomyocyte repopulation by extracardiac progenitors in transplanted human hearts. Circ Res 2002;90:634–40.

11. Reinecke H, Zhang M, Bartosek T, Murry CE. Survival, integration, and differentiation of cardiomyocyte grafts: a study in normal and injured rat hearts. Circulation 1999;100:193–202.

12. Chiu RC, Zibaitis A, Kao RL. Cellular cardiomyoplasty: myocardial regeneration with satellite cell implantation. Ann Thorac Surg 1995;60:12–8.

13. Marelli D, Desrosiers C, el-Alfy M, Kao RL, Chiu RC. Cell transplantation for myocardial repair: an experimental approach. Cell Transplant 1992;1:383–90.

14. Scorsin M, Hagege AA, Marotte F, et al. Does transplantation of cardiomyocytes improve function of infarcted myocardium? Circulation 1997;96:II188–93.

15. Reffelmann T, Dow JS, Dai W, Hale SL, Simkhovich BZ, Kloner RA. Transplantation of neonatal cardiomyocytes after permanent coronary artery occlusion increases regional blood flow of infarcted myocardium. J Mol Cell Cardiol 2003;35:607–13.

16. Roell W, Lu ZJ, Bloch W, et al. Cellular cardiomyoplasty improves survival after myocardial injury. Circulation 2002;105:2435–41.

17. Leor J, Patterson M, Quinones MJ, Kedes LH, Kloner RA. Transplantation of fetal myocardial tissue into the infarcted myocardium of rat. A potential method for repair of infarcted myocardium? Circulation 1996;94:II332–6.

18. Zhang M, Methot D, Poppa V, Fujio Y, Walsh K, Murry CE. Cardiomyocyte grafting for cardiac repair: graft cell death and anti-death strategies. J Mol Cell Cardiol 2001;33:907–21.

19. Taylor DA, Atkins BZ, Hungspreugs P, et al. Regenerating functional myocardium: improved performance after skeletal myoblast transplantation. Nat Med 1998;4:929–33.

20. Menasche P, Hagege AA, Scorsin M, et al. Myoblast transplantation for heart failure. Lancet 2001;357:279–80.

21. Hagege AA, Marolleau JP, Vilquin JT, et al. Skeletal myoblast transplantation in ischemic heart failure: long-term follow-up of the first phase I cohort of patients. Circulation 2006;114:I108–13.

22. Rubart M, Soonpaa MH, Nakajima H, Field LJ. Spontaneous and evoked intracellular calcium transients in donor-derived myocytes following intracardiac myoblast transplantation. J Clin Invest 2004;114:775–83.

23. Dib N, Michler RE, Pagani FD, et al. Safety and feasibility of autologous myoblast transplantation in patients with ischemic cardiomyopathy: four-year follow-up. Circulation 2005;112:1748–55.

24. Pagani FD, DerSimonian H, Zawadzka A, et al. Autologous skeletal myoblasts transplanted to ischemia-damaged myocardium in humans. Histological analysis of cell survival and differentiation. J Am Coll Cardiol 2003;41:879–88.

25. Siminiak T, Kalawski R, Fiszer D, et al. Autologous skeletal myoblast transplantation for the treatment of postinfarction myocardial injury: phase I clinical study with 12 months of follow-up. Am Heart J 2004;148:531–7.

26. Behfar A, Zingman LV, Hodgson DM, et al. Stem cell differentiation requires a paracrine pathway in the heart. FASEB J 2002;16:1558–66.
27. Dai W, Field LJ, Rubart M, et al. Survival and maturation of human embryonic stem cell-derived cardiomyocytes in rat hearts. J Mol Cell Cardiol 2007;43:504–16.
28. Tomescot A, Leschik J, Bellamy V, et al. Differentiation in vivo of cardiac committed human embryonic stem cells in postmyocardial infarcted rats. Stem Cells 2007;25:2200–5.
29. Menard C, Hagege AA, Agbulut O, et al. Transplantation of cardiac-committed mouse embryonic stem cells to infarcted sheep myocardium: a preclinical study. Lancet 2005;366:1005–12.
30. Behfar A, Perez-Terzic C, Faustino RS, et al. Cardiopoietic programming of embryonic stem cells for tumor-free heart repair. J Exp Med 2007;204:405–20.
31. Bonnevie L, Bel A, Sabbah L, et al. Is xenotransplantation of embryonic stem cells a realistic option? Transplantation 2007;83:333–5.
32. Byrne JA, Pedersen DA, Clepper LL, et al. Producing primate embryonic stem cells by somatic cell nuclear transfer. Nature 2007;450:497–502.
33. Guan K, Wagner S, Unsold B, et al. Generation of functional cardiomyocytes from adult mouse spermatogonial stem cells. Circ Res 2007;100:1615–25.
34. Guan K, Nayernia K, Maier LS, et al. Pluripotency of spermatogonial stem cells from adult mouse testis. Nature 2006;440:1199–203.
35. Kocher AA, Schuster MD, Szabolcs MJ, et al. Neovascularization of ischemic myocardium by human bone-marrow-derived angioblasts prevents cardiomyocyte apoptosis, reduces remodeling and improves cardiac function. Nat Med 2001;7:430–6.
36. Orlic D, Kajstura J, Chimenti S, et al. Bone marrow cells regenerate infarcted myocardium. Nature 2001;410:701–5.
37. Murry CE, Soonpaa MH, Reinecke H, et al. Haematopoietic stem cells do not transdifferentiate into cardiac myocytes in myocardial infarcts. Nature 2004;428:664–8.
38. Rota M, Kajstura J, Hosoda T, et al. Bone marrow cells adopt the cardiomyogenic fate in vivo. Proc Natl Acad Sci U S A 2007;104:17783–8.
39. Friedenstein AJ, Petrakova KV, Kurolesova AI, Frolova GP. Heterotopic of bone marrow. Analysis of precursor cells for osteogenic and hematopoietic tissues. Transplantation 1968;6:230–47.
40. Pittenger MF, Mackay AM, Beck SC, et al. Multilineage potential of adult human mesenchymal stem cells. Science 1999;284:143–7.
41. Prockop DJ. Marrow stromal cells as stem cells for nonhematopoietic tissues. Science 1997;276:71–4.
42. Caplan AI. Mesenchymal stem cells. J Orthop Res 1991;9:641–50.
43. Le Blanc K, Ringden O. Immunomodulation by mesenchymal stem cells and clinical experience. J Int Med 2007;262:509–25.
44. Aggarwal S, Pittenger MF. Human mesenchymal stem cells modulate allogeneic immune cell responses. Blood 2005;105:1815–22.
45. Makino S, Fukuda K, Miyoshi S, et al. Cardiomyocytes can be generated from marrow stromal cells in vitro. J Clin Invest 1999;103:697–705.
46. Hakuno D, Fukuda K, Makino S, et al. Bone marrow-derived regenerated cardiomyocytes (CMG Cells) express functional adrenergic and muscarinic receptors. Circulation 2002;105:380–6.
47. Hattan N, Kawaguchi H, Ando K, et al. Purified cardiomyocytes from bone marrow mesenchymal stem cells produce stable intracardiac grafts in mice. Cardiovasc Res 2005;65:334–44.
48. Airey JA, Almeida-Porada G, Colletti EJ, et al. Human mesenchymal stem cells form Purkinje fibers in fetal sheep heart. Circulation 2004;109:1401–7.
49. Toma C, Pittenger MF, Cahill KS, Byrne BJ, Kessler PD. Human mesenchymal stem cells differentiate to a cardiomyocyte phenotype in the adult murine heart. Circulation 2002;105:93–8.
50. Jiang Y, Jahagirdar BN, Reinhardt RL, et al. Pluripotency of mesenchymal stem cells derived from adult marrow. Nature 2002;418:41–9.
51. Yoon YS, Wecker A, Heyd L, et al. Clonally expanded novel multipotent stem cells from human bone marrow regenerate myocardium after myocardial infarction. J Clin Invest 2005;115:326–38.

52. Liao R, Pfister O, Jain M, Mouquet F. The bone marrow–cardiac axis of myocardial regeneration. Prog Cardiovasc Dis 2007;50:18–30.
53. Sandri M, Adams V, Gielen S, et al. Effects of exercise and ischemia on mobilization and functional activation of blood-derived progenitor cells in patients with ischemic syndromes: results of 3 randomized studies. Circulation 2005;111:3391–9.
54. Schmidt-Lucke C, Rossig L, Fichtlscherer S, et al. Reduced number of circulating endothelial progenitor cells predicts future cardiovascular events: proof of concept for the clinical importance of endogenous vascular repair. Circulation 2005;111:2981–7.
55. Hill JM, Zalos G, Halcox JP, et al. Circulating endothelial progenitor cells, vascular function, and cardiovascular risk. N Engl J Med 2003;348:593–600.
56. Heeschen C, Lehmann R, Honold J, et al. Profoundly reduced neovascularization capacity of bone marrow mononuclear cells derived from patients with chronic ischemic heart disease. Circulation 2004;109:1615–22.
57. Ma N, Ladilov Y, Moebius JM, et al. Intramyocardial delivery of human CD133+ cells in a SCID mouse cryoinjury model: bone marrow vs. cord blood-derived cells. Cardiovasc Res 2006;71:158–69.
58. Ma N, Ladilov Y, Kaminski A, et al. Umbilical cord blood cell transplantation for myocardial regeneration. Transplant Proc 2006;38:771–3.
59. Ma N, Stamm C, Kaminski A, et al. Human cord blood cells induce angiogenesis following myocardial infarction in NOD/SCID-mice. Cardiovasc Res 2005;66:45–54.
60. Kogler G, Sensken S, Airey JA, et al. A new human somatic stem cell from placental cord blood with intrinsic pluripotent differentiation potential. J Exp Med 2004;200:123–35.
61. Messas E, Bel A, Morichetti MC, et al. Autologous myoblast transplantation for chronic ischemic mitral regurgitation. J Am Coll Cardiol 2006;47:2086–93.
62. Klein HM, Ghodsizad A, Borowski A, et al. Autologous bone marrow-derived stem cell therapy in combination with TMLR. A novel therapeutic option for endstage coronary heart disease: report on 2 cases. Heart Surg Forum 2004;7:E416–9.
63. Hofmann M, Wollert KC, Meyer GP, et al. Monitoring of bone marrow cell homing into the infarcted human myocardium. Circulation 2005;111:2198–202.
64. Kraitchman DL, Heldman AW, Atalar E, et al. In vivo magnetic resonance imaging of mesenchymal stem cells in myocardial infarction. Circulation 2003;107:2290–3.
65. van Laake LW, Passier R, Monshouwer-Kloots J, et al. Monitoring of cell therapy and assessment of cardiac function using magnetic resonance imaging in a mouse model of myocardial infarction. Nat Protocols 2007;2:2551–67.
66. Ghodsizad A, Ruhparwar A, Marktanner R, et al. Autologous transplantation of CD133+ BM-derived stem cells as a therapeutic option for dilatative cardiomyopathy. Cytotherapy 2006;8:308–10.
67. Yau TM, Kim C, Ng D, et al. Increasing transplanted cell survival with cell-based angiogenic gene therapy. Ann Thorac Surg 2005;80:1779–86.
68. Li W, Ma N, Ong LL, et al. Bcl-2 engineered MSCs inhibited apoptosis and improved heart function. Stem Cells 2007;25:2118–27.
69. Mangi AA, Noiseux N, Kong D, et al. Mesenchymal stem cells modified with Akt prevent remodeling and restore performance of infarcted hearts. Nat Med 2003;9:1195–201.
70. Sasaki K, Heeschen C, Aicher A, et al. Ex vivo pretreatment of bone marrow mononuclear cells with endothelial NO synthase enhancer AVE9488 enhances their functional activity for cell therapy. Proc Natl Acad Sci U S A 2006;103:14537–41.
71. Spyridopoulos I, Haendeler J, Urbich C, et al. Statins enhance migratory capacity by upregulation of the telomere repeat-binding factor TRF2 in endothelial progenitor cells. Circulation 2004;110:3136–42.
72. Maurel A, Azarnoush K, Sabbah L, et al. Can cold or heat shock improve skeletal myoblast engraftment in infarcted myocardium? Transplantation 2005;80:660–5.
73. Choi YH, Stamm C, Hammer PE, et al. Cardiac conduction through engineered tissue. Am J Pathol 2006;169:72–85.

74. Griesel C, Heuft HG, Herrmann D, et al. Good manufacturing practice-compliant validation and preparation of BM cells for the therapy of acute myocardial infarction. Cytotherapy 2007;9:35–43.

75. Seeger FH, Tonn T, Krzossok N, Zeiher AM, Dimmeler S. Cell isolation procedures matter: a comparison of different isolation protocols of bone marrow mononuclear cells used for cell therapy in patients with acute myocardial infarction. Eur Heart J 2007;28:766–72.

76. Strauer BE, Brehm M, Zeus T, et al. Repair of infarcted myocardium by autologous intracoronary mononuclear bone marrow cell transplantation in humans. Circulation 2002;106:1913–8.

77. Strauer BE, Brehm M, Zeus T, et al. Intracoronary, human autologous stem cell transplantation for myocardial regeneration following myocardial infarction. Dtsch Med Wochenschr 2001;126:932–8.

78. Wollert KC, Meyer GP, Lotz J, et al. Intracoronary autologous bone-marrow cell transfer after myocardial infarction: the BOOST randomised controlled clinical trial. Lancet 2004;364:141–8.

79. Meyer GP, Wollert KC, Lotz J, et al. Intracoronary bone marrow cell transfer after myocardial infarction: eighteen months' follow-up data from the randomized, controlled BOOST (BOne marrOw transfer to enhance ST-elevation infarct regeneration) trial. Circulation 2006;113:1287–94.

80. Schaefer A, Meyer GP, Fuchs M, et al. Impact of intracoronary bone marrow cell transfer on diastolic function in patients after acute myocardial infarction: results from the BOOST trial. Eur Heart J 2006;27:929–35.

81. Assmus B, Fischer-Rasokat U, Honold J, et al. Transcoronary transplantation of functionally competent BMCs is associated with a decrease in natriuretic peptide serum levels and improved survival of patients with chronic postinfarction heart failure: results of the TOPCARE-CHD Registry. Circ Res 2007;100:1234–41.

82. Assmus B, Walter DH, Lehmann R, et al. Intracoronary infusion of progenitor cells is not associated with aggravated restenosis development or atherosclerotic disease progression in patients with acute myocardial infarction. Eur Heart J 2006;27:2989–95.

83. Assmus B, Honold J, Schachinger V, et al. Transcoronary transplantation of progenitor cells after myocardial infarction. N Engl J Med 2006;355:1222–32.

84. Schachinger V, Assmus B, Honold J, et al. Normalization of coronary blood flow in the infarct-related artery after intracoronary progenitor cell therapy: intracoronary Doppler substudy of the TOPCARE-AMI trial. Clin Res Cardiol 2006;95:13–22.

85. Schachinger V, Tonn T, Dimmeler S, Zeiher AM. Bone-marrow-derived progenitor cell therapy in need of proof of concept: design of the REPAIR-AMI trial. Nat Clin Pract Cardiovasc Med 2006;3 Suppl 1:S23–8.

86. Schachinger V, Assmus B, Britten MB, et al. Transplantation of progenitor cells and regeneration enhancement in acute myocardial infarction: final one-year results of the TOPCARE-AMI trial. J Am Coll Cardiol 2004;44:1690–9.

87. Britten MB, Abolmaali ND, Assmus B, et al. Infarct remodeling after intracoronary progenitor cell treatment in patients with acute myocardial infarction (TOPCARE-AMI): mechanistic insights from serial contrast-enhanced magnetic resonance imaging. Circulation 2003;108:2212–8.

88. Erbs S, Linke A, Schachinger V, et al. Restoration of microvascular function in the infarct-related artery by intracoronary transplantation of bone marrow progenitor cells in patients with acute myocardial infarction: the Doppler Substudy of the Reinfusion of Enriched Progenitor Cells and Infarct Remodeling in Acute Myocardial Infarction (REPAIR-AMI) trial. Circulation 2007;116:366–74.

89. Schachinger V, Erbs S, Elsasser A, et al. Improved clinical outcome after intracoronary administration of bone-marrow-derived progenitor cells in acute myocardial infarction: final 1-year results of the REPAIR-AMI trial. Eur Heart J 2006;27:2775–83.

90. Janssens S, Dubois C, Bogaert J, et al. Autologous bone marrow-derived stem-cell transfer in patients with ST-segment elevation myocardial infarction: double-blind, randomised controlled trial. Lancet 2006;367:113–21.

91. Lunde K, Solheim S, Aakhus S, et al. Intracoronary injection of mononuclear bone marrow cells in acute myocardial infarction. N Engl J Med 2006;355:1199–209.

92. Schachinger V, Erbs S, Elsasser A, et al. Intracoronary bone marrow-derived progenitor cells in acute myocardial infarction. N Engl J Med 2006;355:1210–21.

93. Lunde K, Solheim S, Aakhus S, et al. Exercise capacity and quality of life after intracoronary injection of autologous mononuclear bone marrow cells in acute myocardial infarction: results from the Autologous Stem cell Transplantation in Acute Myocardial Infarction (ASTAMI) randomized controlled trial. Am Heart J 2007;154:710:e1–8.

94. Perin EC, Dohmann HF, Borojevic R, et al. Improved exercise capacity and ischemia 6 and 12 months after transendocardial injection of autologous bone marrow mononuclear cells for ischemic cardiomyopathy. Circulation 2004;110:II213–8.

95. Brehm M, Strauer BE. Stem cell therapy in postinfarction chronic coronary heart disease. Nat Clin Pract Cardiovasc Med 2006;3 Suppl 1:S101–4.

96. Strauer BE, Brehm M, Zeus T, et al. Regeneration of human infarcted heart muscle by intracoronary autologous bone marrow cell transplantation in chronic coronary artery disease: the IACT Study. J Am Coll Cardiol 2005;46:1651–8.

97. Galinanes M, Loubani M, Davies J, Chin D, Pasi J, Bell PR. Autotransplantation of unmanipulated bone marrow into scarred myocardium is safe and enhances cardiac function in humans. Cell Transplant 2004;13:7–13.

98. Erbs S, Linke A, Adams V, et al. Transplantation of blood-derived progenitor cells after recanalization of chronic coronary artery occlusion: first randomized and placebo-controlled study. Circ Res 2005;97:756–62.

99. Erbs S, Linke A, Schuler G, Hambrecht R. Intracoronary administration of circulating blood-derived progenitor cells after recanalization of chronic coronary artery occlusion improves endothelial function. Circ Res 2006;98:e48.

100. Patel AN, Geffner L, Vina RF, et al. Surgical treatment for congestive heart failure with autologous adult stem cell transplantation: a prospective randomized study. J Thorac Cardiovasc Surg 2005;130:1631–8.

101. Stamm C, Kleine HD, Westphal B, et al. CABG and bone marrow stem cell transplantation after myocardial infarction. Thorac Cardiovasc Surg 2004;52:152–8.

102. Stamm C, Westphal B, Kleine HD, et al. Autologous bone-marrow stem-cell transplantation for myocardial regeneration. Lancet 2003;361:45–6.

103. Stamm C, Kleine HD, Choi YH, et al. Intramyocardial delivery of CD133+ bone marrow cells and coronary artery bypass grafting for chronic ischemic heart disease: safety and efficacy studies. J Thorac Cardiovasc Surg 2007;133:717–25.

104. Amado LC, Saliaris AP, Schuleri KH, et al. Cardiac repair with intramyocardial injection of allogeneic mesenchymal stem cells after myocardial infarction. Proc Natl Acad Sci U S A 2005;102:11474–9.

105. Grinnemo KH, Mansson A, Dellgren G, et al. Xenoreactivity and engraftment of human mesenchymal stem cells transplanted into infarcted rat myocardium. J Thorac Cardiovasc Surg 2004;127:1293–300.

106. Matsumoto T, Kano K, Kondo D, et al. Mature adipocyte-derived dedifferentiated fat cells exhibit multilineage potential. J Cell Physiol 2007;215:210–22.

107. Orlic D, Kajstura J, Chimenti S, et al. Mobilized bone marrow cells repair the infarcted heart, improving function and survival. Proc Natl Acad Sci U S A 2001;98:10344–9.

108. Honold J, Lehmann R, Heeschen C, et al. Effects of granulocyte colony simulating factor on functional activities of endothelial progenitor cells in patients with chronic ischemic heart disease. Arterioscler Thromb Vasc Biol 2006;26:2238–43.

109. Ince H, Stamm C, Nienaber CA. Cell-based therapies after myocardial injury. Curr Treat Options Cardiovasc Med 2006;8:484–95.

110. Nagaya N, Kangawa K, Itoh T, et al. Transplantation of mesenchymal stem cells improves cardiac function in a rat model of dilated cardiomyopathy. Circulation 2005;112:1128–35.

111. Tambara K, Premaratne GU, Sakaguchi G, et al. Administration of control-released hepatocyte growth factor enhances the efficacy of skeletal myoblast transplantation in rat infarcted

hearts by greatly increasing both quantity and quality of the graft. Circulation 2005;112:I129–34.

112. Gnecchi M, He H, Liang OD, et al. Paracrine action accounts for marked protection of ischemic heart by Akt-modified mesenchymal stem cells. Nat Med 2005;11:367–8.

113. Aicher A, Heeschen C, Sasaki K, Urbich C, Zeiher AM, Dimmeler S. Low-energy shock wave for enhancing recruitment of endothelial progenitor cells: a new modality to increase efficacy of cell therapy in chronic hind limb ischemia. Circulation 2006;114:2823–30.

114. Nurzynska D, Di Meglio F, Castaldo C, et al. Shock waves activate in vitro cultured progenitors and precursors of cardiac cell lineages from the human heart. Ultrasound Med Biol 2007;34:334–42.

115. Christman KL, Vardanian AJ, Fang Q, Sievers RE, Fok HH, Lee RJ. Injectable fibrin scaffold improves cell transplant survival, reduces infarct expansion, and induces neovasculature formation in ischemic myocardium. J Am Coll Cardiol 2004;44:654–60.

116. Torella D, Ellison GM, Karakikes I, Nadal-Ginard B. Resident cardiac stem cells. Cell Mol Life Sci 2007;64:661–73.

117. Barile L, Messina E, Giacomello A, Marban E. Endogenous cardiac stem cells. Prog Cardiovasc Dis 2007;50:31–48.

118. Tomita Y, Matsumura K, Wakamatsu Y, et al. Cardiac neural crest cells contribute to the dormant multipotent stem cell in the mammalian heart. J Cell Biol 2005;170.1135–46.

119. Mouquet F, Pfister O, Jain M, et al. Restoration of cardiac progenitor cells after myocardial infarction by self-proliferation and selective homing of bone marrow-derived stem cells. Circ Res 2005;97:1090–2.

120. Urbanek K, Cesselli D, Rota M, et al. Stem cell niches in the adult mouse heart. Proc Natl Acad Sci U S A 2006;103:9226–31.

121. Beltrami AP, Barlucchi L, Torella D, et al. Adult cardiac stem cells are multipotent and support myocardial regeneration. Cell 2003;114:763–76

122. Oh H, Bradfute SB, Gallardo TD, et al. Cardiac progenitor cells from adult myocardium: homing, differentiation, and fusion after infarction. Proc Natl Acad Sci U S A 2003;100:12313–8.

123. Oyama T, Nagai T, Wada H, et al. Cardiac side population cells have a potential to migrate and differentiate into cardiomyocytes in vitro and in vivo. J Cell Biol 2007;176:329–41.

124. Matsuura K, Nagai T, Nishigaki N, et al. Adult cardiac Sca-1-positive cells differentiate into beating cardiomyocytes. J Biol Chem 2004;279:11384–91.

125. Pfister O, Mouquet F, Jain M, et al. CD31- but Not CD31+ cardiac side population cells exhibit functional cardiomyogenic differentiation. Circ Res 2005;97:52–61.

126. Torella D, Ellison GM, Mendez-Ferrer S, Ibanez B, Nadal-Ginard B. Resident human cardiac stem cells: role in cardiac cellular homeostasis and potential for myocardial regeneration. Nat Clin Pract Cardiovasc Med 2006;3 Suppl 1:S8–13.

127. Bearzi C, Rota M, Hosoda T, et al. Human cardiac stem cells. Proc Natl Acad Sci U S A 2007;104:14068–73.

128. Messina E, De Angelis L, Frati G, et al. Isolation and expansion of adult cardiac stem cells from human and murine heart. Circ Res 2004;95:911–21.

129. Smith RR, Barile L, Cho HC, et al. Regenerative potential of cardiosphere-derived cells expanded from percutaneous endomyocardial biopsy specimens. Circulation 2007;115:896–908.

130. Laugwitz KL, Moretti A, Lam J, et al. Postnatal isl1+ cardioblasts enter fully differentiated cardiomyocyte lineages. Nature 2005;433:647–53.

131. Hanna J, Wernig M, Markoulaki S, et al. Treatment of sickle cell anemia mouse model with iPS cells generated from autologous skin. Science 2007;318:1920–3.

132. Meissner A, Wernig M, Jaenisch R. Direct reprogramming of genetically unmodified fibroblasts into pluripotent stem cells. Nat Biotechnol 2007;25:1177–81.

133. Wernig M, Meissner A, Foreman R, et al. In vitro reprogramming of fibroblasts into a pluripotent ES-cell-like state. Nature 2007;448:318–24.

134. Yu J, Vodyanik MA, Smuga-Otto K, et al. Induced pluripotent stem cell lines derived from human somatic cells. Science 2007;318:1917–20.

135. Pasumarthi KB, Nakajima H, Nakajima HO, Soonpaa MH, Field LJ. Targeted expression of cyclin D2 results in cardiomyocyte DNA synthesis and infarct regression in transgenic mice. Circ Res 2005;96:110–8.
136. Soonpaa MH, Koh GY, Pajak L, et al. Cyclin D1 overexpression promotes cardiomyocyte DNA synthesis and multinucleation in transgenic mice. J Clin Invest 1997;99:2644–54.
137. Nakajima H, Nakajima HO, Tsai SC, Field LJ. Expression of mutant p193 and p53 permits cardiomyocyte cell cycle reentry after myocardial infarction in transgenic mice. Circ Res 2004;94:1606–14.
138. Lafontant PJ, Field LJ. The cardiomyocyte cell cycle. Novartis Foundation symposium 2006;274:196–207; discussion 8–13, 72–6.
139. Cheng RK, Asai T, Tang H, et al. Cyclin A2 induces cardiac regeneration after myocardial infarction and prevents heart failure. Circ Res 2007;100:1741–8.
140. Woo YJ, Panlilio CM, Cheng RK, et al. Myocardial regeneration therapy for ischemic cardiomyopathy with cyclin A2. J Thorac Cardiovasc Surg 2007;133:927–33.
141. Brehm M, Strauer BE. Successful therapy of patients in therapy-resistant cardiogenic shock with intracoronary, autologous bone marrow stem cell transplantation. Dtsch Med Wochenschr 2007;132:1944–8.

Embryonic Stem Cells, Cardiomyoplasty, and the Risk of Teratoma Formation

Tomo Saric, Lukas P. Frenzel, Azra Fatima, Manoj K. Gupta, and Jürgen Hescheler

Abstract Embryonic, fetal, and adult stem cells have been extensively tested in the past decade as candidates for cell replacement therapy of severe heart failure. Among them, embryonic stem (ES) cells have been regarded as an especially valuable source of therapeutic cells due to their unlimited growth in culture and well-established cardiogenic potential in vitro. ES cell preparations that were tested in animal models of heart infarction ranged from fully undifferentiated and cardiac-committed ES cells to partially or highly enriched ES cell-derived cardiomyocytes. Here we critically review the current literature on use of fully undifferentiated ES cells for cardiac repair, elaborate on the tumorigenic risk of ES cells and pluripotent cells in general, and summarize strategies for elimination of this threat as an important step toward translation of ES cell-based therapies to clinic. This discussion is also highly relevant for clinical applicability of newly developed autologous ES cell-like stem cells, so-called induced pluripotent stem cells, which circumvent ethical and, to some extent, immunological concerns linked to the use of blastocyst-derived ES cells, but still possess high tumorigenic potential.

Keywords Embryonic stem cell • Cardiomyocyte • Differentiation • Teratoma formation • Transplantation

Introduction

Cardiovascular diseases are the most frequent cause of death in adults and the main noninfectious cause of death in children in the United States and Western Europe (1). In particular, severe heart failure caused by an irreversible loss of

T. Saric (✉) L.P. Frenzel, A. Fatima, M.K. Gupta, and J. Hescheler
Medical Center, Institute for Neurophysiology, University of Cologne,
Robert-Koch-Str. 39, 50931, Cologne, Germany, and Center for Molecular Medicine Cologne,
University of Cologne, Germany
e-mail: tomo. saric@uni-koeln.de

H. Bharavand (ed.), *Trende in Stem Cell Biology and Technology*,
DOI 10.1007/978-1-60327-905-5_14
© Humana Press, a Part of Springer Science + Business Media, LLC 2009

cardiomyocytes has a bad prognosis regardless of the underlying disease. Although the available treatment with pharmacological and surgical means and use of left ventricular assist devices is only of limited help, heart transplantation is currently the only effective therapy providing 1- and 5-year graft survival in 85.5% and 70.6% of cases, respectively (2). However, due to organ shortage, chronic graft rejection, and toxicity of the continual immunosuppressive therapy, there is an urgent need for alternative treatments. One alternative approach for improving the contractile function of a failing heart is transplantation of various types of cells such as fetal heart cells, skeletal myoblasts, multipotent adult stem cells isolated from bone marrow, as well as embryonic stem (ES) cell–derived cardiomyocytes (reviewed in Laflamme and Murry (3)). The use of cardiomyocytes isolated from aborted fetuses for human therapy is problematic because of ethical concerns, limited accessibility, and risk of immune rejection of transplanted cells by histoincompatible recipients. Bone marrow cells and skeletal myoblasts have the great advantage of being available from autologous sources and have therefore already been used in clinical trials in patients with myocardial infarction (reviewed in Murry et al. (4)). However, these cells cannot differentiate into cardiac myocytes (5–7), are incapable of integrating electrically with host myocardium (8, 9), and have only modest therapeutic benefit (10–12). In addition, their use may cause serious side effects such as ventricular tachycardia (8) and bone formation from intramyocardially implanted bone marrow mesenchymal stem cells (13). Although it is now becoming apparent that the beneficial effect on heart function observed with implanted bone marrow cells is most likely mediated by factors acting in paracrine manner through antiapoptotic and pro-angiogenic mechanisms and not, as previously suggested, by stable engraftment of transplanted cells, their fusion with host cardiomyocytes or transdifferentiation into cardiac phenotype (14–17), adult stem cells, or skeletal myoblasts alone cannot be considered as an ideal and the only source of cells for cardiac repair because they are incapable of replacing lost contractile elements.

Pluripotent embryonic stem cells represent an alternative source of clinically useful cardiomyocytes because they are easily accessible and expandable in culture, have broad developmental potential, and have high capacity to reproducibly differentiate into spontaneously beating cardiac cells in vitro. Although no attempt to repair damaged heart in humans with ES cell–derived cardiomyocytes has been made yet due to ethical objections to use of ES cells in general and lack of safe and immunologically acceptable cardiomyocyte preparations at clinically sufficient amounts, many studies have tested the validity of this therapeutic approach in small and large animal models of heart disease. Cell types that were utilized in these experiments ranged from highly purified or enriched ES cell–derived cardiomyocytes to fully undifferentiated pluripotent ES cells that were transplanted intramyocardially after injury (Fig. 1). Normally, when undifferentiated ES cells are injected into adult tissues they grow and differentiate uncontrollably to form tumors, called teratomas. Remarkably, in a few studies the injection

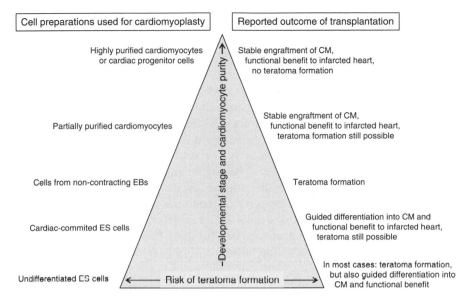

Fig. 1 Overview of different embryonic stem (ES) cell preparations that have been used for cardiac repair and reported outcomes of their transplantation into intact or infarcted hearts

of pluripotent ES cells into myocardium did not lead to expected teratoma formation but rather to their predominant differentiation into cardiomyocytes and functional benefit to the injured heart *(18–23)*.

When mouse or human ES cells are cultured in the absence of factors that normally maintain their pluripotency, such as mitotically inactivated murine embryonic fibroblasts or leukemia inhibitory factor (LIF), they spontaneously develop in vitro into spherical structures called embryoid bodies (EBs). These multicellular aggregates resemble early postimplantation embryos and contain a mixture of differentiating cells of endodermal, ectodermal, and mesodermal origin. In individual EBs these cells are organized in clusters of specific cell types as demonstrated by regional expression of lineage-specific transcripts in in situ hybridizations or of fluorescent protein markers driven by tissue specific promoters *(24, 25)*. Although this exact broad potential of ES cells to differentiate in vitro to practically all tissues found in an adult organism classifies them as a most promising source of cells for cell replacement therapies, their inherent propensity to form teratomas in vivo at the same time greatly impairs this applicability. In this review, we discuss the risk of teratoma formation from transplanted ES cells and their derivatives in view of their utility for cardiac repair in humans and summarize current approaches for elimination of the tumorigenic risk associated with ES cell–based therapies.

Spontaneously Occurring Teratomas and Teratocarcinomas

Teratomas are rare benign tumors, which are usually found in the ovary and occasionally in extragonadal sites, such as the base of the spine in newborn infants. These tumors arise from oocytes, which are activated in the absence of fertilization in a process known as parthenogenesis and develop into a mass composed of randomly distributed areas of differentiated structures resembling various adult tissues and, in some cases, even organs. Their highly malignant counterparts, teratocarcinomas, occur in the testis from abnormal germ cells within the seminiferous tubules and belong to a group of nonseminomatous germ-cell tumors. In addition to heterogeneous assembly of different somatic tissues, which is the common property of teratomas and teratocarcinomas, the latter also contain undifferentiated cells called embryonal carcinoma (EC) cells, which are regarded as pluripotent stem cells of these tumors. In addition, terato-carcinomas sometimes contain structures that closely resemble an early embryo at the gastrulation stage, so-called embryoid bodies. This same term has been later borrowed to describe solid or cystic spherical cell aggregates into which ES cells differentiate in vitro.

The unusual appearance of teratomas (*teratos* is Greek for monster) has attracted the attention of people since antiquity. Until about 50 years ago, the biology of these tumors was difficult to study because of the rare occurrence of such tumors in experimental animals and a lack of established permanent cell lines. However, in 1954 Stevens and Little *(26)* reported that males of mouse strain 129 develop spontaneous testicular teratomas and teratocarcinomas with an incidence of about 1%. Soon afterward the methods were developed for producing murine teratomas experimentally by explanting genital ridges of fetuses between 11 and 13.5 days of development *(27)* or mouse egg cylinders at about 7 days of development to extrauterine sites *(28)*. Interestingly, teratocarcinomas could not be generated by such techniques in other animal species for reasons that are still not sufficiently understood. However, recent studies showed that intrasplenic or kidney-subcapsular implantation of early embryonic liver-, pancreas- and lung-precursor tissues from pig into severe combined immunodeficient (SCID) mice can result in formation of teratoma-like structures if tissues were obtained at an early enough gestational stage *(29)*. For example, pig embryonic liver-precursor tissue obtained from E21 and E24 donors formed teratomas in 37% of SCID mice 6 weeks after implantation under the kidney capsule. These tumors contained cartilage, bone fragments, columnar epithelia lining cystic structures, hepatocytes, and striated muscle fibers. In contrast, no teratoma formation but only organ-specific growth was detected in 23 mice implanted with E28 liver-precursor tissue. Similar findings were also obtained for kidney, pancreas, and lung precursor tissues with the difference that the earliest teratoma-free gestational age differed markedly depending on the organ and species being investigated *(29, 30)*. These data indicate that even embryonic tissues that emerge after the egg-cylinder stage contain multipotent or pluripotent stem cells that can give rise to teratoma-like structures when explanted ectopically.

The ability to experimentally induce teratoma formation in mice, which could be repeatedly transplanted in successive recipients without loosing their pluripotent character, stimulated further developments in the early years of research in this field. Among the most important ones was the demonstration that a single cell derived from a retransplantable tumor can differentiate into diverse cell types found in teratocarcinomas *(31)*. This finding firmly established that these tumors contain pluripotent EC stem cells, which are embryonal in nature, can divide indefinitely, and possess the ability to differentiate into multiple somatic cell types. Subsequent establishment of permanent murine *(32)* and human *(33, 34)* EC cell lines in 1960s and 1970s greatly facilitated the experimentation with this model system and led to discoveries that were later crucial for generation of murine and human ES cell lines. For example, the finding that some EC cell lines required culture on feeder layers of mouse fibroblasts to maintain their pluripotency *(35)* and that their differentiation can be induced by removal from feeder cells and formation of EBs in suspension cultures *(36)* are still routinely utilized today for generation, propagation, and differentiation of mouse *(37, 38)* and human ES cells *(39, 40)*. The similarity between the growth characteristics and developmental potential of EC and ES cells points to their close embryonic origin. This is also highlighted by the similarity in the expression of various cell surface markers between these cell types, all of which have been first identified on either undifferentiated mouse (e.g., stage-specific embryonic antigen 1, SSEA1) or human (e.g., SSEA3, SSEA4, TRA-1-60, and TRA-1-81) EC cells. For a detailed account on a history of teratocarcinoma research and the developments that enabled the establishment of ES cell lines the reader is referred to excellent reviews by Davor Solter *(41)* and Peter Andrews *(42)*.

Teratomas Formed from Transplanted ES Cells

Undifferentiated ES cells routinely develop into teratomas when injected into different anatomical sites of syngeneic or immunodeficient recipients *(43–50)*. Teratoma formation is an important proof of pluripotency of each new ES cell line and is an essential part of a routine quality control of ES cells propagated in culture *(37–39)*. For this purpose, undifferentiated ES cells are prepared as single cell suspensions (or clumps in case of human ES cells) in serum-free medium or phosphate-buffered saline and injected either into the rear leg or thigh muscle, under the kidney capsule, or into the testis capsule of a mouse. Animals are then regularly monitored for tumor growth over a period of several weeks. An ES cell line is regarded pluripotent if differentiated tissues derived from all three germ layers are found in teratomas by histological examination of hematoxylin and eosin-stained sections. Typically, various mature tissues derived from ectoderm (neural and stratified epithelium, ganglia, glia, retina), endoderm (bronchial and gastrointestinal epithelium, thyroid, and salivary glands), and mesoderm (striated and smooth muscle, bone, cartilage) can be easily identified in both human and murine teratomas *(44, 46, 51, 52)*. Using a mixture of three different human ES cell lines for induction of teratomas under the kidney capsule of immunodeficient mice, Blum and Benvenisty *(49)* have demonstrated by means of microsatellite and fluorescence in situ hybridization (FISH) analysis that human teratomas are polyclonal even in discrete differentiated structures such as epithelial tubes within the teratoma. These data indicate that not a single cell but multiple pluripotent cells give rise to different mature tissues within a teratoma. Using the bromodeoxyuridine (BrdU) incorporation assay the same authors further showed that teratomas as old as eight weeks are rapidly proliferating and that the proliferation is scattered throughout the tumor *(49)*. Noteworthy, incorporation of BrdU was detected in the differentiated structures, and there was no evidence for undifferentiated human ES cells in any area of the tumor. Gertow et al. *(44)* were also unable to detect cells expressing any of typical pluripotent ES cell markers as well as any malignant tissues in any of the human tumor samples tested. This is in contrast to teratocarcinomas formed from mouse ES cells and naturally occurring human teratocarcinomas, where multiple foci coexpressing the proliferation marker Ki-67 and the stem cell marker Oct4 were detected *(53)*. Possible explanation for this intriguing difference is the absence of selective pressure on human ES cell–derived teratomas, because they normally grow in immunodeficient mice and are not exposed to immunosurveillance, as is the case with naturally occurring malignant tumors. To tackle this question, it would be important to determine if the foci of undifferentiated ES cells are equally prevalent in teratomas derived from mouse ES cells injected into immunodeficient, syngeneic, and allogeneic animals.

Factors that Affect the Formation of ES Cell–Derived Teratomas

The growth kinetic and tumor phenotype of teratoma may differ significantly depending on the species, genetic background, developmental and immunological status of the recipient, site of ES cell injection, and number and type of injected cells *(21, 45, 48–50, 54)*.

Species of the Host

The influence of the animal species on the ability of ES cells to form teratoma was demonstrated by transplantation of undifferentiated murine D3 ES cells constitutively expressing enhanced green fluorescent protein (eGFP) into the healthy or ischemically injured brain of immunosuppressed rats or immunocompetent syngeneic mice *(55)*. Upon injection of 80,000 ES cells at the hemisphere opposite the ischemic injury, in rats cells migrated along the corpus callosum toward the damaged tissue and differentiated into neurons in the border zone of the lesion. Small tumors were detected histologically only rarely in 9% of transplanted rats with brain injury, whereas no tumorigenesis was observed in intact rats. In contrast, 50,000 ES cells implanted into the healthy or infarcted brain of syngeneic mouse produced highly malignant teratocarcinomas at the site of implantation in more than 90% of animals 3 weeks after implantation and did not migrate to the site of injury. In this study tumors arose in the mouse model even from 500 implanted ES cells or from 10,000 ES cells that were predifferentiated in vitro to neural progenitor cells prior to transplantation. This study also indicates that xenotransplanted and homologously transplanted ES cells may possess radically different teratoma-forming capacities and calls for caution when results of preclinical studies are interpreted and extrapolated to humans. However, in another study the incidence of teratoma from xenotransplanted ES cells was higher. Utilizing traumatic brain injury model Riess et al. *(56)* found that transplantation of 100,000 undifferentiated murine ES cells beneath the damaged brain cortex of immunosuppressed Sprague-Dawley rats leads to teratoma formation in two of seven xenotransplanted animals during the observation period of 7 weeks. In the remaining five rats the ES cell administration improved neurological motor function without teratoma formation. These findings illustrate that under certain circumstances the microenvironment of the injured tissue can play an instructive role in driving the differentiation of ES cells toward specialized cells of that tissue, which could exert therapeutic benefit, but they also point out that the use of undifferentiated ES cells will always be accompanied with the high risk of teratoma formation independent of the host.

Immune System of the Recipient

The humoral and cellular components of specific and nonspecific branches of the immune system of the host exert a major influence on the ability of ES cells to engraft and form teratomas. Xenogeneic transplantation of undifferentiated ES cells across species normally results in acute graft rejection and no teratoma formation unless recipients are immunosuppressed for the whole duration of the experiment or hereditarily immunodeficient *(47, 57, 58)*. Interestingly, the xenorejection of human ES cells in mice *(47, 57)* and mouse ES cells in baboons *(58)* was not mediated by xenoantibodies but, respectively, by T cells or by both T and NK cells. Although under usual circumstance ES cells do not engraft permanently when transplanted across histocompatibility barriers into immunocompetent allogeneic recipients due to immune rejection, teratoma growth has been documented even in this context *(50, 59, 60)*. However, although the initial time course of teratoma development was similar in syngeneic and allogeneic mice, after 3–8 weeks allogeneic teratomas were immunologically rejected while the syngeneic ones still persisted *(50, 60)*. In these studies, allogeneic grafts induced both the humoral and cellular immune responses, which led to rejection of teratomas. These studies refuted the notion that ES cells possess immune privileged properties. The evidence for this assertion comes from a number of in vitro and in vivo observations. In vitro evidence encompasses, first, the observation that ES cells and their derivatives are relatively resistant to lysis by NK cells or cytotoxic T cells *(47, 61–64)*. Second, ES cells were unable to directly stimulate proliferation of alloreactive T cells *(63, 65)*. Furthermore, they were capable of directly suppressing third-party allogeneic dendritic cell-mediated T-cell proliferation *(65)*, abrogating ongoing alloresponses in mixed lymphocyte reactions *(63)* and completely preventing T-cell cytotoxicity against allogeneic concanavalin A blasts *(62)*. In vivo evidence for immunosuppressive activity of ES cells stems from a recent study showing that permanent engraftment of fully allogeneic teratomas can be achieved, provided that a sufficient dose of undifferentiated ES cells is implanted. Koch et al. *(66)* showed that five million murine ES cells allotransplanted into immunocompetent hosts formed teratomas that were tolerated due to immunosuppressive effect of transplanted ES cells. This effect was mediated by the transforming growth factor β (TGF-β) secreted by ES cells. A couple of additional, somewhat disputed in vivo studies suggested that allogeneic ES cells possess immunomodulatory and tolerogenic activity when administered intraportally or intravenously. When administered through this route, undifferentiated allogeneic murine ES cells populated the thymus, spleen, and liver of sublethally irradiated mice and induced mixed hematopoietic chimerism without forming teratomas in 98% of animals *(62, 63)*. However, these observations could not be confirmed by independent investigators *(51)*. The resolution of current controversies about immunological properties of ES cells and their derivatives requires further in-depth investigations.

The humoral component of the nonspecific immune response seems also to affect the fate of transplanted undifferentiated ES cells. A recent study by Koch et al.*(48)* provided evidence that teratoma formation can be prevented by complement system. In vitro, ES cells were shown to be susceptible to lysis by complement in heterologous and homologous sera. The involvement of the complement in preventing tumor growth in vivo was demonstrated by the finding that one million subcutaneously injected ES cells formed teratomas in syngeneic mice that were deficient in the complement component C3 with faster kinetics than in syngeneic wild-type mice. Moreover, when only 100,000 cells were transplanted, teratomas were generated only in C3 knock-out mice but not in wild-type mice. Antibodies, which participate in the activation of the classical complement pathway, did not appear to be involved in control of teratoma formation in this syngeneic system. The complement pathway involved in lysis of ES cells in this study was found to be the antibody-independent alternative pathway. This pathway was activated on the surface of ES cells because the normal inhibitory function of factor H from serum was diminished due to decreased expression of sialic acid on the surface of ES cells *(48)*. Sialic acid is normally required for the inhibitory function of factor H. Interestingly, complement system did not affect the engraftment of ES cell–derived cardiomyocytes, which expressed higher levels of sialic acid *(48)*. These data indicate that both specific and nonspecific elements of the immune system can pose a barrier to teratoma formation as long as the number of injected ES cells is kept below a certain threshold. This conclusion is also supported by the finding that an equal number of injected cells leads to higher teratoma incidence in immunodeficient nude mice than in immunocompetent syngeneic mice *(45)*.

Number of Injected Cells

Many ES cells initially die upon transplantation partially due to immunological mechanisms (even in syngeneic or immunodeficient animals) and partially due to apoptosis induced by the inability of many cells to receive and respond to required survival signals *(67)*. Therefore, teratoma incidence and growth kinetics are greatly dependent on a dose of injected ES cells. According to this assumption, tumors are formed only when ES cell dose is above the certain threshold. For example, Nussbaum et al. *(50)* detected no teratomas 3 weeks after injection of 500–50,000 ES cells into uninjured hearts of immunocompetent syngeneic mice using conventional histology. However, 50% of mice developed teratomas when 100,000 cells were injected, while tumor formation was observed in 80% and 100% of animals that received 250,000 and 500,000 undifferentiated ES cells, respectively. Behfar et al. *(21)* have also shown that the teratoma forming capacity of injected ES cells within myocardium greatly depends on the number of injected cells: if 3×10^5 wild-type murine CGR8 ES cells were injected into intact mouse heart, these ES cells exclusively differentiated into cardiomyocytes without forming teratomas (see the section "Cardiac Repair with ES Cells"). In contrast, when the same cells were delivered into intact hearts at increasing doses of one or three million cells per heart, teratoma formation was observed in 18% and 68% of mice, respectively *(21)*. Significantly, different minimal ES cell doses that still allowed for teratoma formation was reported by Cao et al. *(68)*. These investigators generated a transgenic murine ES cell line to constitutively express fusion reporter protein that consists of firefly luciferase and eGFP to enable longitudinal monitoring of ES cell engraftment and teratoma formation in vivo without sacrificing animals. In order to determine the minimal number of ES cells that still can cause teratomas, they injected subcutaneously in Matrigel (BD Biosciences, San Jose, CA) varying numbers of ES cells (in the range from 1 to 10,000 cells) in a mixture with irradiated murine embryonic fibroblasts to make a final total cell number of 100,000 per injection site. These studies demonstrated that even a dose of 500–1,000 ES cells (but not 1, 10, or 100 cells) can induce tumor growth in syngeneic hosts that was detectable by bioluminescence in vivo imaging already at day 7 after implantation and continued to grow over a time period of 3 months. Injection of 10,000 ES cells (without fibroblasts) into healthy myocardium of nude athymic rats revealed that teratoma formed and could be monitored for a period of 10 months in living animals (68). It remains unclear in the study by Nussbaum et al. (50) if teratoma would be detected in animals that received 500–50,000 ES cells after longer observation periods. In work by Lawrenz et al. (45), the minimal number of ES cells required for teratoma formation was shown to be even lower: only two ES cells administered together with two million nontumorigenic MRC-5 cells were capable of giving rise to teratomas in 60% of nude mice 11 weeks after subcutaneous injection in Matrigel. When 20 or 2,000 ES cells were injected with MRC-5 cells teratomas were formed in 100% of animals 2 and 6 weeks postimplantation, respectively. Interestingly, the incidence of tumors that were generated with 2 and 20 cells injected under the kidney capsule of nude mice was lower as compared to subcutaneously injected cells. In addition, no teratomas were formed when two ES cells were injected subcutaneously into syngeneic immunocompetent mice over the observation period of 20 weeks (45). These results indicate that under clinically relevant conditions, when pure populations of ES cell derivatives are used for transplantation, few ES cell contaminants, if cotransplanted, could lead to tumor formation and that the minimal number of ES cells required for teratogenesis may depend on the site of ES cell injection, immunological status of the host, and type and number of differentiated cells being transplanted. This emphasizes the importance of designing careful protocols for purification of desired cell types devoid of pluripotent contaminants. The risk of teratoma formation would be especially increased in immunosuppressed recipients, which already have a severalfold increased risk of developing tumors due to adverse effects of immunosuppressive drugs (69).

Site of ES Cell Injection

As already mentioned above, the incidence of teratoma formation from pluripotent ES cells can depend on the local environment into which the cells are grafted. However, local cues can also affect the ability of transplanted ES cells to differentiate along the specific cell lineages represented in teratomas. Not only can transplanted ES cells under certain conditions be guided locally to differentiate into mature phenotypes present at the implantation site without forming teratomas, but also the teratomas that are formed from ES cells injected at different anatomical locations appear to differ in their growth properties and composition of differentiated tissues. Cooke et al. (54) reported that 500,000 human ES cells transplanted into the liver of nude mice produce large, fast growing teratomas within a 3- to 4-week period. In contrast, an identical number of ES cells placed subcutaneously failed to produce palpable tumors beneath the skin within the same period of time. Human ES cell–derived teratomas in liver frequently contained fluid-filled cavities. However, large cysts are also occasionally observed when human ES cells are transplanted under the kidney or testis capsule (46). Wakitani et al. (70) reported that teratomas produced from murine ES cells differ phenotypically if generated in the knee joint or subcutaneously, with the former containing more cartilaginous elements than the latter. However, aside from these two reports the question of whether the developmental potential of ES cells is influenced by the graft site still remains open because previous studies did not find any evidence for this conclusion with ES cells implanted into testis or kidney, intraperitoneally or subcutaneously. However, these studies did not perform detailed histological comparisons of different teratomas so that small differences could not be ruled out.

Developmental Status of the Host

The developmental status of the host may also play an important role in determining the fate of transplanted cells. Grafting of ES cells into adult tissues may result in a different outcome compared with grafting into embryonic tissues because environmental cues to which ES cells will respond in such tissues may greatly differ. This is vividly illustrated by the instructive environment of the oocyte or early blastocyst that directs differentiation of injected ES cells to contribute to all mature tissues in adult animal without forming teratomas (71). Embryonic environment of the host is also capable of modulating differentiation of ES cells in a heterologous system, as reported by Goldstein et al. (72) for human ES cells transplanted into the organogenesis-stage chick embryo. In this system human ES cells survived and proliferated in the chicken embryo and differentiated either en masse to specific types of cells or into phenotypes typically found in chicken tissues to which they had migrated. Human cells were found to be well integrated into chicken tissues and to contribute to tissue architecture. These data suggest that early embryonic environment possesses strong instructive signals for directed differentiation of ES cells, which are to a large extent lost in adult tissues where tumorigenic potential of ES cells predominates.

Developmental Status of ES Cells

As discussed above, undifferentiated ES cells are invariably tumorigenic in adult tissues if transplanted under optimal conditions. However, tumorigenic potential is retained even in the progenitor cells of early embryonic tissues, which is evident when these cells are explanted ectopically, as discussed above (29, 30). Therefore, it is not surprising that even differentiated ES cell derivatives such as whole EBs or enriched progenitor cells derived from ES cells still retain their tumorigenic potential (55, 73–78). This is most likely due to the presence of contaminating pluripotent ES cells in these preparations even at late stages of differentiation (79) and/or tumorigenic propensity of early tissue progenitors. In one example, human

ES cells predifferentiated in vitro for 16 days formed severe intracerebral teratomas in rat model of Parkinson's disease, while cells after 20–23 days of differentiation did not generate teratomas *(74)*. In our experience, small number of pluripotent stem cells capable of forming teratomas still persists in ES cell–derived cardiomyocytes after one week of selection with an antibiotic that should kill all cells in EBs except cardiomyocytes, which expresses the antibiotic resistance gene in a cardiospecific manner *(75)* (T. Saric unpublished observations). Only prolonged selection could achieve teratoma-free transplantation of pure populations of selected cardiomyocytes *(75)*. As discussed above, another important issue is the localization of injection. Direct subcapsular transplantation of cells obtained from murine day 9 and day 15 EBs into liver of syngeneic mice did not result in tumors, whereas injection of the same cells through the portal vein via the spleen gave rise to teratomas *(79)*. These results strongly argue for the requirement of complete elimination of ES cells from all cell preparations that are intended for clinical application.

Cardiac Repair with ES Cells

In common perceptions of ES cell–based tissue repair strategies, pluripotent ES cells must be first differentiated in vitro into specific cell lineages, which are then transplanted alone or in combination with other mature or progenitor cell types to directly contribute to tissue regeneration through proliferation and functional integration. However, reports on the developmental plasticity of adult stem cells in general and of adult bone marrow stem cells in particular that were shown to be capable of transdifferentiation into cardiomyocytes if placed into myocardium *(80)* inspired some investigators to determine how stem cells respond to host environment and differentiate toward specific cell phenotype. Instead of using adult stem cells to answer this question, researchers, encouraged by earlier reports that undifferentiated ES cells can specifically differentiate into neural lineages if transplanted into the injured spinal cord *(81)*, set out to determine whether ES cells could also be committed to differentiate specifically toward cardiac lineage if placed into adult cardiac tissue environment and to identify factors that may mediate these effects *(18, 21)*. These and the studies of other groups that directly or indirectly address this question are summarized in Table 1.

Cardiac Repair with Undifferentiated ES Cells

In their first report on this subject in 2002, Behfar et al. *(18)* presented intriguing data that undifferentiated murine CGR8 ES cells could differentiate predominantly into cardiomyocytes when transplanted into intact hearts of immunocompetent syngeneic mice (0.5×10^6 cells/heart). In none of these transplantations was there

Table 1 Transplantation of undifferentiated or cardiac-committed ES cells into healthy or infarcted heart for cardiac repairw

Author (ref)	Type of implanted cells	No. of implanted cells per heart	Transplantation type and disease model	Follow-up period	Outcome of cell transplantation
Group A					
Behfar et al., 2002 (*18*)	Mouse CGR8 ES cells	0.5 × 10⁶	Syngeneic mouse, healthy heart	4 weeks	ES cells differentiated into CMs, no teratoma, no immune infiltration
	CGR8 ES cells expressing Noggin or TGFβRII	0.5 × 10⁶	Syngeneic mouse, healthy heart	4 weeks	Teratoma formation in all cases, no cardiogenic differentiation
	Mouse CGR8 ES cells	0.3 × 10⁶ (4 weeks after infarction at three different locations)	Xenogeneic rat, infarcted heart, no immunosuppression	5 weeks	ES cells differentiated into CMs, functional improvement of the infarcted heart, no teratoma formation, no immune infiltration
Hodgson et al., 2004 (*19*)	Mouse CGR8 ES cells	0.3 × 10⁶ (8 weeks after infarction at three different locations)	Xenogeneic rat, infarcted heart, no immunosuppression	3–12 weeks	ES cells differentiated into CMs in the heart, functional improvement of the infarcted heart was also present at 12 weeks, no teratoma formation, no immune infiltration
Singla et al., 2006 (*22*)	Mouse R1 or HM1 ES cells (originate from 129 mouse strain)	3 × 10⁴ (3 × 10,000 cells at locations, 3–5 h after infarction)	C57BL6 mouse, same H2 haplotype but minor MHC incompatibility is possible, infarcted heart	2 weeks	No teratoma observed but follow-up period was short and ES cell number low; injected ES cells differentiated into CM, endothelial and smooth muscle cells, and improved heart function

(continued)

Table 1 (continued)

Author (ref)	Type of implanted cells	No. of implanted cells per heart	Transplantation type and disease model	Follow-up period	Outcome of cell transplantation
Behfar et al., 2007 (*21*)	Mouse CGR8 ES cells	0.3×10^6	Syngeneic/wild-type mouse, healthy heart	?	No teratoma
		1×10^6			18% teratoma
		3×10^6			68% teratoma
	Mouse CGR8 ES cells	3×10^6	Syngeneic/TNF-α-transgenic mouse, healthy heart	?	No teratoma due to cardiac-restricted expression of TNF-α
	CGR8 ES cells expressing Noggin or TGFβRII	3×10^5	Syngeneic/wild-type mouse, healthy heart	?	100% teratoma due to blockade of TGF-β or BMP signaling
	Mouse CGR8 ES cells	3×10^6 (8 weeks after infarction)	Xenogeneic/rat, infarcted heart, no immunosuppression	8 weeks	70% teratoma
	CGR8 ES cell-derived cardiopoietic progenitors	3×10^6 (8 weeks after infarction)	Xenogeneic/rat, infarcted heart, no immunosuppression	8 weeks	No teratoma, proliferation and integration of progenitor cell-derived CM, functional recovery of the heart
Singla et al., 2007 (*23*)	Mouse R1 or HM1 ES cells (originate from 129 mouse strain)	3×10^4 ($3 \times 10,000$ cells at three locations, 3–5 h after infarction)	C57BL6 mouse, same H-2 haplotype but minor MHC incompatibility is possible, infarcted heart	2 weeks	No teratoma observed but follow-up period was short and ES cell number low, injected cells inhibited apoptosis and hypertrophy of host CM and cardiac remodeling/fibrosis
Xie et al., 2007 (*20*)	Human H1 ES cells	0.5×10^6 (30 min after infarction)	Xenogeneic/rat, infarcted heart, no immunosuppression	8 weeks	Transplanted cells detected in 5 of 11 rats, no teratoma; ES differentiation into CM claimed but not stringently demonstrated

Group B

Ménard et al., 2005 (*83*)	Cardiac-committed mouse CGR8 ES cells	30×10^6 BMP2-treated cells at 25 sites (2 weeks after infarction)	Xenogeneic/sheep, infarcted heart, with and without immunosuppression	4 weeks	Engraftment of cells in immunosuppressed as well as in immunocompetent hearts, differentiation into mature CM accompanied with the functional benefit
Bonnevie et al., 2007 (*58*)	Cardiac committed mouse CGR8 ES cells	40×10^6 BMP2-treated at 20 sites (2 weeks after occlusion)	Xenogeneic/baboon, coronary artery occlusion-reperfusion, without immunosuppression	8 weeks	Mouse cells could not be identified, immune rejection was mediated by T and NK cells, no functional improvement in cell transplanted group
Tomescot et al., 2007 (*82*)	Cardiac committed human HUES-1 or I6 ES cells	3×10^6 BMP2 + SU5402-treated cells (15 days after infarction)	Xenogeneic/rat, infarcted heart, with immunosuppression	8 weeks	Cardiac committed ES cells differentiate into CM without teratoma formation; human CM were detected after 2 months by RT-PCR and ICH

Group C

Kofidis et al., 2004 (*84*)	Mouse D3 ES cells	0.25×10^6 ES cells ± IGF1	Allogeneic/Balb/c, LAD ligation, without immunosuppression	2 weeks	ES cells formed viable grafts and improved heart function, IGF-1 promoted differentiation of ES cells into CM, but large portions of graft were not CM
Kofidis et al., 2005 (*85*)	Mouse D3 ES cells	0.25×10^6 ES cells ± VEGF, FGF, TGF-β	Allogeneic/Balb/c, LAD ligation, without immunosuppression	2 weeks	TGF-β promoted in vivo differentiation of ES cells to CM but large portions of graft were not CM, ES cell transplantation improved myocardial function
Kofidis et al., 2005 (*59*)	Mouse D3 ES cells	0.25×10^6 (following LAD ligation)	Syngeneic/129sv, infarcted heart; Allogeneic/Balb/c, infarcted heart; SCID mouse	1–4 weeks; 1–4 weeks; 1–4 weeks	At 2 weeks all animals showed large grafts that were reduced by week 4, grafts were composed of different cell types but authors state that teratomas were not observed, allogeneic grafts induced humoral and cellular immune response

(continued)

Table 1 (continued)

Author (ref)	Type of implanted cells	No. of implanted cells per heart	Transplantation type and disease model	Follow-up period	Outcome of cell transplantation
Swijnenburg et al., 2005 (**60**)	Mouse D3 ES cells	1×10^6 (5 min after infarction)	Syngeneic/129sv, infarcted heart Allogeneic/Balb/c, infarcted heart	1–8 weeks	Teratomas occurred after 4 weeks in all mice; by 8 weeks teratomas were rejected in allogeneic but not in syngeneic mice
Kolossov et al., 2006 (**75**)	Mouse D3 ES cells	1×10^5	Syngeneic/129sv, infarcted heart	4 weeks	Teratoma in all syngeneic mice ($n = 13$)
	Mouse D3 ES cells	1×10^5	Allogeneic/? infarcted heart, no immunosuppression	4 weeks	Teratoma in two of three allogeneic mice (final outcome not specified)
	Microdissected beating areas	1×10^5	Syngeneic/129sv, infarcted heart	4 weeks	Teratoma in five of six mice with delayed onset
	Puromycin-selected CM (>99% pure)	1×10^5 (with MEF)	Syngeneic/129sv, infarcted heart	>100 days	Long-term engraftment of CM, no teratoma in 95% of mice ($n = 60$); malignant fibrous histiocytoma in two of three mice with tumors
Cao et al., 2007 (**68**)	Mouse D3 ES cells	1×10^6 healthy heart	Syngeneic/mouse 129Sv, intact heart, immunocompetent	5 weeks	Teratomas in the heart and extracardiac tissues, in vivo differentiation of ES cell toward CM only at low frequency
	Mouse D3 ES cells	1×10^4 healthy heart	Xenogeneic/nude athymic rat, intact heart	1–10 months	Teratoma found 10 months after ES cell injection
Nussbaum et al., 2007 (**50**)	Mouse C57BL6 ES cells	0.5×10^6	Syngeneic/mouse, healthy heart	3 weeks	100% teratoma, directed differentiation of ES cells toward CM was not observed
	Mouse C57BL6 ES cells	0.5×10^6	Allogeneic/Balb/c mouse, healthy heart, no immunosuppression	3 weeks	100% teratoma, but all were rejected after 5 weeks
	Mouse CGR8 ES cells (129sv mouse)	0.5×10^6	C57BL6 mouse, healthy heart, minor MHC incompatibility is possible	3 weeks	Injected cells did not survive, neither ES cells nor CM were detected in the heart

Reference	Cells	Dose	Model	Time	Results
Caspi et al., 2007 (76)	Human H9.2 ES cell derivatives, nonbeating	1.5×10^6 (7–8 days after infarction)	Xenogenic/rat, intact and infarcted heart, with immunosuppression	4 weeks	Teratoma formation in six of ten intact and three of six infarcted hearts
	Human H9.2 ES cell–derived beating CM clusters	1.5×10^6 (7–8 days after infarction)		4–8 weeks	CM survived and integrated with host CM, functional improvement of infarcted heart, no teratomas detected
Leor et al., 2007 (77)	Human ES cells	0.5–1×10^6	Xenogenic/athymic nude rat, intact or infarcted heart	2–4 weeks	Teratoma formation occurs, no directed differentiation of ES cells into CM
	Human EB-derived cells 4–9 days of differentiation	0.5–1×10^6		2–4 weeks	Predifferentiated cells of EBs also formed teratomas and did not give rise to CM
	Microdissected human ES cell-derived beating CM, day 10–20 of differentiation	0.1 mm^2 pieces (unspecified dose)		2–4 weeks	Islands of disorganized myofibers and fibrosis detected, teratoma in one case, transplanted cells prevented postinfarct dysfunction and scar thinning

Abbreviations: CM cardiomyocytes, ES cells embryonic stem cells, MHC major histocompatibility complex; TNF tumor necrosis factor; BMP bone morphogenetic protein; TGF tumor growth factor; SCID severe combined immunodeficiency; EBs embryoid bodies. LAD left anterior descending artery, SU5402 synthetic FGF-2 receptor antagonist; RT-PCR reverse transcription-polymerase chain reaction; ICH Imunocytochemistry; IGF insulinlike growth factor; VEGF vascular endothelial growth factor; FGF fibroblast growth factor, ICH imunocytochemistry

References in this table are separated into three groups. The references in Group A claim that ES cells transplanted into the heart preferentially differentiate into cardiomyocytes without forming teratomas. Group B contains references that used cardiac-committed ES cells for transplantation. For that purpose undifferentiated ES cells were pretreated with growth factors 1–2 days prior to transplantation. The references listed in Group C refute the observations presented by studies in the first two groups.

evidence of teratoma formation or inflammatory reaction. This guided differentiation of ES cells was prevented by blocking the TGF-β and bone morphogenetic protein 2 (BMP-2) receptor-mediated signaling through transgenic overexpression of BMP-2 inhibitor Noggin or dominant negative TGF-β receptor II in ES cells. Implantation of such transgenic cells into the heart of wild-type syngeneic mice resulted in teratoma formation in one-third of transplanted animals 4 weeks after cell delivery *(18)*. Interestingly, exclusive teratoma-free cardiogenic differentiation and functionally beneficial integration of TGF-β responsive wild-type murine ES cells was also seen in infarcted hearts of immunocompetent rats (3 × 10^5 cells were injected along the border zone of the infracted area at three different locations). Surprisingly, xenotransplanted cells were not immunologically rejected in rat hearts, and no signs of immune rejection were seen histologically after follow-up of five weeks *(18)*. Later, the same group reported that functional benefit of transplanted ES cells in infarcted rat hearts remained sustained over 12 weeks of continuous follow-up, and that even after this prolonged period no evidence for xenograft rejection or tumor formation could be obtained *(19)*. These early reports supported the hypothesis that undifferentiated ES cells may directly contribute to myocardial healing through directed differentiation within the host myocardium without the need of in vitro predifferentiation. In addition, these studies emphasized the importance of intact TGF-β and BMP-2 signaling pathways for proper cardiogenic differentiation of ES cells. This conclusion was further corroborated by findings that TGF-β and BMP-2 induce cardiac commitment of ES cells in vitro when added for 24 h in low-serum and LIF-containing medium *(18)*. In a recent follow-up study, Behfar et al. *(21)* have corroborated their initial data and shown that the tumorigenic potential of intramyocardially injected CGR8 ES cells (see above) could be completely averted by transplanting wild-type ES cells into hearts of transgenic mice with cardiac-specific expression of tumor necrosis factor α (TNF-α). In this case, implantation of even three million ES cells into TNF-α transgenic hearts did not result in neoplastic growth. Instead, transplanted ES cells were guided by locally produced TNF-α to differentiate into cardiomyocytes and functionally integrate with host myocardium. These data strongly suggest that specific environmental signals indeed play an important role in determining the fate of pluripotent ES cells in vivo, and that this knowledge may be utilized to induce the commitment of ES cells toward desired cell lineage both in vivo as well as in vitro. The cardiogenic TNF-α action could also be recapitulated in EBs in vitro and was found to require an intact endoderm. To identify the endodermal cardiogenic factors, visceral endodermlike cells were stimulated in vitro with or without TNF-α, and genomic and proteomic data were obtained from treated and untreated samples then analyzed and compared. This approach identified an array of TNF-α–induced secreted growth factors as cardiogenic candidates, among them TGF-β1, BMP-1, BMP-2, BMP-4, vascular endothelial growth factor A (VEGF-A), interleukin 6 (IL-6), epidermal growth factor (EGF), fibroblast growth factor 2 (FGF-2) and FGF-4, haptoglobin, colony-stimulating factor 1 (CSF-1), nerve growth factor β (NGF-β), and insulinlike growth factor 1 (IGF-1) and IGF-2 *(21)*. This analysis further identified p38 mitogen activated protein kinase as a central component of the signaling network established in endoderm after TNF-α treatment. When the cocktail composed of 11 recombinant cardiogenic factors identified in this screen was applied to ES cells in vitro, they underwent differentiation into intermediate cardiopoietic progenitor cells and finally to mature cardiomyocytes, which could be transplanted at high loads (three million cells per heart) without being tumorigenic (21). These elegant studies provide insight into molecular mechanisms that drive differentiation of ES cells into cardiomyocytes within EBs in vitro and may act within host myocardium in vivo.

Independent studies of Singla and coworkers (22, 23) corroborated these intriguing observations of guided cardiac differentiation of pluripotent ES cells within the heart using 16 times fewer ES cells than Behfar and coworkers (18, 21). Singla et al. (22, 23) transplanted 3×10^4 undifferentiated R1 or HM1 mouse ES cells (derived from two different 129 mouse substrains) 3–5 h after coronary artery ligation into the ischemic area of immunocompetent C57BL/6 mice. R1 ES cells were engineered to constitutively express β-galactosidase, whereas HM1 cells were stably transfected with an eGFP vector. Evaluation of heart function and histological examination of transplanted tissues was assessed two weeks later. These analyses revealed that injected ES cells survived in about 80% of animals despite minor histocompatibility differences between donor cells and recipient mice (129Sv vs. C57BL/6). Transplanted ES cells reportedly differentiated into cardiomyocytes, endothelial cells, and smooth muscle cells without generating teratomas within this short observation period (22). This process was associated with the functional improvement of the infarcted heart, reduced apoptosis and hypertrophy of host cardiomyocytes, and diminished cardiac remodeling and fibrosis (22, 23). Transplanted cells were found mainly in and around the area of injury, with almost no cells in the intact myocardium. Interestingly, when ES cells were transplanted into healthy heart without infarction, only rare cells could be detected two weeks later. The authors proposed that besides generation of new myocardium the above mentioned beneficial effects can also be attributed to soluble factors released by ES cells such as antiapoptotic proteins cystatin C, osteopontin and clusterin, and the antifibrotic tissue inhibitor of metalloproteinase 1 (TIMP-1) (23). These molecules could be detected in conditioned medium prepared from ES cell cultures in vitro. However, it is unclear if the amount of secreted factors from only 3×10^4 transplanted ES cells in this study would be sufficient to exert the proposed effects. The supposed absence of teratoma in this study should be regarded with caution because a much longer follow-up period would be required to reliably determine the occurrence of teratoma from only 3×10^4 injected cells. In another study that corroborates these data, Xie et al. (20) transplanted undifferentiated GFP-expressing human H1 ES cells into left ventricular infracted area (0.5×10^6 cells per heart) of immunocompetent nonimmunosuppressed rats. Using laser-capture microdissection to isolate GFP-positive areas in infarcted hearts and reverse transcription-polymerase chain reaction (RT-PCR) to detect the expression of human cardiac-specific transcripts two months after cell injection, cells expressing human GATA-4, Nkx2.5 and cardiac troponin I were observed in 5 of 11 rats. No teratoma formation was observed. Although these data also indicate that pluripotent human ES cells can be instructed by local environment to differentiate into cardiac lineage, cardiac differentiation of human ES cells was not stringently demonstrated by additional methods such as immunohistochemistry (20).

Cardiac Repair with Cardiac-Committed ES Cells

In order to enhance the cardiogenesis and reduce tumorigenesis of ES cells, several groups have used cardiac-committed ES cells for transplantation. Tomescot et al. (82) exposed human HUES-1 and I6 ES cells to BMP-2 and synthetic FGF receptor inhibitor SU5402 for 48 hours to induce the expression of mesodermal (Tbx6, FoxH1, Isl1) and cardiac genes (Mef2c, Nkx2.5, α-actin). These precommitted cells were then injected into the infarcted heart of immunosuppressed Wistar rats (3×10^6 cells injected 15 days after infarction). RT-PCR and immunohisto-chemistry performed two months later revealed that 85% of sections contained differentiated human cardiomyocytes. However, the cell engraftment was limited with human cardiomyocytes populating only a small percentage (2.4–3.6%) of the scar area. Neither signs of teratoma formation in the heart or elsewhere in the body nor evidence for inflammation were found in this study.

In another study by Ménard et al. (83), cardiac-committed mouse CGR8 ES cells were implanted into infarcted myocardium of both immunosuppressed and immu-nocompetent sheep. The cells used in this study were engineered to express β-ga-lactosidase under the control of cardiac-specific Nkx2.5 promoter and enhanced yellow fluorescent protein (eYFP) under the control of the α-actin promoter. To commit ES cells toward a cardiac lineage before transplantation, undifferentiated ES cells were treated for 36 h with BMP-2. Two weeks after infarct induction car-diac-committed cells (30×10^6) were implanted in serum-free culture medium at 25 sites (1.2×10^6 cells per site) along the borders and at the center of the infracted area. One month after injection of ES cells the hearts were examined. Detection of β-galactosidase- and eYFP-positive cells in the scar area of infarcted hearts indi-cated that cardiac-committed mouse ES cells differentiated into cardiomyocytes. Interestingly, engraftment of cells was observed in immunosuppressed as well as in nonsuppressed sheep without formation of teratomas, and this was accompanied by improvement of heart function (83). In contrast to these intriguing findings, another study failed to identify cardiac-committed mouse ES cells by X-gal histology or immunofluorescence when they were transplanted into immunocompetent baboons two month after transplantation (58). Based on the composition of the inflamma-tory infiltrate, xenografted cells were rejected in this case presumably by T and NK cells. These data underscore the importance of donor–recipient compatibility in determining the outcome of cellular cardiomyoplasty and warn from direct extrapo-lation of results obtained in one experimental model to another, in particular the human one.

Evidence Against the Exclusive Cardiogenic Potential of ES Cells Injected into the Heart

The intriguing nature of the above delineated results is in their claim that fully undifferentiated ES cells do not form teratomas but rather differentiate into cardiomyocytes even when transplanted at relatively high numbers into intact or infarcted hearts. Nussbaum et al. *(50)* undertook a series of experiments to re-examine these claims. Utilizing mouse pluripotent ES cell lines R1, CGR8, and C57BL/6, which were also used in the above mentioned studies, these investigators found no evidence for teratoma-free cardiac differentiation of myocardially injected ES cells. Four weeks after cell injection (5×10^5 ES cells per heart) teratomas occurred in all nude, syngeneic, and allogeneic animals. Although teratomas persisted in syngeneic animals, those that occurred in allogeneic animals were eventually rejected. Cardiomyocyte content was compared in teratomas formed in the hind limb and in the heart from the same ES cell line, but no preferential differentiation of ES cells toward cardiac myocytes could be detected in the heart as compared to skeletal muscle ($2.1 \pm 0.5\%$ and $1.1 \pm 0.3\%$ of tumor mass contained cardiomyocytes in the hind limb and heart teratomas, respectively). In this study, heart infarction also did not promote ES cell differentiation toward cardiomyocytes or prevent teratoma formation. In addition, neoplastic growth was also observed when murine ES cells were predifferentiated in vitro prior to transplantation. The same results were also obtained with predifferentiated human ES cells *(76)*. Predifferentiation of murine ES cells with single cardiogenic factors in the study by Behfar et al. *(21)* also did not abolish their tumorigenic propensity when these cells were implanted into the heart. In agreement with these results, teratoma formation was also detected by one group in all syngeneic (129sv) and allogeneic (Balb/c) animals that received undifferentiated mouse D3 ES cells (derived from 129S2/SvPas mouse strain) at doses of 0.25×10^6*(59)* or 1×10^6 cells per infarcted heart *(60)*. No evidence for preferential differentiation of ES cells toward cardiomyocytes was reported, and the notion of immune privilege was refuted because teratomas in allogeneic animals were rejected 4–8 weeks after cell injection *(59, 60)*. Kofidis et al. *(84, 85)* tested whether local coadministration of growth factors together with undifferentiated D3 ES cells could enhance their engraftment and host organ-specific differentiation. These investigators transplanted 0.25×10^6 murine D3 ES cells expressing eGFP into left anterior descending artery–ligated hearts of immunocompetent allogeneic Balb/c mice together with recombinant mouse IGF-1, VEGF$_{164}$, FGF-2, or TGF-β1 and after 2 weeks evaluated the degree of ES cell engraftment into infarct area, expression of cardiac markers by GFP-positive cells, and the heart function by echocardiography. These studies revealed that ES cells formed viable grafts in the form of dense conglomerates, which improved the heart function. Within the graft the cardiac gap junction protein connexin 43 was expressed by up to 25–42% and α-sarcomeric actin by 20–87% of GFP-positive graft area in IGF-1- and TGF-β-treated groups. In contrast, only 4–8% of GFP-positive cells expressed α-sarcomeric actin in the group treated with ES cells only *(84)*. Although these data indicate that certain growth factors can promote differentiation of ES cells into cardiomyocytes, and it is important to note that not all cells in the graft adopted cardiomyocyte phenotype (cellular atypia and nuclear polymorphism were present in other regions of the graft), as it was the case in the studies of Behfar et al. *(18)*. It is also very likely that coadministration of growth factors with ES cells would not prevent their neoplastic growth if the animals were observed for longer periods of time than only for 2 weeks, because the same ES cells injected without growth factors formed teratomas and were rejected in allogeneic recipients after 4–8 weeks *(59, 60)*.

Teratoma formation following ES cell transplantation into intact or injured heart without guided cardiomyocytes differentiation has also been reported by a number of other studies. Large tumor mass was found in myocardium of a majority of syngeneic mice 68 days after injection of 1×10^5 mouse ES cells *(75)*. In this study teratomas formed in 80% of mice even when microdissected beating areas were transplanted, illustrating the need for obtaining pure populations of cells for teratoma-free engraftment. Tumorigenic potential of ES cells was also demonstrated by Cao et al. *(52, 68)* utilizing transplantation of 1×10^6 mouse ES cells into intact syngeneic mouse hearts or different doses of mouse ES cells (10,000 or 1×10^7) into intact nude rat hearts. In animals transplanted intramyocardially with one million ES cells, both cardiac and extracardiac teratomas could be easily detected 3–5 weeks after implantation. In contrast, using human ES cells no evidence of teratoma formation was found 2–4 weeks after transplantation of 0.5–1.0×10^6 undifferentiated H9.2 or I6 ES cells into intact or infarcted nude rats ($n = 7$) or after implantation of the same number of cells derived from 4- to 8-day-old EBs, $n = 10$ *(77)*. Although in the majority of cases human cells could be identified in rat hearts independent of their level of differentiation, human ES cells or EB-derived cells did not form new myocardium. Human cardiomyocytes were detected only in hearts that were implanted with microdissected clusters of beating cardiomyocytes *(77)*. In this group of animals myocardial dysfunction of infarcted hearts could be prevented, but in 1 of 23 rats teratoma was found. In contrast, when human ES cells were differentiated into cardiomyocytes in vitro and then enriched by microdissection, as described above, or by Percoll (GE Healthcare), gradient centrifugation to contain about 13–15% cardiomyocytes, teratoma formation was not observed in most studies in the period of 4 weeks, and stable engraftment of cardiomyocytes with functional benefit was achieved in immunosuppressed or immunodeficient rats *(76, 86, 87)*. Similar results were obtained with eGFP-expressing cell preparations derived from murine D3 ES cells that contained approximately 26% of cardiomyocytes *(88)*. In this study, 3×10^5 enriched cardiomyocytes were prepared in Joklik modified medium (Gibco BRL Life Technologies, Grand Island, NY) and injected at three different sites into infarcted hearts of immunocompetent non-suppressed Wistar rats 20 min after induction of infarction. After the follow-up of 32 weeks, transplanted cells could still be identified in the hearts despite being xenogeneic as demonstrated by their eGFP fluorescence. Moreover, the number of blood vessels was higher and the ventricular function was improved in injured hearts that received cell transplants as compared with controls. It is surprising that cell preparations containing only 13–26% of defined cell population did not give rise to teratoma more frequently. Although at lower tumorigenic risk than undifferentiated ES cells, the transplantation of enriched but still not absolutely pure cell preparations may lead to occurrence of teratoma especially after longer observation periods.

Strategies for Preventing Teratoma Formation from ES Cells

The possibility that, under appropriate conditions, only a few cotransplanted ES cells could form teratoma strongly indicates that cells used for replacement therapy will have to be absolutely void of undifferentiated ES cells. Therefore, only those cells that have lost their tumorigenic potential, such as somatic precursors or fully differentiated cells, will be suitable for therapeutic transplantation. There are several strategies for producing pure cell populations for teratoma-free tissue repair.

Directed Differentiation In Vitro

In this strategy, differentiation of ES cells into a desired somatic cell type is conducted in a stepwise manner in strictly controlled culture conditions containing defined differentiation and survival factors. The advantage of this method is that no genetic modification of donor cells is required. However, this approach is complicated because exact factors and culture conditions must be known in order for it to be successful. The most recent example of successful use of this approach is enrichment of cardiomyocytes from murine and human ES cells using defined growth factors. Behfar et al. *(21)* have demonstrated that TNF-α promotes cardiac differentiation of undifferentiated murine ES cells into EBs in vitro as well as in the heart of TNF-α-transgenic mice in vivo. The pro-cardiogenic action of this cytokine was mediated by secreted endodermal cardio-inductive signals including TGFβ1, BMP-2, BMP-4, activin-A, VEGF-A, IL-6, FGF-2, FGF-4, IGF-1, IGF-2, and EGF. These growth factors, applied as a recombinant cocktail to ES cells, have a capacity to generate cardiopoietic progenitor cells in vitro, which were characterized by loss of oncogenic and pluripotency markers like Oct-4, BRCA1, Ect4, and Dek and expression of cardiogenic markers such as Nkx2.5, MEF2C, BMP-1, BMP-2, VEGF-A, and GATA4. In addition, these cells could differentiate into cardiomyocytes if cultivated further in the presence of the cardiogenic cocktail mentioned above. However, removal of these factors after four days of stimulation resulted in continued engagement of cardiopoietic progenitor cells in the cell cycle without differentiation into cardiomyocytes. Moreover, when three million of these cardiopoietic cells were implanted into infarcted hearts of wild-type mice they proliferated and generated mature cardiomyocytes that improved the heart function as demonstrated by decrease in the circumferential area during systole and increase of the fractional heart output by 35%. No teratomas were observed in these experiments 8 weeks after implantation *(21)*. This study sets the basis for directed and scalable production of cardiomyocytes for heart repair under highly controlled culture conditions. However, high costs related to the use of many recombinant cytokines and growth factors may hinder practical application of this approach. Similar success has been recently reported for the production of cardiomyocytes from human ES cells using only two factors. Laflamme et al. *(89)* treated high-density monolayer cultures of human ES cells that were preconditioned for 6 days on Matrigel-coated plates with activin A for 24 h followed by BMP4 for an additional 4 days. Cells were then cultivated for an additional 2–3 weeks in the absence of cytokines until beating cardiomyocytes emerged. This procedure yielded more than 30% cardiomyocytes, which could be further enriched by discontinuous Percoll gradient centrifugation to a purity of 71–95% in independent experiments. The authors have calculated that each starting human ES cell gave rise to approximately three pure cardiomyocytes. When these cells were transplanted into infarcted hearts of athymic rats (ten millions of cells per heart) no teratomas could be observed during the observation period of 4 weeks. Although the purity of up to 95% that could be attained in this relatively simple protocol is impressive, it still may not be satisfactory for clinical applications because it is unlikely that the risk of teratoma formation is completely eliminated with this level of purity. Therefore, directed differentiation approaches will have to be combined with other strategies to achieve complete elimination of pluripotent stem cells from therapeutic cell preparations.

Positive Lineage Selection

The concept of lineage selection involves genetic modification of ES cells to express a selectable marker, which is expressed in differentiated cells in a cell type–specific manner. Approaches that build on this strategy utilize positive selection techniques based on antibiotic resistance, immunoselection, and fluorescence activated cell sorting (sometimes combined with density gradient centrifugation) and had already been successfully applied in generation of highly enriched cardiomyocytes *(75, 90–94)* and several other cell types *(95–98)*. Using stable transgenic murine ES cell line that expresses a reporter gene (GFP) and a selection marker (puromycin resistance) under the transcriptional control of a cardiac-specific promoter α-myosin heavy chain, Kolossov et al. *(75)* achieved a high degree of purification (>99%) and long-term teratoma-free engraftment of puromycin-selected ES cell–derived cardiomyocytes that improved the contractile function of infracted heart. Similarly, injection of highly purified (99%) neomycin-selected human cardiomyocytes into SCID mice (five million per injection site in form of cardiac clusters) did not cause any teratoma formation 13 weeks after injection, whereas the same number of nonselected but differentiated human ES cells frequently formed teratomas *(78)*. Lineage selection has also been recently applied for generation of human ES cell–derived cardiomyocytes.

Huber et al. *(92)* used lentiviral vector to generate transgenic human ES cell lines expressing eGFP under the control of human myosin light chain 2v (MLC2v) promoter. Following fluorescence activated cell sorting of cells obtained from differentiating EBs, cardiomyocytes could be obtained with purity of greater than 95% and viability of greater than 85%. eGFP-positive cell clusters microdissected from beating EBs formed stable myocardial grafts 4 weeks following transplantation into intact hearts of immunosuppressed rats. Most recently, the selection strategy of Kolossov et al. *(75)*, employing a bicistronic reporter composed of eGFP and puromycin-*N*-acetyltransferase under the control of α-myosin heavy chain promoter, was applied for positive selection of human HUES-7 ES cell–derived cardiomyocytes. The purity of cardiomyocytes after puromycin selection was 91.5 ± 4.3% *(94)*.

Negative Selection

The third approach consists of genetic engineering strategies for negative selection of unwanted undifferentiated ES cells, which may be present as contaminants in purified differentiated cells of interest. Negative selection of ES cells could be achieved by expression of a "suicide" gene, such as the herpes simplex virus thymidine kinase gene, in ES cells under the control of a constitutive promoter. This would render ES cells and all their derivatives sensitive to the drug ganciclovir, which could be administered as an additional safety measure in case teratoma occurs upon transplantation. With this strategy established, murine and human teratomas could be efficiently ablated *(43, 99, 100)*. Besides serving as a safety measure, negative selection strategy was recently utilized for enrichment of cardiomyocytes from human ES cells *(94)*. In this approach, the thymidine kinase gene was expressed in proliferating cells that could be negatively selected by exposing differentiating EBs to ganciclovir. Maximally attainable purity of cardiomyocytes utilizing this method was 33.4 ± 2.1%. In conclusion, by combining positive and negative selection strategies with targeted differentiation protocols to generate desired cells, it will be possible to improve the safety to the level where transplantations of ES cell–derived tissues can be performed without the risk of teratoma formation.

Conclusion

Pluripotent ES cells represent an alternative source of clinically useful cardiomyocytes. However, serious ethical concerns, the possibility of immune rejection of allotransplanted cells, low yield and purity of cardiomyocytes, and risk for teratoma formation from contaminating ES cells hindered the translation of ES cell–based therapies in the clinic since they were first generated. Some of these obstacles, namely ethical concerns and immune rejection, seem to have been overcome with recent achievements in reprogramming adult somatic cells into an ES cell–like pluripotent state *(101–110)*. These so-called induced pluripotent stem (iPS) cells were generated by transient overexpression of only four "stemness" factors in adult differentiated cells. This approach enabled their conversion into pluripotent cells, having properties that are highly similar to blastocyst-derived ES cells, including the ability to form teratomas in vivo. These cells raise hope that immunologically compatible patient-tailored cell replacement therapies may become reality in the near future. However, the tumorigenic potential of iPS cells as well as their traditional ES cell counterparts still represent serious obstacles to their clinical use. Even as few as a couple ES cells may give rise to teratoma formation if injected within an appropriate cellular context and if observed for the sufficient length of time. For this reason the reports claiming that fully undifferentiated ES cells can be implanted into the heart without the risk of teratoma formation are very surprising. Since other investigators could not find any evidence to support these observations and also showed that teratoma forms even from partially purified differentiated ES cell derivatives, it is unclear what the reason for these conflicting results may be. In some cases the lack of teratoma formation may be due to the low dose of cells and short follow-up period *(22, 23)*, while in other cases this may be due to peculiarity of the ES cell line used *(18, 19, 21)* or inappropriate methodology *(20)*.

ES cell lines used to demonstrate the cardioinductive properties of the local heart environment were murine CGR8 *(18, 19, 21)* and R1 or HM1 ES cells *(22, 23)*. Human H1 ES cells were applied in only one study to claim the same properties *(20)*. Studies that refute these results mostly used murine D3, R1, or C57BL/6 ES cells. Murine CGR8 ES cells are normally maintained under feeder-free conditions as flat colonies and express very low levels of stem cell surface marker SSEA-1 as compared to D3, R1, or HM1 ES cells that are propagated on feeder cells as 3D colonies (T. Saric, unpublished observation). The work of Nussbaum et al. *(50)* supports the idea that CGR8 cells may behave differently from other ES cell lines in vivo. Transplantation of 0.5×10^6 murine C57BL/6 ES cells into infarcted heart in this study resulted in teratoma formation and did not lead to their preferential differentiation into cardiomyocytes or other major cell types found in the heart, but after injection of the same number of CGR8 ES cells neither ES cells nor cardiomyocytes were detected in the heart 3 weeks after implantation. It may well be that this cell line has lower survival and tumorigenic capacity when transplanted into heart in comparison with other murine ES cell lines. Another reason for the inability of CGR8 ES cells to engraft may be of an immunological nature. Namely, CGR8 ES cells that originate from the 129P2/Ola mouse strain have been transplanted into C57BL/6 mice *(50)*. Although 129P2/Ola and C57BL/6 mouse strains share the same major histocompatibility haplotype (H-2^b), it is possible that ES cells were prevented from engrafting appropriately due to their incompatibility in minor histocompatibility molecules. The same constellation of ES cells (129sv) and the host (C57BL/6) was also present in studies by Singla et al. *(22, 23)*. The exact strain of the recipient mice in studies by Behfar et al. *(18, 21)* has unfortunately not been clearly reported. Beneficial effects on heart function that were reported in some of these studies may be the result of paracrine effects and not of cells that differentiated from injected ES cells. The evidence for the presence of ES cell–derived cardiomyocytes in vivo was usually provided by detection of cells giving green or cyan fluorescent protein (CFP) signal that was interpreted as eGFP or eCFP, but which could also be mistaken for autofluorescence of the host myocardium *(50)*. These issues remain to be examined more carefully.

Although the strategy of transplanting fully undifferentiated ES cells will most likely not be clinically applicable, studies employing this approach stimulated the discussion about the influence of specific tissue environment on tumorigenicity of ES cells, instigated new investigations on environmental cues that may be involved in determining the developmental fate of transplanted stem cells in vitro, and addressed the question of whether factors released by transplanted cells may beneficially affect the function of the recipient organ in a paracrine manner. In an elegant recent study, the transgenic expression of a single cytokine TNF-α in the heart exerted a profound effect on the fate of transplanted ES cells and drove them to cardiogenic lineage *(21)*. This study emphasizes local factors that may exert profound influence on transplanted cells. Elucidating the donor cell and recipient tissue signals that hinder or support stable engraftment of ES/iPS cell derivatives will greatly aid in translating this therapeutic approach into the clinic. If the cardioinductive effect of TNF-α can be confirmed by independent investigators, this strategy could be used to minimize the tumorigenic risk of transplanted cells. In human system, where no transgenic individuals will be available, one could envisage that transplantation of human ES cell–derived cardiomyocytes may be combined with transplantation of TNF α expressing transgenic cells of some other non-ES cell type (e.g., fibroblasts or endothelial cells) to help prevent tumorigenesis from contaminating ES cells. However, it remains to be seen if the expression of this stress cytokine would have some other unwanted side effects. To enable accurate assessment of neoplastic danger stemming from ES cell transplantation in humans, it will be important to design clinically relevant studies with appropriate animal disease models for much larger cohorts and with sufficiently long follow-up periods. In the clinical setting, a single case of teratoma development in a patient treated with ES cell derivatives would represent a serious setback to the whole field, similar to patient deaths in early phase I gene therapy clinical trials. Therefore, transplantation of undifferentiated ES cells cannot be regarded as realistic therapeutic strategy but rather as useful experimental model. Appropriate techniques must be employed to achieve the required purity and safety of pluripotent stem cell derivatives before they can be administered to patients.

Acknowledgments This work was supported by grants from The German Research Society (DFG) to T.Š. (Grant SA 1382/2-1), The Federal Ministry of Education and Research (BMBF) to H.J. and T.Š. (Grant 01GN0541), and from Köln Fortune Program to T.Š. and L.P.F.

References

1. Thom T, Haase N, Rosamund W, et al. Heart disease and stroke statistics—2006 update: a report from the American Heart Association Statistics Committee and Stroke Statistics Subcommittee. Circulation 2006;113:e85–151.
2. Lechler RI, Sykes M, Thomson AW, Turka LA. Organ transplantation—how much of the promise has been realized. Nat Med 2005;11:605–13.
3. Laflamme MA, Murry CE. Regenerating the heart. Nat Biotechnol 2005;23:845–56.
4. Murry CE, Field LJ, Menasche P. Cell-based cardiac repair: reflections at the 10-year point. Circulation 2005;112:3174–83.
5. Murry CE, Soonpaa MH, Reinecke H, et al. Haematopoietic stem cells do not transdifferentiate into cardiac myocytes in myocardial infarcts. Nature 2004;428:664–8.
6. Nygren JM, Jovinge S, Breitbach M, et al. Bone marrow-derived hematopoietic cells generate cardiomyocytes at a low frequency through cell fusion, but not transdifferentiation. Nat Med 2004;10:494–501.
7. Reinecke H, Poppa V, Murry CE. Skeletal muscle stem cells do not transdifferentiate into cardiomyocytes after cardiac grafting. J Mol Cell Cardiol 2002;34:241–9.
8. Roell W, Lewalter T, Sasse P, et al. Engraftment of connexin 43-expressing cells prevents post-infarct arrhythmia. Nature 2007;450:819–24.
9. Leobon B, Garcin I, Menasche P, Vilquin JT, Audinat E, Charpak S. Myoblasts transplanted into rat infarcted myocardium are functionally isolated from their host. Proc Natl Acad Sci U S A 2003;100:7808–11.
10. Wollert KC, Meyer GP, Lotz J, et al. Intracoronary autologous bone-marrow cell transfer after myocardial infarction: the BOOST randomised controlled clinical trial. Lancet 2004;364:141–8.
11. Janssens S, Dubois C, Bogaert J, et al. Autologous bone marrow-derived stem-cell transfer in patients with ST-segment elevation myocardial infarction: double-blind, randomised controlled trial. Lancet 2006;367:113–21.
12. Smits PC, van Geuns RJ, Poldermans D, et al. Catheter-based intramyocardial injection of autologous skeletal myoblasts as a primary treatment of ischemic heart failure: clinical experience with six-month follow-up. J Am Coll Cardiol 2003;42:2063–9.
13. Breitbach M, Bostani T, Roell W, et al. Potential risks of bone marrow cell transplantation into infarcted hearts. Blood 2007;110:1362–9.
14. Noiseux N, Gnecchi M, Lopez-Ilasaca M, et al. Mesenchymal stem cells overexpressing Akt dramatically repair infarcted myocardium and improve cardiac function despite infrequent cellular fusion or differentiation. Mol Ther 2006;14:840–50.
15. Iso Y, Spees JL, Serrano C, et al. Multipotent human stromal cells improve cardiac function after myocardial infarction in mice without long-term engraftment. Biochem Biophys Res Commun 2007;354:700–6.
16. Xu M, Uemura R, Dai Y, Wang Y, Pasha Z, Ashraf M. In vitro and in vivo effects of bone marrow stem cells on cardiac structure and function. J Mol Cell Cardiol 2007;42:441–8.
17. Mazhari R, Hare JM. Mechanisms of action of mesenchymal stem cells in cardiac repair: potential influences on the cardiac stem cell niche. Nat Clin Pract Cardiovasc Med 2007;4 Suppl 1:S21–6.
18. Behfar A, Zingman LV, Hodgson DM, et al. Stem cell differentiation requires a paracrine pathway in the heart. FASEB J 2002;16:1558–66.
19. Hodgson DM, Behfar A, Zingman LV, et al. Stable benefit of embryonic stem cell therapy in myocardial infarction. Am J Physiol Heart Circ Physiol 2004;287:H471–9.
20. Xie CQ, Zhang J, Xiao Y, et al. Transplantation of human undifferentiated embryonic stem cells into a myocardial infarction rat model. Stem Cells Dev 2007;16:25–9.
21. Behfar A, Perez-Terzic C, Faustino RS, et al. Cardiopoietic programming of embryonic stem cells for tumor-free heart repair. J Exp Med 2007;204:405–20.

22. Singla DK, Hacker TA, Ma L, et al. Transplantation of embryonic stem cells into the infarcted mouse heart: formation of multiple cell types. J Mol Cell Cardiol 2006;40:195–200.
23. Singla DK, Lyons GE, Kamp TJ. Transplanted embryonic stem cells following mouse myocardial infarction inhibit apoptosis and cardiac remodeling. Am J Physiol Heart Circ Physiol 2007;293:H1308–14.
24. Itskovitz-Eldor J, Schuldiner M, Karsenti D, et al. Differentiation of human embryonic stem cells into embryoid bodies compromising the three embryonic germ layers. Mol Med 2000;6:88–95.
25. Lavon N, Yanuka O, Benvenisty N. Differentiation and isolation of hepatic-like cells from human embryonic stem cells. Differentiation 2004;72:230–8.
26. Stevens LC, Little CC. Spontaneous testicular teratomas in an inbred strain of mice. Proc Natl Acad Sci U S A 1954;40:1080–7.
27. Stevens LC. Experimental production of testicular teratomas in mice. Proc Natl Acad Sci U S A 1964;52:654–61.
28. Solter D, Skreb N, Damjanov I. Extrauterine growth of mouse egg-cylinders results in malignant teratoma. Nature 1970;227:503–4.
29. Eventov-Friedman S, Katchman H, Shezen E, et al. Embryonic pig liver, pancreas, and lung as a source for transplantation: optimal organogenesis without teratoma depends on distinct time windows. Proc Natl Acad Sci U S A 2005;102:2928–33.
30. Dekel B, Burakova T, Arditti FD, et al. Human and porcine early kidney precursors as a new source for transplantation. Nat Med 2003;9:53–60.
31. Kleinsmith LJ, Pierce GB Jr. Multipotentiality of single embryonal carcinoma cells. Cancer Res 1964;24:1544–51.
32. Finch BW, Ephrussi B. Retention of multiple developmental potentialities by cells of a mouse testicular teratocarcinoma during prolonged culture in vitro and their extinction upon hybridization with cells of permanent lines. Proc Natl Acad Sci U S A 1967;57:615–21.
33. Fogh J, Trempe G. New human tumor cell lines. In Human Tumor Cells In Vitro (ed. J Fogh). Plenum, New York, 1975:115–59.
34. Hogan B, Fellous M, Avner P, Jacob F. Isolation of a human teratoma cell line which expresses F9 antigen. Nature 1977;270:515–8.
35. Martin GR, Evans MJ. The morphology and growth of a pluripotent teratocarcinoma cell line and its derivatives in tissue culture. Cell 1974;2:163–72.
36. Martin GR, Evans MJ. Differentiation of clonal lines of teratocarcinoma cells: formation of embryoid bodies in vitro. Proc Natl Acad Sci U S A 1975;72:1441–5.
37. Martin GR. Isolation of a pluripotent cell line from early mouse embryos cultured in medium conditioned by teratocarcinoma stem cells. Proc Natl Acad Sci U S A 1981;78:7634–8.
38. Evans MJ, Kaufman MH. Establishment in culture of pluripotential cells from mouse embryos. Nature 1981;292:154–6.
39. Thomson JA, Itskovitz-Eldor J, Shapiro SS, et al. Embryonic stem cell lines derived from human blastocysts. Science 1998;282:1145–7.
40. Reubinoff BE, Pera MF, Fong CY, Trounson A, Bongso A. Embryonic stem cell lines from human blastocysts: somatic differentiation in vitro. Nat Biotechnol 2000;18:399–404.
41. Solter D. From teratocarcinomas to embryonic stem cells and beyond: a history of embryonic stem cell research. Nat Rev Genet 2006;7:319–27.
42. Andrews PW. From teratocarcinomas to embryonic stem cells. Philos Trans R Soc Lond B Biol Sci 2002;357:405–17.
43. Schuldiner M, Itskovitz-Eldor J, Benvenisty N. Selective ablation of human embryonic stem cells expressing a "suicide" gene. Stem Cells 2003;21:257–65.
44. Gertow K, Wolbank S, Rozell B, et al. Organized development from human embryonic stem cells after injection into immunodeficient mice. Stem Cells Dev 2004;13:421–35.
45. Lawrenz B, Schiller H, Willbold E, Ruediger M, Muhs A, Esser S. Highly sensitive biosafety model for stem-cell-derived grafts. Cytotherapy 2004;6:212–22.
46. Przyborski SA. Differentiation of human embryonic stem cells after transplantation in immune-deficient mice. Stem Cells 2005;23:1242–50.

47. Drukker M, Katchman H, Katz G, et al. Human embryonic stem cells and their differentiated derivatives are less susceptible to immune rejection than adult cells. Stem Cells 2006;24:221–9.
48. Koch CA, Jordan CE, Platt JL. Complement-dependent control of teratoma formation by embryonic stem cells. J Immunol 2006;177:4803–9.
49. Blum B, Benvenisty N. Clonal analysis of human embryonic stem cell differentiation into teratomas. Stem Cells 2007;25:1924–30.
50. Nussbaum J, Minami E, Laflamme MA, et al. Transplantation of undifferentiated murine embryonic stem cells in the heart: teratoma formation and immune response. FASEB J 2007;21:1345–57.
51. Magliocca JF, Held IK, Odorico JS. Undifferentiated murine embryonic stem cells cannot induce portal tolerance but may possess immune privilege secondary to reduced major histocompatibility complex antigen expression. Stem Cells Dev 2006;15:707–17.
52. Cao F, Lin S, Xie X, et al. In vivo visualization of embryonic stem cell survival, proliferation, and migration after cardiac delivery. Circulation 2006;113:1005–14.
53. Gidekel S, Pizov G, Bergman Y, Pikarsky E. Oct-3/4 is a dose-dependent oncogenic fate determinant. Cancer Cell 2003;4:361–70.
54. Cooke MJ, Stojkovic M, Przyborski SA. Growth of teratomas derived from human pluripotent stem cells is influenced by the graft site. Stem Cells Dev 2006;15:254–9.
55. Erdö F, Buhrle C, Blunk J, et al. Host-dependent tumorigenesis of embryonic stem cell transplantation in experimental stroke. J Cereb Blood Flow Metab 2003;23:780–5.
56. Riess P, Molcanyi M, Bentz K, et al. Embryonic stem cell transplantation after experimental traumatic brain injury dramatically improves neurological outcome, but may cause tumors. J Neurotrauma 2007;24:216–25.
57. Grinnemo KH, Kumagai-Braesch M, Mansson-Broberg A, et al. Human embryonic stem cells are immunogenic in allogeneic and xenogeneic settings. Reprod Biomed Online 2006;13:712–24.
58. Bonnevie L, Bel A, Sabbah L, et al. Is xenotransplantation of embryonic stem cells a realistic option. Transplantation 2007;83:333–5.
59. Kofidis T, de Bruin JL, Tanaka M, et al. They are not stealthy in the heart: embryonic stem cells trigger cell infiltration, humoral and T-lymphocyte-based host immune response. Eur J Cardiothorac Surg 2005;28:461–6.
60. Swijnenburg RJ, Tanaka M, Vogel H, et al. Embryonic stem cell immunogenicity increases upon differentiation after transplantation into ischemic myocardium. Circulation 2005;112:I166–72.
61. Drukker M, Katz G, Urbach A, et al. Characterization of the expression of MHC proteins in human embryonic stem cells. Proc Natl Acad Sci U S A 2002;99:9864–9.
62. Fabricius D, Bonde S, Zavazava N. Induction of stable mixed chimerism by embryonic stem cells requires functional Fas/FasL engagement. Transplantation 2005;79:1040–4.
63. Bonde S, Zavazava N. Immunogenicity and engraftment of mouse embryonic stem cells in allogeneic recipients. Stem Cells 2006;24:2192–201.
64. Abdullah Z, Saric T, Kashkar H, et al. Serpin-6 expression protects embryonic stem cells from lysis by antigen-specific CTL. J Immunol 2007;178:3390–9.
65. Li L, Baroja ML, Majumdar A, et al. Human embryonic stem cells possess immune-privileged properties. Stem Cells 2004;22:448–56.
66. Koch CA, Geraldes P, Platt JL. Immunosuppression by embryonic stem cells. Stem Cells 2007;26:89–98.
67. Zvibel I, Smets F, Soriano H. Anoikis: roadblock to cell transplantation. Cell Transplant 2002;11:621–30.
68. Cao F, Van Der Bogt KE, Sadrzadeh A, et al. Spatial and temporal kinetics of teratoma formation from murine embryonic stem cell transplantation. Stem Cells Dev 2007;16:883–91.
69. Penn I. Post-transplant malignancy: the role of immunosuppression. Drug Saf 2000;23:101–13.

70. Wakitani S, Takaoka K, Hattori T, et al. Embryonic stem cells injected into the mouse knee joint form teratomas and subsequently destroy the joint. Rheumatology (Oxford) 2003;42:162–5.
71. Hochedlinger K, Jaenisch R. Nuclear reprogramming and pluripotency. Nature 2006;441:1061–7.
72. Goldstein RS, Drukker M, Reubinoff BE, et al. Integration and differentiation of human embryonic stem cells transplanted to the chick embryo. Dev Dyn 2002;225:80–6.
73. Chinzei R, Tanaka Y, Shimizu-Saito K, et al. Embryoid-body cells derived from a mouse embryonic stem cell line show differentiation into functional hepatocytes. Hepatology 2002;36:22–9.
74. Brederlau A, Correia AS, Anisimov SV, et al. Transplantation of human embryonic stem cell-derived cells to a rat model of Parkinson's disease: effect of in vitro differentiation on graft survival and teratoma formation. Stem Cells 2006;24:1433–40.
75. Kolossov E, Bostani T, Roell W, et al. Engraftment of engineered ES cell-derived cardiomyocytes but not BM cells restores contractile function to the infarcted myocardium. J Exp Med 2006;203:2315–27.
76. Caspi O, Huber I, Kehat I, et al. Transplantation of human embryonic stem cell-derived cardiomyocytes improves myocardial performance in infarcted rat hearts. J Am Coll Cardiol 2007;50:1884–93.
77. Leor J, Gerecht S, Cohen S, et al. Human embryonic stem cell transplantation to repair the infarcted myocardium. Heart 2007;93:1278–84.
78. Zweigerdt R. The art of cobbling a running pump-Will human embryonic stem cells mend broken hearts. Semin Cell Dev Biol 2007;18:794–804.
79. Teramoto K, Hara Y, Kumashiro Y, et al. Teratoma formation and hepatocyte differentiation in mouse liver transplanted with mouse embryonic stem cell-derived embryoid bodies. Transplant Proc 2005;37:285–6.
80. Orlic D, Kajstura J, Chimenti S, et al. Bone marrow cells regenerate infarcted myocardium. Nature 2001;410:701–5.
81. McDonald JW, Liu XZ, Qu Y, et al. Transplanted embryonic stem cells survive, differentiate and promote recovery in injured rat spinal cord. Nat Med 1999;5:1410–2.
82. Tomescot A, Leschik J, Bellamy V, et al. Differentiation in vivo of cardiac committed human embryonic stem cells in postmyocardial infarcted rats. Stem Cells 2007;25:2200–5.
83. Ménard C, Hagege AA, Agbulut O, et al. Transplantation of cardiac-committed mouse embryonic stem cells to infarcted sheep myocardium: a preclinical study. Lancet 2005;366:1005–12.
84. Kofidis T, de Bruin JL, Yamane T, et al. Insulin-like growth factor promotes engraftment, differentiation, and functional improvement after transfer of embryonic stem cells for myocardial restoration. Stem Cells 2004;22:1239–45.
85. Kofidis T, de Bruin JL, Yamane T, et al. Stimulation of paracrine pathways with growth factors enhances embryonic stem cell engraftment and host-specific differentiation in the heart after ischemic myocardial injury. Circulation 2005;111:2486–93.
86. Laflamme MA, Gold J, Xu C, et al. Formation of human myocardium in the rat heart from human embryonic stem cells. Am J Pathol 2005;167:663–71.
87. Dai W, Field LJ, Rubart M, et al. Survival and maturation of human embryonic stem cell-derived cardiomyocytes in rat hearts. Mol Cell Cardiol 2007;43:504–16.
88. Min JY, Yang Y, Sullivan MF, et al. Long-term improvement of cardiac function in rats after infarction by transplantation of embryonic stem cells. J Thorac Cardiovasc Surg 2003;125:361–9.
89. Laflamme MA, Chen KY, Naumova AV, et al. Cardiomyocytes derived from human embryonic stem cells in pro-survival factors enhance function of infarcted rat hearts. Nat Biotechnol 2007;25:1015–24.
90. Müller M, Fleischmann BK, Selbert S, et al. Selection of ventricular-like cardiomyocytes from ES cells in vitro. FASEB J 2000;14:2540–8.

91. Klug MG, Soonpaa MH, Koh GY, Field LJ. Genetically selected cardiomyocytes from differentiating embryonic stem cells form stable intracardiac grafts. J Clin Invest 1996;98:216–24.

92. Huber I, Itzhaki I, Caspi O, et al. Identification and selection of cardiomyocytes during human embryonic stem cell differentiation. FASEB J 2007;21:2551–63.

93. Xu C, Police S, Hassanipour M, Gold JD. Cardiac bodies: a novel culture method for enrichment of cardiomyocytes derived from human embryonic stem cells. Stem Cells Dev 2006;15:631–9.

94. Anderson D, Self T, Mellor IR, Goh G, Hill SJ, Denning C. Transgenic enrichment of cardiomyocytes from human embryonic stem cells. Mol Ther 2007;15:2027–36.

95. Li Z, Wu JC, Sheikh AY, et al. Differentiation, survival, and function of embryonic stem cell derived endothelial cells for ischemic heart disease. Circulation 2007;116:I46–54.

96. Li M, Pevny L, Lovell-Badge R, Smith A. Generation of purified neural precursors from embryonic stem cells by lineage selection. Curr Biol 1998;8:971–4.

97. Wernig M, Tucker KL, Gornik V, et al. Tau EGFP embryonic stem cells: an efficient tool for neuronal lineage selection and transplantation. J Neurosci Res 2002;69:918–24.

98. Soria B, Roche E, Berna G, Leon-Quinto T, Reig JA, Martin F. Insulin-secreting cells derived from embryonic stem cells normalize glycemia in streptozotocin-induced diabetic mice. Diabetes 2000;49:157–62.

99. Cao F, Drukker M, Lin S, et al. Molecular imaging of embryonic stem cell misbehavior and suicide gene ablation. Cloning Stem Cells 2007;9:107–17.

100. Jung J, Hackett NR, Pergolizzi RG, Pierre-Destine L, Krause A, Crystal RG. Ablation of tumor-derived stem cells transplanted to the central nervous system by genetic modification of embryonic stem cells with a suicide gene. Hum Gene Ther 2007;18:1182–92.

101. Takahashi K, Yamanaka S. Induction of pluripotent stem cells from mouse embryonic and adult fibroblast cultures by defined factors. Cell 2006;126:663–76.

102. Takahashi K, Tanabe K, Ohnuki M, et al. Induction of pluripotent stem cells from adult human fibroblasts by defined factors. Cell 2007;131:861–72.

103. Wernig M, Meissner A, Foreman R, et al. In vitro reprogramming of fibroblasts into a pluripotent ES-cell-like state. Nature 2007;448:318–24.

104. Meissner A, Wernig M, Jaenisch R. Direct reprogramming of genetically unmodified fibroblasts into pluripotent stem cells. Nat Biotechnol 2007;25:1177–81.

105. Okita K, Ichisaka T, Yamanaka S. Generation of germline-competent induced pluripotent stem cells. Nature 2007;448:313–7.

106. Maherali N, Sridharan R, Xie W, et al. Directly reprogrammed fibroblasts show global epigenetic remodeling and widespread tissue contribution. Cell Stem Cell 2007;1:55–70.

107. Qin D, Li W, Zhang J, Pei D. Direct generation of ES-like cells from unmodified mouse embryonic fibroblasts by Oct4/Sox2/Myc/Klf4. Cell Res 2007;17:959–62.

108. Yu J, Vodyanik MA, Smuga-Otto K, Antosiewicz-Bourget J, Frane JL, Tian S. Induced pluripotent stem cell lines derived from human somatic cells. Science 2007;318:1917–20.

109. Hanna J, Wernig M, Markoulaki S, Sun CW, Meissner A, Cassady JP. Treatment of sickle cell anemia mouse model with iPS cells generated from autologous skin. Science 2007;318:1920–3.

110. Nakagawa M, Koyanagi M, Tanabe K, et al. Generation of induced pluripotent stem cells without Myc from mouse and human fibroblasts. Nat Biotechnol 2008;26:101–6.

Neural Differentiation of Human Embryonic Stem Cells and Their Potential Application in a Therapy for Sensorineural Hearing Loss

Objoon Trachoo and Marcelo N. Rivolta

Abstract Deafness is a condition produced primarily by the loss of hair cells and their associated neurons in the spiral ganglion of the inner ear. Hair cells can be replaced under certain circumstances with a prosthetic device, the cochlear implant. However, neurons have to be preserved and functional for the cochlear implant to succeed. The generation of neural stem cells (NSCs) and the control of neural differentiation from human embryonic stem cells (hESCs) have opened new doors for therapies for several neurological conditions, and people suffering from hearing impairments could potentially benefit from these recent developments. However, not all neurons are equivalent. The best-suited neurons for a particular application would be those that have been derived following protocols that closely match the development in vivo. The three main sources of neurons during early embryonic development are the neural plate, the neural crest, and the ectodermal placodes. Understanding how these structures are formed and what the signals involved in the process are will facilitate and should guide the generation of methods to control hESC differentiation. In this chapter we review several protocols used to generate neural precursors from hESCs, including initial attempts to establish otic placodal precursors, and we discuss their potential application in the development of a new therapy for deafness.

Keywords Human embryonic stem cells • Neural stem cells • Otic progenitors • Auditory neurons • Deafness • Cochlear implant

O. Trachoo and M.N. Rivolta (✉)
Centre for Stem Cell Biology and Department of Biomedical Sciences,
University of Sheffield, Addison Building, Western Bank, Sheffield S10 2TN, UK
e-mail: m.n.rivolta@sheffield.ac.uk

H. Baharvand (ed.) *Trends in Stem Cell Biology and Technology*,
DOI 10.1007/ 978-1-60327-905-5_15,
© Humana Press, a Part of Springer Science+Business Media, LLC 2009

Introduction

Human embryonic stem cells are derived from the inner cell mass of blastocysts and hold great therapeutic potential given their capacity for self-renewal and the ability to give rise to specific cell types from the three germ layers *(1)*. For these reasons hESCs can provide useful tools for the study of early developmental biology in humans, to model diseases, to advance drug discovery, and to develop potentially ground-breaking new cell-based therapies. In recent years, specialized neural progenitors have been derived from human embryonic stem cells (hESCs) and the generation of these precursors could open new possibilities for the treatment of neurological disorders and neurodegenerative diseases such as sensorineural hearing loss, stroke, multiple sclerosis, amyotrophic lateral sclerosis, Huntington's, Parkinson's, and Alzheimer's diseases *(2)*. hESC-derived neural progenitors could be used to replace the damaged cells or tissues and then be allowed to differentiate into new functional cell population of the nervous system. At the present, the generation of protocols for neural differentiation from hESCs is moving forward at a rapid pace, and several laboratories are focusing on how to obtain the specific cell types, followed by later efforts to translate the basic advances into the clinic as therapies for these particular disorders. From all the neurological conditions named above, deafness is an ideal one to be targeted by a cell-based therapy. It is a major public health issue, with almost nine million adults in the United Kingdom affected by hearing problems. Congenital deafness has an incidence of 1/1,000 of the newborn population worldwide *(3)*. The causes of deafness are various and the range of severity is broad. In more than 80% of deaf people the fundamental problem, regardless of the cause, translates as the degeneration and death of hair cells and their associated spiral ganglion neurons in the cochlea of the inner ear *(4)*. Neurons and hair cell progenitors are only produced during embryonic development and perhaps, perinatally *(5, 6)*. Later in life, these cells are not replaced, making the hearing loss irreversible. There is no cure for deafness, although, in certain situations, the sensory function of the inner ear can be restored by a cochlear implant (CI). CI can only replace the function of hair cells, the first sensory element of the auditory pathway *(7)*. However, they need intact, healthy neurons to transmit the auditory information to the brain. Without neurons, most of the currently available CIs will not function *(8, 9)*. A potential therapeutic approach to the loss of sensory cells would be to replace them by transplantation of exogenous, in vitro maintained stem cells. These transplantations could also provide a means of delivering supporting neurotrophic factors to promote further the survival of neurons. Moreover, the delivery of neural progenitors at the time of cochlear implantation could extend the applicability and success rate of the current CI approach. Although part of the efforts to develop a stem cell–based therapy for deafness aims to restore hair cells, the replacement of neurons would appear more feasible in the initial stages. This chapter will concentrate primarily in the strategies and potential applications of hESC to replace auditory neurons.

Several methods have been developed to generate neuronal progenitors and neurons from hESCs. Many of these protocols are based on the use of empirical factors without a clear rational for their employment. The generation of functional neurons that could be used for a regenerative therapy, however, is more likely to be achieved by trying to mimic the normal events that occur during development in vivo. The closer we can resemble the native environment and reproduce the sequential stages of differentiation, the better and more physiologically able the final cell types would be. Therefore, the understanding of the processes involved in the development of the nervous system should be at the core of the design of protocols to induce neural progenitors and to support neuronal differentiation. With this idea in mind we start this chapter describing some basic outlines on the developmental biology of stem cells in the embryonic nervous system, we then move to discuss some of the current methods used for the differentiation of hESCs toward neural stem cells (NSCs), and finally, we concentrate in how we can apply hESCs-derived neural precursors to the development of a therapy for sensorineural hearing loss.

Developmental Biology of the Nervous System

Due to ethical constraints and technical limitations to study early neurogenesis in humans, most of our knowledge has originated from studying model organisms (mammalian and nonmammalian) such as the frog, the zebrafish, the chick, and the mouse. After gastrulation, specific organogenesis of the nervous system begins with the formation of the neural plate. During vertebrate neurulation, three main structures derived from the ectoderm contain cells with neurogenic potential we could loosely denominate NSCs. First, the neural plate further develops into the neural tube, producing NSCs that will provide neuronal-restricted progenitors, astrocytes, and oligodendrocytes of the central nervous system (CNS) *(10)*. Second, the neural crest, a ridge of cells flanking the rostrocaudal length of the neural tube, form the peripheral nervous system (PNS), autonomic neurons, pigmented cells of the skin, and craniofacial connective tissue structures *(11)*, and, third, the ectodermal placodes that line the outside of the neural plate lateral to neural crests form cranial sensory ganglia and adenohypophysis *(12)*. Finally, the remaining nonneural ectoderm will give rise to the skin (Fig. 1) *(13)*.

During neurulation, the embryonic ectodermal cells are responsive to the inductive interactions from surrounding mesodermal and/or endodermal cells *(14, 15)*. Neural induction requires the inhibition of epidermal cell fate and promotion of a neural fate within the ectoderm. Bone morphogenetic protein (BMP) family signaling molecules play a role in epidermal fate promotion by triggering the expression of ventralizing transcriptional factors and genes expressed by presumptive epidermis *(16–18)*. From studies in *Xenopus*, the underlying mesoderm appears to have a role as an organizer in regulating the gradient of BMP activity. The production of BMP inhibitors, such as noggin, by the dorsal mesoderm opposing the effect of BMPs secreted by the ectoderm generates localized regions of low, intermediate, and high BMP activity that will translate into neural plate, neural crest, and epidermis, respectively *(19)*. The exposure of *Xenopus* animal cap explants in culture to

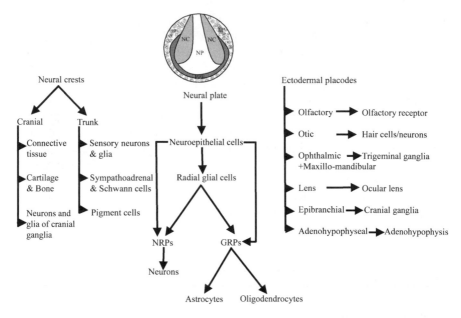

Fig. 1 Schematic drawing and diagram of multiple neural and nonneural derivatives derived from various origins during embryonic neural development. A dorsal view of an early embryo shows the formation of neural plate, neural crest, preplacodal ectoderm, and epidermis. Embryonic neuroepithelial cells from the neural plate can progress directly to become neuronal-restricted progenitors (NRPs) and glial-restricted progenitors (GRPs), or extend projections and form radial glial cells of the subventricular zone, to continue production of progenitors in fetal and adult life. Neural crests are located laterally to the neural plate and migrate to cranial and trunk area and play a role in the formation of cranial sensory ganglia, peripheral and autonomic nervous system, as well as other nonneural derivatives. Ectodermal placodes give rise to specific placodal precursors that progress into specialized cranial sensory system, lens and adenohypophysis. *NP* neural plate, *NC* neural crest, *PPE* preplacodal ectoderm, *EP* epidermis

decreasing concentrations of the BMP antagonist noggin can induce differentiation of neural plate, neural crests, or early developed placodes. Up-regulation of the expression of Six1 and Eya1, markers of early placodal development, was observed when explants were cultured in the lowest concentration of noggin, while progressively higher concentrations induce the expression of FoxD3 (neural crests marker) and then Sox2 (marker of neural plate) *(20)*. These findings suggest that induction of neural plate, neural crests, and placodes require different concentrations of BMP antagonists, which correspond to their respective distances from an endogenous source of these factors at the dorsal midline *(21)*.

In amniotes embryos, however, neural plate formation requires fibroblast growth factors (FGFs) to repress BMP mRNA activity and progress to the neural fate *(14)*. On the other hand, while FGFs, either alone or in combination with BMP antagonists (i.e., noggin, chordin, and follistatin), secreted from the dorsal mesoderm and anterior ectoderm are required, they are not sufficient to induce neural fate *(14, 22–24)*.

In addition, positive and negative Wnt signaling is needed to make the selection between the epidermal and neural fate. It has been demonstrated in chick embryos that continual Wnt signaling can inhibit the response of epiblast cells to FGFs, allowing the expression of BMPs to direct an epidermal fate. Reciprocally, the lack of exposure to Wnt signals promotes the neural fate induction by permitting FGFs to act *(25)*. In summary, the integration of several signaling molecules, such as BMPs, FGFs, and Wnts, is essential for early neural development.

As described, neural plate, neural crests, and placodes can generate the cells with neural potential. Mainly, the proliferation and differentiation of NSCs and progeny depend on their choices between symmetrical and asymmetrical cell divisions, which are in balance at the different stages of development. Symmetrical divisions are found primarily at the earliest stage of neurogenesis, with the elaboration of equivalent daughter cells resulting in the maintenance of NSCs as a reservoir *(26, 27)*. By contrast, at later stages, two different daughter cells are generated by asymmetrical divisions, which regulate the correct number of terminally differentiated cells such as neurons and glia at the appropriate time and sites. The mechanism of asymmetric cell division during neurogenesis appears to be evolutionarily conserved *(10, 28)*. Nevertheless, different regions of the nervous system are under the influence of various ligands and signaling molecules that are secreted by specific embryonic tissues and organizing centers.

Neuroepithelial Cells of the Neural Tube and Subventricular Zone Give Rise to the Cells of CNS

Embryonic NSCs arise from the neuroepithelial (NEP) cells of the neural plate region by the end of neurulation. Undifferentiated NEP cells are located inside the epithelia of the neural tube, giving rise to all neurons, astroglial, oligodendroglial, and even ependymal cells, of the CNS *(10)*. NEP cells elongate, extending from the ventricular (apical) to the pial (basal) surface of the early neural tube, maintaining their multipotent capacity and expressing some specific glial markers such as glial fibrillary acidic protein (GFAP) and astrocyte-specific glutamate transporter (GLAST) *(29, 30)*. Given their morphology and profile of markers, they are referred to as radial glia cells. The second germinal subventricular zone (SVZ) arises after young neurons migrate from the ventricular zone (VZ) to continue diving symmetrically during the midgestational period as they migrate to their cortical destination *(31)*. Later, during perinatal and early postnatal life, asymmetrical division occurs to generate the population of astrocytes and oligodendrocytes while NSCs in the VZ disappear. NSCs in the SVZ of some areas of the brain, such as the dentate gyrus of the hippocampus, can survive and maintain their stem cell capacity until adult life *(10, 31, 32)*. To date, four types of cells in the SVZ have been described: A, B, and C are the cells with active proliferative capacity, while E cells or ependymal cells do not proliferate. The B cells are the true stem cells the would give rise to C cells, which are the transit amplifying population (TAP) and can also generate the

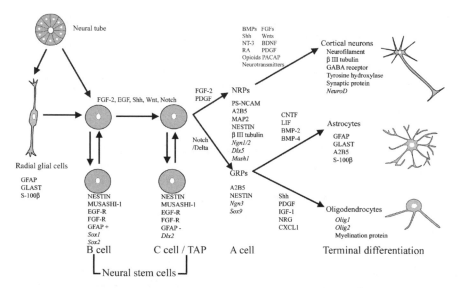

Fig. 2 Schematic drawing showing the specification of neural stem cells (NSCs) isolated from neuroepithelial (NEP) cells lying in the neural tube, including important signaling molecules that play a role in each step of development as well as relevant markers. B cells are quiescent NSCs that can maintain self-renewal and express neuroepithelial stem cell markers and glial fibrillary acidic protein (GFAP). Subsequently, they progress into actively dividing C cells characterized as transient amplifying progenitors (TAPs) in which GFAP is absent and *Dlx2* is up-regulated. TAPs give rise to A cells, which are specific neuronal or glial-restricted progenitors that required different signaling for their particular lineages to produce neurons, astrocytes, and oligodendrocytes

A cells (neuron-restricted precursors) and glial-restricted precursors. The A cells are young neurons that are able to continue their proliferation before committing to their specific neuronal phenotypes *(33)*. The specific markers of each stage of the development are listed in Fig. 2.

Several signaling molecules play roles in the proliferation and differentiation of NSCs. FGFs and epidermal growth factor (EGF) are well known to regulate the proliferation of NSCs in vivo and in vitro. Ligands of the FGF family are widely expressed in the developing brain during embryogenesis *(34)*. FGF2 appears to have a main role in the regulation of NSCs proliferation at early embryo stages, while at a later stage NSCs can proliferate in either FGF or EGF *(35)*. EGF is stimulated by FGF2 and inhibited by BMP4. During the final stages of neural development, BMPs also promote NSCs proliferation and apoptosis through the activation of bone morphogenetic protein receptor-1A (BMPR1A), while final differentiation is induced through bone morphogenetic protein receptor-1B (BMPR1B) *(36)*. The other signaling molecules that take part in the proliferation of NSCs include insulin-like growth factor 1 (IGF1), Sonic hedgehog (Shh), and Wnt. IGF1 is essential for the proliferation mediated by FGF2 and EGF in cultured NSCs, and there is evidence that neurosphere generation is IGF1 dependent in a dose-responsive manner *(37)*.

Shh is a signaling protein expressed by the ventrally located notochord and floor plate, which plays a role in the induction and patterning process of the vertebrate and invertebrate embryos *(38)*. Apart from those effects, it can regulate the proliferation of NSCs in vitro and in vivo *(39–41)*. Wnt proteins are also a large group of signaling molecules that can stimulate the proliferation of NSCs. The mitogenic activity belongs to the dorsal midline Wnts (Wnt1 and Wnt3a), whereas ventral Wnts (Wnt3, Wnt4, Wnt7a, and Wnt7b) do not affect proliferation *(42, 43)*. However, Wnt3, Wnt7a, and Wnt7b can stimulate the generation of neurosphere in vitro *(44)*. Recently, there is evidence that Wnt3a regulates survival, expansion, and maintenance of NSCs derived from hESCs in vitro. When endogenous Wnt signaling is inhibited, hESC-derived neurosphere formation and size appear to be significantly reduced *(45)*. The other important molecule is retinoic acid (RA), which has a role in embryonic neurogenesis especially in posteriorizing CNS tissues *(46)*. RA can bind to its binding protein and is transported to specific RA receptors in the nucleus, resulting in transcription of target genes *(47)*.

Final differentiation of NSCs to committed specific lineages is dependent on several extrinsic signals that promote the exit from cell cycle. Those signaling molecules include cytokines (BMPs and stem cell factor [SCF]), peptides (pituitary adenylate cyclase-activating peptide [PACAP] and opioids), neurotrophic factors (neurotrophin-3 [NT3] and brain-derived neurotrophic factor [BDNF]), and neurotransmitters (glutamate, g-aminobutyric acid [GABA], and dopamine) *(13)*.

PNS Precursors Arise from the Development of Neural Crests

The PNS is primarily derived from a population of neural crest cells that arise at the interface between nonneural ectoderm (presumptive epidermis) and the dorsal region of neural plate within the CNS and migrate to the peripheral regions *(48, 49)*. Neural crest cells have demonstrated the capacity for self-renewal in vivo, appear to have stem cell ability to differentiate in several cell types, and express their specific markers including FoxD3, SNAIL, dHAND, Sox9, and Sox10 *(20, 50, 51)*. During the development of CNS, different populations of neural crest cells arise from different levels of the neural axis and can give rise to diverse cell types, both neuronal and nonneuronal *(52)*. The neural crest cells at cranial levels contribute to the differentiation into cranial sensory ganglia, connective tissue, bone, and cartilage elements of the face and skull; while neural crests at the trunk level would give rise to neurons and supporting cells of PNS, pigmented cells, and sympathoadrenal system *(50)*, as shown in Fig. 3. Neural crest induction requires several signaling molecules, which are secreted by the ectodermal interaction border and mesoderm. BMPs, expressed throughout the ectoderm, have been demonstrated to modulate neural crest induction, and its effect at intermediate concentrations is highly conserved among vertebrates *(52, 53)*. BMP4 and BMP7 have also been implicated as downstream targets of Notch signaling, which participates in neural crest induction in avians *(54–56)*. In vitro, BMP4/7 are able to induce neural crest from neural plate culture *(54)*.

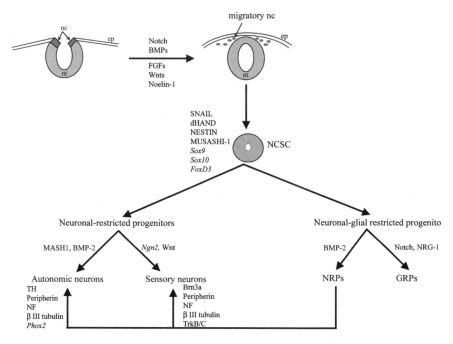

Fig. 3 Schematic drawing of neural crest stem cells (NCSCs) in the development of neuronal- and glial-restricted progenitors, showing their markers and critical signaling molecules involved in their specification. NCSCs migrate after the complete formation of neural tube, and a population of neural derivatives can be differentiated into autonomic, sensory neurons, and other glial progenitors

However, the inhibition of BMP alone by BMP antagonistic molecules cannot solely explain the entire phenomenon of neural crest induction *(53, 57)*. Wnts, FGFs, Notch, and Noelin-1 are all signaling molecules expressed by both the ectoderm and mesoderm and involved in the triggering of neural crest formation, even when BMP signaling is inhibited *(57–60)*. From the phenotypes described to date, no complete absence of neural crest formation has been reported in any species for Wnt, FGF, or Notch-mutant embryos *(52, 61–63)*.

Neural crest stem cells (NCSCs) have been isolated from trunk neural crest cells in the rat and mouse embryo, demonstrating their self-renewal and pluripotent capacity to generate diverse cell types, i.e., neurons, glia, myofibroblast, and melanocytes *(11, 64, 65)*. Pluripotent NCSCs can commit into both neural and nonneural progenitors, as described previously. hESCs have been directed to differentiate into NCSCs and their derivatives by coculturing with the mouse stromal line PA6 without the addition of BMP4 treatment *(51)*. During PNS development, NCSCs have to choose their fate between neurogenesis and gliogenesis. Final differentiation of specific lineage is also regulated by a couple of signaling pathways. Notch and Neuregulin 1 (Nrg1) appear to be potent inducers for glial cell fate, while Neurogenin2 (Ngn2) and Wnt elicit their effects for sensory neuron differentiation. On the other hand, the regulation of autonomic neurons differentiation is under the influence of the activation of the Mash1 gene, which is the target for BMP2 signaling *(11, 52)*.

Cranial Sensory Organs Develop from Ectodermal Placodes

The cranial sensory structures and the adenohypophysis derive from specialized ectodermal thickenings of the embryonic vertebrate head called placodes *(12, 66)*. It is not well understood how placodes can be induced during early development. An accepted model suggests that placodes begin to form in a common territory called preplacodal ectoderm (PPE), which is an undifferentiated nonneural ectoderm that is able to generate and subdivide into various specific types of placodes *(67)*. The otic placode gives rise to the specialized cells of both inner ear organs, auditory and vestibular, including mechanosensory hair cells, supporting cells, and spiral ganglion neurons. Olfactory placodes can generate neurons and glia of the olfactory tract and also gonatotropin-releasing hormone secreting neurons. Lens placodes can give rise to the crystalline-containing cells of the lens, without other neural derivatives. Hypophyseal placodes give rise to Rathke's pouch, which further differentiates into adenohypophysis. The ophthalmic and maxillomandibular placodes together give rise to the trigeminal (Gasserian) ganglia. Other distal sensory ganglia, i.e., facial (geniculate), vagal (nodose)m and glossopharyngeal (petrosal), are derived from epibranchial placodes *(12, 21, 66, 68)*.

The induction of PPE has been implicated as a response to the appropriate levels of BMPs, Wnt, and FGFs signals, similar to the neural plate and neural crest development described previously. Transcription factors Six and Eya, have been identified as critical in establishing the PPE identity *(21, 68–71)*. Six1/2, Six4/5 and Eya promote the generic aspects of placode development, but the expression patterns are not identical among different species of vertebrate *(20, 66)*. Once the specific placode is established as a separate domain from the unique PPE, the differentiation process goes on as shown in Fig. 4.

Regarding the ear, FGF signaling is fundamental in setting up the initial otic placodal identity from the PPE. Among the FGF ligands, FGF3, FGF8, and FGF10 have critical roles in the development of the inner ear, spanning from the induction of the otic placode and otic vesicle formation to the differentiation to the mature cells types *(72, 73)*. The mesodermally expressed FGF10 together FGF3 expressed by the neural tube are involved in the initial induction of the otic placode *(72, 73)*. FGF8 is expressed by the endoderm and exerts an inductive effect on the otic placode *(74)*, although probably indirectly, by stimulating the production of FGF10 by the mesenchyme *(75)*. FGF8 can also modulate terminal cell differentiation of pillar cells *(76)*. At later stages, FGF1 and FGF2 can promote the neurite outgrowth of mouse spiral ganglion neurons *(77, 78)*. BMP signaling is also important in ear development. The balance between BMP4 and BMP antagonist signals does

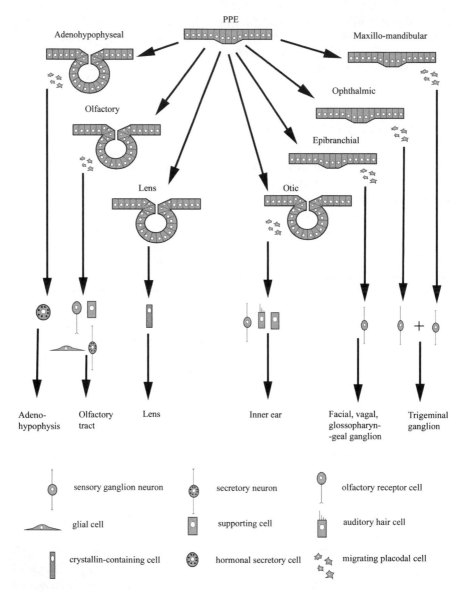

Fig. 4 Schematic drawing of mammalian ectodermal placodes giving rise to various types of specialized neural derivatives. To produce specific cell types, the initial preplacodal ectoderm (PPE) becomes partitioned into thicken, identifiable placodes. These invaginate to form a pit first and later a vesicle, cells can simply delaminate and migrate to a target location or use a combination of both mechanisms

influence hair cell survival and specification. The final number of hair cells is reduced by raised levels of BMP4, whereas the final size of the sensory patches and the number of specified hair cells is increased by the BMP4 antagonist, noggin *(79)*.

Different Methods to Induce Neural Differentiation In Vitro

Concepts of In Vitro Differentiation

The progression of ESCs into becoming a differentiated neuron should involve four major, distinct stages: competence, specification, commitment, and differentiation. Cells are competent when they are able to become neural precursors if they are exposed to the right combination of signals, theoretically similar to the ones present during the development of the nervous system in vivo. It is a stage of "potential capability." Specification refers to the acquisition of neural progenitor properties after having received the right stimulation, but they are still capable of responding to signals that can repress a neural character. It is basically a reversible state. The next stage, commitment, describes a situation where the cells have adopted the characteristics of neuronal precursors and will not change or lose them, even in the presence of signals that normally would repress the neural features. It is, then, an irreversible state. Finally, they will exit cell cycle to differentiate into a postmitotic cell type of their lineage such as neurons or glial cells *(14)*.

Methods of differentiation of neural progenitors from hESCs can be described as following two basic strategies: those that will have an "inductive" approach, exposing the cells to signals that mimic the environment that produces neuroectoderm in the embryo, and those that will follow a "passive" strategy, plating cells at low density in serum-free medium to devoid them on any signal that would repress their "neural program" and therefore evoke a default mechanism for NSC differentiation *(80)*. To resemble the native environment ESCs are allowed to form aggregates, known as embryoid bodies (EBs). These multicellular structures resemble those of the inner cell mass of the blastocyts and can be differentiated into all three primary germ layers. The gene expression of these EBs has been reported to be similar to the developing embryo *(81)*. When mice ESCs are placed in suspension culture in the absence of leukemia inhibitory factor (LIF) (which maintains pluripotency in mouse but not human ESCs), aggregates that comprises an outer hypoblastic-like cell layer and epiblastic-like core are formed within 2–4 days *(82)*. As occurs in embryonic development, the epiblastic-like core is able to give rise to ectoderm, mesoderm, and definitive endoderm; while the outer layer generates an extraembryonic visceral endoderm *(83)*. Signals, such as BMP antagonists, FGFs, and Wnt, from a region of anterior extraembryonic visceral endoderm (AVE) are well known to regulate anterior neuroectodermal differentiation *(80, 84, 85)*. At the more caudal region of the embryo, these signals are supplied by the notochord, but there is no structure in the EB suggested to resemble this function *(86)*. Thus far, the key to establish protocols for hESCs differentiation into neural progenitors and their specific lineages by signaling through EBs is the presence of the right signaling molecules, produced by equivalent structures to those found in vivo.

The other approach for directing differentiation toward NSCs is to evoke a default mechanism by depriving the ESCs of both cell–cell interaction and signaling. In principle, this approach does not involve the formation of EBs, but ESCs are plated at low density, and therefore the signaling from an outer layer of EBs that functions as AVE is absent *(80)*. Cells are cultured in a serum-free and nutrient-poor neurobasal medium that has been formulated for culturing pure population of NSCs. This medium mainly supports the survival of NSCs, selects out the apoptotic nonneural derivatives, and excludes the signaling molecules that are present in the serum *(87)*. Several protocols use the combination of these two approaches by promoting EBs, sometimes in the presence of growth factors, to be followed by neural lineage-specific selection under a defined medium *(80, 88)*.

The first evidence of neural differentiation in hESCs came out of the initial transplantation studies of these cells into severe combined immunodeficiency (SCID) mice, as they could produce teratomas containing all germ layers. After 4 months of transplantation, rosettes of neuroepithelium and neural ganglia were formed *(1)*. hESCs could be induced to have spontaneous differentiation by removing them from substrate and growing them in suspension culture in the presence of serum *(1, 89)*. At this stage, hESCs formed an EB composed of an ectodermal inner layer and endodermal outer layer similar to mice ESCs. After 4 days of culture in suspension, Oct4 expression was down-regulated and cell types of all primary layers arose *(89)*. When these cells were dissociated, plated in monolayer, and exposed to growth factors that promote their specific neural fate, neural progenitors and their lineages were induced *(90)*.

Methods of Differentiation of hESCs Toward NSCs and Neural Progenitors

Retinoic Acid Induction

By combining the use of EBs and RA treatment, hESCs were induced to form neural derivatives that expressed a dopamine receptor, DRD1, and two serotonin receptors, 5HT21 and 5HT5A *(91)*. In a parallel study, it was shown that RA treatment can generate a population of cells expressing the cell surface markers polysialylated neural cell adhesion molecule (PS-NCAM) and A2B5. PS-NCAM is expressed in multiple neuronal phenotypes, including dopamine-responsive neurons, but not found in glia, while A2B5 is expressed in both neuronal and glial derivatives *(92)*.

FGF2 Induction

This protocol mimics the function of FGF2 triggering neural induction during embryo development. Upon aggregation of EBs in the presence of FGF2, large numbers of neural tubelike structures or neural rosettes were detected in the culture. These neural rosette formations were isolated by selective enzymatic digestion, which led to the preferential detachment of the rosette islands and left the surrounding flat cells attached.

The selected rosette cells expressed markers of neuroepithelial cells, being positive for nestin, musashi-1, and PS-NCAM. In a following step, the isolated neuroepithelial cells were expanded in suspension culture and formed neurospheres, which subsequently, upon withdrawal of FGF2, gave rise to neurons, astrocytes, and oligodendrocytes (93).

Overgrow Culture

In this protocol, EBs were not required for neural differentiation. This study was performed by overgrowing hESCs in culture as an adherent monolayer. This led to spontaneous differentiation at the edges of hESC colonies; these neural precursors were subsequently isolated from the undifferentiated colonies by mechanical dissociation and grown in suspension as neurospheres in the presence of FGF2 and EGF. Under these conditions, neurons, astrocytes, and oligodendrocytes were produced (89, 94).

MedII Conditioned Medium

This method was originally established by treating mice ESCs with MedII conditioned medium, which is collected from the cultured hepatocellular carcinoma cell line HepG2 (95, 96). This conditioned medium caused the conversion of mouse EBs to early primitive ectoderm-like (EPL) cells, which appeared to be similar in morphology, transcriptional profiles, cytokine responsiveness, and differentiation potential to primitive ectoderm in vivo (96). In hESCs, a population of neural rosettes was found in either serum-free EBs or adherent cultures when they were exposed to MedII for 7–10 days (97, 98). It has been demonstrated that these neural progenitors expressed nestin and musashi-1, which subsequently gave rise to A2B5-positive and PS-NCAM-postive cells, postmitotic neurons, and astrocytes (98). A subpopulation of these neurons expressed tyrosine hydroxylase (TH), a dopaminergic marker (97).

Inhibition of BMP Signaling

As previously described, BMP signaling can inhibit the induction of neural cell fate from ectoderm during embryo development. Several protocols have used BMP antagonists to maintain the formation of neural rosettes in the process of differentiation. When hESCs were cultured in the presence of noggin, both in suspension and adherent culture, the commitment of hESCs to extraembryonic endoderm was blocked and neuroectodermal formation was enhanced, resulting in the generation of neural progenitors. Oct4 expression was shown to be down-regulated during differentiation, while neural marker expression increased during noggin treatment. These neural progenitors could be further derived into astrocytes, oligodendrocytes, and mature electrophysiologically functional neurons (99–101).

Stromal-Derived Inducing Activity

It has been reported that neurons are efficiently induced from mice ESCs when they are cocultured with the mouse stromal cell line PA6 *(102)*, which produces an unidentified stromal-derived inducing activity (SDIA). This system would appear to induce differentiation of dopaminergic neurons without the use of EBs or RA. In hESC, SDIA can also induce dopaminergic neurons *(103)*; however, it would appear that the coculturing of naive hESCs and PA6 could also direct differentiation into neural crest precursors and peripheral sensory neuronlike cells. Under these conditions, several neural crest markers, i.e., SNAIL, dHAND, and Sox9, were upregulated after 1 week of culture. Within the third week, neural crest marker genes were down-regulated, and a population expressed peripherin combined with Brn3a and tyrosine hydroxylase, which represented sensory and sympathetic neurons, respectively *(51)*.

To date, several protocols to induce neural differentiation in hESCs have been reported and are shown and summarized in Fig. 5. Some laboratories attempt to

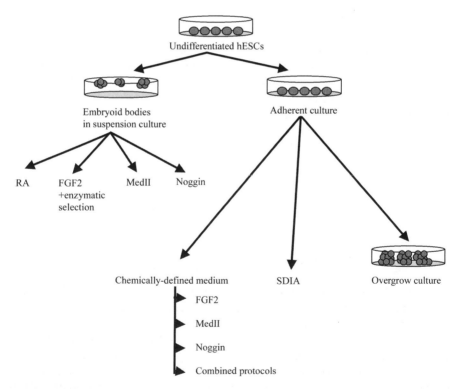

Fig. 5 Summary of current protocols used for differentiation of human embryonic stem cells (hESCs) toward neural progenitors. Two main strategies include the activation of neural differentiation pathway through embryoid body formation in suspension culture and through adherent culture in the presence of chemically defined medium

combine the differentiation methods in chemically defined medium (88, 104). After neural progenitors are obtained from the inductive steps, efforts are directed to support further differentiation into specific cell types; i.e., motor neurons, sensory neurons, dopaminergic neurons, and so on, by using specific growth factors that again would mimic the environment of the human embryonic nervous system.

Translation of hESC-Derived Neural Precursors to Clinic: The Attempt to Develop the Tools for Regeneration of Sensorineural Hearing Loss

Hair cells act as a primary sensory receptor that is innervated by sensory afferent neurons in the spiral ganglion, which in turn give rise to the fibers of the auditory nerve that propagate the electrical impulse to the auditory cortex (3, 4). Hair cells and spiral ganglion neurons are, therefore, the two main targets for a potential stem cell–based therapy for sensorineural hearing loss. As highlighted in the introduction, the only available therapy for deafness is the CI. The possibility of combining a CI with stem cell–derived neurons offers the potential of making the prosthetic device available to a larger segment of patients, and it could represent the initial and most viable application of stem cells in the shorter term. The development of techniques that could apply hESCs to the treatment of sensorineural hearing loss should then concentrate in two objectives: first, how to differentiate hESCs toward hair cells and spiral ganglion neurons and, second, how to transplant the ESC-derived sensory neural progenitors in vivo.

A few reports have explored the potential of murine ESCs to produce auditory derivatives. The combination of EB formation and exposure to IGF1, EGF, and basic fibroblast growth factor (bFGF) produced ear progenitors (105, 106); although by using this protocol the generation of hair cell–like types, rather than neurons, has been reported. Neural progenitors, generated as a monolayer culture in the presence of bFGF, were transplanted into deaf gerbils, showing grafting and generation of neurites (107). Mice ESCs have also been transplanted into mouse ears after treatment with SDIA. As explained before, this activity, obtained by growing the ESCs on PA6 feeder cells, promotes neural differentiation (102). Cells survived for 4 weeks and expressed the neuronal marker bIII-tubulin, but no hair cell markers. Differentiation was not complete, since cells were still proliferating and expressing SSEA3, a marker of the undifferentiated state (108). However, in a subsequent study, the ESCs-derived neurons showed an improvement of functionality in guinea pigs, as measured by auditory brainstem evoked responses (109). In an independent study, Hildebrand et al. (110) transplanted partially differentiated mouse ESCs into deafened guinea pigs cochleae. Although the cells survived in the cochlea, they failed to support functional recovery. Studies by a different group have described that untreated mouse ESCs transplanted into the vestibulocochlear nerve migrated centrally into the brainstem (111, 112). The survival of ESCs differentiated into TUJ1[+] cells was improved by cotransplanting them with fetal dorsal root ganglion

(DRG) neurons *(113)*. Moreover, coculturing EBs with cochlear explants facilitated the formation of neuron-like cells *(114)*.

Despite of the efforts described above to direct differentiation of murine ESCs into ear cell types, very little has been developed using hESCs. As mentioned before, both sensory hair cells and spiral ganglion neurons are derived from the ectodermal otic placode and share a common precursor *(115, 116)*. Given the placodal origin of the auditory neurons, we believed that an ideal protocol to obtain auditory sensory neurons should involve the generation of placodal progenitors. Our laboratory has devised a protocol that allows the production, from hESCs, of auditory neurons as well as hair cells. By using the signals involved in placodal induction we have been able to obtain both of these cells types *(117)*. It is also possible that useful sensory neurons could be obtained by differentiating hESCs into neural crest cells. Although strictly speaking auditory neurons are not derived in vivo from the neural crest *(118)*, sensory neurons that share many of the molecular markers *(119)* with the otic ones are of neural crest origin, and it is possible that they would function correctly once implanted in the right context. The use of protocols to induce neural crest progenitors from hESCs and their application to the ear remains to be explored. Regardless of the protocol to be used, the development of hESCs-derived sensory cells for clinical purposes should very likely happen under serum-free conditions, since serum-treated cells or coculture with mouse cell lines could pose a risk for transfer of pathogens across species.

A major issue to consider would be the stage of differentiation at which cells should be delivered in vivo: pluripotential hESCs, multipotential NSCs, or terminally differentiated hair cells and neurons? Evidence suggests that hESCs will readily incorporate into their host environment, but they could result in malignant tumor formation. When hESCs were transplanted into animal models, they generated teratomas *(1, 89)*. Transplantation of multipotential or differentiated cells should not encounter this problem *(120)*. On the other hand, postmitotic hair cells and spiral ganglion neurons, having lost their proliferating capacity, may not graft easily in the host organ. Therefore, predifferentiation, committed progenitors might be the best choice for therapeutic application. However, there is a risk in relying too much on the local environment or niche in vivo to complete the differentiation process, as the degree of pathology in the inner ear is likely to vary in each individual and might exert very different effects on the final differentiation of transplanted cells *(121)*.

Another challenge for transplantation lies in the route of delivery to the target site. In the particular case of the inner ear, the cochlea is encased by bone and is very difficult to access. Few experimental approaches tried in rodents have injected cells into the modiolus (central axis of cochlea) *(109, 111, 122, 123)*, the scala tympani (which is one of the fluid-filled compartment of the cochlea containing perilymph) *(113, 124, 125)*, or into Rosenthal's canal (where spiral ganglion neurons are enclosed) *(121)*; but all of these methods require surgical intervention that might cause damage to the cochlea. Few studies attempted to transplant NSCs directly to the auditory nerve that had minimal trauma, resulting in a longer survival period of transplanted cells and yielding satisfactory expression of neuronal markers such as bIII tubulin. These particular cells could extend their processes toward the hair cell target *(107, 112, 126)*.

All these routes will prove very difficult for hair cell restoration, but in order to deliver neural precursors, the most likely approach will be to do it simultaneously during cochlear implantation into the scala tympani, through a cochleostomy in the vicinity of the round window *(127)*.

Conclusion

Neural differentiation from hESCs is a complex process that should follow the principles of the developmental biology of the embryonic nervous system. NSCs in vivo arise mainly from the neural plate, neural crest, and the ectodermal placodes. In this chapter we have discussed several published protocols that attempted to generate neural progenitors from hESCs and subsequently tried to differentiate them into the specific cell types, aiming to produce the functional neural derivatives that can replace the damaged cells in various pathological disorders. Some of these protocols could be adapted or translated into applications for hearing disorders. Although a generic neural precursor may not be ideal for a deafness therapy, the generation of otic placodal progenitors as well as the creation of sensory neurons from NCSCs are highly viable alternatives. The combination of hESC-derived auditory neurons with cochlear implants may be a real therapeutic alternative in a not too distant future. However, they are not the only application of this technology. Great benefit will come from the generation of a model system that could be used for functional studies, drug testing, and discovery. The promise of stem cells in the ear goes far beyond cell replacement.

Acknowledgments This work has been supported by grants from the Royal National Institute for Deaf People (RNID) and Deafness Research UK (DRUK). MNR was also supported by a Wellcome Trust VIP Award. OT is supported by a PhD studentship from the Ramathibodi Hospital, Mahidol University, Bangkok, Thailand.

References

1. Thomson JA, Itskovitz-Eldor J, Shapiro SS, et al. Embryonic stem cell lines derived from human blastocysts. Science 1998;282:1145–7.
2. Lindvall O, Kokaia Z. Stem cells for the treatment of neurological disorders. Nature 2006;441:1094–6.
3. Morton CC. Genetics, genomics and gene discovery in the auditory system. Hum Mol Genet 2002;11:1229–40.
4. Holley MC. Keynote review: the auditory system, hearing loss and potential targets for drug development. Drug Discov Today 2005;10:1269–82.
5. Oshima K, Grimm CM, Corrales CE, et al. Differential distribution of stem cells in the auditory and vestibular organs of the inner ear. J Assoc Res Otolaryngol 2007;8:18–31.
6. Ruben RJ. Development of the inner ear of the mouse: a radioautographic study of terminal mitoses. Acta Otolaryngol 1967;220 Suppl:1–44.

7. Waltzman SB. Cochlear implants: current status. Expert Rev Med Devices 2006;3:647–55.
8. Hall RD. Estimation of surviving spiral ganglion cells in the deaf rat using the electrically evoked auditory brainstem response. Hear Res 1990;49:155–68.
9. Pettingill LN, Richardson RT, Wise AK, O'Leary SJ, Shepherd RK. Neurotrophic factors and neural prostheses: potential clinical applications based upon findings in the auditory system. IEEE Trans Biomed Eng 2007;54:1138–48.
10. Temple S. The development of neural stem cells. Nature 2001;414:112–7.
11. Dupin E, Calloni G, Real C, Goncalves-Trentin A, Le Douarin NM. Neural crest progenitors and stem cells. C R Biol 2007;330:521–9.
12. Streit A. The preplacodal region: an ectodermal domain with multipotential progenitors that contribute to sense organs and cranial sensory ganglia. Int J Dev Biol 2007;51:447–61.
13. Jalali A, Bonaguidi, M, Hamill, C, Kessler, JA. Multipotent stem cells in the embryonic nervous system. In: Rao MS, editor. Neural development and stem cells, 2nd ed. Humana Press, Totowa, NJ, 2006:67–95.
14. Wilson SI, Edlund T. Neural induction: Toward a unifying mechanism. Nat Neurosci 2001;4 Suppl:1161–8.
15. Kuroda H, Wessely O, De Robertis EM. Neural induction in Xenopus: requirement for ectodermal and endomesodermal signals via Chordin, Noggin, beta-Catenin, and Cerberus. PLoS Biol 2004;2:E92.
16. Suzuki A, Kaneko E, Ueno N, Hemmati-Brivanlou A. Regulation of epidermal induction by BMP2 and BMP7 signaling. Dev Biol 1997;189:112–22.
17. Tribulo C, Aybar MJ, Nguyen VH, Mullins MC, Mayor R. Regulation of Msx genes by a Bmp gradient is essential for neural crest specification. Development 2003;130:6441–52.
18. De Robertis EM, Kuroda H. Dorsal–ventral patterning and neural induction in Xenopus embryos. Annu Rev Cell Dev Biol 2004;20:285–308.
19. Marchant L, Linker C, Ruiz P, Guerrero N, Mayor R. The inductive properties of mesoderm suggest that the neural crest cells are specified by a BMP gradient. Dev Biol 1998;198:319–29.
20. Brugmann SA, Pandur PD, Kenyon KL, Pignoni F, Moody SA. Six1 promotes a placodal fate within the lateral neurogenic ectoderm by functioning as both a transcriptional activator and repressor. Development 2004;131:5871–81.
21. Brugmann SA, Moody SA. Induction and specification of the vertebrate ectodermal placodes: precursors of the cranial sensory organs. Biol Cell 2005;97:303–19.
22. Liu W, Ren C, Shi J, et al. Characterization of the functionally related sites in the neural inducing gene noggin. Biochem Biophys Res Commun 2000;270:293–7.
23. Larrain J, Bachiller D, Lu B, Agius E, Piccolo S, De Robertis EM. BMP-binding modules in chordin: a model for signalling regulation in the extracellular space. Development 2000;127:821–30.
24. Zimmerman LB, De Jesus-Escobar JM, Harland RM. The Spemann organizer signal noggin binds and inactivates bone morphogenetic protein 4. Cell 1996;86:599–606.
25. Wilson SI, Rydstrom A, Trimborn T, et al. The status of Wnt signalling regulates neural and epidermal fates in the chick embryo. Nature 2001;411:325–30.
26. Chenn A, McConnell SK. Cleavage orientation and the asymmetric inheritance of Notch1 immunoreactivity in mammalian neurogenesis. Cell 1995;82:631–41.
27. Lu B, Jan L, Jan YN. Control of cell divisions in the nervous system: symmetry and asymmetry. Annu Rev Neurosci 2000;23:531–56.
28. Temple S. Defining neural stem cells and their role in normal development of the nervous system. In: Rao MS, editor. Neural development and stem cells, 2nd ed. Humana Press, Totowa, NJ, 2006:1–28.
29. Morest DK, Silver J. Precursors of neurons, neuroglia, and ependymal cells in the CNS: what are they? where are they from? how do they get where they are going? Glia 2003;43:6–18.
30. Kriegstein AR, Gotz M. Radial glia diversity: a matter of cell fate. Glia 2003;43:37–43.
31. Noctor SC, Martinez-Cerdeno V, Ivic L, Kriegstein AR. Cortical neurons arise in symmetric and asymmetric division zones and migrate through specific phases. Nat Neurosci 2004;7:136–44.

32. Temple S, Alvarez-Buylla A. Stem cells in the adult mammalian central nervous system. Curr Opin Neurobiol 1999;9:135–41.
33. Gangemi RM, Perera M, Corte G. Regulatory genes controlling cell fate choice in embryonic and adult neural stem cells. J Neurochem 2004;89:286–306.
34. Emoto N, Gonzalez AM, Walicke PA, et al. Basic fibroblast growth factor (FGF) in the central nervous system: Identification of specific loci of basic FGF expression in the rat brain. Growth Factors 1989;2:21–9.
35. Zhu G, Mehler MF, Mabie PC, Kessler JA. Developmental changes in progenitor cell responsiveness to cytokines. J Neurosci Res 1999;56:131–45.
36. Liu SY, Zhang ZY, Song YC, et al. SVZa neural stem cells differentiate into distinct lineages in response to BMP4. Exp Neurol 2004;190:109–21.
37. Arsenijevic Y, Weiss S, Schneider B, Aebischer P. Insulin-like growth factor-I is necessary for neural stem cell proliferation and demonstrates distinct actions of epidermal growth factor and fibroblast growth factor-2. J Neurosci 2001;21:7194–202.
38. Goodrich LV, Scott MP. Hedgehog and patched in neural development and disease. Neuron 1998;21:1243–57.
39. Rowitch DH, Benoit S-J, Lee SM, Flax JD, Snyder EY, McMahon AP. Sonic hedgehog regulates proliferation and inhibits differentiation of CNS precursor cells. J Neurosci 1999;19:8954–65.
40. Zhu G, Mehler MF, Zhao J, Yu Yung S, Kessler JA. Sonic hedgehog and BMP2 exert opposing actions on proliferation and differentiation of embryonic neural progenitor cells. Dev Biol 1999;215:118–29.
41. Wechsler-Reya RJ, Scott MP. Control of neuronal precursor proliferation in the cerebellum by Sonic hedgehog. Neuron 1999;22:103–14.
42. Lee SM, Tole S, Grove E, McMahon AP. A local Wnt-3a signal is required for development of the mammalian hippocampus. Development 2000;127:457–67.
43. Megason SG, McMahon AP. A mitogen gradient of dorsal midline Wnts organizes growth in the CNS. Development 2002;129:2087–98.
44. Viti J, Gulacsi A, Lillien L. Wnt regulation of progenitor maturation in the cortex depends on Shh or fibroblast growth factor 2. J Neurosci 2003;23:5919–27.
45. Davidson KC, Jamshidi P, Daly R, Hearn MT, Pera MF, Dottori M. Wnt3a regulates survival, expansion, and maintenance of neural progenitors derived from human embryonic stem cells. Mol Cell Neurosci 2007;36:408–15.
46. Durston AJ, Timmermans JP, Hage WJ, et al. Retinoic acid causes an anteroposterior transformation in the developing central nervous system. Nature 1989;340:140–4.
47. Chambon P. A decade of molecular biology of retinoic acid receptors. FASEB J 1996;10:940–54.
48. Selleck MA, Bronner-Fraser M. Origins of the avian neural crest: the role of neural plate-epidermal interactions. Development 1995;121:525–38.
49. Suzuki HR, Kirby ML. Absence of neural crest cell regeneration from the postotic neural tube. Dev Biol 1997;184:222–33.
50. Morrison SJ, White PM, Zock C, Anderson DJ. Prospective identification, isolation by flow cytometry, and in vivo self-renewal of multipotent mammalian neural crest stem cells. Cell 1999;96:737–49.
51. Pomp O, Brokhman I, Ben-Dor I, Reubinoff B, Goldstein RS. Generation of peripheral sensory and sympathetic neurons and neural crest cells from human embryonic stem cells. Stem Cells 2005;23:923–30.
52. Crane JF, Trainor PA. Neural crest stem and progenitor cells. Annu Rev Cell Dev Biol 2006;22:267–86.
53. Kanzler B, Foreman RK, Labosky PA, Mallo M. BMP signaling is essential for development of skeletogenic and neurogenic cranial neural crest. Development 2000;127:1095–104.
54. Liem KF Jr, Tremml G, Roelink H, Jessell TM. Dorsal differentiation of neural plate cells induced by BMP-mediated signals from epidermal ectoderm. Cell 1995;82:969–79.
55. Selleck MA, Garcia-Castro MI, Artinger KB, Bronner-Fraser M. Effects of Shh and Noggin on neural crest formation demonstrate that BMP is required in the neural tube but not ectoderm. Development 1998;125:4919–30.

56. Wakamatsu Y, Maynard TM, Weston JA. Fate determination of neural crest cells by NOTCH-mediated lateral inhibition and asymmetrical cell division during gangliogenesis. Development 2000;127:2811–21.
57. LaBonne C, Bronner-Fraser M. Induction and patterning of the neural crest, a stem cell-like precursor population. J Neurobiol 1998;36:175–89.
58. Saint-Jeannet JP, He X, Varmus HE, Dawid IB. Regulation of dorsal fate in the neuraxis by Wnt-1 and Wnt-3a. Proc Natl Acad Sci U S A 1997;94:13713–8.
59. Chang C, Hemmati-Brivanlou A. Neural crest induction by Xwnt7B in Xenopus. Dev Biol 1998;194:129–34.
60. Hemmati H, Moreno TA, Bronner-Fraser M. PNS precursor cells in development and cancer. In: Rao MS, editor. Neural development and stem cells, 2nd ed. Humana Press, Totowa, NJ, 2006:189–217.
61. Garcia-Castro MI, Marcelle C, Bronner-Fraser M. Ectodermal Wnt function as a neural crest inducer. Science 2002;297:848–51.
62. Monsoro-Burq AH, Fletcher RB, Harland RM. Neural crest induction by paraxial mesoderm in Xenopus embryos requires FGF signals. Development 2003;130:3111–24.
63. Lewis JL, Bonner J, Modrell M, et al. Reiterated Wnt signaling during zebrafish neural crest development. Development 2004;131:1299–308.
64. Stemple DL, Anderson DJ. Isolation of a stem cell for neurons and glia from the mammalian neural crest. Cell 1992;71:973–85.
65. Trentin A, Glavieux-Pardanaud C, Le Douarin NM, Dupin E. Self-renewal capacity is a wide-spread property of various types of neural crest precursor cells. Proc Natl Acad Sci U S A 2004;101:4495–500.
66. Schlosser G. Induction and specification of cranial placodes. Dev Biol 2006;294:303–51.
67. Baker CV, Bronner-Fraser M. Establishing neuronal identity in vertebrate neurogenic placodes. Development 2000;127:3045–56.
68. Streit A. Early development of the cranial sensory nervous system: from a common field to individual placodes. Dev Biol 2004;276:1–15.
69. Pandur PD, Moody SA. Xenopus Six1 gene is expressed in neurogenic cranial placodes and maintained in the differentiating lateral lines. Mech Dev 2000;96:253–7.
70. Ghanbari H, Seo HC, Fjose A, Brandli AW. Molecular cloning and embryonic expression of Xenopus Six homeobox genes. Mech Dev 2001;101:271–7.
71. Schlosser G, Ahrens K. Molecular anatomy of placode development in Xenopus laevis. Dev Biol 2004;271:439–66.
72. Alvarez Y, Alonso MT, Vendrell V, et al. Requirements for FGF3 and FGF10 during inner ear formation. Development 2003;130:6329–38.
73. Wright TJ, Mansour SL. Fgf3 and Fgf10 are required for mouse otic placode induction. Development 2003;130:3379–90.
74. Zelarayan LC, Vendrell V, Alvarez Y, et al. Differential requirements for FGF3, FGF8 and FGF10 during inner ear development. Dev Biol 2007;308:379–91.
75. Ladher RK, Wright TJ, Moon AM, Mansour SL, Schoenwolf GC. FGF8 initiates inner ear induction in chick and mouse. Genes Dev 2005;19:603–13.
76. Mueller KL, Jacques BE, Kelley MW. Fibroblast growth factor signaling regulates pillar cell development in the organ of corti. J Neurosci 2002;22:9368–77.
77. Hossain WA, Morest DK. Fibroblast growth factors (FGF-1, FGF-2) promote migration and neurite growth of mouse cochlear ganglion cells in vitro: immunohistochemistry and antibody perturbation. J Neurosci Res 2000;62:40–55.
78. Nicholl AJ, Kneebone A, Davies D, et al. Differentiation of an auditory neuronal cell line suitable for cell transplantation. Eur J Neurosci 2005;22:343–53.
79. Pujades C, Kamaid A, Alsina B, Giraldez F. BMP-signaling regulates the generation of hair-cells. Dev Biol 2006;292:55–67.
80. Cai C, Grabel L. Directing the differentiation of embryonic stem cells to neural stem cells. Dev Dyn 2007;236:3255–66.
81. Gajovic S, St-Onge L, Yokota Y, Gruss P. Retinoic acid mediates Pax6 expression during in vitro differentiation of embryonic stem cells. Differentiation 1997;62:187–92.

82. Smith AG, Heath JK, Donaldson DD, et al. Inhibition of pluripotential embryonic stem cell differentiation by purified polypeptides. Nature 1988;336:688–90.
83. Rodda SJ, Kavanagh SJ, Rathjen J, Rathjen PD. Embryonic stem cell differentiation and the analysis of mammalian development. Int J Dev Biol 2002;46:449–58.
84. Rathjen J, Haines BP, Hudson KM, Nesci A, Dunn S, Rathjen PD. Directed differentiation of pluripotent cells to neural lineages: homogeneous formation and differentiation of a neurectoderm population. Development 2002;129:2649–61.
85. Maye P, Becker S, Siemen H, et al. Hedgehog signaling is required for the differentiation of ES cells into neurectoderm. Dev Biol 2004;265:276–90.
86. Stern CD. Neural induction: old problem, new findings, yet more questions. Development 2005;132:2007–21.
87. Reynolds BA, Weiss S. Clonal and population analyses demonstrate that an EGF-responsive mammalian embryonic CNS precursor is a stem cell. Dev Biol 1996;175:1–13.
88. Li XJ, Zhang SC. In vitro differentiation of neural precursors from human embryonic stem cells. Methods Mol Biol 2006;331:169–77.
89. Reubinoff BE, Pera MF, Fong CY, Trounson A, Bongso A. Embryonic stem cell lines from human blastocysts: somatic differentiation in vitro. Nat Biotechnol 2000;18:399–404.
90. Schuldiner M, Yanuka O, Itskovitz Eldor J, Melton DA, Benvenisty N. Effects of eight growth factors on the differentiation of cells derived from human embryonic stem cells. Proc Natl Acad Sci U S A 2000;97:11307–12.
91. Schuldiner M, Eiges R, Eden A, et al. Induced neuronal differentiation of human embryonic stem cells. Brain Res 2001;913:201–5.
92. Carpenter MK, Inokuma MS, Denham J, Mujtaba T, Chiu CP, Rao MS. Enrichment of neurons and neural precursors from human embryonic stem cells. Exp Neurol 2001;172:383–97.
93. Zhang SC, Wernig M, Duncan ID, Brustle O, Thomson JA. In vitro differentiation of transplantable neural precursors from human embryonic stem cells. Nat Biotechnol 2001;19:1129–33.
94. Reubinoff BE, Itsykson P, Turetsky T, et al. Neural progenitors from human embryonic stem cells. Nat Biotechnol 2001;19:1134–40.
95. Rathjen J, Lake JA, Bettess MD, Washington JM, Chapman G, Rathjen PD. Formation of a primitive ectoderm like cell population, EPL cells, from ES cells in response to biologically derived factors. J Cell Sci 1999;112(Pt 5):601–12.
96. Lake J, Rathjen J, Remiszewski J, Rathjen PD. Reversible programming of pluripotent cell differentiation. J Cell Sci 2000;113(Pt 3):555–66.
97. Schulz TC, Palmarini GM, Noggle SA, Weiler DA, Mitalipova MM, Condie BG. Directed neuronal differentiation of human embryonic stem cells. BMC Neurosci 2003;4:27.
98. Shin S, Mitalipova M, Noggle S, et al. Long-term proliferation of human embryonic stem cell-derived neuroepithelial cells using defined adherent culture conditions. Stem Cells 2006;24:125–38.
99. Itsykson P, Ilouz N, Turetsky T, et al. Derivation of neural precursors from human embryonic stem cells in the presence of noggin. Mol Cell Neurosci 2005;30:24–36.
100. Gerrard L, Rodgers L, Cui W. Differentiation of human embryonic stem cells to neural lineages in adherent culture by blocking bone morphogenetic protein signaling. Stem Cells 2005;23:1234–41.
101. Wilson PG, Stice SS. Development and differentiation of neural rosettes derived from human embryonic stem cells. Stem Cell Rev 2006;2:67–77.
102. Kawasaki H, Mizuseki K, Nishikawa S, et al. Induction of midbrain dopaminergic neurons from ES cells by stromal cell-derived inducing activity. Neuron 2000;28:31–40.
103. Brederlau A, Correia AS, Anisimov SV, et al. Transplantation of human embryonic stem cell-derived cells to a rat model of Parkinson's disease: effect of in vitro differentiation on graft survival and teratoma formation. Stem Cells 2006;24:1433–40.
104. Baharvand H, Mehrjardi NZ, Hatami M, Kiani S, Rao M, Haghighi MM. Neural differentiation from human embryonic stem cells in a defined adherent culture condition. Int J Dev Biol 2007;51:371–8.
105. Li H, Roblin G, Liu H, Heller S. Generation of hair cells by stepwise differentiation of embryonic stem cells. Proc Natl Acad Sci U S A 2003;100:13495–500.

106. Rivolta MN, Li H, Heller S. Generation of inner ear cell types from embryonic stem cells. Methods Mol Biol 2006;330:71–92.

107. Corrales CE, Pan L, Li H, Liberman MC, Heller S, Edge AS. Engraftment and differentiation of embryonic stem cell-derived neural progenitor cells in the cochlear nerve trunk: growth of processes into the organ of Corti. J Neurobiol 2006;66:1489–500.

108. Sakamoto T, Nakagawa T, Endo T, et al. Fates of mouse embryonic stem cells transplanted into the inner ears of adult mice and embryonic chickens. Acta Otolaryngol Suppl 2004;551:48–52.

109. Okano T, Nakagawa T, Endo T, et al. Engraftment of embryonic stem cell-derived neurons into the cochlear modiolus. Neuroreport 2005;16:1919–22.

110. Hildebrand MS, Dahl HH, Hardman J, Coleman B, Shepherd RK, de Silva MG. Survival of partially differentiated mouse embryonic stem cells in the scala media of the guinea pig cochlea. J Assoc Res Otolaryngol 2005;6:341–54.

111. Hu Z, Ulfendahl M, Olivius NP. Central migration of neuronal tissue and embryonic stem cells following transplantation along the adult auditory nerve. Brain Res 2004;1026:68–73.

112. Regala C, Duan M, Zou J, Salminen M, Olivius P. Xenografted fetal dorsal root ganglion, embryonic stem cell and adult neural stem cell survival following implantation into the adult vestibulocochlear nerve. Exp Neurol 2005;193:326–33.

113. Hu Z, Andang M, Ni D, Ulfendahl M. Neural cograft stimulates the survival and differentiation of embryonic stem cells in the adult mammalian auditory system. Brain Res 2005;1051:137–44.

114. Coleman B, Fallon JB, Pettingill LN, de Silva MG, Shepherd RK. Auditory hair cell explant co-cultures promote the differentiation of stem cells into bipolar neurons. Exp Cell Res 2007;313:232–43.

115. Satoh T, Fekete DM. Clonal analysis of the relationships between mechanosensory cells and the neurons that innervate them in the chicken ear. Development 2005;132:1687–97.

116. Raft S, Koundakjian EJ, Quinones H, et al. Cross-regulation of Ngn1 and Math1 coordinates the production of neurons and sensory hair cells during inner ear development. Development 2007;134:4405–15.

117. Chen W, Moore H, Andrews PW, Rivolta MN. Isolation and characterization of human auditory stem cells and multipotent progenitors. Presented at Third Annual Meeting of the International Society for Stem Cell Research, San Francisco, CA, 2005.

118. Lang H, Fekete DM. Lineage analysis in the chicken inner ear shows differences in clonal dispersion for epithelial, neuronal, and mesenchymal cells. Dev Biol 2001;234:120–37.

119. Anderson DJ. Lineages and transcription factors in the specification of vertebrate primary sensory neurons. Curr Opin Neurobiol 1999;9:517–24.

120. Sell S. Stem cell origin of cancer and differentiation therapy. Crit Rev Oncol Hematol 2004;51:1–28.

121. Coleman B, de Silva MG, Shepherd RK. Concise review: The potential of stem cells for auditory neuron generation and replacement. Stem Cells 2007;25:2685–94.

122. Naito Y, Nakamura T, Nakagawa T, et al. Transplantation of bone marrow stromal cells into the cochlea of chinchillas. Neuroreport 2004;15:1–4.

123. Tamura T, Nakagawa T, Iguchi F, et al. Transplantation of neural stem cells into the modiolus of mouse cochleae injured by cisplatin. Acta Otolaryngol Suppl 2004;551:65–8.

124. Hu Z, Wei D, Johansson CB, et al. Survival and neural differentiation of adult neural stem cells transplanted into the mature inner ear. Exp Cell Res 2005;302:40–7.

125. Coleman B, Hardman J, Coco A, et al. Fate of embryonic stem cells transplanted into the deafened mammalian cochlea. Cell Transplant 2006;15:369–80.

126. Sekiya T, Kojima K, Matsumoto M, Kim TS, Tamura T, Ito J. Cell transplantation to the auditory nerve and cochlear duct. Exp Neurol 2006;198:12–24.

127. Adunka OF, Radeloff A, Gstoettner WK, Pillsbury HC, Buchman CA. Scala Tympani cochleostomy II: Topography and histology. Laryngoscope 2007;117:2195–200.

Stem Cell Transplantation Supports the Repair of Injured Olfactory Neuroepithelium After Permanent Lesion

Valeria Franceschini, Simone Bettini, Riccardo Saccardi, and Roberto P. Revoltella

Abstract We investigated whether human cord blood-selected CD133[+] stem cells (HSC) may engraft the olfactory mucosa and contribute to restoration of neuro-olfactory epithelium (NE) in *nod-scid* mice damaged by dichlobenil. The herbicide dichlobenil selectively causes necrosis of the dorsomedial part of the NE and underlying mucosa, while the lateral part of the olfactory region remains undamaged. The aim of this research was to demonstrate that HSC stimulate self-renewal of neuronal stem cells and promote their differentiation into bipolar olfactory neurons to replace the injured NE. By PCR, we tested the presence of three human-specific microsatellites (CODIS; Combined DNS Index System), used as DNA markers for traceability of the engrafted cells, demonstrating their presence in various tissues of the host, including the olfactory mucosa, 1 month after transplantation. By immunohistochemistry and lectin staining, we demonstrated that, in injured mice, HSC contributed to stimulating residual endogenous olfactory neurons, promoting recovery of the original phenotype of the NE, in contrast to the lack of spontaneous regeneration in similar injured areas always seen in the nontransplanted control mice. Multiple colour fluorescence in situ hybridisation (M-FISH) analysis detected seven human genomic sequences present in different chromosomes and provided further evidence of positive prolonged engraftment of chimeric cells in the olfactory mucosa. This study provides the first evidence that transplanted HSC migrating to the neuro-olfactory mucosa may contribute to NE structure restoration with resumption of the sensorineural olfactory loss.

Keywords Olfaction • Neuro-olfactory epithelium • Stem cell • Transplantation • Tissue regeneration

V. Franceschini, S. Bettini, R. Saccardi, and R.P. Revoltella (✉)
Foundation "Stem Cells & Life" onlus; IPCF-CNR, Molecular Modelling Lab.,
Via G. Moruzzi 1, 56124, Pisa, Italy.
e-mail: roberto.revoltella@itb.cnr.it e-mail: rrevoltella@yahoo.it

H. Baharvand (ed.), *Trends in Stem Cell Biology and Technology*,
DOI 10.1007/ 978-1-60327-905-5_16,
© Humana Press, a Part of Springer Science+Business Media, LLC 2009

Introduction

The Neuro-olfactory Mucosa and Its Neuroepithelium

The initial event in primary odorant recognition occurs at the level of the olfactory neuroepithelium (NE), whereas processing the sensory information occurs in the olfactory bulb and higher cortical centres. Figure 1 gives a schematic representation of the olfactory NE. This is a pseudostratified epithelium lying on the convoluted turbinates within the posterior region of the nasal cavity. From the apical surface to the basal lamina, the sensory epithelium is comprised of sustentacular cells, mature and immature olfactory sensory neurons, globose basal cells, and horizontal basal cells. The sustentacular cells are nonneuronal supporting cells capped by microvilli *(1)*. They express many biotransformation enzymes, suggesting that these cells serve in detoxification *(2, 3)*. Moreover the sustentacular cells act as phagocyte, eliminating dead olfactory neurons *(4, 5)*. The olfactory receptors are elongated, columnar bipolar cells with an apical dendrite ending with a knob bearing 12 or more cilia that provide an enlarged membrane surface for interaction with odorants. The dendritic knob and the cilia are the only parts of the sensory neuron exposed to the external environment. When moving from the air space of the nasal cavity to olfactory receptor sites, odorant molecules must first diffuse through the mucus layer covering the olfactory epithelium. The mucus provides the milieu in which receptor activation itself is presumed to occur. Each olfactory neuron has a thin, unmyelinated axon that exits the epithelium basally to join with fascicles of the olfactory nerve, running accompanied by ensheathing glia to the main olfactory

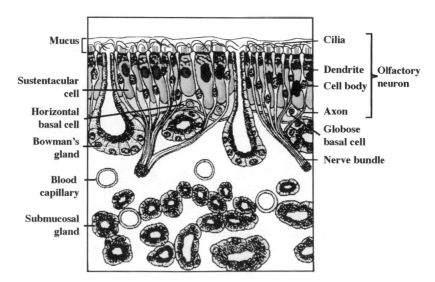

Fig. 1 Schematic diagram of the normal olfactory epithelium

bulb. Here the primary olfactory projections ramify extensively and, synapsing on richly branched dendritic processes of neurons situated in the bulb, constitute entangled spherical plexuses named glomeruli *(1)*.

The olfactory neurogenesis originates from the globose basal cells, the major population of proliferating cells located close to the basal lamina whose daughter cells differentiate, maturate, and form new olfactory receptors *(6)*. Numerous submucosa glands (Bowman's glands) contribute to produce the mucus.

Sensory olfactory cells are directly exposed to a wide range of environmental factors. During inhalation, there is local exposure of the nasal tissue to chemicals and fine particles that may affect the longevity of the olfactory neurons *(7)*. As a consequence of their unprotected position, they are continuously replaced throughout life *(8, 9)*.

Environmental Interactions

An increasing number of chemicals are known to preferentially induce extensive morphological changes in the olfactory epithelium, not only following inhalation. In an aquatic environment, chemical wastes, such as heavy metals produced by industrial and agricultural activities, give rise to toxic effects on olfaction, known to mediate a wide number of behavioural repertoires such as food search, intra- and interspecific interactions, and reproduction *(10, 11)*. In rodents, studies on the local effects of olfactory toxicant such as Triton X-100 or methylbromide, have demonstrated an increased rate of neurogenesis, leading to a complete regeneration of the NE *(12)*. In mice, a single intraperitoneal injection of methimazole (50 mg/kg), an antithyroid drug that can induce loss of taste and smell in humans, induces necrosis throughout the olfactory region (epithelium), whereas the nasal respiratory mucosa is spared. This lesion is followed by a rapid regeneration of the tissue and 2 weeks later the appearance of NE is almost normal *(13)*. On the contrary, a single intraperitoneal injection of the herbicide dichlobenil causes necrosis of the dorsomedial part of the NE and underlying lamina propria, whereas the lateral part of the olfactory region remains undamaged. Necrosis is not followed by regeneration, and 6 months after exposure to dichlobenil, the olfactory dorsomedial region shows a respiratory metaplasia with invaginations into a fibrotic lamina propria *(14, 15)*. This epithelium is almost completely devoid of Bowman's glands. Bowman's glands are known to produce and secrete a large number of growth or differentiation factors, including glial cell line-derived neurotrophic factor *(16)*. On the basis of these findings, they suggested that an intact lamina propria is a prerequisite for the regeneration of the olfactory NE after toxicant-induced injury.

Stem Cells and Neuronal Differentiation

Recently, several studies have reported that when cultured in an appropriate medium, human and rodent mesenchymal bone marrow cells, as well as cord blood–derived cells and adipose tissue–derived stem cells, undergo differentiation along classical mesenchymal lineages: adipogenesis, chondrogenesis, and osteogenesis, as shown by the expression of several lineage-specific genes and proteins *(17, 18)*. These stem cells, under particular culturing conditions, can differentiate to nonmesenchymal lineages. Culture in the presence of dexamethasone, hydrocortisone, or 5-azacytidine results in a time-dependent pattern of expression of muscle-related genes consistent with normal myogenesis *(19)*. Furthermore, these stem cells can be induced to express markers consistent with a neuronal phenotype, suggesting an ectodermal potential. Treatment of rat and mouse mesenchymal bone marrow cells and adipose tissue–derived stem cells with β-mercaptoethanol results in rapid transition of cells to a neuronal morphology and expression of neuron-specific proteins such as nestin, neuron-specific enolase, and neuron-specific protein, all of which are early markers of the neuronal lineage. Expression of markers characteristic of mature neurons has not been described *(18)*.

The present study investigated the capacity of CD133$^+$ stem cells selectively obtained from human umbilical cord blood (UCB) in recovery of mouse olfactory NE after degeneration induced by dichlobenil injections.

Materials and Methods

Human Umbilical Cord Blood CD133$^+$ Cells

UCB was obtained from informed consenting donors (Bone Marrow Transplantation Unit, Careggi Hospital, Florence). The mononuclear cell fraction was isolated from UCB by density gradient centrifugation (Ficoll Hypaque, Amersham Biosciences Europe GmbH, Freiburg, DE). CD133$^+$ cells were obtained by magnetic isolation with MACS® MicroBeads (Miltenyi Biotech, GmbH, Gladbach, DE) conjugated to a monoclonal mouse antihuman CD133 antibody (clone AC133, which recognizes epitope CD133/1), using CD133$^+$ Cell Isolation Kits (Miltenyi Biotec), following the manufacturer's procedure. The efficiency of CD133$^+$ cell purification was verified by flow-cytometry counterstaining with CD133/2 (293C3)-PE (Miltenyi Biotec). The percentage of CD133$^+$ cells was on average 90% or more.

Experimental Protocols

All animal experiments were performed according to protocols approved by the Institutional Review Board and the Ethics Committee of the Institute for

Research against Tumors (IST) of Genua, Italy, and the National Research Council of Italy (CNR). Twenty-one 2-month-old female inbred *nod-scid* mice (Charles River Laboratories, Wilmington, MA) were used for this study. Animals were housed in microisolators and fed sterile food and acidified water to discourage bacterial growth. All mice were injected intraperitoneal on day 0 with dichlobenil (Fluka, Buchs SG, CH) (50 mg/kg body weight) dissolved in dimethylsulfoxide (DMSO) (1 µL/g) *(15)*; 2 days later mice were total body X-ray irradiated (300 rad/mouse) and after 5 more days they were injected again with a lower dose of dichlobenil (20 mg/kg). Two days later, nine mice (Group A) were transplanted by intravenous tail injection with human UCB-selected CD133$^+$ stem cells isolated from only a single donor (≤70,000 cells/mouse). Twelve mice (Group B) nontransplanted but given dichlobenil were used only as controls. Four of them were sacrificed 4 days after dichlobenil treatment.

Thirty-one days after transplantation animals were euthanized using gaseous carbon dioxide and the following organs were explanted: NE, olfactory bulb, liver, skin, and spleen. Each explant was divided into two halves. One was immediately immersed in ice, frozen in liquid nitrogen, and then kept at −24°C for molecular biology analysis; the other half was fixed in Glyo-fixx (Shandon Lipshaw, Pittsburgh, PA), subsequently paraffin-embedded, and then evaluated by histology and fluorescence in situ hybridisation (FISH) analysis.

The right side of the nasal region of each animal was immersion-fixed in Glyo-fixx for 24 h, decalcified in 0.25 M ethylenediaminetetraacetic acid (EDTA) (Fluka, Buchs SG, CH) in 0.1 M Na$^+$ phosphate buffer (pH 7.4) for 10 days, and embedded in Paraplast plus (Sherwood Medical, MO; melting point 55–57°C). Coronal serial sections of 5–7 µm were collected on silane-coated slides. The sections were used for histological, immunohistochemical, and lectin staining procedures.

Histochemistry and Immunological Staining

Tables 1 and 2 list the lectins and antibodies used and their dilutions and specificity (20). Lectins (Sigma, St. Louis, MO) and immunoperoxidase staining were performed according to the manufacturer's instructions for each lectin and antibody used. Antigen retrieval was performed as needed with citrate buffer at pH 6.0. Lectin and antigen detection was performed with 3.3 diaminobenzidine (DAB, Sigma). All sections were counterstained with hematoxylin. Lectin controls included competitive inhibition with the appropriate sugar (100–200 mM) for 1 h at room temperature (RT), buffered saline in place of the lectin, and the use of a known positive control tissue. The specificity of the immunostaining was verified by incubating sections without primary antibodies, replaced by 3% normal goat serum. All controls were negative.

Table 1 List of lectins used in this study. Carbohydrate binding specificities are from Van Damme et al. (20)

Lectin	Carbohydrate specificity	Dilution (µg/mL)	Inhibitory sugar (molarity)	Source
Glycine max agglutinin (SBA)	N-Acetyl-D-galacto-samine > α-D-galactose	10	GalNAc (200 mM)	Sigma Chemical
Bandeirea simplicifolia agglutinin I (BSA-I)	α-d-Gal > α-D-galNAc	10	Gal (200 mM)	Sigma Chemical
Bandeirea simplicifolia agglutinin isolectin B₄ (BSA-I-B₄)	α-Galactose	10	Gal (500 mM)	Sigma Chemical
Dolichos biflorus agglutinin (DBA)	α-N-acetyl-d-galactosamine	10	GalNAc (200 mM)	Sigma Chemical
Ricinus communis agglutinin I (RCA-I)	β-Galactose	10	Gal (200 mM)	Sigma Chemical
Ulex europaeus agglutinin I (UEA-I)	α-l-Fucose	10	L-Fucose (100 mM)	Sigma Chemical
Lycopersicum esculentum agglutinin (LEA)	N-Acetyl-D-glucosamine	10	Chitin hydroxylase (0.6 mg/mL)	Vector Laboratories

Table 2 List of primary antibodies used in this study

Primary antibodies	Clones	Dilution	Source
Anti PCNA	PC10	1:500	Sigma Chemical
Anti β-tubulin isotype III	SDL.3D10	1:500	Sigma Chemical
Anti GAP-43	GAP-7B10	1:500	Sigma Chemical
Anti PGP 9.5	Polyclonal	1:300	DAKO Cytomation

Molecular Biology Analysis

Polymerase Chain Reaction

Total DNA (0.5–1 mg) was extracted from each organ with QIAamp DNA mini kit (Qiagen GmbH, Hilden, DE) and analyzed by polymerase chain reaction (PCR) for tissue traceability in order to confirm that the DNA engraftment was due exclusively to the donor human UBC CD133⁺ cells DNA, excluding potential artefacts due to sample or operational contaminations. Samples of all Group A and Group B mice were analyzed for the presence of three human-specific microsatellites (short tandem repeats, STRs) (D8S1179, D18S51, D21S11) (Table 3) from the Combined DNA Index System (CODIS) commonly used for paternity testing and other forensic applications *(21)*. STR amplification was performed as a nested PCR, using the primers indicated in Table 3 for the first step and the second step of amplification, and the same conditions for each couple of primers: 94°C denaturation (10 min); 94°C (1 min), 57°C annealing (1 min) for, 72°C extension (1 min), all repeated for 35 cycles. A final extension at 60°C for 30 min was performed. Each PCR was performed in duplicate. All forward primers were labelled with 6-FAM fluorochrome.

Two of 30 µL from each PCR were subsequently run on 3100 ABI Prism Genetic Analyzer (Applied Biosystems, Foster City, CA) and positive peaks of the expected length were subsequently purified (Wizard SV gel and PCR cleanup system, Promega, Madison, WI) and verified by sequencing in order to confirm the specificity of the human DNA sequences.

Multiple-Fluorescence In Situ Hybridisation Analysis

Glyo-fixx fixed, paraffin-embedded tissues tested by PCR and proved to contain human DNA, were subjected to multiple fluorescence in situ hybridisation (M-FISH) analysis with combinations of seven probes, each labelled with a different fluorochrome: one pan-centromeric, one pan-telomeric, three cosmidic (locus specific identifier DNA probes), and two telomeric mapping on different chromosomes were employed in this analysis: LSI-p16 (Urovision®, Vysis, ABBOTT Laboratories, Abbott Park, IL); Tel 12 (12p) and Tel 18 (18p) (Vysis ABBOTT Laboratories); LSI-LPL, LSI-c-myc (Provision®, Vysis, ABBOTT Laboratories). Serial tissue

Table 3 Characteristics of short tandem repeats (STR) loci investigated and list of the primers. Primers for a nested polymerase chain reaction (PCR) are indicated (first step, internal primers; second step, external primers)

STR locus	Chromosomal location	GeneBank accession number	Primer sequence (all primers A labelled with 6-FAM in 5′)
D18S51	18q21.3	L18333	First step: A-CAAACCCGACTACCAGCAAC
			B-GAGCCATGTTCATGCCACTG
			Second step: A-GCCATCGCACTTCACTCTGA
			B-AAGGTGGACATGTTGGCTTC
D21S11	21q11-21	M84567	First step: A-ATATGTGAGTCAATTCCCCAAG
			B-TGTATTAGTCAATGTTCTCCAG
			Second step: A-CCCCAAGTGAATTGCCTTCT
			B-AGTCAATGTTCTCCAGAGACAGAC
D8S1179	8q	AF250877	First step: A-TTTTTGTATTTCATGTGTACATTCG
			B-CGTATCCCATTGCGTGAATATG

sections of 4–5 μm thickness were mounted on silane-coated slides and backed overnight at 45–50°C. The slides were deparaffinized in xylene at RT, washed in 100% ethanol at RT, air dried, and kept in pretreatment solution (Vysis ABBOTT Laboratories) for 30 min. Slides were subjected to protease digestion for 20–30 min depending on the type of tissue. The slides were dehydrated in ethanol and then air dried. Denatured probes were added to each sample and covered with a cover-slip; slides were then placed in a dark humidity box (Hybrite®, Vysis ABBOTT Laboratories) for 16–18 h at 37°C.

Slides were immersed in 0.5× SSC buffer for 5 min at 75°C and counterstained with DAPI. Morphological analysis and engrafted cell count of M-FISH positive slides were performed using a fluorescence microscope equipped with recommended filters (Olympus BX 51).

Results

Four days after the administration of the olfactory toxicant dichlobenil, the dorsomedial part of the olfactory region presented thin, disorganized epithelium in four Group B mice investigated (Fig. 2a) and the dorsomedial region of the olfactory epithelium of Group B mice that were not transplanted with blood-selected CD133+ stem cells (HSC) was severely damaged. After 31 days in control mice the dorsomedial region was covered by an atypical respiratory-like epithelium with no sign of neuronal regeneration. The basal lamina of the original NE had often disappeared and the underlying mucosa was almost completely devoid of Bowman's glands. A representative example is shown in Fig. 2b. By contrast, the olfactory NE of Group A mice treated with dichlobenil and subsequently transplanted with human HSC showed significant neuronal recovery. Neuronal cells at different differentiation stages were identified by immunological staining for a pattern of specific antigen markers, i.e., GAP-43, a phosphoprotein known to be expressed during axonal growth, PGP 9.5, against protein gene product 9.5, and β-III-tubulin, involved in neuronal differentiation at an extremely early stage of commitment to the olfactory neuron lineage, prior to the extension of neurites. The epithelium covering the dorsomedial olfactory region was characterized by clusters of differentiated neuronal cells, approximately three to four neurons in thickness above the basal cell layer, arranged in a pseudostratified manner (Fig. 2c). The lectin staining performed on subsequent sections showed that the majority of these cells were strongly stained after SBA, BSA-I, BSA-I-B$_4$, DBA, and UEA-I binding. These lectin-labelled cells, as depicted in two representative pictures (Fig. 2d,e), were similar to the olfactory receptor cells present in the undamaged olfactory region or in untreated, untransplanted controls (Fig. 2f). They included few undifferentiated globose basal neuronal cells located close to the basal lamina. However, there were mainly bipolar elongated cells, with their vertically elongated body located in the midregion of the epithelium, with an extension terminating in a knob reaching the epithelial surface. Their lectin staining indicated differences in the pattern and distribution of the

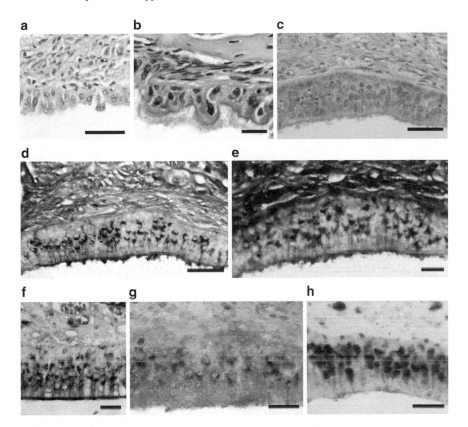

Fig. 2 Regeneration of olfactory epithelium of the dorsomedial olfactory region damaged by the toxicant dichlobenil in *nod-scid* mice transplanted or not with human umbilical cord blood (UCB)-selected CD133$^+$ stem cells. (**a–c**) Histological section; (**d–f**) lectin histochemical staining; (**g–h**) immunostaining. (**a**) At 4 days after the first administration of dichlobenil in a control representative Group B mouse not transplanted, the olfactory epithelium is seriously damaged and appears looser due to the degeneration of the olfactory receptors. (**b**) At 1 month after the treatment with dichlobenil in not transplanted mice the olfactory region is covered by a ciliated respiratory-like epithelium and the underlying mucosa is almost completely devoid of Bowman's glands. (**c**) On the contrary in a representative Group A mouse at 1 month after the treatment with dichlobenil and transplanted with human CD133$^+$ stem cells, the olfactory epithelium has a quite normal appearance. (**d**) Olfactory receptors labelled by BSA-I-B$_4$ agglutinin. (**e**) Olfactory receptors labelled by SBA agglutinin. (**f**) Olfactory neurons in the undamaged olfactory region after BSA-I-B$_4$ binding in a representative Group A mouse. (**g**) GAP43 and (**h**) PGP9.5 expression in the repaired dorsomedial olfactory epithelium. (Scale bar = 40 μm for (**a, c**); 20 μm for (**b, d, e, f, g, h**))

saccharidic moieties among neuronal cells at different stages of differentiation. Moreover these cells were positive after GAP-43 (Fig. 2g) and PGP 9.5 (Fig. 2h) immunostaining, confirming their different stage of neuronal differentiation. One month after stem cells transplantation no dividing cells were detectable in the NE, by anti–proliferating cell nuclear antigen (PCNA) immunostaining (not shown).

Table 4 Cumulative results from polymerase chain reaction (PCR) amplification of three short tandem repeats and multiple fluorescence in situ hybridisation analysis of seven different human chromosomal DNA sequences in tissues of Group A mice

Days	UCB	Mice	Li	Sp	Sk	ON/NE	OB
31	467	1	−	+	++	++	++
		2	−	+	++	++	++
		3	−	+	−	+	+
	358	4	++	+	−	++	++
		5	−	++	−	++	+
		6	+	+	+	++	++
		7	−	−	+	+	+
		8	−	++	+	++	+
	469	9	−	++	+	++	+

Abbreviations: UCB number of cord blood (individual patient's code); *Li* liver; *Sp* spleen; *Sk* skin; *OM/NE* olfactory mucosa and neuroepithelium; *OB* olfactory bulb. Score: − negative in both analyses; + positive by polymerase chain reaction (PCR) for at least two out of three short tandem repeats (STRs) and positive by multiple fluorescence in situ hybridisation (M-FISH) for at least two of seven probes tested; ++ positive by PCR for all three STRs and positive by M-FISH for at least four of seven probes tested

PCR analysis for the presence of three specific human STRs, used as markers for tracing the fate and the engraftment of human cells in Group A transplanted mice, revealed that in vivo chimeric cells had divided and migrated to different organs (spleen, liver, skin, olfactory bulb) (Table 4), including the injured olfactory mucosa and its NE (Fig. 3a). Positivity for the three human STRs varied at random in each mouse among the different tissues analysed, with significant differences among transplanted mice. Control injured but untransplanted Group B mice always failed to reveal their presence (not shown).

M-FISH revealed the presence of rare chimeric human cells (≤2% positive cells counted in five subsequent sections) in the olfactory mucosa of the transplanted mice, around small capillaries and gland structures, or dispersed among mesenchymal or stromal cells; positive cells were also found, but less frequently, above the basal lamina surrounding the bipolar neurons in the upper NE. M-FISH did not always reveal the presence in the olfactory mucosa of all the seven different human STR examined (Table 4), possibly due to different angle of cutting or different proportions of positive cell engraftment in the various tissues as well as a different assay sensitivity for the detection of the seven fluorescent probes.

Discussion

In this study we demonstrated that human UCB-selected CD133⁺ stem cells transplantation into Group A *nod-scid* mice injected with dichlobenil stimulated the repair and regeneration of the injured olfactory NE. In contrast, there was no

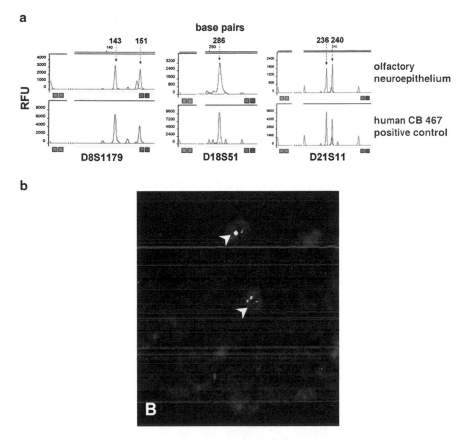

Fig. 3 Detection of chimerism in Group A mice analysed 31 days following transplantation. (**a**) polymerase chain reaction (PCR): electropherograms of three human short tandem repeats (STRs) (D8S1179, D18S51, D21S51) in the olfactory bulb of a representative mouse (mouse no. 4) and in donor human umbilical cord blood (UCB)-selected CD133⁺ cells as positive control. *RFU* relative fluorescent unit. Positive control: human UCB 467. *Arrows* indicate the length of the donor human alleles for the heterozygous D8S1179 and D21S51 STRs, and of one allele for the homozygous D18S51 STR. The peaks are identical to those obtained from donor CD133⁺ stem cells. (**b**) Multiple fluorescence in situ hybridisation (M-FISH) analysis with seven probes mapping on different human chromosomes, each labelled with a different fluorochrome (see the section "Materials and Methods"). NE section (mouse no. 4) reveals two positive cells (*arrows*)

spontaneous regeneration in similar injured areas in control Group B mice that were treated with dichlobenil but not transplanted with HSC.

By histology in the dorsomedial region at 1 month from transplantation, we observed clusters of differentiated neuronal cells arranged in a pseudostratified epithelium in all Group A mice. These cells were intensively labelled by all the lectins used except LEA and RCA-I, and also labelled by anti-β-III-tubulin, anti-GAP-43, and anti-PGP 9.5 antibodies. The expression of GAP-43 in neurons correlates with axonal elongation and synaptogenesis during embryonic development and regeneration in the adult central nervous system *(39)*. Moreover PGP 9.5 immunoreactivity is demonstrated to be a useful marker for investigation of the olfactory and vomeronasal receptor neurons not only in the early developmental stage but also in the adult *(22–24)*. The patterns of lectin binding revealed by these cells suggests that these neurons were characterized by a high density of α-D-galactose, α-*N*-acetyl-D-galactosamine, and α-fucose residues. Sugar residues on cell surface are known to play a key role in cellular function, differentiation, degeneration, and regeneration of the olfactory neurons mediating axon–axon or axon–matrix interactions *(25–31)*.

The presence of these regenerating olfactory neurons in the dorsomedial region raises the question of which precursor cell(s) they originate from, since dichlobenil treatment causes permanent lesions in this olfactory area. Using PCR analysis for the presence of three human STR we excluded potential artefacts due to sample or operational contaminations. Additionally STRs were used as reliable markers for tracing host cells in different tissues of the mice transplanted with HSC.

We detected the presence of these DNA markers in different organs after engraftment, with random distribution among the animals. Their presence was more frequent and at higher levels in certain tissues of the injured mice, mainly in the olfactory NE, olfactory bulb, but less in liver, skin, and spleen. In vitro studies have indicated that UBC-selected CD133⁺ cells include a subset of stem or precursor cells that may exhibit a pluripotent phenotype: when cultured in appropriate medium they are able to differentiate to endothelial or neural cells *(32, 33)*. It is possible that in vivo CD133⁺ circulating HSC, once migrated in a suitable microenvironment, generated progenies that exhibited different features, for example of the mesenchymal or the neuronal or the endothelial lineage. However, the very low level of human chimeric cells detected by FISH in mice 1 month following HSC transplantation, suggests the alternative possibility that engrafting of the transplanted CD133⁺ cells could be enhanced by cell–cell contacts and the release of growth factors from mesenchymal or stromal cells and macrophages that are abundant and activated in a damaged olfactory tissue. Studies on cell death and neurogenesis in the olfactory epithelium indicated that macrophages play a key role in mesenchymal stem cell self-renewal and differentiation *(34)*. Within 3–5 days following bilateral olfactory bulb ablation, macrophages resident in the olfactory NE could be stimulated to enhance the synthesis and release of chemokines, recruiting additional macrophages that in turn stimulated self-renewal, growth, and differentiation of residual mesenchymal-like multipotent progenitor cells present in the olfactory mucosa. Besides being able to phagocytize dead or damaged host cells, the recruited macrophages may be stimulated to enhance the secretion of numerous bioactive molecules, including several chemokines, cytokines, such as leukaemia inhibitory factor (LIF), and other growth or differentiation factors active for different tissue lineages and for stimulating locally residual mesenchymal-like cells *(34–36)*. LIF induces proliferation of the globose basal cells, the olfactory receptor cell progenitors of the NE *(34, 37)*. Moreover, evidence indicates that, while differentiating in three-dimensional culture systems, embryonic stem cells can produce endogenous extracellular matrix proteins, cell–cell adhesion molecules, cell–surface receptors, and lectins and their glycoligands, forming a microenvironment, a *niche*, able to positively influence stem cell behaviour and their surrounding environment *(38)*. In vivo, creating such an appropriate three-dimensional microenvironment, modulated by exogenous soluble factors, HSC may constitute a template for adequate tissue repair and regeneration.

In conclusion, these findings strongly favour the hypothesis that in our mouse model, transplanted human HSC may contribute to piloting the repair of an injured NE, emerging as a potential strategy for therapy.

Acknowledgments The authors are grateful to Dr. Isabella Andreini for her valuable suggestions and generous help. Supporting grants to RPR by: C.N.R.: RSTL 2007; Italian Ministry of Health (Project "Stem 2001"; Istituto Zooprofilattico Sperimentale Lazio e Toscana, I.F. 2005–2007); Italian Ministry of University and Research (M.I.U.R.) (FIRB: "New Medical Engineering" and "Technologies in Oncology"); Joint Project "Kontakt" between the Ministries of Foreign Affairs of Italy and Czech Republic; Foundation "Stem Cells & Life" Pisa, Italy. Supporting Grants to VF by M.I.U.R.

References

1. Farbman AI. Cell biology of olfaction. Cambridge University Press, Cambridge, 1992.
2. Ding XX, Coon MJ. Purification and characterization of two unique forms of cytochrome P 450 from rabbit nasal microsomes. Biochemistry 1988;27:8330–7.
3. Chen Y, Getchell ML, Ding X, et al. Immunolocalization of two cytochrome P450 isozymes in rat nasal chemosensory tissue. Neuroreport 1992;3:749–52.
4. Suzuki Y, Schafer J, Farbman AI. Phagocytic cells in the rat olfactory epithelium after bulbectomy. Exp Neurol 1995;136:225–33.
5. Suzuki Y, Takeda M, Farbman AI. Supporting cells as phagocytes in the olfactory epithelium after bulbectomy. J Comp Neurol 1996;376:509–17.
6. Huard JM, Schwob JE. Cell cycle of globose basal cells in rat olfactory epithelium. Dev Dyn 1995;203:17–26.
7. Calderon-Garciduenas L, Rodriguez-Alcaraz A, Villarreal-Calderon A, et al. Nasal epithelium as a sentinel for airborne environmental pollution. Toxicol Sci 1998;46:352–64.
8. Graziadei, PPC, Monti Graziadei GA. Continuous nerve cell renewal in the olfactory system. In: Jacobson M, editor. Handbook of sensory physiology. Vol. 9 Development of sensory systems. Springer, New York, 1978:55–83.
9. Farbman AI. Olfactory neurogenesis: genetic or environmental controls? Trends Neurosci 1990;13:362–5.
10. Beyers DW, Farmer MS. Effects of copper on olfaction of Colorado pikeminnow. Environ Toxicol Chem 2001;20:907–12.
11. Baldwin DH, Sandhal JF, Labenia JS, et al. Sublethal effect of copper on coho salmon: impacts on nonoverlapping receptor pathways in the peripheral olfactory nervous system. Environ Toxicol Chem 2003;22:2266–74.
12. Schwob JE, Youngentob SL, Mezza RC. Reconstitution of the rat olfactory epithelium after methyl bromide-induced lesion. J Comp Neurol 1995;359:15–37.
13. Bergman U, Brittebo EB. Methimazole toxicity in rodents: covalent binding in the olfactory mucosa and detection of glial fibrillary acidic protein in the olfactory bulb. Toxicol Appl Pharmacol 1999;155:190–200.
14. Brandt I, Brittebo EB, Feil VJ, et al. Irreversible binding and toxicity of the herbicide dichlorobenil (2,6-dichlorobenzonitrile) in the olfactory mucosa of mice. Toxicol Appl Pharmacol 1990;103:491–501.

15. Bergman U, Ostergren A, Gustafson A-L, et al. Differential effects of olfactory toxicants on olfactory regeneration. Arch Toxicol 2002;76:104–12.

16. Buckland ME, Cunningham AM. Alterations in expression of the neurotrophic factors glial cell line-derived neurotrophic factor, ciliary neurotrophic factor and brain-derived neurotrophicfactor, in the target-deprived olfactory neuroepithelium. Neuroscience 1999;90:333–47.

17. Pittenger MF, Mackay AM, Beck SC, et al. Multilineage potential of adult human mesenchymal stem cells. Science 1999;284:143–7.

18. Strem BM, Hicok KC, Zhu M, et al. Multipotential differentiation of adipose tissue-derived stem cells. Keio J Med 2005;54:132–41.

19. Mizuno H, Zuk PA, Zhu M, et al. Myogenic differentiation by human processed lipoaspirate cells. Plast Reconstr Surg 2002;109:199–209.

20. Van Damme EJM, Peumans WJ, Pusztai A, et al. Handbook of plant lectins: properties and biomedical applications. Wiley, Chichester, England, 1998.

21. Ricci U, Sani I, Guarducci S, et al. Infrared fluorescent automated detection of thirteen short tandem repeat polymorphisms and one gender-determining system of the CODIS core system. Electrophoresis 2000;213:564–70.

22. Taniguchi K, Saito H, Okamura M, et al. Immunohistochemical demonstration of protein gene product 9.5 (PGP 9.5) in the primary olfactory system of the rat. Neurosci Lett 1993;156:24–6.

23. Johnson EW, Eller PM, Jafek BW. Protein gene product 9.5 in the developing and mature rat vomeronasal organ. Dev Brain Res 1994;78:259–64.

24. Schofield JN, Day INM, Thompson RJ, et al. PGP 9.5, a ubiquitin C-terminal hydrolase; pattern of mRNA and protein expression during neural development in the mouse. Dev Brain Res 1995;85:229–38.

25. Key B, Akeson RA. Olfactory neurons express a unique glycosylated form of the neural cell adhesion molecule NCAM. J Cell Biol 1990;110:1729–43.

26. Key B, Akeson RA. Immunochemical markers for the frog olfactory neuroepithelium. Dev Brain Res 1990;57:103–17.

27. Breer H. Molecular reaction cascade in olfactory signal transduction. J Steroid Biochem Mol Biol 1991;39:621–5.

28. Franceschini V, Lazzari M, Revoltella RP, et al. Histochemical study by lectin binding of surface glycoconjugates in the developing olfactory system of rat. Int J Dev Neurosci 1994;12:197–206.

29. Lipscomb BW, Treolar HB, Klehoff J, et al. Cell surface carbohydrates and glomerular targeting of olfactory sensory neuron axons in the mouse. J Comp Neurol 2003;467:22–31.

30. Henion TR, Raitcheva D, Grosholz R, et al. β1,3-N-Acetylglucosaminyltransferase 1 glycosilation is required for axon pathfinding by olfactory sensory neurons. J Neurosci 2005;25:1894–903.

31. St. John J, Key B. A model for axonal navigation based on glycocodes in the primary olfactory system. Chem Senses 2005;30 Suppl 1:i123–4.

32. Hao HN, Zhao J, Thomas RL, et al. Fetal human hemopoietic stem cells can differentiate sequentially into neural stem cells and then astrocytes in vitro. J Hematother Stem Cell Res 2003;12:23–32.

33. Baal N, Reisinger K, Jahr H, et al. Expression of transcription factor Oct-4 and other embryonic genes in CD133 positive cells from human umbilical cord blood. Thromb Haemost 2004;92:767–75.

34. Nan B, Getchell ML, Partin JV, et al. Leukemia inhibitory factor, interleukin-6, and their receptors are expressed transiently in the olfactory mucosa after target ablation. J Comp Neurol 2001;435:60–77.

35. Getchell TV, Shah DS, Partin JV, et al. Leukemia inhibitory factor mRNA expression is upregulated in macrophages and olfactory receptor neurons after target ablation. J Neurosci Res 2002;67:246–54.

36. Getchell TV, Subhedar NK, Shah DS, et al. Chemokine regulation of macrophage recruitment into the olfactory epithelium following target ablation: involvement of macrophage inflammatory protein-1 and monocyte chemoattractant protein-1. J Neurosci Res 2002;70:784–93.

37. Bauer S, Rasika S, Han J, et al. Leukemia inhibitory factor is a key signal for injury-induced neurogenesis in the adult mouse olfactory epithelium. J Neurosci 2003;23:1792–803.
38. Michelini M, Franceschini V, Sihui Chen S, et al. Primate embryonic stem cells create their own *niche* while differentiating in three-dimensional culture systems. Cell Prolif 2006;39:217–29.
39. Awenagha O, Campbell G, Bird MM. Distribution of GAP-43, β-III tubulin and F-actin in developing and regenerating axons and their growth cones in vitro, following neurotrophin treatment. J Neurocyt 2003;32:1077–89.

Strategies Toward Beta-Cell Replacement

Enrique Roche, Nestor Vicente-Salar, Maribel Arribas,
and Beatriz Paredes

Abstract Embryonic and adult stem cells are considered to be potential sources of insulin-secreting cells to be transplanted into type 1 and advanced stages of type 2 diabetic patients. After years of study, the key determinants necessary for the differentiation process are beginning to be fully characterised, and several protocols have been published. However, investigators still have to face several problems before finding a therapeutic application, such as to increase the amount of insulin produced by the final cell product, to examine the processing of the hormone in these protocols, and to confirm the existence of a stimulus-coupled secretory process. Concerning transplantation, we must also pay attention to implant survival, tumour formation, and immune rejection. Mimicking endocrine pancreas development in vitro seems to yield the best results in terms of obtaining insulin-secreting cells from embryonic stem cells. To this end, definitive endoderm precursors have been generated. The final cell product contained amounts of insulin and phenotypic traits similar to mature β cells. However, these cells seemed to be immature, since they did not respond to stimulatory concentrations of extracellular glucose and coexpressed two hormones in the same cell (insulin glucagon, insulin somatostatin). Therefore, further improvements are required. Concerning adult stem cells, the possibility of identifying pancreatic precursors or of reprogramming extrapancreatic-derived cells are key possibilities that may circumvent some of the problems that appear when using embryonic stem cells. However, current protocols are not capable of obtaining a functional cell product useful for therapeutic purposes. The identification of signals that operate in vivo in the different niches could help in the design of more adequate strategies.

Keywords Diabetes • Stem cells • Cell therapy

E. Roche (✉), N. Vicente-Salar, M. Arribas, and B. Paredes
Instituto de Bioingenieria, Universidad Miguel Hernandez, Alicante, Spain
e-mail: eroche@usmh.es

H. Baharvand (ed.), *Trends in Stem Cell Biology and Technology*,
DOI 10.1007/ 978-1-60327-905-5_17,
© Humana Press, a Part of Springer Science + Business Media, LLC 2009

299

Introduction

Diabetes is a degenerative pathology that has different causes but displays the same symptom: high levels of circulating nutrients (hyperglycolipemia) *(1)*. Although there are different forms, diabetes is typically classified as type 1 and type 2. Type 1 is caused by an autoimmune attack that results in a rapid β cell (the cell that produces insulin in the organism) destruction at early stages in the life of affected individuals *(2)*. Type 2 diabetes has a multifactorial origin in which different determinants influence at different levels, such as genetic predisposition, diet, and lifestyle. Type 2 diabetes develops in two phases: insulin resistance and insulin deficiency. In the insulin resistance phase, the target tissues do not detect the hormone, which results in the β cell increasing the secretion *(1)*. The second phase results from an exhaustion of the pancreatic insulin-secreting cells that culminates in the destruction of this cell type through apoptotic mechanisms. The excess of circulating calorinergic nutrients is instrumental in activating the suicidal mechanisms (glucolipotoxicity hypothesis) *(3)*. This type of diabetes occurs in older individuals and is strongly associated with the development of obesity and metabolic syndrome *(1)*.

Whereas type 1 diabetes is a fulminate disease with a very complicated prognosis, type 2 is preventable, where changes in diet and lifestyle can delay and even prevent the development of the disease. The lack of specific markers or functional proof complicates the early detection of the disease, usually being diagnosed once the patient is completely affected by the disorder. In this context, exogenous insulin injection is the only effective method to normalize hyperglycaemia. However, this is not the only requirement to fulfill for affected patients. Balanced diets, proper exercise, and even precise pharmacological intervention can complete the hormone injections. Altogether, this requires motivated patients in order to maintain such a regimented lifestyle. However, hormone injection cannot mimic the secretory function performed by mature β cells, usually resulting in the development of secondary complications that affect eyes (retinopathy), kidneys (nephropathy), nerves (neuropathy), and the cardiovascular system *(1)*.

Although pharmacological agents and insulin formulations have reached a high degree of specificity, the cure for diabetes is still far in the future. The logic therapeutic intervention may consist in the replacement of the damaged tissue, i.e., pancreatic β-cell replacement using transplantation technologies. This is achieved through the complete dual transplantation of pancreas and kidney or by implanting the insulin-producing structures, called islets of Langerhans, which represent only 1% of the total mass of the pancreas. Over recent years, key advances have been made in islet transplantation *(4)*, although this strategy still has many problems to face, such as graft immune compatibility, side effects of immunosuppressors in β-cell viability, adjustment of the correct number of islets, implant survival, and scarcity of the biomaterial *(5–10)*. The latter case indicates that alternative sources of β cells need to be further investigated to offer new therapeutic alternatives.

In this sense, it is theoretically possible to obtain customized β cells from embryonic stem cells (ESCs) or adult stem cells (ASCs). Both cell types bear two key properties: self-renewal and differentiation to diverse cell fates *(11)*. In addition, drugs can be developed to stimulate the replication of existing β cells and recover the lost islet cell mass. However, pancreas regeneration in a diabetic environment has to first solve the problem associated with the immune or apoptotic destruction of pancreatic β cells that surpasses the reparative capacity of pancreatic stem cells, resulting in a deficient number of functional cells.

Embryonic stem cells are obtained from the inner cell mass of blastocysts. Human embryonic stem cells are maintained in culture on feeder layers of inactivated embryonic fibroblasts *(11, 12)*, which supposedly secrete specific factors that maintain these cells in an undifferentiated state *(13)*. Ectoderm-committed cells tend to accumulate over passages constraining the original plasticity of primary stem cells *(14)*. These ectodermal precursor cells tend to present chromosomal aberrations and very often are oblivious to the agents used in the differentiation protocols, maintaining its tumorigenic potential that culminates in teratoma formation when transplanted in animal models *(15)*. Furthermore, ESCs present immunocompatibility problems that severely constrain their clinical applications *(5)*. Nevertheless, ESCs are a good testing material in order to check the efficiency of extracellular factors and culture conditions that could be instrumental in specific protocols with ASCs.

The spontaneous expression of specific transcription factors involved in pancreas development during ESC cultures has encouraged some groups to hypothesize that some steps of islet development could be reproduced in vitro *(16, 17)* These spontaneous differentiation programmes were triggered in ESCs when they were transferred from adherent monolayers to bacteriological Petri dishes, forming particular cell aggregates called embryoid bodies (EBs). The activation of differentiation programmes is confirmed by the expression of genetic markers from primitive and definitive endoderm, ectoderm, and mesoderm *(18)*. It is believed that nutrients, oxygen, and growth factor gradients established inside EB structures are determinant in activating cell differentiation processes. Cell-to-cell interactions, as well as paracrine and biophysical interactions, also have to be taken into account.

ASCs do not pose problems related to immune rejection (provided that donor and recipient is the same person) and apparently under normal circumstances display a committed differentiation to specific cell types. This usually occurs when ASCs are maintained in particular tissue niches, where the cells receive the proper signals and establish adequate cell-to-cell interactions to cover tissue turnover and repopulation *(19, 20)*. However, ASCs from nonpancreatic tissues, such as liver, intestine, and bone marrow, have shown the potential to transdifferentiate into insulin-secreting cells *(21–25)*. However, the molecular mechanisms underlying such processes remain largely unknown. The precise identification of ASCs in the different tissues by cell surface markers and the detection of specific niches will be new scientific challenges to confront in the future.

Although there is still much work to do, several advances have been achieved in this field, providing key data to design a definitive protocol in the future. This chapter will summarize the key work that has been performed in the bioengineering of both ESCs and ASCs toward insulin-secreting cells.

Insulin-Producing Cells from Embryonic Stem Cells

The possibility that mouse ESCs can give rise to insulin-producing cells was first reported in our laboratory and subsequently confirmed in human cells *(26, 27)*. The presence of insulin-positive cells in those protocols was based in the spontaneous expression of the insulin gene in EB structures or in the outgrowth phase of the culture. The implant of the final cell product in the spleen of streptozotocin diabetic mice resulted in euglycaemia recovery *(26)*. However, the variable amounts of insulin produced by the different clones constrained the reproducibility of the protocol, suggesting the necessity of designing coaxial strategies to obtain insulin-producing cells from ESCs.

In this sense, Lumelsky et al. *(28)* proposed a new protocol based in the idea that nestin-positive neurons and pancreatic β cells share many features, allowing the possibility of obtaining insulin-positive cells from ectoderm-derived cells. Indeed, certain neuroectoderm cells are capable of expressing insulin as well as β-cell specific transcription factors, proteins of the glucose-sensing machinery, and components of the secretory pathway. However, the functional capacity of neurons and β cells differs markedly. Insulin-positive neuroectodermal cells express the insulin II gene, which does not respond to the same regulators as the insulin I gene, the gene that is expressed in pancreatic β cells. In addition, the proinsulin produced by neuroectoderm-derived cells

Table 1 Different protocols used to differentiate embryonic stem cells toward definitive endoderm precursors, liver, and insulin-producing cells

Author (ref)	Cell type	Culture conditions
Lumelsky et al. *(28)*	Mouse (R1)	*Undifferentiated:* DMEM, 1,400 U/mL LIF, 100 mM NEAAs, 0.55 mM βME, L-glutamine and 15% FCS *Differentiation:* (1) EB without LIF (4 days) and 15% FCS (2) Outgrowth with ITSF. (insulin/trasferrin/selenium/fibronectin) medium (6–7 days) serum free selection nestin-positive cells (3) Outgrowth with N2 medium containing ? B27 and 10 ng/mL bFGF, serum free expansion (4) Outgrowth with N2 medium containing B27 and 10 mM nicotinamide without bFGF, serum free formation of insulin-secreting clusters
Jones et al. *(29)*	Mouse (HR with 1.114 Gtar-βgal)	*Undifferentiated:* BHK21, 100 U/mL LIF and 10% FCS *Differentiation:* (1) EB (2 × 10⁴ cells/mL). BHK21 and 10% FCS (5 days) (2) Outgrowth in gelatin (1 EB/well Mw24) (4–10 days) (step 1 same media)
Shiroi et al. *(30)*	Mouse (EB3; HR with Oct ¾ – Blasticidin S)	*Undifferentiated:* DMEM (4.5 mg/L d-glucose), 0.1 mM βME, 0.1 mM NEAAs, 1 mM sodium pyruvate, 1,000 U/mL LIF and 10% FBS *Differentiation:* (1) HD (500 cells/20 µL) without LIF (5 days) (2) Outgrowth in gelatin (20 EBs/dish) 10% FBS (23 days)

(continued)

Table 1 (continued)

Author (ref)	Cell type	Culture conditions
Kubo et al. (31)	Mouse (E14.1; HR with Bry-GFP)	*Undifferentiated:* Feeder layer. DMEM, 1% LIF, 1.5 × 10⁻⁴ M MTG and 15% FCS *Differentiation:* (A): (1.1) – EB (10³–8 × 10⁴ cells/mL). IMDM, 2 mM Glutamine, 0.5 mM ascorbic acid, 4.5 × 10⁻⁴ M MTG, 5% PFHM-II, 200 µg/mL transferrin and 15% FCS (2.5 days) (1.2) – EB. IMDM, 2 mM glutamine, 0.5 mM ascorbic acid, 4.5 × 10⁻⁴ M MTG and 15% KSR (4 days) (B): (1.1) – EB (10³–8 × 10⁴ cells/mL). IMDM, 2 mM glutamine, 0.5 mM ascorbic acid, 4.5 × 10⁻⁴ M MTG, 5% PFHM-II, 200 µg/mL transferrin and 15% FCS (2.5 days) (1.2) – EB. StemPro 34 medium, 2 mM glutamine, 0.5 mM ascorbic acid, 4.5 × 10⁻⁴ M MTG, 1% Kit ligand and 15% KSR (2 days) (1.3) – EB. IMDM, 2 mM glutamine, 0.5 mM ascorbic acid, 4.5 × 10⁻⁴ M MTG, 100 ng/mL activin A and 15% KSR
Ku et al. (32)	Mouse (R1, E14.1 and CCE)	*Undifferentiated:* Feeder layer. DMEM, ? LIF, 3 × 10⁻⁴ M MTG and 15% FCS *Differentiation:* (1) EB (5–2 × 10³ cells/mL). IMDM, 50 µg/mL ascorbic acid, 6 × 10⁻³ M MTG and 15% FCS (2 days) (2) EB. IMDM, 50 µg/mL ascorbic acid, 6 × 10⁻⁴ M MTG and 15% FCS or KSR (4 days) (3.1) EB or outgrowth in gelatin. DMEM/F12 (1:1) with or without 10 ng/mL FGF2 and 15% KSR (5–11 days) (3.2) EB coming from step 2 with 15% FCS. DMEM/F12 (1:1), 10 mM nicotinamide, 0.1 nM exendin-4, 10 ng/mL activin βB and 15% KSR (13 days)
Choi et al. (103)	Mouse (129/SvJ)	*Undifferentiated:* DMEM, 0.1 mM βME, 1,000 U/mL LIF, 1× NEAAs and 15% FBS *Differentiation:* (A) EB. DMEM, 0.1 mM βME, 1× NEAAs and 15% FBS (54 days) (B): (1) – EB. DMEM, 0.1 mM βME, 1× NEAAs and 15% FBS (14 days) (2) – EB disaggregated and culture in gelatin. 100 ng/mL aFGF. 15% FBS (2 days) (3) – 20 ng/mL HGF. 15% FBS (3 days) (4) – Replated on Matrigel matrix. 20 ng/mL HGF, 10 ng/mL oncostatin M, 10⁻⁷ M dexamethasone, ITS (5 mg/mL insulin, 5 mg/mL transferrin and 5 µg/mL selenious acid). 15% FBS (3 days)
D'Amour et al. (33)	Human (H7 and H9)	*Undifferentiated:* Feeder layer. DMEM/F12, 1 mM NEAAs, 0.55 mM βME, 4 ng/mL FGF2 and 20% KSR (sometimes 10 ng/mL activin A is added to maintain undifferentiated state) *Differentiation:* (A) Monolayer. RPMI. 100 ng/mL activin A and different concentrations of FBS (0.5–10%) (5 days) (B) Monolayer. RPMI. 100 ng/mL activin A or 100 ng/mL BMP4, 5 µM SU5402 and FBS (0% 1 day; 0.2% 1 day and 2% 2 days)

(continued)

Table 1 (continued)

Author (ref)	Cell type	Culture conditions
Ishii et al. *(34)*	Mouse (ES C57BL/6 transfected with AFP-GFP)	*Undifferentiated:* Feeder layer. DMEM, 0.1 mM βME, ? NEAAs, 1 mM sodium pyruvate, 1,000 U/mL LIF and 20% FBS *Differentiation:* (1) Monolayer (2×10^4 cells/cm^2) with collagen type I. DMEM, 2 mM L-glutamine, 1 mM sodium pyruvate and 10% KSR (1.A) 10 μM RA and 1,000 U/mL LIF (7 days) (1.B) 20 ng/mL bFGF and 20 ng/mL dHGF (7 days) (1.C) 1,000 U/mL LIF, 10 μM RA (first 2 days), 20 ng/mL bFGF and 20 ng/mL dHGF (next 5 days) (1.C.1) 20 ng/mL bFGF and 20 ng/mL dHGF (1 day) (1.C.2) 10 ng/mL oncostatin M (1 day) (2) AFP-GFP cell sorting (3) Monolayer. Coculture (2.5×10^4 cells/cm^2) AFP-GFP$^+$ cells with Thy1$^+$ mesenchymal cells. DMEM, 1 mM sodium pyruvate, 10 mM nicotinamide, 2 mM L-ascorbic acid phosphate, ? insulin–transferrin–selenium, 1×10^{-7} M dexamethasone, 20 ng/mL dHGF, 10 ng/mL oncostatin M and 10% FBS (7 days)
Milne et al. *(35)*	Mouse (CCE and D3)	*Undifferentiated:* DMEM (25 mM glucose), 2 mM glutamine, ? NEAAs, 0.1 mM βME, 1,000 U/mL LIF and 15% FBS *Differentiation:* Monolayer (high density?) (4–12 days) with or without LIF
Tada et al. *(36)*	Mouse (EB5 with HR Oct4-Blasticidin S and Goosecoid-GFP)	*Undifferentiated:* Gelatin. G-MEM, 0.1 mM NEAAs, 1 mM sodium pyruvate, 0.1 mM βME, 1,000 U/mL LIF, 20 μg/mL Blasticidin S, 1% FCS and 10% KSR *Differentiation:* (A) Monolayer. Type IV collagen-coated 10-cm dishes (1×10^5 cells/dish). O3 medium, 0.1% BSA, 50 μM βME and/or 10 ng/mL activin A, 10 ng/mL BMP4 and 1,000 ng/mL nodal. Serum free (4 days) (B) EB (3×10^4 cells/dishes 6 cm). O3 medium, 0.1% BSA, 50 μM βME and/or 10 ng/mL activin A, 10 ng/mL BMP4 and 1,000 ng/mL Nodal. Serum free (4 days) (C) Monolayer. Sorting Gsc$^+$ECD$^+$ and Gsc$^+$ECD^{low}. Type IV collagen-coated 10-cm dishes (1×10^5 cells/dish). O3 medium, 0.1% BSA, 50 μM βME and 10 ng/mL activin A. Serum free (4–6 days)
Yasunaga et al. *(37)*	Mouse (EB5 with HR Goosecoid-GFP; Sox17-hCD25 and Oct4-Blasticidin S)	*Undifferentiated:* Gelatin. G-MEM, 0.1 mM NEAAs, 1 mM sodium pyruvate, 0.1 mM βME, 1,000 U/mL LIF, 20 μg/mL Blasticidin S, 1% FCS and 10% KSR *Differentiation:* (A) Definitive endoderm. Monolayer. Type IV collagen-coated 10-cm dishes. O3 medium and 10 ng/mL activin A. Serum free (6 days) (B) Visceral endoderm. Monolayer (10^5 cells/mL). Gelatin or human fibronectin coated dishes. O3 medium. Serum free (6 days) (C) Monolayer. Sorting Gsc$^+$ECD$^+$Sox17$^+$ and Gsc$^-$ECD$^+$Sox17$^+$. Type I collagen-coated. O3 medium, 20 ng/mL EGF, 20 ng/mL BMP4^{12}, 20 ng/mL aFGF and 5 ng/mL bFGF. Serum free (4–6 days)

(continued)

Table 1 (continued)

Author (ref)	Cell type	Culture conditions
D'Amour et al. *(38)*	Human (CyT203)	*Undifferentiated:* Feeder layer. DMEM[1]/F12, 1 mM NEAAs, 0.55 mM βME, 4 ng/mL FGF2 and 20% KSR (sometimes 10 ng/mL activin A is added to maintain undifferentiated state) *Differentiation:* (1) Monolayer. RPMI, 100 ng/mL activin A, 25 ng/mL Wnt3a (first 2 days) and FBS (0% 2 days; 0.2% 2 days) (2) Monolayer. RPMI, 50 ng/mL hFGF10, 0.25 μM KAAD-cyclopamine and 2% FBS (4 days) (3) Monolayer. DMEM, 50 ng/mL hFGF10, 0.25 μM KAAD-cyclopamine, 2m μM RA and 1% B27 (4 days) (4) Monolayer. DMEM, 1 μM DAPT, 50 ng/mL exendin 4 and 1% B27 (3 days) (5) Monolayer. CMRL, 50 ng/mL exendin 4, 50 ng/mL IGF1, 50 ng/mLHGF and 1% B27 (>3 days)
Gadue et al. *(39)*	Mouse (E14.1; HR with Bry-GFP and Foxa2-hCD4)	*Undifferentiated:* 50% Neurobasal medium, 50% DMEM/F12, 0.5× N2, 0.5× B27, 0.05% BSA, ? LIF, 10 ng/mL BMP4 and 1.5×10^{-4} M MTG. Free serum (without feeder) or containing ? serum (with feeder layer) *Differentiation:* (1) EB (1.5×10^{5} cells/mL). 75% IMDM, 25% Ham's F12 medium, 0.5× N2, 0.5× B27, 0.05% BSA, 2 mM glutamine, 0.5 mM ascorbic acid and 4.5×10^{-4} M MTG. Free serum or containing ? serum (2 days) (2) EB dissociated and reaggregated. 1 or 25 ng/mL activin A, 100 ng/mL Wnt3, 150 ng/mL DKK1 or 10 μM SB. Free serum (3–4 days)
Gouon-Evans et al. *(40)*	Mouse (E14.1; HR with Bry-GFP and Foxa2-hCD4)	*Undifferentiated:* 50% Neurobasal medium, 50% DMEM/F12, 0.5× N2, 0.5× B27, 0.05% BSA, ? LIF, 10 ng/mL BMP4 and 1.5×10^{-4} M MTG. Free serum *Differentiation:* (1) EB (9,000 cells/mL for step (2.A) or 20,000 cells/mL for step (2.B). 75% IMDM, 25% Ham's F12 medium, 0.5× N2, 0.5× B27, 0.05% BSA, 2 mM Glutamine, 0.5 mM ascorbic acid and 4.5×10^{-4} M MTG. Free serum (2 days) (2.A) EB. 50 ng/mL activin A (2 days) (2.B) EB dissociated and reaggregated. 50 ng/mL activin A (2 days) (2.B.1) EB dissociated, populations isolated by cell sorting and reaggregated (250,000 cells/mL). Combinations of 50 ng/mL BMP4, 10 ng/mL bFGF, 50 ng/mL activin A, 10 ng/mL VEGF (2 days) (2.B.2) Outgrowth. It appears adherent colonies and suspended aggregated. Combinations of 50 ng/mL BMP4, 10 ng/mL bFGF, 50 ng/mL activin A, 10 ng/mL VEGF (6 days)

(continued)

Table 1 (continued)

Author (ref)	Cell type	Culture conditions
Jiang et al. *(41)*	Human (H1 and H9)	*Undifferentiated:* Feeder layer. DMEM (4.5 mg/L D-glucose), 1 mM glutamine, 0.1 mM βME, 1% NEAAs and 20% FBS *Differentiation:* (1) Monolayer. CDM (1:1 IMDM:F12 NUT-MIX), insulin–transferrin–selenium-A (1:100), 450 μM MTG, 5 mg/mL albumin fraction V or X-vivo 10 with 55 μM βME and 0.1% albumin fraction V. Free serum (2 days) (2) Monolayer. CDM and 50 ng/mL activin A (4 days) (3) Monolayer. CDM and 10^{-6} M RA (4 days) (4) Monolayer. DMEM/F12 1:1, insulin–transferrin–selenium-A (1:100), 2 mg/mL albumin fraction V and 10 ng/mL bFGF (3 days) (5) Monolayer. DMEM/F12 1:1, insulin–transferrin–selenium-A (1:100), 2 mg/mL albumin fraction V, 10 ng/mL bFGF and 10 mM nicotinamide (5 days) (6) Suspension. DMEM/F12 1:1, insulin–transferrin–selenium-A (1:100), 2 mg/mL albumin fraction V, 10 ng/mL bFGF and 10 mM nicotinamide (2 days)
Nakanishi et al. *(42)*	Mouse (E14 and CMTI-1)	*Undifferentiated:* Feeder layer. DMEM (4.5 mg/L D-glucose), ? NEAAs, 0.001% βME, 1,500 U/mL LIF and 15% FBS *Differentiation:* (1) EB without LIF and 15% KSR (4 days) (2) EBs with 15% KSR (2 days), 10–50 ng/mL activin A and 0.001–1 μM all trans RA (exocrine 10 ng/mL activin A and 0.1 μM RA; INS II 25 ng/mL activin A and 0.1 μM RA) (3) Outgrowth with gelatin and 10% KSR (6–12 days)

Abbreviations: aFGF acidic fibroblast growth factor; *AFP-GFP* alpha fetoprotein-green fluorescence protein; *bFGF* basic fibroblast growth factor; *BMP4* bone morphogenetic protein 4; *Bry-GFP* brachyury-green fluorescence protein; *BSA* bovine serum albumin; *DAPT* Notch pathway inhibitor; *dHGF* deleted form of hepatocyte growth factor; *DMEM* Dulbecco's modified Eagle's medium; *EBs* embryoid bodies; *EGF* epidermal growth factor; *FBS* fetal bovine serum; *FCS* fetal calf serum; *FGF2* fibroblast growth factor 2; *G-MEM* Glasgow minimum essential medium; *Gsc-ECD* Goosecoid-E cadherin; *HD* hanging drop; *hFGF10* human fibroblast growth factor 10; *HGF* hepatocyte growth factor; *HR* homologous recombination; *IGF1* insulinlike growth factor 1; *IMDM* Iscove's modified Dulbecco's medium; *KSR* knock-out serum replacement; *LIF* leukemia inhibitory factor; *MTG* 1-thioglycerol; *NEAAs* nonessential amino acids; *PFHM* protein free hybridoma medium; *RA* all transretinoic acid; *SB* inhibitor SB-431542; *SU5402* inhibitor of FGFr1 (fibroblast growth factor receptor 1); *VEGF* vascular endothelial growth factor; *βME* beta-mercaptoethanol; *?* unknown concentration

is not fully processed, and the amounts of hormone are still far from the amounts found in the secretory vesicles of pancreatic β cells. Although the reproducibility of this protocol was its best warranty, the modifications to circumvent the aforementioned problems were not sufficient *(43–46)*.

In addition, teratoma formation after implantation was a key problem in the published protocols, thereby limiting their therapeutic potential. This is due to cells that escape from the differentiation processes, remaining undifferentiated and not responding to specific differentiation factors present in the culture medium. These reluctant cells displayed continuous expression of stem cell markers (i.e., Oct3/4) and presented aberrant chromosomal numbers, an altered pattern of oncogene expression, and a high degree of BrdU (bromodeoxyuridine) incorporation *(15, 47)*.

Recently, a protocol proposed a new rationale to obtain insulin-secreting cells from ESCs by using a differentiation strategy that recapitulates in vitro pancreas ontogeny in ESC monolayers. The cells were incubated in the presence of specific compounds in five different steps in which the expression of specific genes was analyzed. The key point in this protocol was the enrichment of the human ESC monolayers in definitive endoderm precursors. As a result, 80% of the culture was definitive endoderm-derived cells *(48)*, compared to the 2.7% seen in other reports using EB structures *(32)*.

The main obstacle addressed in this protocol concerns the distinction between definitive and primitive endoderm, since both lineages perform very similar functions in different contexts during development and share many markers, including insulin (insulin II in mice) *(35)*. Therefore, the use of a combination of various markers to establish lineage association of the resulting cells through the different stages of the protocol has been a useful criterion for the isolation of definitive endoderm-committed cells *(33, 37)*. It is important to indicate that definitive endoderm could derive in vivo from a progenitor cell population that is capable of also giving rise to mesoderm and thereby has been called mesendoderm *(39, 49, 50)*. The application of this rationale to bioengineering protocols resulted in obtaining definitive endoderm and insulin-positive cells from brachyury-selected cells *(31)*. Nevertheless, it has to be mentioned that brachyury gene expression is not exclusively restricted to mesendoderm, being also expressed in visceral endoderm *(51)*. On the other hand, Yasunaga et al. *(37)* reported the isolation of mesendodermic precursors based on the expression of Gsc (goosecoid) and Sox17/CD25. Gsc is a mesoderm marker, while Sox17/CD25 is a well-known marker of definitive and visceral endoderm. Therefore, the endoderm-committed cell line must be Gsc$^+$/Sox17$^+$, whereas Gsc$^-$/Sox17$^+$ cells are most likely committed to visceral endoderm *(33, 36, 37)* (Fig. 1).

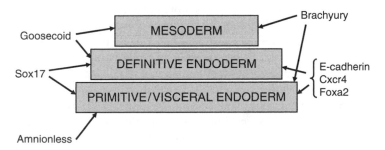

Fig. 1 Scheme of gene marker combinations used sto identify mesoderm and primitive and definitive endoderms. It is important to note that mesoderm and definitive endoderm derive from a common precursor called the mesendoderm. Insulin gene is expressed in primitive endoderm (insulin II in rodents) and in definitive endoderm (insulin I in rodents). Endocrine pancreas derives from definitive endoderm

Particular culture conditions are required to guide the progenitors toward definitive endoderm and thus toward endocrine pancreatic cells. To favour definitive endoderm precursors, ESCs were cultured in the absence of serum *(37, 52, 53)* and in the presence of activin A, which modulates intracellular events by binding to Nodal receptors (Nodal is a transforming growth factor β [TGF-β] family member) *(37, 39, 54)*. Afterward, the addition of specific growth factors in five well-defined stages drove the endoderm-committed monolayers to pancreatic fates *(38)*. These include primitive gut endoderm (fibroblast growth factor 10 [FGF10] + cyclopamine), posterior foregut (FGF10 + cyclopamine + retinoic acid), pancreatic endoderm and endocrine precursors (Notch-pathway inhibitor DAPT + exendin-4), and finally hormone-expressing cells (exendin-4 + insulin growth factor 1 [IGF1] + hepatocyte growth factor [HGF]).

A key question is to know whether the classical approach via EB formation allows spontaneous differentiation to definitive endoderm or if it requires the addition of specific factors. In the aforementioned protocol, the use of monolayers allowed the direct access of extracellular factors to all cells, assessing the reproducibility of the protocol *(38)*. However, it is difficult to control all the microenvironmental determinants during differentiation via EBs, which explains the variability in the yields of insulin-positive cells obtained in different protocols *(18, 55)*. One possible advantage of using protocols of differentiation via EBs is the presence of different cell types that could produce key factors and thereby mimic certain steps of pancreatic development using paracrine mechanisms. Nevertheless, this point remains to be fully demonstrated. The presence of glucagon mRNA in EBs by spontaneous differentiation could suggest the presence of definitive endoderm in these cell aggregates *(18)*. However, a focused phenotypic and functional characterization of glucagon-positive cells in EB structures remains to be performed. In any case, the presence of glucagon is not in agreement with the production of Shh (Sonic hedgehog) by EBs, repressing any possible endocrine pancreas differentiation. Although it is unknown if this observation can be extended to all EBs or if it is only observed in a few EB structures, the inhibition of Shh can positively improve the yield of insulin-producing cells *(56)*.

Baetge's protocol is actually a reference in diabetes cell bioengineering *(38)*. However, the transfer of this strategy to a more therapeutic context requires solving some key questions. The differentiation protocol presents an artificial situation in which pancreatic progenitors are obtained in 15 days. In vivo the same process takes about four times longer. One explanation could be that under the restricted conditions in the culture dish, differentiation might be enhanced over proliferation, which usually occurs during mammalian organogenesis. This could suggest that some key steps are not accurately performed in the culture plate, resulting in immature cells. The typical traits of the resulting cell product are: (a) proinsulin is not fully processed to mature insulin, (b) low secretory response to extracellular glucose concentrations, (c) coexpression of insulin and glucagon or insulin and somatostatin in the same cell type, and (d) no expression of the transcription factor MafA, which is required for the progression to a mature phenotype *(38)*. On the other hand, and of great importance, the resulting cells possessed the highest insulin contents described so far, close to those found in mature β cells.

Therefore, the unsolved aspects of this protocol open the possibility of searching for new determinants that could complete this in vitro differentiation protocol in order to obtain a more functional cell. The characterization of the microenvironment or niche in which the pancreas develops could help in this context. Indeed, when human ESCs were cotransplanted with mouse foetal dorsal pancreas, they expressed processed insulin as well as many pancreatic transcription factors *(57)*. It can be hypothesized that the niche where pancreas develops should contain a mixture of key molecules derived from the same endocrine pancreas, as well as from surrounding tissues. Complexity increases if we consider that this particular niche is subject to dynamic processes in which concentration, composition, and time of action of the different determinants change according to particular patterns specified by the orchestrated developmental programmes *(58)*.

Although the idea seems interesting in theory, it needs to be sustained by consistent experiments. The first study addressing this point presents deep inconsistencies with a lack of controls in the experimental design *(59)*. Therefore, conclusions raised in the study were completely useless for bioengineering protocols. This study did not report the insulin gene that was expressed (I or II), nor mentioned the processing of the hormone, and the commitment to a specific cell fate was also obviated. In addition, the resulting cells released 20% of the insulin content when stimulated with glucose, showing a very rare and unusual secretion pattern. Finally, the large amounts of BrdU incorporated into the cells strongly indicated that the final cell obtained could produce tumours, which is ignored by the authors by simply sacrificing the transplanted animals 15 days postimplant.

In conclusion, aside from useless protocols like this one *(59)*, notable improvements have been performed in this field. A notable amount of data has been accumulated in this respect, and therefore the discovery of a definitive protocol is simply a matter of time. Table 1 presents some additional examples that could be useful for designing more complete protocols to drive ESCs to endoderm-derived cells, including cells committed into the hepatic and the pancreatic pathways. In this context, innovative ideas require adequate financial support for their development and for the advancement in this field. In this line, national agencies must be aware of the positive results found in some protocols that are apparently evaluated by "competent" reviewers but that do not add significant improvements to the stem cell field.

Insulin-Producing Cells from Adult Stem Cells

Conversely to other endoderm-derived organs, such as the liver, pancreas is not capable of regenerating when it suffers a partial ablation. Similar observations have been made in the brain and heart. However, a quiescent stem cell population has been described in these organs, opening the possibility for developing future tissue regeneration strategies. If this observation could be extended to the pancreatic tissue, this should provide a source of precursors committed to islet cells. However, despite the large amount of research being performed, the location and identification of such a pancreatic stem cell population remains elusive. Several candidates have been proposed, including ductal cells, exocrine-associated cells, pancreatic mesenchymal cells, as well as β-cell self-regeneration.

Pancreatic and hepatic ducts seem to bear a precursor population that can be bioengineered in vitro to obtain insulin-producing cells *(60, 61)*. The observation of islets budding from pancreatic ducts has suggested the existence of islet precursors in this tissue. However, the identification of the ductal cell among the cell types present in the ducts capable of differentiating toward insulin-positive cells remains elusive. This candidate cell has received several names, such as cultivated human islet buds (CHIBs) *(60)*, islet pluripotent stem cells (IPSCs) *(61)*, nestin-positive islet-derived progenitors [NIPs], also found to be associated with islets) *(62)*, and nonendocrine pancreatic epithelial precursor (NEPEC) *(63)*. Certain gene markers, such as CK19, PDX-1, or nestin, seem to be expressed in these precursors *(64, 65)*. Interestingly, duct epithelium can be easily purified from the rest of the pancreatic cell types, such as islets and exocrine tissue, due to its exceptional adherent capacity to cell culture surfaces *(60)*. This property could favour ex vivo bioengineering, thereby offering a new potential application of human ductal tissue obtained from cadaveric pancreata. Several agents could favour islet neogenesis from the ducts, such as islet neogenesis-associated protein (INGAP) peptide or all-transretinoic acid, although the operating mechanism has not been fully characterized *(66, 67)*. Finally, insulin-positive cells derived from ductal precursors produce modest amounts of insulin, and their stimulus-coupled secretory response is still very poor, thereby requiring further improvements.

On the other hand, it has been proposed that a nestin-positive isolated precursor (NIP) could be associated to islet structures *(62)*. Nestin is an intermediate filament protein in the nervous system and in mesenchymal cells. In this context, bioengineered mesenchymal stem cells from pancreatic or extrapancreatic tissues could represent alternative sources to obtain insulin-secreting cells. However, a definitive in vitro differentiation protocol has not been developed yet, more likely due to the heterogeneity of this cell population *(68)*. Aside from their location, the pathways from which new β cells appear are unknown. The proposed mechanism, called the epithelial-to-mesenchymal-to-epithelial transition, implies the dedifferentiation of islets cells to mesenchymal cells in order to redifferentiate into islets *(69, 70)*. However, whether this mechanism, which has been described only in vitro, occurs in vivo is a debated question *(71–75)*. On the other hand, pancreatic exocrine cells represent another source for insulin-positive cells, although several authors claim the opposite *(76, 77)*.

Finally, it seems that new β cells can arise from preexisting β cells. Although this turnover strategy could be fully operative during normal life physiological adaptations, such as pregnancy *(78, 79)*, in overt diabetes with accelerated β-cell destruction, this mechanism does not seem to be sufficient for β-cell replacement *(80, 81)*. This self-replicating capacity of pancreatic β cells has been demonstrated by convincing cell-tracing experiments *(80)*. However, adult β cells display low proliferation rates (less than 1% of β-cell mass/day) compared to the embryonic counterparts, which display turnovers of 10% of β-cell mass/day. This replicative capacity diminishes with age to values lower than 0.1% of β-cell mass/day.

Additional knowledge is necessary in order to decipher the molecular mechanisms of the different determinants that can alter β-cell cycle. This would help to design intervention strategies to recover β-cell loss in diabetic pathology.

Therefore, the main application of pancreatic ASCs in the context of regenerative medicine in diabetes should be the in vitro expansion of islets, either from the same patient or from cadaveric donors. Ideally, islets should replicate in order to obtain sufficient amounts and be functional in order to rescue the patient from the disease when they are reimplanted. At best, the accumulated knowledge should give key information in order to design pharmacological agents that could stimulate in vivo islet/β-cell turnover. Several agents present promising results for diabetes treatment, including exendin, a pharmacological product derived from glucagon-like peptide (GLP-1), β-cellulin, nicotinamide, gastrin, epidermal growth factor 1 (EGF-1), and thyroid hormone *(82)*.

In conclusion, in healthy organisms, β-cell mass could be maintained by neogenesis from pancreatic stem cells and by self-replication of β cells. In pathological situations where insulin production is deficient, β-cell function adapts through hypertrophy (increased cell size) and replication (increased cell number). However, persistent damage usually favours cell death over proliferation/hypertrophy. The location of candidate pancreatic niches for stem cells (ducts, islets, acini) and the characteristics of the cellular progenitors (epithelium, mesenchyme, exocrine tissue, β cells) remain active areas of research *(78)*.

Bioengineering extrapancreatic ASCs is supported by the accumulated experience of ectopic insulin expression in several nonpancreatic tissues *(83)*. In this sense, bone marrow represents an important ASC population in the organism that offers the possibility of transdifferentiating into insulin-secreting cells. However, the reports concerning this possibility have not clarified the operating mechanism. In this context, a direct, but modest, transdifferentiation process has been proposed *(22)* that has not been reproduced by others *(84–88)*. Bone marrow harbours a heterogeneous population of stem cells, including haematopoietic, endothelial, and mesenchymal, among others, that could stimulate insulin production through different mechanisms. In this context, bone marrow-derived endothelial progenitors can indirectly favour islet regeneration through angiogenesis *(23)*.

Mesenchymal cells are also present in bone marrow as well as in other locations, such as adipose tissue. Bone marrow mesenchymal cells are located in the stroma and display the ability to differentiate into mesoderm-derived tissues, such as bone, cartilage, and adipoblasts. In addition, in vitro experiments have shown that bone marrow mesenchymal cells can give rise to insulin-producing cells by adding specific factors to the culture medium or by transfection with β-cell–specific DNA constructs *(89–92)*. In this context, it has been described that $CD29^+/CD44^+/CD106^+$ bone marrow–purified mesenchymal cells can differentiate to insulin-positive cells *(93)*. The question remains if other mesenchymal stem cell populations reprogrammed to pancreatic fates can display higher values of insulin production and secretion. Interestingly, mesenchymal stem cells from adipose tissue are very similar to bone marrow–derived mesenchymal cells. Mesenchymal cells from adipose tissue can express insulin and pancreatic genes when they are cultured in the presence of basic fibroblast growth factor (bFGF) and high glucose concentrations *(94)*. Altogether, these reports indicate an interesting plasticity for mesenchymal stem cells, which deserves further investigation.

Precise phenotyping of precursors would be mandatory in all published reports in order to avoid problems related with reproducibility when these experiments are carried out in other laboratories. For instance, a poorly characterised population of circulating human monocytes could be coaxed to express the insulin gene *(95)*. However, some authors of this work publicly claimed in scientific forums the difficulties to reproduce these published results. On the other hand, insulin-positive cells can also be obtained from a human stem cell population derived from peripheral blood that is positive for Oct4, Nanog, and the hematopietic markers CD9, CD45, and CD117, as well as negative for CD34 and markers of the monocyte or lymphocyte lineages *(96)*.

Similar to pancreas, liver derives from the upper foregut definitive endoderm and displays a robust self-renewing potential. In this context, oval hepatic stem cells are capable of differentiating to hepatocytes, bile duct epithelium, and insulin-positive cells as well *(97, 98)*. A hepatic mesenchymal stem cell population has been isolated from adult human liver with positive expression for CD29, CD73, CD44, CD90, nestin, and vimentin, and the ability to differentiate into pancreatic endocrine fates. In addition, the overexpression of the pancreatic transcription factor Pdx-1 allows the transdifferentiation of human foetal hepatic stem cells *(99)* and adult human and mouse hepatic cells into insulin-producing cells *(24, 83, 100)*. However, Pdx-1–transformed cells could additionally induce the differentiation to pancreatic exocrine tissue, causing hepatic destruction. Transfection with other transcription factors (i.e., NeuroD combined with betacellulin) that are more specific of the endocrine lineage of the pancreas seems to circumvent this problem *(25)*.

Following the same rationale than in liver, endoderm-derived endocrine cells of the intestinal epithelium might be good candidates for ectopic insulin expression and secretion. Indeed, endocrine intestinal cells share many functional traits with β cells, including the nutrient-sensor machinery and the regulated pathway for hormone secretion *(101, 102)*. However, the obtained results, although promising, still require important improvements.

Altogether, ASCs from bone marrow, adipose tissue, liver, and intestine open new possibilities for autologous cell therapy in diabetes. To characterize, isolate, and explore the molecular mechanisms involved in the differentiation processes is fundamental to provide consistent protocols to the scientific community.

Conclusion

Research on ESCs and ASCs as potential sources for the treatment of diabetes requires substantial improvements before transferring to a clinical field. In vitro culture conditions need to be assessed in order to set up reproducible protocols. It is also necessary to investigate the endocrine pancreas physiology and development in order to accumulate knowledge to afford in vitro problems. Immune rejection, implant survival, and tumour formation are serious limitations that can edge the development of future therapeutic strategies in humans. Finally, the mechanisms operating in the transdifferentiation of extrapancreatic ASCs need to be investigated in more detail. In conclusion, the objective is to obtain a cell that produces sufficient amounts of proinsulin, is competent in correctly processing the hormone, and can be secreted in response to several secretagogues (mainly glucose) in a regulated manner. In addition, this cell needs to be capable of surviving in an appropriate body niche, neither giving rise to tumours nor to inducing immune rejection *(5)*.

References

1. DeFronzo RA, Ferrannini E, Keen H, et al. International textbook of diabetes mellitus, 3th ed. Wiley, Chichester, UK, 2004.
2. Isermann B, Ritzel R, Zorn M, Schilling T, Nawroth PP. Autoantibodies in diabetes mellitus: current utility and perspectives. Exp Clin Endocrinol Diabetes 2007;115:483–90.
3. Prentki M, Joly E, El-Assaad W, Roduit R. Malonyl-CoA signaling, lipid partitioning, and glucolipotoxicity: role in beta-cell adaptation and failure in the etiology of diabetes. Diabetes 2002;51:S405–13.
4. Shapiro AM, Lakey JR, Ryan EA, et al. Islet transplantation in seven patients with type 1 diabetes mellitus using a glucocorticoid-free immunosuppressive regimen. N Engl J Med 2000;343:230–8.
5. Roche E, Reig JA, Campos A, et al. Insulin-secreting cells derived from stem cells: clinical perspectives, hypes and hopes. Transplant Immunol 2005;15:113–29.
6. Ryan EA, Paty BW, Senior PA, et al. Five-year follow-up after clinical islet transplantation. Diabetes 2005;54:2060–9.
7. Hermann M, Margreiter R, Hengster P. Molecular and cellular key players in human islet transplantation. J Cell Mol Med 2007;11:398–415.
8. Laybutt DR, Hawkins YC, Lock J, et al. Influence of diabetes on the loss of β-cell differentiation after islet transplantation in rats. Diabetologia 2007;50:2117–25.
9. Marzorati S, Pileggi A, Ricordi C. Allogeneic islet transplantation. Expert Opin Biol Ther 2007;7:1627–45.
10. Kaestner KH. Beta cell transplantation and immunosuppression: can't live with it, can't live without it. J Clin Invest 2007;117:2380–2.
11. Smith AG. Embryo-derived stem cells of mice and men. Annu Rev Cell Dev Biol 2001;17:435–62.
12. Smith AG. Culture and differentiation of embryonic stem cells. J Tissue Culture Methods 1991;13:89–94.
13. Prowse AB, McQuade LR, Bryant KJ, Marcal H, Gray PP. Identification of potential pluripotency determinants for human embryonic stem cells following proteomic analysis of human and mouse fibroblast conditioned media. J Proteome Res 2007;6:3796–807.
14. Roche E, Sepulcre P, Reig JA, Santana A, Soria B. Ectodermal commitment of insulin-producing cells derived from mouse embryonic stem cells. FASEB J 2005;19:1341–3.
15. Ensenat-Waser R, Santana A, Vicente-Salar N, et al. Isolation and characterization of residual undifferentiated mouse embryonic stem cells from embryoid body cultures by fluorescence tracking. In Vitro Cell Develop Biol Animal 2006;42:115–23.
16. Kahan BW, Jacobson LM, Hullett DA, et al. Pancreatic precursors and differentiated islet cell types from murine embryonic stem cells. An in vitro model to study islet differentiation. Diabetes 2003;52:2016–24.

17. Skoudy A, Rovira M, Savatier P, et al. Transforming growth factor (TGF)β, fibroblast growth factor (FGF) and retinoid signaling pathways promote pancreatic exocrine gene expression in mouse embryonic stem cells. Biochem J 2004;379:749–56.

18. Ensenat-Waser R, Santana A, Paredes B, et al. Embryonic stem cell processing in obtaining insulin-producing cells: a technical review. Cell Preserv Technol 2006;4:278–89.

19. Watt FM, Hogan BL. Out of Eden: stem cells and their niches. Science 2000;287:1427–30.

20. Barrilleaux B, Phinney DG, Prockop DJ, O'Connor KC. Ex vivo engineering of living tissues with adult stem cells. Tissue Eng 2006;12:3007–19.

21. Fellous TG, Guppy NJ, Brittan M, Alison MR. Cellular pathways to β-cell replacement. Diabetes/Metab Res Rev 2007;23:87–99.

22. Ianus A, Holz GG, Theise ND, Hussain MA. In vivo derivation of glucose-competent pancreatic endocrine cells from bone marrow without evidence of cell fusion. J Clin Invest 2003; 111:843–50.

23. Hess D, Li L, Martin M, et al. Bone marrow-derived stem cells initiate pancreatic regeneration. Nat Biotechnol 2003;21:763–70.

24. Ferber S, Halkin A, Cohen H, et al. Pancreatic and duodenal homeobox gene 1 induces expression of insulin genes in liver and ameliorates streptozotocin-induced hyperglycemia. Nat Med 2000;6:568–72.

25. Kojima H, Fujimiya M, Matsumura K, et al. NeuroD-betacellulin gene therapy induces islet neogenesis in the liver and reverses diabetes in mice. Nat Med 2003;9:596–603.

26. Soria B, Roche E, Bernat G, Leon-Quinto T, Reig JA, Martin F. Insulin-secreting cells derived from embryonic stem cells normalize glycemia in streptozotocin-induced diabetic mice. Diabetes 2000;49:157–62.

27. Assady S, Maor G, Amit M, Itskovitz-Eldor J, Skorecki KL, Tzukerman M. Insulin production by human embryonic stem cells. Diabetes 2001;50:1691–7.

28. Lumelsky N, Blondel O, Laeng P, et al. Differentiation of embryonic stem cells to insulin-secreting structures similar to pancreatic islets. Science 2001;292:1389–94.

29. Jones EA, Tosh D, Wilson DI, Lindsay S, Forrester LM. Hepatic differentiation of murine embryonic stem cells. Exp Cell Res 2002;272:15–22.

30. Shiroi A, Yoshikawa M, Yokota H, et al. Identification of insulin-producing cells derived from embryonic stem cells by zinc-chelating dithizone. Stem Cells 2002;20:284–92.

31. Kubo A, Shinozaki K, Shannon JM, et al. Development of definitive endoderm from embryonic stem cells in culture. Development 2004;131:1651–62.

32. Ku HT, Zhang N, Kubo A, et al. Committing embryonic stem cells to early endocrine pancreas in vitro. Stem Cells 2004;22:1205–17.

33. D'Amour KA, Agulnick AD, Eliazer S, Kelly OG, Kroon E, Baetge EE. Efficient differentiation of human embryonic stem cells to definitive endoderm. Nat Biotechnol 2005;23:1534–41.

34. Ishii T, Yasuchika K, Fujii H, et al. In vitro differentiation and maturation of mouse embryonic stem cells into hepatocytes. Exp Cell Res 2005;309:68–77.

35. Milne HM, Burns CJ, Kitsou-Mylona I, et al. Generation of insulin-expressing cells from mouse embryonic stem cells. Biochem Biophys Res Commun 2005;328:399–403.

36. Tada S, Era T, Furusawa C, et al. Characterization of mesendoderm: a diverging point of the definitive endoderm and mesoderm in embryonic stem cell differentiation culture. Development 2005;132:4363–74.

37. Yasunaga M, Tada S, Torikai-Nishikawa S, et al. Induction and monitoring of definitive and visceral endoderm differentiation of mouse ES cells. Nat Biotechnol 2005;23:1542–50.

38. D'Amour KA, Bang AG, Eliazer S, et al. Production of pancreatic hormone-expressing endocrine cells from human embryonic stem cells. Nat Biotechnol 2006;24:1392–401.

39. Gadue P, Huber TL, Paddison PJ, Keller GM. Wnt and TGF-β signaling are required for the induction of an in vitro model of primitive streak formation using embryonic stem cells. Proc Natl Acad Sci U S A 2006;103:16806–11.

40. Gouon-Evans V, Boussemart L, Gadue P, et al. BMP-4 is required for hepatic specification of mouse embryonic stem cell-derived definitive endoderm. Nat Biotechnol 2006;24:1402–11.

41. Jiang W, Shi Y, Zhao D, et al. In vitro derivation of functional insulin-producing cells from human embryonic stem cells. Cell Res 2007;17:333–44.

42. Nakanishi M, Hamazaki TS, Komazaki S, Okochi H, Asashima M. Pancreatic tissue formation from murine embryonic stem cells in vitro. Differentiation 2007;75:1–11.

43. Rajagopal J, Anderson WJ, Kume S, Martinez OI, Melton DA. Insulin staining of ES cell progeny from insulin uptake. Science 2003;299:363.

44. Hori Y, Rulifson IC, Tsai BC, Heit JJ, Cahoy JD, Kim SK. Growth inhibitors promote differentiation of insulin-producing tissue from embryonic stem cells. Proc Natl Acad Sci U S A 2002;99:16105–10.

45. Moritoh Y, Yamato E, Yasui Y, Miyazaki S, Miyazaki J. Analysis of insulin-producing cells during in vitro differentiation from feeder-free embryonic stem cells. Diabetes 2003;52:1163–8.

46. Blyszczuk P, Czyz J, Kania G, et al. Expression of Pax4 in embryonic stem cells promotes differentiation of nestin-positive progenitor and insulin-producing cells. Proc Natl Acad Sci U S A 2003;100:998–1003.

47. Santana A, Enseñat-Waser R, Arribas MI, Reig JA, Roche E. Insulin-producing cells derived from stem cells: recent progress and future directions. J Cell Mol Med 2006;10:866–83.

48. Madsen OD, Serup P. Towards cell therapy for diabetes. Nat Biotechnol 2006;24:1481–3.

49. Shook D, Keller R. Mechanisms, mechanics and function of epithelial–mesenchymal transitions in early development. Mech Dev 2003;120:1351–83.

50. Yamanaka Y, Ralston A, Stephenson RO, Rossant J. Cell and molecular regulation of the mouse blastocyst. Dev Dyn 2006;235:2301–14

51. Inman KE, Downs KM. Localization of brachyury (T) in embryonic and extraembryonic tissues during mouse gastrulation. Gene Expr Patterns 2006;6:783–93.

52. Wang L, Schulz TC, Sherrer ES, et al. Self-renewal of human embryonic stem cells requires insulin-like growth factor-1 receptor and ERBB2 receptor signaling. Blood 2007; 110:4111–9.

53. Nguyen TT, Sheppard AM, Kaye PL, Noakes PG. IGF-1 and insulin activate mitogen-activated protein kinase via the type 1 IGF receptor in mouse embryonic stem cells. Reproduction 2007;134:41–9.

54. McLean AB, D'Amour KA, Jones KL, et al. Activin A efficiently specifies definitive endoderm from human embryonic stem cells only when phosphatidylinositol 3-kinase signaling is suppressed. Stem Cells 2007;25:29–38.

55. Schroeder IS, Rolletschek A, Blyszczuk P, Kania G, Wobus AM. Differentiation of mouse embryonic stem cells to insulin-producing cells. Nat Protocols 2006;1:495–507.

56. Mfopou JK, De Groote V, Xu X, Heimberg H, Bouwens L. Sonic hedgehog and other soluble factors from differentiating embryoid bodies inhibit pancreas development. Stem Cells 2007;25:1156–65.

57. Brolén GKC, Heins N, Edsbagge J, Semb H. Signals from the embryonic mouse pancreas induce differentiation of human embryonic stem cells into insulin-producing β-cell-like cells. Diabetes 2005;54:2867–74

58. Rivas-Carrillo JD, Okitsu T, Tanaka N, Kobayashi N. Pancreas development and beta-cell differentiation of embryonic stem cells. Curr Med Chem 2007;14:1573–8.

59. Vaca P, Martin F, Vegara-Meseguer JM, Rovira JM, Berna G, Soria B. Induction of differentiation of embryonic stem cells into insulin secreting cells by fetal soluble factors. Stem Cells 2006;24:258–65.

60. Bonner-Weir S, Taneja M, Weir GC, et al. In vitro cultivation of human islets from expanded ductal tissue. Proc Natl Acad Sci U S A 2000;97:7999–8004.

61. Ramiya VK, Maraist M, Arfors KE, Schatz DA, Peck AB, Cornelius JG. Reversal of insulin-dependent diabetes using islets generated in vitro from pancreatic stem cells. Nat Med 2000;6:278–82.

62. Zulewski H, Abraham EJ, Gerlach MJ, et al. Multipotential nestin-positive stem cells isolated from adult pancreatic islets differentiate ex vivo into pancreatic endocrine, exocrine and hepatic phenotypes. Diabetes 2001;50:521–33.

63. Hao E, Tyrberg B, Itkin-Ansari P, et al. Beta-cell differentiation from nonendocrine epithelial cells of the adult human pancreas. Nat Med 2006;12:310–6.

64. Taguchi M, Yamaguchi T, Otsuki M. Induction of PDX-1-positive cells in the main duct during regeneration after acute necrotizing pancreatitis in rats. J Pathol 2002;197:638–46.

65. Kritzik MR, Jones E, Chen Z, et al. PDX-1 and Msx-2 expression in the regenerating and developing pancreas. J Endocrinol 1999;163:523–30.
66. Roche E, Jones J, Arribas MI, Leon-Quinto T, Soria B. Role of small bioorganic molecules in stem cell differentiation to insulin-producing cells. Bioorg Med Chem 2006;14:6466–74.
67. Rafaeloff R, Pittenger GL, Barlow SW, et al. Cloning and sequencing of the pancreatic islet neogenesis associated protein (INGAP) gene and its expression in islet neogenesis in hamsters. J Clin Invest 1997;99:2100–9.
68. Zulewski H. Stem cells with potential to generate insulin-producing cells in man. Swiss Med Wkly 2006;136:647–54.
69. Gershengorn MC, Hardikar AA, Hardikar A, Geras-Raaka E, Marcus-Samuels B, Raaka BM. Epithelial-to-mesenchymal transition generates proliferative human islet precursor cells. Science 2004;306:2261–4.
70. Davani B, Ikonomou L, Raaka BM, et al. Human islet-derived precursor cells are mesenchymal stromal cells that differentiate and mature to hormone-expressing cells in vivo. Stem Cells 2007;25:3215.
71. Atouf F, Park CH, Pechhold K, Ta M, Choi Y, Lumelsky NL. No evidence for mouse pancreatic beta-cell epithelial-mesenchymal transition in vitro. Diabetes 2007;56:699–702.
72. Chase LG, Ulloa-Montoya F, Kidder BL, Verfaillie CM. Islet-derived fibroblast-like cells are not derived via epithelial–mesenchymal transition from Pdx-1 or insulin-positive cells. Diabetes 2007;56:3–7.
73. Eberhardt M, Salmon P, von Mach MA, et al. Multipotential nestin and Isl-1 positive mesenchymal stem cells isolated from human pancreatic islets. Biochem Biophys Res Commun 2006;345:1167–76.
74. Weinberg N, Ouziel-Yahalom L, Knoller S, Efrat S, Dor Y. Lineage tracing evidence for in vitro dedifferentiation but rare proliferation of mouse pancreatic beta-cells. Diabetes 2007;56:1299–304.
75. D'Alessandro JS, Lu K, Fung BP, Colman A, Clarke DL. Rapid and efficient in vitro generation of pancreatic islet progenitor cells from nonendocrine epithelial cells in the adult human pancreas. Stem Cells Dev 2007;16:75–89.
76. Minami K, Okuno M, Miyawaki K, et al. Lineage tracing and characterization of insulin-secreting cells generated from adult pancreatic acinar cells. Proc Natl Acad Sci U S A 2005;102:15116–21.
77. Desai BM, Oliver-Krasinski J, De Leon DD, et al. Preexisting pancreatic acinar cells contribute to acinar cell, but not islet beta cell regeneration. J Clin Invest 2007;117:971–7.
78. Ackermann AM, Gannon M. Molecular regulation of pancreatic β-cell mass development, maintenance, and expansion. J Mol Endocrinol 2007;38:193–206.
79. Murtaugh LC. Pancreas and beta-cell development: from the actual to the possible. Development 2007;134:427–38.
80. Dor Y, Brown J, Martinez OI, Melton DA. Adult pancreatic β-cells are formed by self-duplication rather than stem-cell differentiation. Nature 2004;429:41–6.
81. Kodama S, Kuhtreiber W, Fujimura S, Dale EA, Faustman DL. Islet regeneration during the reversal of autoimmune diabetes in NOD mice. Science 2003;302:1223–7.
82. Banerjee M, Kanitkar M, Bhonde RR. Approaches towards endogenous pancreatic regeneration. Rev Diab Stud 2005;2:165–76.
83. Efrat S. Prospects for gene therapy of insulin-dependent diabetes mellitus. Diabetologia 1998;41:1401–9.
84. Choi JB, Uchino H, Azuma K, et al. Little evidence of transdifferentiation of bone marrow-derived cells into pancreatic beta cells. Diabetologia 2003;46:1366–74.
85. Lechner A, Yang Y-Q, Blacken RA, Wang L, Nolan AL, Habener JF. No evidence for significant transdifferentiation of bone marrow into pancreatic β-cells in vivo. Diabetes 2004;53:616–23.
86. Hasegawa Y, Ogihara T, Yamada T, et al. Bone marrow (BM) transplantation promotes beta-cell regeneration after acute injury through BM cell mobilization. Endocrinology 2007;148:2006–15.
87. Butler AE, Huang A, Rao PN, et al. Hematopoietic stem cells derived from adult donors are not a source of pancreatic beta-cells in adult nondiabetic humans. Diabetes 2007;56:1810–6.

88. Lavazais E, Pogu S, Sai P, Martignat L. Cytokine mobilization of bone marrow cells and pancreatic lesion do not improve streptozotocin-induced diabetes in mice by transdifferentiation of bone marrow cells into insulin-producing cells. Diabetes Metab 2007;33:68–78.

89. D'Ippolito G, Diabira S, Howard GA, Menei P, Roos BA, Schiller PC. Marrow-isolated adult multilineage inducible (MIAMI) cells, a unique population of postnatal young and old human cells with extensive expansion and differentiation potential. J Cell Sci 2004;117:2971–81.

90. Tayaramma T, Ma B, Rohde M, Mayer H. Chromatin-remodeling factors allow differentiation of bone marrow cells into insulin-producing cells. Stem Cells 2006;24:2858–67.

91. Moriscot C, de Fraipont F, Richard MJ, et al. Human bone marrow mesenchymal stem cells can express insulin and key transcription factors of the endocrine pancreas developmental pathway upon genetic and/or microenvironmental manipulation in vitro. Stem Cells 2005;23:594–604.

92. Karnieli O, Izhar-Prato Y, Bulvik S, Efrat S. Generation of insulin-producing cells from human bone marrow mesenchymal stem cells by genetic manipulation. Stem Cells 2007;25:2837–44.

93. Yu S, Li C, Xin-guo H, et al. Differentiation of bone marrow-derived mesenchymal stem cells from diabetic patients into insulin-producing cells in vitro. Chin Med J (Engl) 2007;120:771 6.

94. Timper K, Seboek D, Eberhardt M, et al. Human adipose tissue-derived mesenchymal stem cells differentiate into insulin, somatostatin, and glucagon expressing cells. Biochem Biophys Res Commun 2006;341:1135–40.

95. Ruhnke M, Ungefroren H, Nussler A, et al. Differentiation of in vitro-modified human peripheral blood monocytes into hepatocyte-like and pancreatic islet-like cells. Gastroenterology 2005;128:1774–86.

96. Zhao Y, Huang Z, Lazzarini P, Wang Y, Di A, Chen M. A unique human blood-derived cell population displays high potential for producing insulin. Biochem Biophys Res Commun 2007;360:205–11.

97. Yang L, Li S, Hatch H, et al. In vitro trans-differentiation of adult hepatic stem cells into pancreatic endocrine hormone-producing cells. Proc Natl Acad Sci U S A 2002;99:8078–83.

98. Kim S, Shin JS, Kim HJ, Fisher RC, Lee MJ, Kim CW. Streptozotocin-induced diabetes can be reversed by hepatic oval cell activation through hepatic transdifferentiation and pancreatic islet regeneration. Lab Invest 2007;87:702–12.

99. Zalzman M, Gupta S, Giri RK, et al. Reversal of hyperglycemia in mice by using human expandable insulin-producing cells differentiated from fetal liver progenitor cells. Proc Natl Acad Sci U S A 2003;100:7253–8.

100. Shternhall-Ron K, Quintana FJ, Perl S, et al. Ectopic PDX-1 expression in liver ameliorates type 1 diabetes. J Autoimmun 2007;28:134–42.

101. Kojima H, Nakamura T, Fujita Y, et al. Combined expression of pancreatic duodenal homeobox 1 and islet factor 1 induces immature enterocytes to produce insulin. Diabetes 2002;51:1398–408.

102. Han J, Lee HH, Kwon H, Shin S, Yoon JW, Jun HS. Engineered enteroendocrine cells secrete insulin in response to glucose and reverse hyperglycemia in diabetic mice. Mol Ther 2007;15:1195–202.

103. Choi D, Lee HJ, Jee S, et al. In vitro differentiation of mouse embryonic stem cells: enrichment of endodermal cells in the embryoid body. Stem Cells 2005;23:817–27.

Corneal Epithelial Stem Cells and Their Therapeutic Application

Sai Kolli, Majlinda Lako, Francisco Figueiredo, and Sajjad Ahmad

Abstract The cornea is the clear window at the front of the eye and its clarity is vital for the transmission of light to the retina at the back of the eye for visual perception. The surface of the cornea is made up of an epithelium, which is continuous with that of the surrounding conjunctiva. The transition between the corneal and conjunctival epithelia is formed by the limbal epithelium. The limbal epithelium has two particular important functions. First, it harbours the corneal epithelial stem cells (CESCs), also known as limbal stem cells (LSCs). These stem cells (SCs) provide a reservoir for corneal epithelial cells, which are needed to replace those lost continuously from the corneal surface. Second, it acts as a barrier to prevent the phenotypically and functionally different conjunctival epithelium from encroaching onto the corneal surface, which would impair the transparency of the cornea and lead to visual loss.

LSC deficiency (LSCD) is a disease characterised by the loss or dysfunction of CESCs. It results from a variety of causes such as chemical or thermal burns, contact len–related eye disease, hereditary disorders (such as aniridia and ectodermal dysplasia), iatrogenic causes (such as surgery, radiotherapy, and cryotherapy), and inflammatory eye diseases (such as Stevens-Johnson's syndrome and ocular cicatricial pemphigoid). In LSCD, the conjunctival epithelium and its underlying blood vessels encroach onto the surface of the cornea, resulting in significant visual impairment. Additionally, the corneal epithelium fails to heal normally, resulting in recurrent epithelial breakdown associated with constant pain and photophobia. The understanding of CESC biology and its clinical application has allowed the development of treatment strategies for this blinding and painful condition.

The mainstay of treatment in severe LSCD is the transplantation of large pieces of healthy limbal tissue. This tissue can be obtained from the other eye of the patient (if healthy), or the healthy eye of a living related donor or cadaveric donor. Existing techniques are not ideal due to the quantity of tissue required and the

S. Kolli (✉), M. Lako, F. Figueiredo, and S. Ahmad
North East England Stem Cell Institute, Newcastle University, United Kingdom &
Department of Ophthalmology, Royal Victoria Infirmary, Newcastle upon Tyne, United Kingdom
e-mail: drsaikolli@gmail.com

H. Bharavand (ed.), *Trende in Stem Cell Biology and Technology*,
DOI 10.1007/978-1-60327-905-5_18
© Humana Press, a Part of Springer Science + Business Media, LLC 2009

additional need for immunosuppression in allograft recipients. Recently it has been proposed that much smaller pieces of limbal tissue containing CESCs can be cultured in the laboratory, and this ex vivo expanded tissue can then be transplanted to the eye with LSCD. In a significant number of with LSCD, the disease is total and bilateral, which precludes any expansion of existing CESCs. The possibility of using alternative autologous sources of epithelial SCs to regenerate the corneal surface is also now the subject of laboratory study and clinical application.

Our knowledge of CESC biology is rapidly growing. However, to fully benefit from their therapeutic use, the ability to accurately identify them by specific marker expression and the ability to understand the exact nature of the cellular and molecular mechanisms that maintain their "stemness" is required. Once these markers have been identified and our understanding of the physiological processes involved in maintaining the SC niche are complete, the therapeutic implications will be vast.

Keywords Stem cell • Corneal epithelium • Corneal epithelial stem cell • Limbal stem cell • Limbal stem cell deficiency • Stem cell transplantation • Amniotic membrane • Allograft • Autograft • Cultured limbal stem cells

Introduction

In the field of regenerative medicine, the possibility of using stem cells (SCs) to replace diseased or dysfunctional tissue has always been an ultimate goal. In the case of the cornea, the use of adult SCs in the regeneration of corneal epithelium has now become a clinical reality. The cornea is unique in that it is transparent, superficially located, and contains a SC population that is spatially segregated from its progeny, allowing its separate study. This chapter will discuss corneal epithelial anatomy, corneal epithelial stem cell (CESC) biology, and the application of this biology in the field of regenerative medicine.

Corneal Structure and Function

The cornea is the clear dome-shaped window at the front of the eye (Fig. 1a, b). It is the most important focusing structure in the eye, contributing more than two-thirds of the total refractive power. The focusing of light by the cornea and to a lesser extent, the crystalline lens, allows a sharp image to fall on the retina, which allows for subsequent clear visual perception. Even minor disturbances of the corneal ultrastructure will have marked consequences on its optical properties, leading to devastating visual loss. Corneal disease represents the second most common cause of world blindness after cataract (1). The structure of the cornea is made up of five distinct layers (Fig. 2):

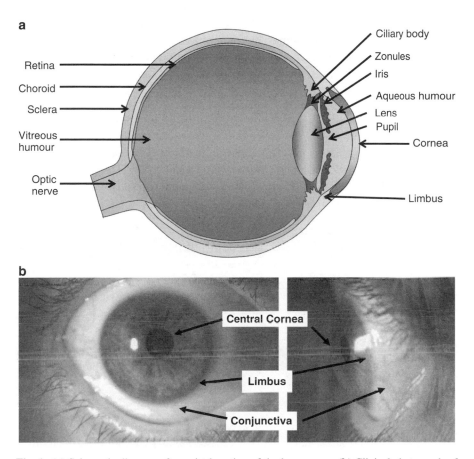

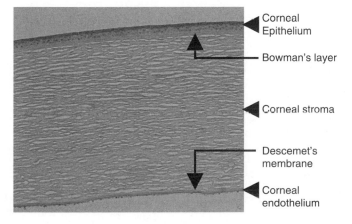

Fig. 1 (**a**) Schematic diagram of a sagittal section of the human eye. (**b**) Clinical photograph of the external view of the anterior segment of the human eye

Fig. 2 Histology of the human cornea stained with haematoxylin and eosin (H&E)

1. *Corneal epithelium* (Fig. 3a, b). This is composed of nonkeratinising stratified squamous epithelial cells that make up approximately 10% of the corneal thickness. It is usually 5–7 layers thick and is composed of a single basal layer of columnar cells, two or three layers of wing cells, and two or three layers of superficial squamous cells. The cells of the corneal epithelium are connected by a variety of intercellular junctions, especially desmosomes (Fig. 3c), and are attached to the underlying basement membrane by hemidesmosomes (Fig. 3d). The most superficial cells of the corneal epithelium are attached by means of tight junctions that prevent the penetration of the tear film and its components. The superficial squamous epithelial cells contain multiple projecting microplicae (Fig. 3e) that are covered in a glycocalyx, allowing a stable interaction with the overlying tear film.

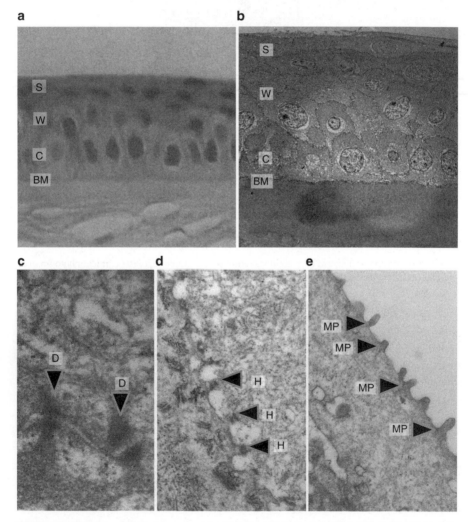

Fig. 3 Ultrastructure of the human corneal epithelium. (**a**) Light microscopy (stained with haematoxylin and eosin [H&E]). (**b–e**) Electron microscopy: *BM* basement membrane, *C* columnar cell monolayer, *W* wing cells, *S* squamous cells, *D* desmosomes, *H* hemidesmosome, *MP* microplicae

Together, these anterior structures form the important optically smooth outer layer of the anterior segment, allowing precise light refraction. Unlike the other layers of the cornea, the epithelium is continuously replaced. To allow this to occur, there is a source of CESCs that resides in the basal epithelium of the corneal limbus. The limbus is the name given to the junction of the corneal and conjunctival epithelium. The conjunctiva extends from the corneoscleral limbus to the mucocutaneous junction of the eyelids. The structure of the conjunctival epithelium differs from that of the cornea in a number of ways. First, it consists of nonkeratinising squamous epithelial cells that are *irregularly* and *loosely* arranged on a basement membrane. Second, it contains goblet cells that are normally never present in corneal epithelium. Goblet cells are unicellular, mucin-secreting cells that make up approximately 10% of the basal conjunctival cell population. Finally, the conjunctival stroma underlying the epithelium is richly vascularised, unlike that of the cornea, which is avascular.

2. *Bowman's layer.* This is an acellular condensation of the outer portion of the corneal stroma consisting mainly of collagen fibres (types I and III) and proteoglycans. Unlike the epithelium, it has no power of regeneration.

3. *Stroma (substantia propria).* The stroma makes up 90% of the thickness of the cornea. It consists of an extracellular matrix made up of collagens (types I, V, and VI) and proteoglycans (decorin, associated with dermatan sulphate, and lumican, associated with keratan sulphate). The collagen fibrils are regularly arranged, which allows for corneal transparency. Keratocytes are scattered throughout the stroma lying between the collagen lamellae. They secrete components of the stroma, which allows for its maintenance and repair.

4. *Descemet's membrane.* This is the basement membrane of the corneal endothelium. Its principal components are collagen type IV and laminin.

5. *Corneal endothelium.* This is made up of a single layer of closely interdigitating hexagonal cells. They have an important role in pumping out water from the corneal stroma, which, together with the regular spacing of collagen lamellae, allows for corneal transparency. Unlike the corneal epithelium, human corneal endothelium does not proliferate in vivo.

The corneal epithelium is the only layer of the cornea that is exposed to the external environment, and, as such, it undergoes a rapid and continuous cell turnover. As corneal epithelial cells are continuously lost from the surface of the eye, they must be continuously replaced to maintain the integrity and function of the cornea. The cells that are ultimately responsible for repopulation of any epithelium are termed SCs.

Corneal Epithelial Stem Cells

Stem Cells

SCs can be defined as relatively undifferentiated cells that have the capacity to self-renew and also generate one or more differentiated daughter cells *(2)*. SCs can be classified by the extent to which they can differentiate into different cell types, such as *totipotent, pluripotent, multipotent,* or *unipotent.*

1. *Totipotent SCs* can differentiate into any cell type in the body, including the placenta. A zygote is an example of a totipotent SC. Cells produced by the first few divisions of the zygote are also totipotent.
2. *Pluripotent SCs* develop about 4 days after fertilisation and can differentiate into any cell type except for totipotent SCs and cells of the placenta. Cells of the inner cell mass are considered pluripotent.
3. *Multipotent SCs* are capable of producing lineages that can differentiate into two or more cell types. For example, haematopoietic SCs can give rise to all the cells found in blood, including red blood cells, white blood cells, and platelets.
4. *Unipotent SCs*, also known as progenitor cells, can produce only one cell type. Epidermal SCs and CESCs fall into this category.

The study of SC biology has recently become one of the most active areas in scientific research. However, to date, no molecular markers have been recognised that definitively identify SCs. Therefore, SCs are usually defined by a group of common characteristics *(3)*:

1. *Long lifespan.* SCs have a long life and have the ability to proliferate indefinitely throughout the lifetime of the organism in which they reside *(4)*.
2. *Undifferentiated.* SCs are poorly differentiated or undifferentiated. The cytoplasm of SCs appears primitive and contains few, if any, differentiation products *(5)*.
3. *Slow cycling.* SCs have a long cell cycle time and thus are slow cycling, indicating low mitotic activity.
4. *High proliferative potential.* Although SCs exhibit extremely low rates of proliferation under steady-state conditions, they are endowed with high proliferative potential, which allows replacement of tissue when the need arises *(4, 5)*.
5. *Self-renewing and asymmetric division.* It is essential for the functioning of the tissue that a constant pool of SC is maintained throughout the lifetime of the organism. This is achieved by asymmetric cell division, giving rise to one daughter cell that remains a SC and a second daughter cell that will go on to differentiate (Fig. 4). These latter cells have been traditionally referred to as transient amplifying cells (TACs) *(6)*. TACs are more committed to differentiation than the SCs and also have a more limited ability to proliferate. The primary purpose of these TACs is to increase the number of cells resulting from each SC division. SCs and TACs fall into the proliferative tissue compartment and are capable of preceding cell mitosis with DNA synthesis. TACs differentiate into postmitotic cells (PMCs), which fall into the nonproliferative (differentiative) tissue compartment and are incapable of further division. PMCs are committed to cellular differentiation and mature into terminally differentiated cells (TDCs), which represent the ultimate expression of the functional aspect of the tissue *(3, 5)*.
6. *Error-free division.* This feature is essential since any genetic error at the level of the SC will permanently pass on to the whole clone of cells, resulting in cellular dysfunction. Several protective mechanisms have been developed to minimise any error made in SC mitosis. First, SCs remain relatively quiescent during steady growth and leave the job of active DNA synthesis and cell amplification to TACs.

Compartment   **Cell Type**

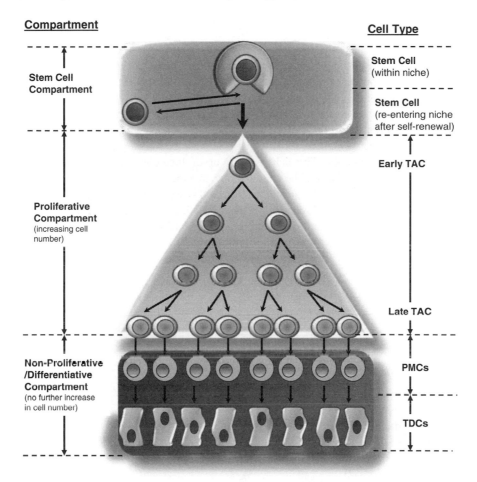

Fig. 4 Schematic diagram of the stem cell hierarchy showing the three cell compartments: the stem cell compartment, the proliferative compartment, and the nonproliferative or differentiative compartment. *SC* stem cell, *TAC* transient amplifying cell, *PMC* postmitotic cell, *TDC* terminally differentiated cell. The SC divides asymmetrically to produce another SC, which reenters the SC compartment and a young TAC, which enters the proliferative compartment. The TAC undergoes a variable number of divisions and matures, thereby allowing expansion in cell numbers. Eventually, the late TAC differentiates into a PMC, which can no longer divide, and enters the nonproliferative compartment, where the cell number is constant and finally matures into the TDC, which has the characteristic phenotype of the host tissue

Therefore, even if an error is made, this will be self-limited since TACs have a limited lifespan. Second, it has been shown that there may be asymmetrical DNA segregation during SC mitosis, suggesting that the SC retains its original genetic message and allows the new copy to be passed on to the TAC (*7*).

Stem Cell Niche

The factors that maintain the "stemness" of a SC are incompletely understood. Both intrinsic SC properties (characteristics inherent to the SC) and extrinsic SC properties (characteristics of the environment surrounding the SC) are thought to play a role *(8)*. The maintenance of "stemness" by extrinsic properties is best explained by the "stem cell niche" model initially proposed by Schofield *(9)*, who suggested that SCs exist in an optimal microenvironment or niche that promotes the maintenance of the SC in an undifferentiated condition. The niche is made up of three components: the SCs themselves, supporting cells within the mesenchyme, and the extracellular matrix produced by both SCs and support cells surrounded by blood vessels *(3)*. Following cell division, only one of the daughter cells can reenter the niche, thus replenishing the SC pool. The other cell cannot reenter the niche and therefore enters a less favourable environment that does not protect the cell from entering the pathway of differentiation. Therefore, this cell becomes a TAC that will eventually differentiate into a TDC (Fig. 5).

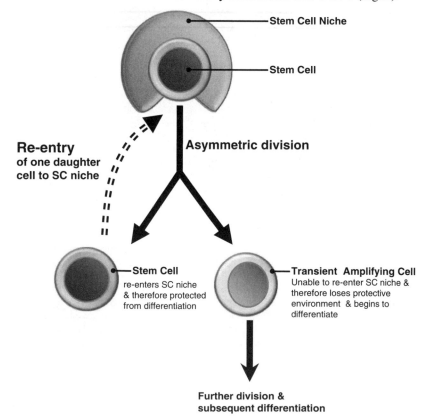

Fig. 5 Diagrammatic representation of Schofield's niche hypothesis. The stem cell (SC) is protected from differentiation by the various components of the SC niche. Following division, only one daughter cell can re-enter the niche thereby preserving SC numbers (*dashed arrow*). The other daughter cell can no longer enter the niche and therefore loses its protective environment and differentiates into a transient amplifying cell

Corneal Epithelial Cell Turnover

The corneal epithelium is exposed to the external environment. There is a continual loss of epithelium through desquamation of the superficial cells centrally. Since the epithelial mass remains relatively constant, there must be continuous replacement of these lost cells through mitotic division. In 1983, prior to our knowledge of the existence of SCs in the cornea, Thoft and Friend *(10)* proposed their X, Y, Z hypothesis to explain the turnover dynamic of the corneal epithelium during normal homeostasis and in response to injuries (Fig. 6). The hypothesis states that the amount of proliferation of basal epithelial cells (*X*), together with the centripetal movement of the peripheral cells (*Y*), is equal to the epithelial cell loss from the surface (*Z*). Corneal epithelial maintenance can then be defined by the equation $X + Y = Z$, which simply states that if the corneal epithelium is to be maintained, cell loss must be balanced by cell replacement.

The cells ultimately responsible for the repopulation of the epithelium are the CESCs, which are also known as limbal stem cells (LSCs) due to their anatomical location. The CESCs are thought to reside in the basal layer of the limbus, the transition zone between the corneal and conjunctival epithelium. Differentiating SCs from the limbus move centripetally onto the cornea to become TACs. These cells, located in the basal layer of the cornea, proliferate and then travel vertically and differentiate into the PMCs of suprabasal corneal epithelium. Further differentiation of PMCs results in the TDCs of the superficial corneal epithelium. Only basal cells of the epithelium are able to divide. Once epithelial cells lose their attachment to the basement membrane, they lose their capacity to divide and become terminally differentiated during migration to the surface of the cornea *(11)*. CESCs at the limbus can also differentiate and migrate superficially rather than centripetally. This process is believed to be important in establishing a barrier to prevent the encroachment of conjunctiva onto the corneal surface *(12)*.

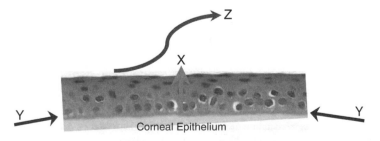

Fig. 6 Diagrammatic representation of Thoft's X, Y, and Z hypothesis. *X* proliferation of basal cells, *Y* centripetal movement of cells, *Z* cell loss from the surface

Evidence for Existence and Location of CESCs

Precise identification, isolation, and characterisation of CESCs have not been possible due to the lack of specific markers. However, there are several pieces of indirect evidence to indicate the presence of CESCs at the limbus:

1. The concept of the limbal location of corneal SC was first specifically suggested by Davanger and Evenson *(13)* in 1971. They studied the healing of corneal epithelial defects in pigmented eyes of guinea pigs. During healing of such defects, they observed pigmented epithelial cell lines that migrated from the limbal region toward the central cornea. They suggested that the palisades of Vogt, a pigmented structure found at the limbus, served as the generative organ for corneal epithelial cells.
2. In 1986 Schermer et al. *(6)* studied the expression of a tissue restricted cytokeratin K3 (CK3), a 64-kDa basic keratin, in rabbit corneal epithelial cells using the monoclonal antibody AE5. The differential expression of keratins allows the separation of cell populations in the cornea according to their level of differentiation. Schermer et al. showed that CK3 represents a marker for an advanced state of corneal epithelial differentiation. They found that CK3 is expressed in *all* the layers of the central corneal epithelium but only in the *suprabasal* layers of the limbal epithelium (Fig. 7). Since basal cells of the limbal epithelium lacked a marker for an advanced stage of corneal epithelial differentiation, limbal basal cells must be more primitive than corneal basal cells.

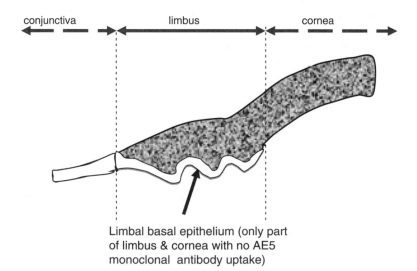

Fig. 7 Diagrammatic representation of the experiment by Schermer et al. All layers of the cornea and limbus except the basal layer of the limbus are stained with the monoclonal antibody (AE5), which is directed against the corneal differentiation marker cytokeratin K3, shown here by the *mottled shading*

This led to the conclusion that CESCs are not uniformly distributed throughout the entire corneal epithelial basal layer, but preferentially located in the limbal epithelial basal layer. Another important finding of their study was that *conjunctival* epithelium is CK3 negative, indicating that corneal epithelium is distinct from conjunctival epithelium and arguing against the previously purported concept of conjunctival transdifferentiation acting as a source of corneal epithelial cell renewal. Subsequently, a similar expression profile has been shown for CK12, a 55-kDa acidic keratin, which is also specific for corneal differentiation *(14)*.

3. SCs have a slower cell cycle and can be detected experimentally as label-retaining cells (LRCs) *(15)*. LRCs can be identified in this way by means of a pulse chase experiment. This involves long-term labelling of all the dividing cells in the tissue with a DNA precursor such as tritiated thymidine (^{3}HT) or bromodeoxyuridine (BrdU). Following a chase period of usually 4–8 weeks, the rapidly dividing cells lose most or all of their labels, while the slow cycling SCs retain the label. In 1989 Lavker et al. *(11)* were the first to show that LRCs (labelled with ^{3}HT) were found exclusively in the basal layer of the limbal epithelium but were completely absent in the central cornea. From studies measuring the percentage of LRCs in the limbal zone, it has been concluded that SCs may represent less than 10% of the total limbal population.

4. SCs have a very high proliferative potential. Several findings suggest the presence of cells with high proliferative potential at the limbal basal epithelium. First, cells from various regions of the cornea can be grown in identical culture conditions and subcultured repeatedly in order to compare the proliferative potential of the different subpopulations. This approach shows that corneal epithelial cells from the limbus grow far better than peripheral and central corneal epithelium *(11, 16, 17)*. In addition, labelling studies have demonstrated that the mitotic index of the corneal epithelium is higher toward the periphery, suggesting that the peripheral corneal basal cells are more active in DNA synthesis *(18)*.

5. Clinical evidence for the limbal location of CESCs comes from the observation that corneal epithelial wounds fail to heal normally after removal of the limbal epithelium *(12, 19, 20)*. In such cases, the abnormal corneal surface is replaced by conjunctival epithelial ingrowth, corneal vascularisation, and chronic inflammation, which are all cardinal features of limbal stem cell deficiency (LSCD).

6. Larger peripheral corneal epithelial wounds heal significantly faster than smaller central wounds. These findings indicate the CESCs are located at the corneal periphery and give rise to dividing cells that migrate centrally *(21, 22)*.

7. Further clinical evidence supporting the limbal location of CESCs comes from the success of limbal transplantation in LSCD. In 1989 Kenyon and Tseng *(23)* demonstrated that limbal epithelial cell transplants can be used to reconstitute a healthy corneal epithelium in patients with LSCD. Similarly, Tsai et al. *(24)* have demonstrated the successful reversal of experimentally induced LSCD with limbal transplants in a rabbit model.

8. Finally, corneal tumours nearly always arise from the limbus rather than the cornea itself *(25)*. Since most tumours are thought to arise from SCs *(26)*, this further suggests the limbal location of CESCs.

Current Model of Corneal and Limbal Epithelial Physiology

Basic SC theory, corneal anatomy, the X, Y, Z hypothesis, and the evidence for the existence of CESCs at the limbus have been presented above. Collectively, this knowledge allows us to understand the current model of corneal and limbal epithelial physiology, which will now be outlined (Fig. 8).

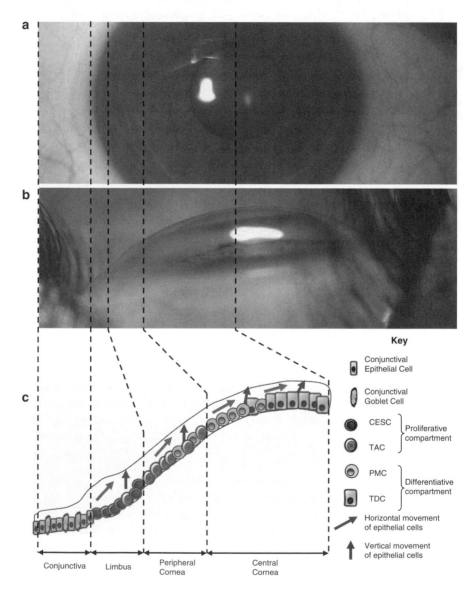

Fig. 8 (**a**) Anterior view of the human anterior segment. (**b**) Corresponding side view. (**c**) Corresponding diagrammatic representation of the model for corneal epithelial cell turnover

CESCs are located exclusively in the limbal basal epithelium. These CESCs are undifferentiated (therefore lack CK3), slow cycling (therefore label retaining), and have high proliferative potential (therefore faster healing capacity). Under steady state conditions, the CESCs are relatively quiescent. The specific location of CESCs at the limbus confers several functional advantages by providing a specialised niche environment, which is discussed in the section "CESC Niche and the Palisades of Vogt." The CESCs give rise to TACs by asymmetric division, as described above. These TACs are found in the basal layer of the corneal epithelium and are CK3 positive, indicating that they are more differentiated than the limbal basal cells. The TACs are larger in number compared to the CESCs but are not slow cycling and have a lower proliferative potential. Unlike CESCs, the TACs have a limited capacity for self-renewal and actively divide in steady-state conditions to maintain the corneal epithelial mass. As with other epithelial systems, the advantage of having a TAC population is that they allow the amplification of each SC division, thereby conserving SC energy and numbers and minimising the accumulation of mutations (27). The young daughter TACs have a high proliferative potential and are located in the peripheral corneal epithelium (28). These TACs migrate toward the central cornea and slowly lose their proliferative potential. Once the proliferative capacity of the TACs has been depleted, they undergo differentiation into PMCs, which mature into TDCs. The latter two cell types form the bulk of the corneal epithelium and display the inherent characteristics of this tissue. The whole process from CESC division to terminal differentiation requires 7–14 days, after which the superficial cells are desquamated into the tear film (29).

CESC Niche and the Palisades of Vogt

The SC niche hypothesis model proposed by Schofield in 1983, as discussed above, suggests the importance of intrinsic properties (characteristics inherent to the cell) and extrinsic properties (characteristics of the SC environment) in maintaining "stemness." The limbus and particularly the palisades of Vogt are thought to contain the CESC niche (8). The palisades of Vogt consist of a series of radially orientated fibrovascular ridges that are concentrated along the superior and inferior limbus (Fig. 9). The region in between the ridges are occupied by the epithelial rete pegs, which consist of 10–15 layers of epithelial cells where the CESCs are thought to reside (30). If the palisades of Vogt at the limbus represent the LSC niche, then the basement membrane and stromal matrix in this zone must be different from that seen in the central cornea to allow the "stemness" of these cells to be maintained. Indeed, the limbal epithelium of the palisades of Vogt has many differences from the central corneal epithelium:

1. The palisades of Vogt consist of a highly folded limbal epithelium projecting into a richly vascularised stroma. The presence of blood vessels is one of the most fundamental differences of the limbus from the cornea. The close proximity of the limbal epithelium to blood vessels allows increased levels of nutrition as well as closer proximity to blood-borne cytokines.

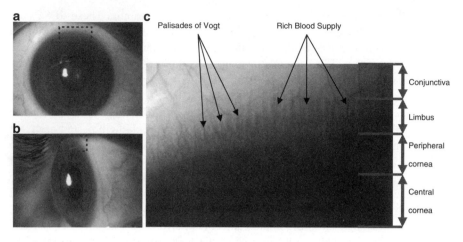

Fig. 9 The human superior limbus. (**a**) Anterior view. (**b**) Side view. (**c**) Magnified view of the area demarcated by the *dotted line* in (**a**) and (**b**) demonstrating the palisades of Vogt and the rich blood supply

2. The limbal epithelium is highly undulating with underlying pegs of stroma extending upward *(31)*. This is seen on the external surface of the eye as radial lines at the limbus. Such an arrangement results in better anchorage of the CESCs and therefore protects them from shearing forces *(4)*.
3. The CESCs of the palisades of Vogt are located at the deepest level of the limbal epithelium. This naturally protects them from external environmental insults, which must penetrate the full thickness of the epithelium before the CESCs are damaged.
4. It is the requirement that the central cornea must be transparent and as such the basal cells of the corneal epithelium lack any pigmentation and are therefore susceptible to ultraviolet (UV) damage. However, basal limbal epithelial cells do not have this restriction and are heavily pigmented due to the presence of melanin deposits within the palisades of Vogt, which confers UV protection to the SC population *(11, 16)*.
5. Indirect immunohistochemical evidence suggests CESCs are more abundant at the superior and inferior limbus compared with the nasal and temporal limbus *(32)*. Such an arrangement would further protect the CESCs because the eyelids cover most of the superior limbus, whereas the nasal and temporal limbus is exposed to the external environment.
6. The ultrastructure of the basement membrane (BM) of CESCs differs significantly from the BM of corneal epithelium. Reported differences include different isoforms of collagen IV and laminin *(33)*. These differences presumably allow differential binding of cytokines, which allows the maintenance of CESCs in their undifferentiated state within the niche *(3)*.
7. CESCs within the palisades of Vogt are uniquely positioned to be influenced by signals from a variety of sources, including adjacent corneal epithelial cells (early TACs), adjacent conjunctival epithelial cells, mesenchymal cells including corneal

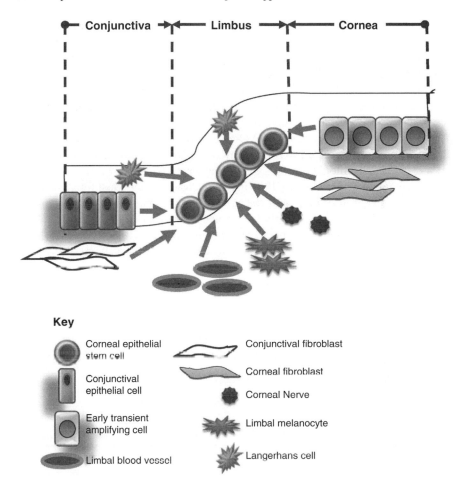

Fig. 10 Diagrammatic representation of the corneal epithelial stem cells (CESC) niche. The CESCs are uniquely positioned to receive cellular signals from a variety of sources (as indicated by the *arrows*) both by direct cell contact and diffusible cytokines

and conjunctival fibroblasts, smooth muscle and vascular endothelium of the multiple limbal blood vessels, corneal nerves, Langerhans cells, and limbal melanocytes (Fig. 10).

CESC Markers

The indirect identification of SCs can be made on the basis of demonstration of their common properties, as discussed above, such as label-retaining capacity and high proliferative potential. However, the direct identification of SCs within a population of cells requires the presence of a specific SC marker. The identification of a uniquely specific CESC marker remains elusive. Many potential markers have been studied over the past few years (Table 1). The role of many of these molecules as CESC markers remains controversial due to conflicting results between researchers. This may be due, in part, to the different methodologies used between studies and also variations in expression of these markers between different species. Although none of the above markers can be considered absolutely CESC specific, certainly the presence, absence, or relative expression of these markers in the corneal epithelium allows the description of a putative SC phenotype *(63)*. In general, these putative markers fall into two categories: first, SC-associated markers that positively identify CESCs, and second, those markers that are associated with corneal *differentiation* and therefore can be thought of as negative markers of CESCs. The most studied of the *positive* putative markers of CESC that are considered the most reliable are the transcription factor p63 *(47, 48)*, the ATP binding cassette transporter protein ABCG2 *(60, 64)*, and more recently the leucine zipper transcription factor C/EBPδ *(51)*. The most studied and reliable of the *negative* markers (corneal differentiation markers) are the keratin pair CK3/12, the cytosolic structural protein involucrin, and the gap junction proteins connexin 43 and connexin 50.

p63

p63 is a nuclear transcription factor that is a member of the p53 and p73 gene family *(65)*. p63 has been suggested as a positive marker for CESCs *(47)*. p63 is expressed

Table 1 Putative corneal epithelial stem cell (CESC) markers (with corresponding references)

	Putative markers	Presence(+) or absence(−) in CESCs (basal limbal epithelium)	References
Cytoskeleton proteins			
Cytokeratins	CK3/12	−	*(6, 14, 34, 35)*
	CK19	+	*(34, 36, 37)*
	Vimentin	+	*(36–38)*
Cytosolic proteins			
Metabolic enzymes	Cytochrome oxidase	+	*(39)*
	Na/K-ATPase	+	*(8)*
	Carbonic anhydrase	+	*(40)*
	α-enolase	+	*(41–43)*
	Protein kinase C-γ	+	*(44)*
Cell-cycle proteins	Cyclin A, D, & E	+	*(45)*
Structural protein	Involucrin	−	*(34, 46)*
Nuclear proteins			
Transcription factors	p63	+	*(47–50)*
	C/EBPδ	+	*(51)*
Cell surface proteins			
Cell–cell and cell–matrix	Connexin 43 & 50	−	*(46, 52, 53)*
interaction molecules	Integrin α9 & β1	+	*(54–56)*
Growth factor receptors	TrKA	+	*(57–59)*
Transporter molecules	ABCG2	+	*(34, 60–62)*

in the basal cells of stratified epithelium such as skin and cornea and plays an essential role in epithelial development and differentiation. This is demonstrated in p63 knock-out mice that have a complete absence of stratified squamous epithelium *(66)*. Similarly, in humans, mutations in the p63 gene can result in the EEC syndrome (ectrodactyly, ectodermal dysplasia, and facial clefts), which is an autosomal dominant condition characterised by the lack of normal stratified epithelium *(67)*.

p63 exists in several isoforms containing a DNA-binding domain and a carboxyl-terminal oligomerisation domain. Isoforms with a transactivating aminoterminal domain are termed TAp63 and are generated by activity of an upstream promoter. Isoforms lacking this transactivating domain are termed ΔNp63 and are produced by a downstream intronic promoter. Both the TAp63 and ΔNp63 transcripts can have alternative splicing of the c-terminal domain, designated α, β, and γ. This gives a total of six possible isoforms. The ΔNp63 isoforms have been shown to be preferentially expressed in the epithelial basal cells of different organs *(68)*. In 2001 Pellegrini et al. *(47)* demonstrated that ΔNp63 was specifically located in the basal layer of the limbus but not in the corneal epithelium of human corneas. Using a combination of clonal analysis and Western blotting, they demonstrated that ΔNp63 was strongly expressed in limbal holoclones (SCs), weakly in meroclones (early TACs), but undetectable in paraclones (late TACs). Immunohistochemistry demonstrated abundant expression in the limbal basal layer but complete absence in the central corneal epithelium. There were very low levels of expression observed in occasional basal cells of the peripheral cornea adjacent to the limbus, which would correspond to the location of early TACs. Similarly, a study investigating the expression of ΔNp63 in human limbal epithelial cells grown on amniotic membrane indicated that p63 was localised to cells with high proliferative potential, including both CESCs and early TACs *(69)*. Chee et al. *(63)* demonstrated a gradient of ΔNp63 signalling across the basal corneal epithelium with the highest signal intensity in the limbus, followed by the peripheral cornea and finally the central cornea. The above data suggest that ΔNp63 is expressed by CESCs *and* early TACs rather than CESCs alone. Di Iorio et al. *(48)* have recently suggested that the ΔNp63α isoform may be a more specific marker of the CESC. They provide data that show the ΔNp63α isoform is confined to the basal layer of the limbus and is up-regulated during and after corneal wounding and richly represented in holoclones. The β and γ isoforms are virtually absent from the resting limbus, are only found suprabasally on corneal wounding, and are absent in holoclones, suggesting that they are involved in epithelial differentiation following corneal regeneration and are not CESC markers.

ATP-Binding Cassette Subfamily G, Member 2 (ABCG2)

SCs from bone marrow and other tissues can be isolated based on their ability to efflux the vital dye Hoechst 33342, giving rise to a side population (SP) on flow cytometry *(70)*. The ability of SCs to exclude this dye is mediated by ABCG2 protein, a member of the ATP-binding cassette transporters. ABCG2 has been proposed as a universal marker of SCs *(71)*. Several recent studies have shown the expression of the ABCG2 protein exclusively in a subset of limbal epithelial cells *(60–62)*. De Paiva et al. *(60)* have localised ABCG2 to the basal cells of the limbal epithelium but absent from limbal suprabasal and corneal epithelia. ABCG2 positive corneal epithelial cells were found to have enriched SC properties, such as slow cycling, higher colony forming efficiencies, and higher p63 expression, when compared with ABCG2 negative cells. These early studies suggest that ABCG2 may represent a specific marker for CESCs.

C/EBPδ

CCAAT/enhancer-binding proteins (C/EBPs) are a family of transcription factors critical for cellular proliferation, differentiation, metabolism, inflammatory response, and many other biological events *(72)*. Six members of the family have been identified (C/EBPα, β, δ, γ, ε, and ζ) and have been shown to be distributed in a variety of tissues *(72)*. C/EBPδ regulates cell cycle specifically in epithelial cells by inducing a G_0/G_1 arrest *(73)*. Barbaro et al. *(51)* have recently shown that C/EBPδ is instrumental in regulating self-renewal and cell cycle length of CESCs. Their studies indicate that a proportion of quiescent CESCs coexpress ΔNp63α (discussed above), C/EBPδ and Bmi-1 under resting conditions. C/EBPδ is found to produce mitotic quiescence of CESCs by forcing cells into the G_0/G_1 phase of the cell cycle. Bmi-1 is responsible for SC renewal in haematopoetic and neural systems *(74, 75)* and is hypothesised to play a similar role in CESCs. In this scheme, it is suggested that upon corneal wounding, a fraction of these CESCs switch off C/EBPδ and Bmi-1 but maintain ΔNp63α, which becomes activated and proliferate to regenerate the corneal epithelium while losing their SC properties and switching from ΔNp63α expression to ΔNp63β and ΔNp63γ expression. Therefore, although both C/EBPδ and ΔNp63α may act as CESC markers, they are responsible for related but distinct processes: ΔNp63α is required for proliferation, whereas self-renewal also requires C/EBPδ *(51)*.

CK3/12

Cytokeratins are intermediate filament keratins found in the intracytoplasmic cytoskeleton of epithelial tissue. There are two types of cytokeratins: the low weight, acidic type I cytokeratins and the high weight, basic, or neutral type II cytokeratins. Cytokeratins are usually found in pairs comprising a type I cytokeratin and a type II cytokeratin. The cytokeratin pair CK3/12 is specifically found in corneal epithelial cells and regarded as markers of corneal differentiation (see the second point in the section "Evidence for Existence and Location of CESCs"). Specific antibodies can be used to stain CK3/12 within a tissue. Such an approach shows a complete absence of CK3/12 in the basal layer of the limbus underlying the undifferentiated nature of these cells. However, suprabasal limbal and all corneal epithelial cells are positive for CK3/12. Therefore, CK3/12 can be considered a reliable *negative* marker for CESCs, which by definition are undifferentiated.

Involucrin

In stratified epithelia, division of cells takes place in the basal layer and once these cells lose contact with the basement membrane, they begin to differentiate. In human epidermal keratinocytes, final differentiation is associated with the formation of an insoluble protein envelope on the cytoplasmic side of the cell plasma membrane. Involucrin has been demonstrated to be a soluble protein precursor of this envelope formed during an early stage of differentiation *(76)*. In the human cornea, the distribution of involucrin has been demonstrated to be similar to CK3/12 in that it is found in all layers of the cornea and superficial layers of the limbus but is completely absent from the basal cell layer of the limbus and is therefore also a marker of corneal differentiation *(34)*. Similarly, in ex vivo expanded CESC cultures, involucrin expression was found to be associated with the larger differentiated cells but relatively absent in the smaller SC-like cells *(77)*.

Connexin 43 and Connexin 50

Gap junctions allow communication between adjacent cells and are physically made up of six transmembrane proteins known as connexins *(78)*. Two members of this family are abundantly expressed in the corneal epithelium, namely connexin 43 (Cx43) and connexin 50 (Cx50) *(13, 78)*. Paucity or complete absence of gap junctions may be an essential property of SCs *(79)*. It is important that SCs are isolated from their adjacent progeny since their intracellular environment is under the control of separate regulatory pathways, so that it would make sense that connexins are absent from CESCs. Indeed, Cx43 is found in the basal layer of the corneal epithelium but is absent from the basal layer of the limbal epithelium where the SCs reside. Limbal basal cells that lack Cx43 are thought to represent SCs, and their subsequent expression of Cx43 denotes their differentiation into TACs *(52)*. Unlike Cx43, Cx50 is expressed in all the *suprabasal* layers of the corneal and limbal epithelium but absent in the basal layer of the limbal epithelium *(53)*. Thus, both Cx43 and Cx50 may be considered markers of corneal differentiation, with Cx43 expressed in TACs and Cx50 expressed at a more advanced stage of differentiation *(52, 80)*.

In conclusion, numerous studies show that CESCs preferentially express several markers, as shown above. However, as yet, there is no evidence to show that one specific marker can reliably identify the CESC population. The coexpression of more than one of the putative markers in CESCs suggests that these cells may express a set of markers rather than a single marker. With this in mind, a reliable approach to recognising CESC cells within a population would be to demonstrate the expression of a selection of a putative marker, such as coexpression of p63 and ABCG2 together with the absence of CK3/12 and Cx43.

Limbal Stem Cell Deficiency

The maintenance of a healthy ocular surface requires a healthy corneal epithelium, which in turn requires a healthy CESC population. The loss or dysfunction of CESCs leads to LSCD, a painful and blinding condition. LSCD is an important cause of corneal vascularisation and scarring. Corneal vascularisation and opacity have been estimated to cause blindness in eight million people worldwide each year *(81)*. In LSCD, reduced vision is due to corneal disease where otherwise the patient's eye is healthy. This would mean that effective treatment of LSCD would restore vision in these patients. Until recently, the long-term results of conventional treatment of severe LSACD have been rather disappointing (See "Keratolimbal Allograft" section). The understanding of the SC theory summarised above has allowed the development of new strategies involving cultured CESCs in treating this condition with encouraging results.

Causes of LSCD

In the majority of individuals, the complement of CESCs is able to maintain a healthy corneal epithelium for life. However, any pathology in which the CESCs are absent or destroyed or in which the SC niche is damaged leads to an inability to maintain a normal stratified corneal epithelium, and the clinical state of LSCD ensues.

The causes of LSCD can be divided into congenital and acquired. Additionally, they can be divided, first, into causes where there is direct damage to and loss of CESCs and, second, into causes where the CESC niche is damaged. The main causes are summarised in Table 2. Although specific aetiologies may preferentially

Table 2 Causes of limbal stem cell deficiency

Primary (hereditary/congenital)	Secondary (acquired)	Idiopathic
Aniridia	Chemical or thermal burns	No specific cause found
Ectodermal dysplasia	Mechanical trauma	
Multiple endocrine deficiency	Stevens-Johnson's syndrome	
	Ocular cicatricial pemphigoid	
	Contact lens–related limbal stem cell deficiency	
	Cytotoxic agents: (5-fluorouracil/ mitomycin C)	
	Inflammatory eye disease	
	Multiple or extensive limbal surgery	
	Limbal cryotherapy	
	Severe microbial infection involving limbus	
	Peripheral ulcerative keratitis	
	Neurotrophic keratopathy	
	Pterygium	
	Radiotherapy	
	Limbal tumours	

damage the SC itself or the SC niche, most affect both of these factors to some degree. Indeed, successful reconstruction of the ocular surface in LSCD requires both restoration of CESC numbers (SC transplants) *and* restoration of the normal SC niche environment (normal lid function, adequate tear function, and the possible adjunctive use of amniotic membrane and autologous serum [AS]).

Symptoms and Signs

The symptoms and signs of LSCD can be understood by considering the main functions of CESCs in normal eyes. First, the ability of the corneal epithelium to repopulate itself is lost, leading to recurrent epithelial breakdown and inflammation that result in the clinical symptoms of ocular pain, redness, photophobia, tearing, blepharospasm, and reduced vision. Second, the barrier function of the CESCs at the limbus is lost, leading to replacement of the ordered corneal epithelium with conjunctival epithelium. This process is known as *conjunctivalisation* and it represents the hallmark of LSCD. The direct loss of CESCs and the conjunctivalisation of the corneal surface lead to a number of clinical signs (Fig. 11):

1. *Abnormal corneal epithelium.* As described earlier, normal corneal epithelial cells are held together with tight junctions that prevent permeability to many hydrophilic molecules such as fluorescein. However, cellular connections between conjunctival epithelial cells are much looser, making the tissue more permeable. This results in the late staining of conjunctivalised corneal epithelium with fluorescein seen in LSCD. In addition, the corneal epithelium loses its normal clarity and takes on a grey hazy appearance. The abnormal epithelium also demonstrates a very short tear film break-up time. Additionally, the loose

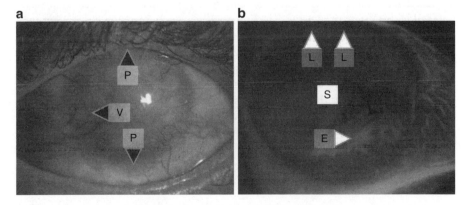

Fig. 11 (**a**) Slit lamp photograph of an eye with limbal stem cell deficiency (LSCD). Note the opaque and scarred cornea and conjunctival vessel dilation. *V* conjunctival vessels on the corneal surface, *P* loss of palisade of Vogt architecture. (**b**) Slit lamp photograph of an eye with LSCD stained with topical fluorescein. *S* diffuse mild corneal staining with fluorescein, *L* late and persistent staining of the abnormal palisades of Vogt, and *E* epithelial defect

connections between the epithelial cells allow entry of leucocytes from the tear film, contributing to redness from the chronic inflammation seen in this condition.

2. *Persistent epithelial defects*. The lack of CESCs results in epithelial defects that are unable to heal. Even conjunctivalised corneal epithelium is susceptible to breakdown due to the lack of hemidesmosomes in conjunctival tissue. These persistent epithelial defects may lead to secondary stromal infiltration and subsequent melting and even perforation of the cornea on occasion, which may ultimately lead to loss of viability of the eye.

3. *Loss of limbal architecture*. Normal limbal architecture with regular rows of palisades of Vogt has been described above. In LSCD, there is loss of the palisades and characteristic late irregular and radial staining of the limbal epithelium due to conjunctival encroachment.

4. *Corneal vascularisation*. The ability of corneal tissue to secrete antiangiogenic factors has recently been elucidated *(82)*. However, conjunctival epithelial cells that encroach on the cornea are unable to produce these antiangiogenic factors, leading to the development of both superficial and deep corneal vascularisation. The extent of the vascularisation will depend on the degree of CESC loss.

5. *Corneal scarring*. Typically, LSCD is associated with chronic epithelial defects that continually heal and breakdown and are associated with low-grade inflammation. Provided that the epithelial defects do not progress to thinning and perforation, there is often remodelling of the affected cornea to form scar tissue, resulting in further visual loss.

Diagnosis

The accurate diagnosis of LSCD is crucial since conventional treatments for scarring corneal disease such as penetrating keratoplasty (full thickness corneal transplant) will fail in this group of patients. The diagnosis is usually made on clinical grounds. In patients with a known cause of LSCD and the characteristic signs of LSCD, namely a conjunctivalised corneal surface, persistent epithelial defects, and loss of palisades of Vogt, a diagnosis is fairly straightforward. However, where the diagnosis is less clear or when surgical treatment is considered, a tissue diagnosis is required.

Although several of the symptoms and signs discussed above may be seen in many ocular surface diseases, the hallmark of LSCD is conjunctivalisation of the corneal surface *(4)*. From a histological perspective, this can be shown definitively by the presence conjunctivalised epithelium on the corneal surface. To perform such a histological examination, the superficial layers of the abnormal corneal epithelium may be removed in the form of impression cytology or a tissue biopsy.

Corneal Impression Cytology and Histopathological Examination

Corneal impression cytology was initially presented as a relatively non-invasive technique that involves obtaining corneal epithelial cell samples by gently pressing cellulose acetate paper onto the ocular surface for a few seconds *(83)*. This has been shown to remove one to three cell layers of the surface epithelium, preserving its morphology and allowing further histological examination *(84)*. Unlike other techniques of sampling, such as corneal biopsy and corneal scrapes, impression cytology preserves morphological information and allows sampling of a large area of corneal epithelium that can be repeated as necessary.

Histologically, the hallmark of LSCD is the presence of goblet cells on the corneal surface. Impression cytology specimens of the corneal surface can be examined by traditional cytology techniques, such as haematoxylin and eosin (H&E) and periodic acid Schiff (PAS) staining. These techniques will clearly show the presence of any goblet cells in the impression cytology sample (Fig. 12a). Studies suggest that when the presence of goblet cells in an impression cytology sample is the only criterion considered, there is a significant false-negative diagnosis of LSCD *(85, 86)*. This is mainly due to the difficulty of distinguishing ectopic conjunctival epithelium and corneal epithelium by conventional cytological means if typical goblet cells are not present on the specimen *(86)*. For this reason, the identification of cytokeratins (CKs) that are specific for corneal and conjunctival epithelium has been suggested as a more accurate method of identification of LSCD. Of the human CKs, only CK3 and CK19 have been demonstrated to discriminate between corneal and conjunctival epithelia. CK3 stains all layers of normal corneal epithelium but does not stain the conjunctival epithelium, whereas CK19 stains the conjunctival but not corneal epithelium *(87)*. Since goblet cells may not always be present in early conjunctivalisation of the cornea or may not be picked up by the impression cytology, Donisi et al. *(86)* suggest that the only possible way to differentiate true corneal epithelium from the conjunctiva is to study the differences in phenotypic expression of CKs. The immunophenotypic definition of cornea derived tissue is

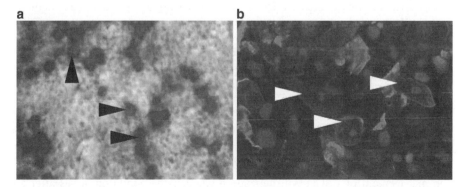

Fig. 12 Impression cytology specimens from a central cornea of a patient with limbal stem cell deficiency (LSCD). (**a**) Corneal impression stained with periodic acid Schiff. The *black arrows* indicate the presence of multiple goblet cells (stained darkly), indicating conjunctivalisation of the corneal surface (i.e., LSCD). (**b**) Corneal impression stained for CK19. The *white arrows* indicate the diffuse staining of the cytoplasm of multiple cells positive for CK19, indicating the presence of conjunctival cells on the surface of the cornea (i.e., LSCD). The round central nucleus of each cell is stained with propidium iodide

CK3$^+$/CK19$^-$ and of conjunctiva-derived tissue is CK3$^-$/CK19$^+$. According to the same authors high CK3$^+$/CK19$^+$ expression could be considered an initial step (recent damage) in the progression of LSCD, whose end stage is the acquisition of a full CK3$^-$/CK19$^+$ phenotype. Taken together, the above information indicates that the ability to demonstrate CK19$^+$ cells on impression cytology specimens is a sensitive way of diagnosing LSCD (Fig. 12b).

Direct sampling of conjunctivalised corneal epithelium can also be obtained by direct surgical biopsy or scraping. This may be performed on a small area for diagnostic purposes or during superficial keratectomy at definitive surgery. In either case, the biopsy specimens are stained routinely with H&E or PAS to demonstrate a multilayered epithelium, with goblet cells indicating a conjunctival phenotype. In addition, blood vessels and leucocytes may also be seen on the histology specimens, resulting from conjunctival vessel invasion and corneal inflammation. The specimen can also be stained with CK19 to demonstrate conjunctival cytokeratin expression on the surface of the cornea.

Confocal Microscopy

The presence of the palisades of Vogt can be used as a clinical marker to indicate the presence of CESCs *(88)*. The ability to examine detailed corneal structure in vivo has been possible with the recent development of white-light confocal microscopy *(89)*. However, the ability to examine the palisades of Vogt with this modality has been limited due to their deep location and image degradation from light scattered by adjacent sclera unless the palisades are particularly prominent *(90)*. These problems have recently been overcome by the use of the laser scanning in vivo confocal microscope *(91)*. Using this modality, limbal epithelial cells have been clearly seen and their decreased density with increasing age has also been demonstrated *(91)*. It has been suggested that this technique may be important in comparing limbal epithelial density in LSCD compared to normal controls. This technique also allows examination of the limbus in its physiological state and avoids the artefacts induced by ex vivo studies and allows multiple examinations of the same tissue over time *(91)*.

Management of Partial LSCD

The management of LSCD will depend on whether the condition arises after an acute injury or is an established case. Assuming the case of LSCD is established, the management will depend on whether the patient has *partial* or *total* LSCD. In partial LSCD, there is still the presence of functioning CESCs. When the visual axis is covered in normal corneal epithelium and the patient is a relatively asymptomatic with good vision, *conservative* (medical) management is indicated *(92)*. However, where there is decreased vision, significant irritation, persistent epithelial defects, and conjunctival encroachment across the visual axis, *surgical* management, including mechanical debridement of conjunctival epithelium from the corneal surface and/or amniotic membrane transplantation, may be indicated *(4)*.

Conservative Management (Medical)

As discussed above, LSCD is often associated with an irregular corneal epithelial surface with decreased tear function, epithelial erosions and secondary infection, chronic inflammation, and persistent epithelial defects. The medical management is tailored toward dealing with each of these problems. The main medical treatments make use of topical medications that are unpreserved so as to prevent further toxicity associated with most topical preservatives and consist of:

1. *Unpreserved ocular lubricants (with punctal plugs)*. Ocular lubricants are used on a frequent basis to alleviate dry eye symptoms and signs. Ocular lubrication can be further increased by the insertion of small plugs (punctal plugs) into the lacrimal punctae of the upper and lower eyelids that work by preventing tear outflow from the conjunctival sac.
2. *Unpreserved topical antibiotics*. In the presence of significant epithelial erosions, broad-spectrum topical antibiotics are useful in preventing the development of secondary infection due to lack of the protective epithelial barrier.
3. *Unpreserved topical steroids*. Chronic inflammation of the ocular surface is a feature of LSCD and can be controlled by a judicious application of topical steroids.

Autologous Serum

Recently, the use of AS drops has been advocated in the medical treatment of many ocular surface disorders including LSCD where conventional treatment is insufficient to allow stabilisation of the ocular surface. The cornea is avascular and therefore must satisfy its metabolic needs by means other than its blood supply. Glucose, electrolytes, and amino acids are supplied by the aqueous humour. However, the tears produced by the lacrimal gland supply the remaining nutrients and are found to be rich in growth factors, vitamins, antimicrobial factors, and neuropeptides, which support proliferation, migration, and differentiation of the ocular surface epithelia *(93, 94)*. Although commercially available artificial tears are able to provide lubrication, maintain an optically smooth surface, and are biomechanically stable, they lack the complex nutrient composition of natural tears that supports ocular surface health. AS is a natural tear substitute that has several advantages over commercially available artificial tear supplements:

1. The composition of AS is very similar to natural tears (Table 3) *(95)*.
2. The pH and osmolality of tears and AS are virtually identical.
3. AS is free of preservatives.
4. AS by definition lacks antigenicity.

Table 3 Comparison of human tears and serum [modified from (95)]

	Tears	Autologous serum
Properties		
pH	7.4	7.4
Osmolality standard deviation (SD)	298 *(10)*	296
Biological components		
EGF (ng/mL)	0.2–3.0	0.5
TGF-β (ng/mL)	2–10	6–33
Vitamin A (mg/mL)	0.02	46
Lysozyme (mg/mL) (SD)	1.4 (0.2)	6
SIgA (μg/mL) (SD)	1,190 (904)	2
Fibronectin (μg/mL)	21	205

Abbreviations: EGF epidermal growth factor, *TGF-β* transforming growth factor β; *SIgA* surface immunoglobulin A, *SD* Standard deviation.

The concept of *serum* should be differentiated from *plasma*, both of which are derived from whole blood. Plasma is derived by removing the cellular components of whole blood by centrifugation in the presence of an anticoagulant. However, serum is produced by centrifugation of whole *clotted* blood. Therefore, unlike plasma, serum contains higher quantities of all the platelet-derived growth factors including epidermal growth factor (EGF), platelet-derived growth factor (PDGF), and transforming growth factor β (TGF-β), which are released from the platelets during clotting.

AS eye drops have been shown to have a clear advantage over artificial tear supplements in the treatment of several conditions including severe dry eye *(96)*, persistent epithelial defects *(94)*, recurrent erosion syndrome *(97)*, and superior limbic keratoconjunctivitis *(98)*. AS has also been successfully used as an adjunctive treatment in ocular surface reconstruction *(99)*.

The final composition of AS eye drops will vary slightly between individuals due to differences in composition of their blood. However, significant differences may occur due to different parameters used in the clotting, coagulation, dilution, and storage stages of the production of AS (Fig. 13):

1. *Clotting stage.* Parameters: duration and temperature
2. *Centrifugation stage.* Parameters: force and duration
3. *Dilution stage.* Parameters: dilution reagent and dilution factor
4. *Storage stage.* Parameters: container, temperature and duration

Many of the published studies fail to mention some or all of the parameters used. There have been no controlled clinical trials evaluating the impact of differences in these production parameters in clinical outcomes to date. Liu et al. *(100)* have tried to establish an optimised protocol for the production of AS eye drops by in vitro study. They examined the effect of various clotting times, centrifugation forces, types of dilution reagent, and differing dilutions on the concentration of growth factors, fibronectin, and vitamins in serum and tested the epitheliotropic capacity of these serum modifications in a cell culture model of corneal epithelial cells. They found that all the above parameters had a significant impact on the composi-

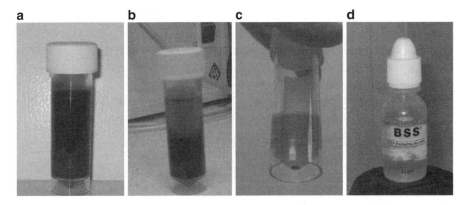

Fig. 13 Manufacture of autologous serum under good manufacturing practice guidelines. (**a**) Appearance of blood after 2 h clotting time. (**b**) Appearance following initial high-speed centrifugation; the supernatant is removed for further centrifugation. (**c**) Appearance following further high-speed centrifugation; the supernatant is removed for further processing. (**d**) Appearance of final product following dilution to 50% with basic salt solution (BSS) and filter sterilisation

tion and epitheliotropic effects of serum. A long clotting time (≥120 min), a sharp centrifugation (3,000 × g for 15 min), and dilution with basic salt solution (BSS) improve the ability of serum eyedrops to support proliferation, migration, and differentiation of corneal epithelial cells *(100)*.

Surgical Treatment of Partial LSCD

In more advanced cases of partial LSCD, especially when the visual axis becomes covered in conjunctival epithelium or in cases where ocular pain and irritation do not respond to medical treatment, surgical intervention is required. The main surgical modalities for the treatment of partial LSCD include sequential conjunctival epitheliectomy and amniotic membrane transplantation. These two surgical modalities may be used alone or in combination.

Sequential Conjunctival Epitheliectomy

The technique of sequential conjunctival epitheliectomy was first described by Dua et al. in 1994 *(92, 101)*. It involves the removal of conjunctival epithelium that has grown onto the cornea in LSCD by mechanical debridement (Fig. 14). This allows the bare cornea to be covered by corneal epithelial cells derived from the remaining healthy limbus. The procedure may be repeated several times until all the abnormal conjunctival epithelium has been replaced by corneal epithelium. Once this has occurred, the barrier function of the peripheral corneal epithelium and limbus would prevent further conjunctivalisation *(5)*. This procedure has been shown to be

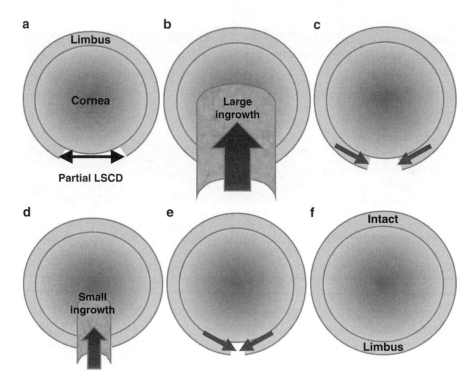

Fig. 14 Diagrammatic representation of sequential sector conjunctival epitheliectomy (SSCE) technique. (**a**) A cornea with partial limbal stem cell deficiency (LSCD) (the *black double headed arrow* indicates an area of corneal epithelial stem cells [CESC] deficiency), therefore the barrier function of the limbus is lost in this area. (**b**) Conjunctiva grows over the area of absent limbal barrier to cover the cornea. (**c**) The conjunctiva is scraped away from the cornea and limbus to allow the adjacent CESCs to fill the gap in the limbal barrier. (**d**) If the limbal barrier is not closed completely, smaller amounts of conjunctiva may grow inward again. (**e**) The conjunctival ingrowth is repeatedly scraped back to ensure the cornea is re-epithelialised by CESC derived cells rather than conjunctival cells. (**f**) Once the limbal barrier is completed with cells from the adjacent CESCs, further conjunctival ingrowth is prevented

successful in treating partial LSCD when used alone *(92)* and in conjunction with amniotic membrane transplantation *(102)*. Dua et al. suggest whichever technique is used in partial LSCD, it is not always important to achieve total corneal epithelial cover for the corneal surface. Good visual results can be obtained provided that the visual axis is covered in corneal epithelial cells, whereas other areas may be covered in cells of conjunctival phenotype *(92)*.

Amniotic Membrane

In recent years, amniotic membrane (AM) transplantation has been widely used for the reconstruction of the damaged ocular surface. AM represents the innermost layer of the placenta. It consists of three main layers: an avascular stromal matrix, a thick basement membrane, and a cuboidal epithelial monolayer (Fig. 15a). AM is prepared from placentas obtained after elective caesarean section. The donors undergo serology screening for HIV, hepatitis B and C, and syphilis. Once the amnion has been isolated from the placenta, it is flattened onto nitrocellulose paper with the epithelium side up *(103)*. The epithelial side ("shiny" side) is *not* sticky unlike the stromal side ("sticky" side), which allows for correct orientation during surgery. In general, AM is cryopreserved in 50% glycerol at −80°C. This results in most of the epithelial cells becoming nonviable and therefore losing their proliferative capability *(104)*. Although the exact mechanism by which cryopreserved AM transplantation encourages ocular surface healing is not known, studies have purported several favourable attributes:

1. *Promotion of corneal epithelialisation.* AM contains high levels of several growth factors including EGF *(105)*. AM can also induce growth factor expression in tissue with which it has contact *(106)*. In addition, the thick basement membrane provides a suitable substrate for epithelial cells to adhere to and expand.
2. *Inhibition of scarring.* The stromal side of AM has a matrix composition that suppresses TGF-β signalling and suppresses the proliferation and myofibroblast differentiation of human corneal and limbal fibroblasts.
3. *Reduction of inflammation.* The stromal matrix of AM reduces inflammation by stimulating inflammatory cells into rapid apoptosis. Reduced inflammation is also mediated by the presence of various protease inhibitors *(107)* and anti-inflammatory cytokines such as interleukin 1RA (IL-1RA) and IL-10 *(108)*.

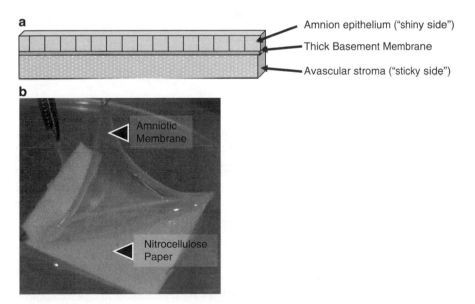

Fig. 15 (**a**) Diagrammatic representation of the three main layers of human amniotic membrane. (**b**) Clinical photograph of human amniotic membrane (attached to nitrocellulose filter paper)

In addition, the barrier function of AM will exclude inflammatory cells and proteins in the tear film from gaining access to the corneal stroma.

4. *Inhibition of vascularisation.* AM is observed clinically to prevent or retard neo-vascularisation. This may be due to its barrier effect, preventing the diffusion of vasculogenic mediators, or due to the presence of anti-angiogenic mediators such as endostatin, thrombospondin-1, and tissue inhibitors of metalloproteases *(109)*.

5. *Antimicrobial action.* Antibacterial effects of amniotic membrane have been shown against a wide variety of organisms *(110)*. This antimicrobial action may be because AM provides an effective physical barrier to infection *(111)* or due to the presence of various antimicrobial factors such as lysozyme, bactricidin, β-lysin, and specific antibodies *(112)*.

6. *Low antigenicity.* AM lacks most of the major histocompatibility agents, therefore exciting little or no immunological reaction *(113)*.

7. *Prolonged lifespan and maintenance of clonogenicity of epithelial progenitor cells (57).*

As a result of these attributes, AM transplantation (AMT) has been used in the management of a multitude of ocular surface diseases. With regard to partial LSCD, AMT alone has been shown to successfully promote re-epithelialisation of the cornea with a corneal epithelial phenotype, as it allows the expansion of the remaining CESCs in vivo *(102)*. When LSCD is total, the combination of AMT with limbal cell transplantation produces a higher success rate than limbal cell transplantation alone *(57)*. AM has been shown to be an ideal matrix for ex vivo preservation and expansion of CESCs and is thought to act as a surrogate SC niche *(114)*. This has allowed the development of ex vivo expansion of CESCs using a human nonantigenic carrier rather than an animal fibroblast feeding system, as described below.

Management of Total LSCD

The only options for the treatment of total LSCD are surgical. In the presence of total LSCD, the CESC population must be restored by limbal epithelial grafts. In addition, it is vital that for any type of CESC transplant to succeed, the environment into which these CESCs are transferred must be made receptive, as described below.

Failure of Penetrating Keratoplasty (Conventional Corneal Transplantation) in Total LSCD

In LSCD, the conjunctivalisation of the corneal surface leads to an opaque cornea, resulting in marked visual loss. Standard surgery for replacement of opaque corneas consists of penetrating keratoplasty (PK) or full thickness corneal transplantation. However, the success of this operation relies on the ability of the host producing a healthy corneal epithelium, which will gradually replace the epithelium of the donated cornea. Since this cannot occur in total LSCD, conjunctivalisation of the PK ensues, resulting in recurrence of visual loss and the other symptoms and signs of LSCD. In addition, the immunologic privilege of the normally avascular cornea is lost and the donor cornea can be rejected. In summary, standard PK will inevitably fail in patients with TOTAL LSCD as confirmed by several studies *(115–117)*. For successful corneal transplantation to work in these patients, the CESC population must first be replaced by the surgical procedure known as *limbal stem cell transplantation*.

Whole Tissue Limbal Grafts

The management of severe ocular surface disease was revolutionised following the discovery of the location and function of CESCs. In 1989 Kenyon and Tseng *(23)* were the first to use limbal transplants to successfully treat LSCD. In summary, they transplanted normal conjunctiva and limbal tissue from the healthy eye to the opposite eye of a patient with unilateral total LSCD. This resulted in reversal of the LSCD and long-term stabilisation of the ocular surface.

Since Kenyon and Tseng's original paper, techniques of limbal transplantation have been modified and advanced. Holland and Schwartz *(118–120)* have proposed an international classification system (described below) for the various limbal transplantation procedures available for the treatment of severe ocular surface disease. The exact technique used will depend on first, whether the LSCD is unilateral (in which case the patient may undergo an autograft) or bilateral (in which case an allograft is required), and second, on any other components of the ocular surface that are damaged. Whichever procedure is employed, the preexisting ocular environment must be made hospitable to allow survival of the CESCs to be grafted. This involves correction of eyelid and eyelash abnormalities, suppression of inflammation, and adequate ocular lubrication *(120)*. Any of the procedures may be combined with AMT, which can be used as a suitable bed on which the CESCs from the limbal grafts can grow or as a biological patch to provide a suitable environment and to prevent conjunctival tissue from growing onto the cornea.

Conjunctival Limbal Autograft

Conjunctival limbal autograft (CLAU) involves transplantation of limbal tissue attached to a conjunctival carrier from the healthy eye to the contralateral SC deficient eye in patients with unilateral LSCD (Fig. 16). It is the procedure of choice in unilateral cases since there is no need for postoperative immunosuppression. The success of CLAU in this group of patients has been documented in several studies *(23, 121, 122)*. A prerequisite for CLAU is that the donor eye must have no condition that predisposes it to LSCD as the amount of limbal tissue required is large (usually two limbal grafts spanning 2-3 clock hours). Obviously, this limits the use

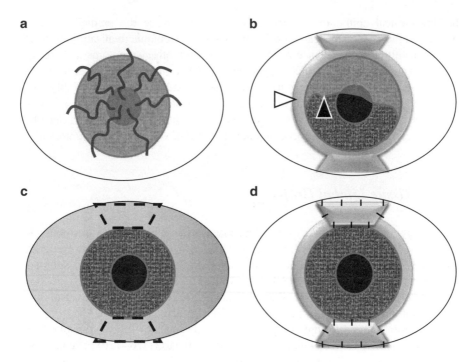

Fig. 16 Diagrammatic representation of the conjunctival limbal autograft (CLAU) technique. (**a**) An eye with total limbal stem cell deficiency (LSCD). The surface of the cornea is covered with conjunctiva derived fibrovascular pannus. (**b**) The first part of the procedure involves removal of all the abnormal corneal epithelium. The *black arrow* indicates the edge of the abnormal corneal epithelium as it is peeled away (superficial keratectomy). Next, a rim of conjunctiva is removed (shown by the *white arrow*) with additional tissue resection at 12 and 6 o'clock to accommodate the limbal grafts. (**c**) Two limbal grafts are harvested from the healthy opposite eye. (**d**) The limbal grafts are sutured into place on the previously prepared recipient bed

of this procedure as most causes of LSCD are bilateral to varying degrees. Since the amount of tissue to be removed is to be minimised, this procedure is most effective in *subtotal* LSCD.

Living Related Conjunctival Limbal Allograft

This procedure is similar to CLAU except the tissue is harvested from the patient's living relative. This means it can be used for patients with bilateral disease. Unlike CLAU, living related conjunctival limbal allograft (lr-CLAL) is an *allograft* rather than an *autograft* procedure. Since the amount of tissue removed is limited to prevent SC exhaustion in the donor, it is again more suitable for subtotal LSCD. Reported 2-year success rates vary from 50 *(123)* to 80% *(124)*. Unlike CLAU, these patients require immunosuppression to maintain graft survival even in the presence of HLA matching if long-term graft survival is to be maximised *(123)*.

Keratolimbal Allograft

This procedure involves the transplantation of limbal tissue attached to a *corneal* carrier harvested from *cadaveric* eyes to the recipient eye (Fig. 17). Since the harvested tissue is taken from cadaveric eyes, large amounts of limbus can be harvested. In fact, the technique suggested by Holland and Schwartz *(120)* involves the use of two stored corneoscleral rings per eye to be treated. Since the available SC population is large, this procedure is suitable for severe LSCD. It is also indicated in patients with unilateral LSCD who do not want to risk damage to the other eye or in patients who have no available living relative. The procedure is *not* suitable in patients with significant conjunctival damage who require a combined conjunctival and keratolimbal allograft (C-KLAL). Initial success rates of between 51 *(125)* and 83% *(126)* have been reported. The use of potent postoperative immunosuppression appears to be crucial for a successful outcome *(127)*. The reports of long-term success rates have been limited, and some workers have shown poor long-term results. Solomon et al. *(128)* reported a stable ocular surface in 77% of patients following KLAL at 1 year but this fell to 24% at 5 years.

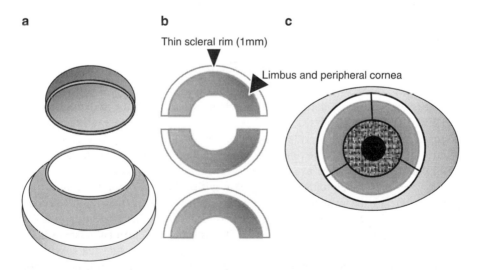

Fig. 17 Diagrammatic representation of the keratolimbal allograft (KLAL) technique. (**a**) Preparation of cadaveric corneal tissue. The central 7–8 mm of the cornea is removed to produce a corneoscleral ring. (**b**) Three lenticules are needed, and these are obtained from two donor corneas. Each lenticule undergoes lamellar dissection as only the anterior one-third is used for transplantation. (**c**) The three limbal crescents are sutured in place with no intervening spaces so that a 360-degree limbal barrier is produced, which will prevent further conjunctival ingrowth

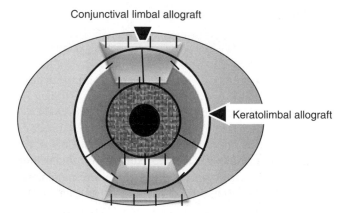

Conjunctival limbal allograft

Keratolimbal allograft

Fig. 18 Diagrammatic representation of the combined conjunctival and keratolimbal allograft (C-KLAL) technique. This represents a combination of the living related conjunctival limbal allograft (lr-CLAL) and KLAL techniques to replace the corneal epithelial stem cells (CESC) population and conjunctival population

Combined Conjunctival and Keratolimbal Allograft

This procedure essentially involves a combination of lr-CLAL and KLAL (Fig. 18). It is the recommended procedure in patients with severe LSCD combined with severe damage to the conjunctiva as exemplified in Stevens-Johnson's syndrome and ocular cicatricial pemphigoid.

Problems Associated with Whole Tissue Limbal Grafts

The most worrisome of the potential problems associated with limbal autografts and living related allografts is that of producing iatrogenic LSCD in the healthy donor eye *(19)*. Unfortunately, to ensure successful limbal autografting, at least 6 mm of limbal tissue must be removed, which has been shown to potentially cause LSCD in the healthy donor eye in animals *(12)* and humans *(129)*.

PK can generally be managed long term with little or no topical immunosuppression due to the avascularity and immune privilege of the central cornea. However, limbal tissue is highly vascularised and contains large quantities of antigen presenting cells and leucocytes. This results in significant susceptibility of limbal grafts to rejection. It has been categorically shown that the survival of limbal allografts requires aggressive systemic immunosuppression *(127, 130)*. This approach is also needed for HLA-matched limbal allografts *(123)*. The initial protocol recommended by Holland and Schwartz *(120)* included oral prednisolone, cyclosporin A, and azathioprine. This has subsequently changed to oral prednisolone, tacrolimus, and mycophenolate. The main problem with such aggressive immunosuppression is that each of the mentioned agents is associated with several serious and potentially life-threatening complications. Therefore, these patients require regular lifelong follow-up by specialists trained in the use of these agents.

Cultured CESC/LSC Grafts

Recently, with the increased knowledge of CESC biology, techniques have been developed to expand *small* limbal biopsies in ex vivo culture for subsequent transplantation with good results *(131–134)*. The underlying theory is that ex vivo cultivated limbal epithelial cultures contain CESCs that after transplantation would persist and regenerate the corneal epithelium and repopulate the CSEC niche in patients with LSCD.

This approach has clear advantages to the use of whole tissue limbal grafts. As discussed above, the main problem associated with conventional whole tissue allografts is the potential for causing donor CESC exhaustion and also allograft rejection. However, if only small amounts of limbal tissue are removed from the patient's normal eye and the CESC population could be expanded ex vivo and transplanted into the diseased eye, both of these problems could be circumvented. In addition, the procedure could be used with caution in patients with bilateral disease provided the less injured eye had partial LSCD. Obviously, in this latter scenario, conventional limbal autografts are absolutely contraindicated due to the amount of tissue required. Since the transplanted tissue is not allogeneic, the need for aggressive immunosuppression with its increased morbidity and mortality is removed.

The concept of using adult SCs for therapeutic purposes has revolutionised treatment options available for many diseases such as haematological malignancy and is now revolutionising the treatment of LSCD. Since these techniques make use of adult SCs, this bypasses the ethical issues inherent in the manipulation of embryonic SCs.

Technique of Ex Vivo Expansion of Limbal Epithelial Cells

In general, the technique starts with the removal of a very small amount of limbal tissue from the contralateral healthy eye (or from the eye of a living relative or cadaveric donor). Typically, a 1 mm^2 limbal biopsy is sufficient to produce enough expanded epithelium to resurface the whole cornea, which minimises the risk of damaging the CESC population in the healthy donor eye or other eye in unilateral LSCD or living related allograft in bilateral LSCD *(131)*. The biopsy is placed on a carrier and fed at appropriate intervals with a defined growth medium and incubated under standard conditions (Fig. 19). Once a sheet of cultured CESCs is produced, they can be transplanted onto the affected cornea once the abnormal fibrovascular tissue has been removed. Once the CESCs have successfully engrafted, a PK can be performed at a later date to remove any residual central corneal scarring if necessary (Fig. 20).

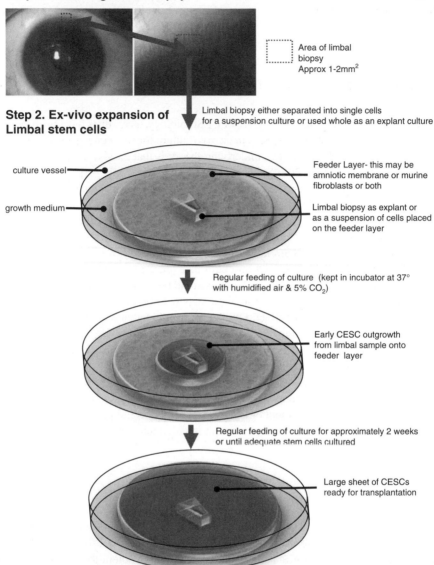

Step 1. Obtaining limbal biopsy

Area of limbal biopsy Approx 1-2mm²

Step 2. Ex-vivo expansion of Limbal stem cells

Limbal biopsy either separated into single cells for a suspension culture or used whole as an explant culture

culture vessel

Feeder Layer- this may be amniotic membrane or murine fibroblasts or both

growth medium

Limbal biopsy as explant or as a suspension of cells placed on the feeder layer

Regular feeding of culture (kept in incubator at 37° with humidified air & 5% CO₂)

Early CESC outgrowth from limbal sample onto feeder layer

Regular feeding of culture for approximately 2 weeks or until adequate stem cells cultured

Large sheet of CESCs ready for transplantation

Fig. 19 Diagrammatic representation of the technique used for ex vivo expansion of a limbal epithelial biopsy

Significant differences of the various stages of the above process have been described by various workers. The limbal explant should ideally be autologous, but ex vivo expansion has also been reported using allogeneic living related tissue *(132, 135)* and allogeneic cadaveric tissue *(134, 136, 137)*. Once the limbal

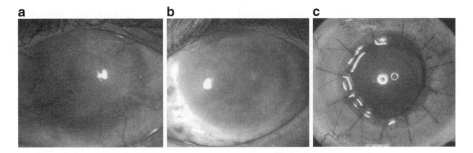

Fig. 20 Clinical photographs of (**a**) an eye with severe limbal stem cell deficiency (LSCD). Note extensive vascularisation of the irregular corneal surface and dense corneal opacification. (**b**) The same eye 1 month following ex vivo expanded corneal epithelial stem cells (CESC) graft. Note the white eye with no inflammation and no vascularisation. The cornea still has significant opacity due to stromal scarring from the initial insult causing the LSCD. (**c**) The same eye following a central corneal graft (performed 1 year after the CESC graft). The clarity of the corneal graft can be seen by the bright red reflex. The corneal graft is successful due to the previous stem cell graft, which allows the donor tissue to be epithelialised

biopsy is obtained, the tissue can be cultured using one of two techniques: the explant technique or the single cell suspension technique. With the explant technique, where the limbal biopsy can be placed directly onto a carrier *(133, 138)*. In the suspension technique, the limbal biopsy is enzymatically digested, typically with a combination of dispase and trypsin to produce a suspension of single cells, which are then inoculated onto a carrier *(139)*. Several carriers or substrates have been used on which to expand the cell population. These include soft contact lenses *(131)*, coated plates (fibronectin, collagen IV, laminin) *(140)*, culture inserts *(139)*, and temperature responsive gels *(141)*. However, by far the most common substrate used for both research and clinical purposes is human AM *(57, 102, 133, 142–145)*.

The use of AM in limbal culture techniques has been adopted as the preferred carrier system for a number of reasons. First, in the absence of AM, limbal epithelial cultures had to be expanded by coculturing with murine 3T3 fibroblasts *(131, 146)*. However, CESC can be preserved and expanded on AM without the need for 3T3 fibroblasts, which reduces the potential risk of using an animal-derived cell line in humans *(114)*. Second, the direct growth of CESCs on AM allows for ease of direct transfer of the CESCs at transplantation due to its thick basement membrane, allowing ease of handling. Since the epithelial cells of cryopreserved AM are not viable, it is nonimmunogenic and does not require the recipient patient to be immunosuppressed *(57)*. The ability of AM to maintain SC characteristics (SC marker retention, LRC capability, and proliferative capacity) in ex vivo culture has led to the suggestion it acts as an effective SC niche *(57, 105, 145, 147)*. Researchers have demonstrated success using both intact AM *(57, 114)* and epithelial denuded AM *(105, 139, 147)* in maintaining CESC characteristics in ex vivo culture.

Once the culture is set up, the culture is fed with a defined culture medium, which varies among investigators but is usually a modification of basic epithelial culture medium. Although some workers have successfully replaced fetal calf serum in the culture medium with AS *(148, 149)*, a completely animal-free product culture system has not yet been achieved clinically *(150)*. The cultured sheets of cells have been shown to contain CESCs by measuring their colony-forming efficiency *(131, 132)* and with immunohistochemistry targeted at various putative CESC markers *(77)*.

The culture period lasts approximately 2–4 weeks before adequate CESCs are available to ensure successful transplantation. Once these cells are available, the patient undergoes superficial keratectomy of the eye with LSCD to remove all abnormal conjunctivalised epithelium followed by the transplantation of the sheet of cultured CESCs. The patient is then closely monitored and given appropriate adjunctive therapy.

Results of Clinical Trials of Ex Vivo Expanded Limbal Transplants

In 1997 Pellegrini et al. *(131)* were the first to show successful reconstitution of a healthy ocular surface in two patients with unilateral LSCD by the transplantation of ex vivo expanded CESCs grown on 3T3 fibroblasts. Tsai et al. *(133)* were the first to show successful restoration of the corneal surface in patients with unilateral LSCD using AM as the carrier for the CESCs. They showed successful re-epithelialisation and maintenance of intact epithelium in all six patients of their study group with a follow up of 15 months. Significant visual improvement occurred in five of the six patients. Since then, a number of clinical trials have been performed that have recently been reviewed *(150)*. In general, these studies show encouraging results that this new technique is successful in restoring healthy corneal epithelium in severe LSCD, resulting in improvement in the ocular surface, improvement of symptoms, and increased visual acuity. Although long-term results of this experimental SC technique are awaited, the evidence so far points to it becoming a viable and preferred option for treating patients with severe LSCD. A summary of the treatment options available for the management of LSCD is shown in Fig. 21.

Alternative Sources of SCs in Total Bilateral LSCD

The ex vivo expansion of CESCs and subsequent transplantation for the treatment of LSCD has the advantage that it only requires a very small amount of healthy limbal tissue (typically 1 mm^2), which means the risk of causing LSCD in the donor eye is negligible. However, there are a proportion of patients with bilateral *total* LSCD who lack any healthy CESC population to biopsy in the first place. Ex vivo expansion of limbal tissue from a living relative or cadaveric donor is a possible

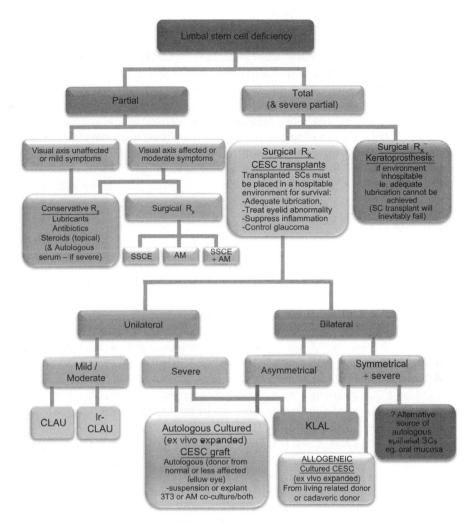

Fig. 21 Algorithm for the management of limbal stem cell deficiency (LSCD). *SC* stem cell, *CESC* corneal epithelial stem cell, *AM* amniotic membrane, *SSCE* sequential sector conjunctival epitheliectomy, *CLAU* conjunctival limbal autograft, *lr-CLAL* living related conjunctival limbal allograft, *KLAL* keratolimbal allograft, *PK* penetrating keratoplasty, R_x treatment. Notes: (1) The key to success with any CESC graft depends on adequate postoperative treatment with antibiotics, steroids, protective contact lens, and autologous serum together with close patient follow-up. (2) Any of the surgical SC techniques may be combined with AM transplantation to maximise success. (3) KLAL and lr-CLAL require adequate systemic immunosuppression. (4) If the LSCD is associated with significant damage or loss of conjunctiva, this population must also be replaced. (5) The CESC graft may be combined or followed at a later date by PK as necessary

option in these patients. Such an approach would still have the problems associated with whole tissue allografts, as discussed above, namely tissue rejection, and the requirement for immunosuppression with its associated potentially serious side effects.

In an attempt to overcome the problems inherent in transplantation of allogeneic tissue, there has been a recent interest in the possibility of using alternative autologous epithelial cells from oral mucosa to replace the CESC population. Nakamura et al. *(151)* have successfully expanded rabbit oral mucosa epithelial SC on AM and subsequently transplanted the cells onto the ocular surface of rabbits with total LSCD. This led to the successful re-epithelialisation of the corneal surface in this animal model. This approach has been applied to humans with encouraging early results *(152, 153)*. The use of alternative sources of cultured epithelial cells provides an exciting possibility of treating this difficult group of patients with blindness from bilateral total LSCD and warrants further study.

Summary

- The cornea is the most important focussing structure in the eye.
- The cornea is covered with a highly specialised stratified squamous epithelium that has a continuous and rapid turnover.
- The corneal epithelium is replenished by epithelial SCs.
- These SCs are anatomically segregated at the corneal limbus and are known as CESCs or LSCs. These CESCs have been shown to exhibit all the features that define an adult SC population.
- Although no *single* specific marker has been identified that can categorically identify CESCs, a number of putative SC have been studied. The relative expression of these positive and negative markers can be used to identify a putative CESC phenotype.
- The role of the CESC niche is critically important in maintaining a healthy SC population, and our understanding of SC niche physiology will be crucial in the successful use of SCs for therapeutic means.
- Absence or acquired loss of CESCs leads to the condition of LSCD.
- LSCD leads to recurrent breakdown of the corneal epithelium and conjunctivalisation of the corneal surface. This results in the patient suffering significant ocular pain and visual loss.
- LSCD can occur unilaterally or bilaterally and can be either partial or total.
- Mild partial LSCD can usually be treated medically. Severe partial LSCD usually requires treatment with AS treatment, amniotic membrane transplantation, or sequential conjunctival epitheliectomy.
- Total LSCD will require surgical treatment in the form of CESC grafts.
- In the majority of treatment centres, CESC grafting involves the transplantation of large whole tissue grafts.
- The majority of whole tissue grafts used for transplantation are allogeneic, and their success depends on long-term potent immunosuppression, which is associated with significant patient morbidity.
- Recently, the ex vivo expansion of small amounts of autologous limbal tissue has been used to successfully treat LSCD while eliminating the need for immunosuppression. Early results of this new group of treatments are very encouraging.

- The technique of SC ex vivo expansion is still in its infancy. However, our understanding of SC biology is rapidly advancing, and this knowledge will undoubtedly allow the further refinement of the current techniques for corneal SC therapy. This will lead to our ultimate goal, namely successful long-term SC engraftment and host tissue regeneration with no associated side effects.

References

1. Whitcher JP, Srinivasan M, Upadhyay MP. Corneal blindness: a global perspective. Bull World Health Organ 2001;79:214–21.
2. Boulton M, Albon J. Stem cells in the eye. Int J Biochem Cell Biol 2004;36:643–57.
3. Stepp MA, Zieske JD. The corneal epithelial stem cell niche. Ocul Surf 2005;3:15–26.
4. Dua HS, Azuara-Blanco A. Limbal stem cells of the corneal epithelium. Surv Ophthalmol 2000;44:415–25.
5. Tseng SC. Concept and application of limbal stem cells. Eye 1989;3(Pt 2):141–57.
6. Schermer A, Galvin S, Sun TT. Differentiation-related expression of a major 64K corneal keratin in vivo and in culture suggests limbal location of corneal epithelial stem cells. J Cell Biol 1986;103:49–62.
7. Potten CS, Hume WJ, Reid P, Cairns J. The segregation of DNA in epithelial stem cells. Cell 1978;15:899–906.
8. Zieske JD. Perpetuation of stem cells in the eye. Eye 1994;8(Pt 2):163–9.
9. Schofield R. The stem cell system. Biomed Pharmacother 1983;37:375–80.
10. Thoft RA, Friend J. The X, Y, Z hypothesis of corneal epithelial maintenance. Invest Ophthalmol Vis Sci 1983;24:1442–3.
11. Lavker RM, Dong G, Cheng SZ, Kudoh K, Cotsarelis G, Sun TT. Relative proliferative rates of limbal and corneal epithelia. Implications of corneal epithelial migration, circadian rhythm, and suprabasally located DNA-synthesizing keratinocytes. Invest Ophthalmol Vis Sci 1991;32:1864–75.
12. Chen JJ, Tseng SC. Abnormal corneal epithelial wound healing in partial-thickness removal of limbal epithelium. Invest Ophthalmol Vis Sci 1991;32:2219–33.
13. Davanger M, Evensen A. Role of the pericorneal papillary structure in renewal of corneal epithelium. Nature 1971;229:560–1.
14. Kurpakus MA, Stock EL, Jones JC. Expression of the 55-kD/64-kD corneal keratins in ocular surface epithelium. Invest Ophthalmol Vis Sci 1990;31:448–56.
15. Bickenbach JR. Identification and behavior of label-retaining cells in oral mucosa and skin. J Dent Res 1981; 60(Spec No C):1611–20.
16. Cotsarelis G, Cheng SZ, Dong G, Sun TT, Lavker RM. Existence of slow-cycling limbal epithelial basal cells that can be preferentially stimulated to proliferate: implications on epithelial stem cells. Cell 1989;57:201–9.
17. Ebato B, Friend J, Thoft RA. Comparison of limbal and peripheral human corneal epithelium in tissue culture. Invest Ophthalmol Vis Sci 1988;29:1533–7.
18. Hanna C, O'Brien JE. Cell production and migration in the epithelial layer of the cornea. Arch Ophthalmol 1960;64:536–9.
19. Chen JJ, Tseng SC. Corneal epithelial wound healing in partial limbal deficiency. Invest Ophthalmol Vis Sci 1990;31:1301–14.
20. Huang AJ, Tseng SC. Corneal epithelial wound healing in the absence of limbal epithelium. Invest Ophthalmol Vis Sci 1991;32:96–105.
21. Matsuda M, Ubels JL, Edelhauser HF. A larger corneal epithelial wound closes at a faster rate. Invest Ophthalmol Vis Sci 1985;26:897–900.
22. Srinivasan BD, Eakins KE. The reepithelialization of rabbit cornea following single and multiple denudation. Exp Eye Res 1979;29:595–600.

23. Kenyon KR, Tseng SC. Limbal autograft transplantation for ocular surface disorders. Ophthalmology 1989;96:709–22; discussion 22–3.
24. Tsai RJ, Sun TT, Tseng SC. Comparison of limbal and conjunctival autograft transplantation in corneal surface reconstruction in rabbits. Ophthalmology 1990;97:446–55.
25. Waring GO 3rd, Roth AM, Ekins MB. Clinical and pathologic description of 17 cases of corneal intraepithelial neoplasia. Am J Ophthalmol 1984;97:547–59.
26. Reya T, Morrison SJ, Clarke MF, Weissman IL. Stem cells, cancer, and cancer stem cells. Nature 2001;414:105–11.
27. Potten CS, Loeffler M. Stem cells: attributes, cycles, spirals, pitfalls and uncertainties. Lessons for and from the crypt. Development 1990;110:1001–20.
28. Lehrer MS, Sun TT, Lavker RM. Strategies of epithelial repair: modulation of stem cell and transit amplifying cell proliferation. J Cell Sci 1998;111(Pt 19):2867–75.
29. Hanna C, Bicknell DS, O'Brien JE. Cell turnover in the adult human eye. Arch Ophthalmol 1961;65:695–8.
30. Townsend WM. The limbal palisades of Vogt. Trans Am Ophthalmol Soc 1991;89:721–56.
31. Gipson IK. The epithelial basement membrane zone of the limbus. Eye 1989;3(Pt 2):132–40.
32. Wiley L, SundarRaj N, Sun TT, Thoft RA. Regional heterogeneity in human corneal and limbal epithelia: an immunohistochemical evaluation. Invest Ophthalmol Vis Sci 1991;32:594–602.
33. Ljubimov AV, Burgeson RE, Butkowski RJ, Michael AF, Sun TT, Kenney MC. Human corneal basement membrane heterogeneity: topographical differences in the expression of type IV collagen and laminin isoforms. Lab Invest 1995;72:461–73.
34. Chen Z, de Paiva CS, Luo L, Kretzer FL, Pflugfelder SC, Li DQ. Characterization of putative stem cell phenotype in human limbal epithelia. Stem Cells 2004;22:355–66.
35. Espana EM, Di Pascuale M, Grueterich M, Solomon A, Tseng SC. Keratolimbal allograft in corneal reconstruction. Eye 2004;18:406–17.
36. Kasper M, Moll R, Stosiek P, Karsten U. Patterns of cytokeratin and vimentin expression in the human eye. Histochemistry 1988;89:369–77.
37. Kasper M. Patterns of cytokeratins and vimentin in guinea pig and mouse eye tissue: evidence for regional variations in intermediate filament expression in limbal epithelium. Acta Histochem 1992;93:319–32.
38. Schlotzer-Schrehardt U, Kruse FE. Identification and characterization of limbal stem cells. Exp Eye Res 2005;81:247–64.
39. Hayashi K, Kenyon KR. Increased cytochrome oxidase activity in alkali-burned corneas. Curr Eye Res 1988;7:131–8.
40. Steuhl KP, Thiel HJ. Histochemical and morphological study of the regenerating corneal epithelium after limbus-to-limbus denudation. Graefes Arch Clin Exp Ophthalmol 1987;225:53–8.
41. Zieske JD, Bukusoglu G, Yankauckas MA. Characterization of a potential marker of corneal epithelial stem cells. Invest Ophthalmol Vis Sci 1992;33:143–52.
42. Zieske JD, Bukusoglu G, Yankauckas MA, Wasson ME, Keutmann HT. Alpha-enolase is restricted to basal cells of stratified squamous epithelium. Dev Biol 1992;151:18–26.
43. Chung EH, DeGregorio PG, Wasson M, Zieske JD. Epithelial regeneration after limbus-to-limbus debridement. Expression of alpha-enolase in stem and transient amplifying cells. Invest Ophthalmol Vis Sci 1995;36:1336–43.
44. Tseng SC, Li DQ. Comparison of protein kinase C subtype expression between normal and aniridic human ocular surfaces: implications for limbal stem cell dysfunction in aniridia. Cornea 1996;15:168–78.
45. Joyce NC, Meklir B, Joyce SJ, Zieske JD. Cell cycle protein expression and proliferative status in human corneal cells. Invest Ophthalmol Vis Sci 1996;37:645–55.
46. Chen Z, Evans WH, Pflugfelder SC, Li DQ. Gap junction protein connexin 43 serves as a negative marker for a stem cell-containing population of human limbal epithelial cells. Stem Cells 2006;24:1265–73.

47. Pellegrini G, Dellambra E, Golisano O, et al. p63 identifies keratinocyte stem cells. Proc Natl Acad Sci U S A 2001;98:3156–61.
48. Di Iorio E, Barbaro V, Ruzza A, Ponzin D, Pellegrini G, De Luca M. Isoforms of DeltaNp63 and the migration of ocular limbal cells in human corneal regeneration. Proc Natl Acad Sci U S A 2005;102:9523–8.
49. Di Iorio E, Barbaro V, Ferrari S, Ortolani C, De Luca M, Pellegrini G. Q-FIHC: quantification of fluorescence immunohistochemistry to analyse p63 isoforms and cell cycle phases in human limbal stem cells. Microsc Res Tech 2006;69:983–91.
50. Arpitha P, Prajna NV, Srinivasan M, Muthukkaruppan V. High expression of p63 combined with a large N/C ratio defines a subset of human limbal epithelial cells: implications on epithelial stem cells. Invest Ophthalmol Vis Sci 2005;46:3631–6.
51. Barbaro V, Testa A, Di Iorio E, Mavilio F, Pellegrini G, De Luca M. C/EBPdelta regulates cell cycle and self-renewal of human limbal stem cells. J Cell Biol 2007;177:1037–49.
52. Matic M, Petrov IN, Chen S, Wang C, Dimitrijevich SD, Wolosin JM. Stem cells of the corneal epithelium lack connexins and metabolite transfer capacity. Differentiation 1997;61:251–60.
53. Dong Y, Roos M, Gruijters T, et al. Differential expression of two gap junction proteins in corneal epithelium. Eur J Cell Biol 1994;64:95–100.
54. Stepp MA, Zhu L, Sheppard D, Cranfill RL. Localized distribution of alpha 9 integrin in the cornea and changes in expression during corneal epithelial cell differentiation. J Histochem Cytochem 1995;43:353–62.
55. Stepp MA, Zhu L. Upregulation of alpha 9 integrin and tenascin during epithelial regeneration after debridement in the cornea. J Histochem Cytochem 1997;45:189–201.
56. Pajoohesh-Ganji A, Ghosh SP, Stepp MA. Regional distribution of alpha9beta1 integrin within the limbus of the mouse ocular surface. Dev Dyn 2004;230:518–28.
57. Grueterich M, Espana EM, Tseng SC. Ex vivo expansion of limbal epithelial stem cells: amniotic membrane serving as a stem cell niche. Surv Ophthalmol 2003;48:631–46.
58. Lambiase A, Bonini S, Micera A, Rama P, Bonini S, Aloe L. Expression of nerve growth factor receptors on the ocular surface in healthy subjects and during manifestation of inflammatory diseases. Invest Ophthalmol Vis Sci 1998;39:1272–5.
59. Touhami A, Grueterich M, Tseng SC. The role of NGF signaling in human limbal epithelium expanded by amniotic membrane culture. Invest Ophthalmol Vis Sci 2002;43:987–94.
60. de Paiva CS, Chen Z, Corrales RM, Pflugfelder SC, Li DQ. ABCG2 transporter identifies a population of clonogenic human limbal epithelial cells. Stem Cells 2005;23:63–73.
61. Watanabe K, Nishida K, Yamato M, et al. Human limbal epithelium contains side population cells expressing the ATP-binding cassette transporter ABCG2. FEBS Lett 2004;565:6–10.
62. Budak MT, Alpdogan OS, Zhou M, Lavker RM, Akinci MA, Wolosin JM. Ocular surface epithelia contain ABCG2-dependent side population cells exhibiting features associated with stem cells. J Cell Sci 2005;118:1715–24.
63. Chee KY, Kicic A, Wiffen SJ. Limbal stem cells: the search for a marker. Clin Experiment Ophthalmol 2006;34:64–73.
64. Vascotto SG, Griffith M. Localization of candidate stem and progenitor cell markers within the human cornea, limbus, and bulbar conjunctiva in vivo and in cell culture. Anat Rec A Discov Mol Cell Evol Biol 2006;288:921–31.
65. Lohrum MA, Vousden KH. Regulation and function of the p53-related proteins: same family, different rules. Trends Cell Biol 2000;10:197–202.
66. Yang A, Schweitzer R, Sun D, et al. p63 is essential for regenerative proliferation in limb, craniofacial and epithelial development. Nature 1999;398:714–8.
67. Celli J, Duijf P, Hamel BC, et al. Heterozygous germline mutations in the p53 homolog p63 are the cause of EEC syndrome. Cell 1999;99:143–53.
68. Yang A, Kaghad M, Wang Y, et al. p63, a p53 homolog at 3q27-29, encodes multiple products with transactivating, death-inducing, and dominant-negative activities. Mol Cell 1998;2:305–16.
69. Hernandez Galindo EE, Theiss C, Steuhl KP, Meller D. Expression of delta Np63 in response to phorbol ester in human limbal epithelial cells expanded on intact human amniotic membrane. Invest Ophthalmol Vis Sci 2003;44:2959–65.

70. Goodell MA, McKinney-Freeman S, Camargo FD. Isolation and characterization of side population cells. Methods Mol Biol 2005;290:343–52.

71. Zhou S, Schuetz JD, Bunting KD, et al. The ABC transporter Bcrp1/ABCG2 is expressed in a wide variety of stem cells and is a molecular determinant of the side-population phenotype. Nat Med 2001;7:1028–34.

72. Lekstrom-Himes J, Xanthopoulos KG. Biological role of the CCAAT/enhancer-binding protein family of transcription factors. J Biol Chem 1998;273:28545–8.

73. O'Rourke JP, Newbound GC, Hutt JA, DeWille J. CCAAT/enhancer-binding protein delta regulates mammary epithelial cell G0 growth arrest and apoptosis. J Biol Chem 1999;274:16582–9.

74. Molofsky AV, He S, Bydon M, Morrison SJ, Pardal R. Bmi-1 promotes neural stem cell self-renewal and neural development but not mouse growth and survival by repressing the p16Ink4a and p19Arf senescence pathways. Genes Dev 2005;19:1432–7.

75. Lessard J, Sauvageau G. Bmi-1 determines the proliferative capacity of normal and leukaemic stem cells. Nature 2003;423:255–60.

76. Watt FM, Green H. Involucrin synthesis is correlated with cell size in human epidermal cultures. J Cell Biol 1981;90:738–42.

77. Kim HS, Jun Song X, de Paiva CS, Chen Z, Pflugfelder SC, Li DQ. Phenotypic characterization of human corneal epithelial cells expanded ex vivo from limbal explant and single cell cultures. Exp Eye Res 2004;79:41–9.

78. Beyer EC, Paul DL, Goodenough DA. Connexin family of gap junction proteins. J Membr Biol 1990;116:187–94.

79. Chang CC, Trosko JE, el-Fouly MH, Gibson-D'Ambrosio RE, D'Ambrosio SM. Contact insensitivity of a subpopulation of normal human fetal kidney epithelial cells and of human carcinoma cell lines. Cancer Res 1987;47:1634–45.

80. Wolosin JM, Xiong X, Schutte M, Stegman Z, Tieng A. Stem cells and differentiation stages in the limbo-corneal epithelium. Prog Retin Eye Res 2000;19:223–55.

81. Lee P, Wang CC, Adamis AP. Ocular neovascularization: an epidemiologic review. Surv Ophthalmol 1998;43:245–69.

82. Ambati BK, Nozaki M, Singh N, et al. Corneal avascularity is due to soluble VEGF receptor-1. Nature 2006;443:993–7.

83. Egbert PR, Lauber S, Maurice DM. A simple conjunctival biopsy. Am J Ophthalmol 1977;84:798–801.

84. Nelson JD. Impression cytology. Cornea 1988;7:71–81.

85. Sacchetti M, Lambiase A, Cortes M, Sgrulletta R, Bonini S, Merlo D. Clinical and cytological findings in limbal stem cell deficiency. Graefes Arch Clin Exp Ophthalmol 2005;243:870–6.

86. Donisi PM, Rama P, Fasolo A, Ponzin D. Analysis of limbal stem cell deficiency by corneal impression cytology. Cornea 2003;22:533–8.

87. Elder MJ, Hiscott P, Dart JK. Intermediate filament expression by normal and diseased human corneal epithelium. Hum Pathol 1997;28:1348–54.

88. Kinoshita S, Adachi W, Sotozono C, et al. Characteristics of the human ocular surface epithelium. Prog Retin Eye Res 2001;20:639–73.

89. Cavanagh HD, Petroll WM, Alizadeh H, He YG, McCulley JP, Jester JV. Clinical and diagnostic use of in vivo confocal microscopy in patients with corneal disease. Ophthalmology 1993;100:1444–54.

90. Kobayashi A, Sugiyama K. In vivo confocal microscopy in a patient with keratopigmentation (corneal tattooing). Cornea 2005;24:238–40.

91. Patel DV, Sherwin T, McGhee CN. Laser scanning in vivo confocal microscopy of the normal human corneoscleral limbus. Invest Ophthalmol Vis Sci 2006;47:2823–7.

92. Dua HS. The conjunctiva in corneal epithelial wound healing. Br J Ophthalmol 1998;82:1407–11.

93. Wilson SE, Liang Q, Kim WJ. Lacrimal gland HGF, KGF, and EGF mRNA levels increase after corneal epithelial wounding. Invest Ophthalmol Vis Sci 1999;40:2185–90.

94. Tsubota K, Goto E, Shimmura S, Shimazaki J. Treatment of persistent corneal epithelial defect by autologous serum application. Ophthalmology 1999;106:1984–9.

95. Geerling G, Maclennan S, Hartwig D. Autologous serum eye drops for ocular surface disorders. Br J Ophthalmol 2004;88:1467–74.
96. Tsubota K, Goto E, Fujita H, et al. Treatment of dry eye by autologous serum application in Sjogren's syndrome. Br J Ophthalmol 1999;83:390–5.
97. del Castillo JM, de la Casa JM, Sardina RC, et al. Treatment of recurrent corneal erosions using autologous serum. Cornea 2002;21:781–3.
98. Goto E, Shimmura S, Shimazaki J, Tsubota K. Treatment of superior limbic keratoconjunctivitis by application of autologous serum. Cornea 2001;20:807–10.
99. Tsubota K, Satake Y, Ohyama M, et al. Surgical reconstruction of the ocular surface in advanced ocular cicatricial pemphigoid and Stevens-Johnson syndrome. Am J Ophthalmol 1996;122:38–52.
100. Liu L, Hartwig D, Harloff S, Herminghaus P, Wedel T, Geerling G. An optimised protocol for the production of autologous serum eyedrops. Graefes Arch Clin Exp Ophthalmol 2005;243:706–14.
101. Dua HS, Gomes JA, Singh A. Corneal epithelial wound healing. Br J Ophthalmol 1994;78:401–8.
102. Tseng SC, Prabhasawat P, Barton K, Gray T, Meller D. Amniotic membrane transplantation with or without limbal allografts for corneal surface reconstruction in patients with limbal stem cell deficiency. Arch Ophthalmol 1998;116:431–41.
103. Lee SH, Tseng SC. Amniotic membrane transplantation for persistent epithelial defects with ulceration. Am J Ophthalmol 1997;123:303–12.
104. Kruse FE, Joussen AM, Rohrschneider K, et al. Cryopreserved human amniotic membrane for ocular surface reconstruction. Graefes Arch Clin Exp Ophthalmol 2000;238:68–75.
105. Koizumi N, Inatomi T, Quantock AJ, Fullwood NJ, Dota A, Kinoshita S. Amniotic membrane as a substrate for cultivating limbal corneal epithelial cells for autologous transplantation in rabbits. Cornea 2000;19:65–71.
106. Choi TH, Tseng SC. In vivo and in vitro demonstration of epithelial cell induced myofibroblast differentiation of keratocytes and an inhibitory effect by amniotic membrane. Cornea 2001;20:197–204.
107. Kim JS, Kim JC, Na BK, Jeong JM, Song CY. Amniotic membrane patching promotes healing and inhibits proteinase activity on wound healing following acute corneal alkali burn. Exp Eye Res 2000;70:329–37.
108. Chacko DM, Das AV, Zhao X, James J, Bhattacharya S, Ahmad I. Transplantation of ocular stem cells: the role of injury in incorporation and differentiation of grafted cells in the retina. Vision Res 2003;43:937–46.
109. Hao Y, Ma DH, Hwang DG, Kim WS, Zhang F. Identification of antiangiogenic and antiinflammatory proteins in human amniotic membrane. Cornea 2000;19:348–52.
110. Kjaergaard N, Hein M, Hyttel L, et al. Antibacterial properties of human amnion and chorion in vitro. Eur J Obstet Gynecol Reprod Biol 2001;94:224–9.
111. Kjaergaard N, Helmig RB, Schonheyder HC, Uldbjerg N, Hansen ES, Madsen H. Chorioamniotic membranes constitute a competent barrier to group B streptococcus in vitro. Eur J Obstet gynecol Reprod Biol 1999;83:165–9.
112. Galask RP, Snyder IS. Antimicrobial factors in amniotic fluid. Am J Obstet Gynecol 1970;106:59–65.
113. Shimmura S, Shimazaki J, Ohashi Y, Tsubota K. Antiinflammatory effects of amniotic membrane transplantation in ocular surface disorders. Cornea 2001;20:408–13.
114. Meller D, Pires RT, Tseng SC. Ex vivo preservation and expansion of human limbal epithelial stem cells on amniotic membrane cultures. Br J Ophthalmol 2002;86:463–71.
115. Brown SI, Bloomfield SE, Pearce DB. A follow-up report on transplantation of the alkali-burned cornea. Am J Ophthalmol 1974;77:538–42.
116. Kremer I, Rajpal RK, Rapuano CJ, Cohen EJ, Laibson PR. Results of penetrating keratoplasty in aniridia. Am J Ophthalmol 1993;115:317–20.
117. Tugal-Tutkun I, Akova YA, Foster CS. Penetrating keratoplasty in cicatrizing conjunctival diseases. Ophthalmology 1995;102:576–85.
118. Holland EJ. Epithelial transplantation for the management of severe ocular surface disease. Trans Am Ophthalmol Soc 1996;94:677–743.

119. Holland EJ, Schwartz GS. The evolution of epithelial transplantation for severe ocular surface disease and a proposed classification system. Cornea 1996;15:549–56.
120. Holland EJ, Schwartz GS. The Paton lecture: ocular surface transplantation: 10 years' experience. Cornea 2004;23:425–31.
121. Dua HS, Azuara-Blanco A. Autologous limbal transplantation in patients with unilateral corneal stem cell deficiency. Br J Ophthalmol 2000;84:273–8.
122. Yao YF, Zhang B, Zhou P, Jiang JK. Autologous limbal grafting combined with deep lamellar keratoplasty in unilateral eye with severe chemical or thermal burn at late stage. Ophthalmology 2002;109:2011–7.
123. Rao SK, Rajagopal R, Sitalakshmi G, Padmanabhan P. Limbal allografting from related live donors for corneal surface reconstruction. Ophthalmology 1999;106:822–8.
124. Daya SM, Ilari FA. Living related conjunctival limbal allograft for the treatment of stem cell deficiency. Ophthalmology 2001;108:126–33; discussion 133–4.
125. Tsubota K, Satake Y, Kaido M, et al. Treatment of severe ocular-surface disorders with corneal epithelial stem-cell transplantation. N Engl J Med 1999;340:1697–703.
126. Dua HS, Azuara-Blanco A. Allo-limbal transplantation in patients with limbal stem cell deficiency. Br J Ophthalmol 1999;83:414–9.
127. Holland EJ, Djalilian AR, Schwartz GS. Management of aniridic keratopathy with keratolimbal allograft: a limbal stem cell transplantation technique. Ophthalmology 2003;110:125–30.
128. Solomon A, Ellies P, Anderson DF, et al. Long-term outcome of keratolimbal allograft with or without penetrating keratoplasty for total limbal stem cell deficiency. Ophthalmology 2002;109:1159–66.
129. Jenkins C, Tuft S, Liu C, Buckley R. Limbal transplantation in the management of chronic contact-lens-associated epitheliopathy. Eye 1993;7(Pt 5):629–33.
130. Sloper CM, Powell RJ, Dua HS. Tacrolimus (FK506) in the management of high-risk corneal and limbal grafts. Ophthalmology 2001;108:1838–44.
131. Pellegrini G, Traverso CE, Franzi AT, Zingirian M, Cancedda R, De Luca M. Long-term restoration of damaged corneal surfaces with autologous cultivated corneal epithelium. Lancet 1997;349:990–3.
132. Schwab IR, Reyes M, Isseroff RR. Successful transplantation of bioengineered tissue replacements in patients with ocular surface disease. Cornea 2000;19:421–6.
133. Tsai RJ, Li LM, Chen JK. Reconstruction of damaged corneas by transplantation of autologous limbal epithelial cells. N Engl J Med 2000;343:86–93.
134. Koizumi N, Inatomi T, Suzuki T, Sotozono C, Kinoshita S. Cultivated corneal epithelial stem cell transplantation in ocular surface disorders. Ophthalmology 2001;108:1569–74.
135. Daya SM, Watson A, Sharpe JR, et al. Outcomes and DNA analysis of ex vivo expanded stem cell allograft for ocular surface reconstruction. Ophthalmology 2005;112:470–7.
136. Shimazaki J, Aiba M, Goto E, Kato N, Shimmura S, Tsubota K. Transplantation of human limbal epithelium cultivated on amniotic membrane for the treatment of severe ocular surface disorders. Ophthalmology 2002;109:1285–90.
137. Nakamura T, Koizumi N, Tsuzuki M, et al. Successful regrafting of cultivated corneal epithelium using amniotic membrane as a carrier in severe ocular surface disease. Cornea 2003;22:70–1.
138. Joseph A, Powell-Richards AO, Shanmuganathan VA, Dua HS. Epithelial cell characteristics of cultured human limbal explants. Br J Ophthalmol 2004;88:393–8.
139. Koizumi N, Cooper LJ, Fullwood NJ, et al. An evaluation of cultivated corneal limbal epithelial cells, using cell-suspension culture. Invest Ophthalmol Vis Sci 2002;43:2114–21.
140. Nakagawa S, Nishida T, Kodama Y, Itoi M. Spreading of cultured corneal epithelial cells on fibronectin and other extracellular matrices. Cornea 1990;9:125–30.
141. Nishida K, Yamato M, Hayashida Y, et al. Functional bioengineered corneal epithelial sheet grafts from corneal stem cells expanded ex vivo on a temperature-responsive cell culture surface. Transplantation 2004;77:379–85.
142. Schwab IR. Cultured corneal epithelia for ocular surface disease. Trans Am Ophthalmol Soc 1999;97:891–986.

143. Wang DY, Hsueh YJ, Yang VC, Chen JK. Propagation and phenotypic preservation of rabbit limbal epithelial cells on amniotic membrane. Invest Ophthalmol Vis Sci 2003;44:4698–704.
144. Sangwan VS, Vemuganti GK, Singh S, Balasubramanian D. Successful reconstruction of damaged ocular outer surface in humans using limbal and conjuctival stem cell culture methods. Biosci Rep 2003;23:169–74.
145. Tseng SC, Meller D, Anderson DF, et al. Ex vivo preservation and expansion of human limbal epithelial stem cells on amniotic membrane for treating corneal diseases with total limbal stem cell deficiency. Adv Exp Med Biol 2002;506:1323–34.
146. Lindberg K, Brown ME, Chaves HV, Kenyon KR, Rheinwald JG. In vitro propagation of human ocular surface epithelial cells for transplantation. Invest Ophthalmol Vis Sci 1993;34:2672–9.
147. Koizumi N, Rigby H, Fullwood NJ, et al. Comparison of intact and denuded amniotic membrane as a substrate for cell-suspension culture of human limbal epithelial cells. Graefes Arch Clin Exp Ophthalmol 2007;245:123–34.
148. Sangwan VS, Matalia HP, Vemuganti GK, et al. Clinical outcome of autologous cultivated limbal epithelium transplantation. Indian J Ophthalmol 2006;54:29–34.
149. Nakamura T, Inatomi T, Sotozono C, et al. Transplantation of autologous serum-derived cultivated corneal epithelial equivalents for the treatment of severe ocular surface disease. Ophthalmology 2006;113:1765–72.
150. Shortt AJ, Secker GA, Notara MD, et al. Transplantation of ex vivo cultured limbal epithelial stem cells: a review of techniques and clinical results. Surv Ophthalmol 2007;52:483–502.
151. Nakamura T, Endo K, Cooper LJ, et al. The successful culture and autologous transplantation of rabbit oral mucosal epithelial cells on amniotic membrane. Invest Ophthalmol Vis Sci 2003;44:106–16.
152. Nakamura T, Inatomi T, Sotozono C, Amemiya T, Kanamura N, Kinoshita S. Transplantation of cultivated autologous oral mucosal epithelial cells in patients with severe ocular surface disorders. Br J Ophthalmol 2004;88:1280–4.
153. Inatomi T, Nakamura T, Koizumi N, Sotozono C, Yokoi N, Kinoshita S. Midterm results on ocular surface reconstruction using cultivated autologous oral mucosal epithelial transplantation. Am J Ophthalmol 2006;141:267–75.

Tissue Engineered Scaffolds for Stem Cells and Regenerative Medicine

Hossein Hosseinkhani and Mohsen Hosseinkhani

Abstract For successful tissue regeneration, the cells constituting the tissue to be regenerated, such as matured, progenitor, or precursor cells, are necessary. Considering the proliferation activity and differentiation potential of cells, stem cells are practically promising. Among them, mesenchymal stem cells (MSCs) have been widely investigated for use by themselves or combined with scaffolds if necessary for promotion of cell proliferation and differentiation. It was found that MSCs have an inherent nature to differentiate into not only osteogenic linage cells but also chondrogenic, myogenic, adipogenic, and neurogenic lineages. MSCs have been experimentally used to demonstrate their in vivo potential to induce the regeneration of mesenchymal tissues. Since it is reported that the cells are effective in inducing the regeneration of tissues other than mesenchym, their feasibility in the cell source for regenerative medicine is highly expected. They are practically isolated from patients themselves. Material design of scaffold for cell proliferation and differentiation is one of the key technologies for tissue engineering. Porous materials with various dimensional structures have been investigated for the cell scaffold because they have a larger surface for cell attachment and proliferation than two-dimensional materials and are preferable to assist the formation of three-dimensional cell constructs, which may resemble the structure and function of body tissues. In addition, the three-dimensional scaffold also plays an important role in the substrate for in vitro cell culture to increase the number of cells as high as clinically applicable. This chapter reviews the basic principles of tissue engineering and the recent developments of stem cells for their potential applications in regenerative medicine.

Keywords Tissue engineering; Scaffolds; Stem cells; Regenerative medicine

H. Hosseinkhani (✉) and M. Hosseinkhani
International Center for Young Scientists (ICYS), National Institute for Materials Science
(NIMS), Tsukuba, Ibaraki 305-0044, Japan
e-mail: hossein.hosseinkhani@nims.go.jp

H. Bharavand (ed.), *Trende in Stem Cell Biology and Technology*,
DOI 10.1007/978-1-60327-905-5_19

367

Introduction

Stem cells are a self-renewing cell type that can be differentiated into other cells. Conventional in vitro models to study differentiation of stem cells use freshly isolated cells grown in two-dimensional (2D) cultures. Clinical trails using in vitro stem cell culture can be expected only when the differentiated stem cells mimic the tissue regeneration in vivo. Therefore, the design of an in vitro three-dimensional (3D) model that mimics the in vivo environment is needed to effectively study its use for regenerative medicine. Biodegradable scaffolds play an important role for tissue-engineered scaffolds for tissue regeneration. Tissue-engineered scaffolds have a significant effect on stem cells' proliferation and differentiation. The application of scaffolding materials together with stem cell technologies is believed to hold enormous potential for applications in tissue regeneration. This chapter emphasizes that tissue-engineered scaffolds represent a viable strategy for the development of certain engineered tissue replacements and tissue regeneration systems using stem cells.

The Source of Stem Cells and Their Therapeutic Application

There are many limitations and problems that remain from current therapies such as autografts, allografts, xenografts, or metal prosthesis for the replacement of tissue defects resulting from tumors, surgical resections, trauma, or aging. Therefore, using stem cells for tissue regeneration represents a new direction toward regenerative medicine. Bone, cartilage, tendon, muscle, fat, and marrow stroma are formed during embryologic development (1). Bone marrow contains stromal or mesenchymal stem cells (MSCs) that have the ability to differentiate into osteoblasts, adipocytes, chondrocytes, or myoblasts (2, 3). Figure 1 shows the main source of stem cells and its differentiation capacity. Embryonic stem (ES) cells (fetal or adult cells) obtained from any germ layers such as ectoderm (epidermal tissues and nerves), mesoderm (muscle, bone, and blood), or endoderm (liver, pancreas, gastrointestinal tract, lungs) can differentiate into any cell type (4). The ability of stem cells to differentiate and replace mature cells is fundamental for regenerative medicine (5). In recent years, a combination of materials engineering and stem cells technology has been used to study tissue regeneration and ultimately mimic the stem cell niche (6). The application of stem cells for regenerative medicine is one of the most attractive research areas in biomedical engineering. There are some research trials under way in replacement therapy using stem cells, such as in infarcted heart, diabetes, and Parkinson's disease (6). Figure 2 indicates the differentiation tree of MSCs toward tissue regeneration by use of MSCs, precursor cells, and blast cells. In the field of regenerative medicine, scientists apply the principles of cell biology and materials engineering to construct biological substitutes that will restore and maintain normal function in injured tissues (7). Researchers have been investigating the fabrication of functional living tissue, or tissue engineering, using cells seeded on highly porous synthetic biodegradable polymer scaffolds as a novel approach toward

Source of Stem Cells

Embryonic Stem Cells

Embryonic stem cells constitute stem cell mass and give rise to a multiple of cell types and tissues.

Compared to embryonic stem cells, adult stem cells are preferable for therapeutic purposes since they are considered safer for implantation, with lesser proliferation capacity and tumorgenecity. They are also easier to differentiate to specific lineages, while ES cells can give a wide range of tissues following local implantation.

Adult Stem Cells

Adult stem cells constitute adult tissues and give rise to differentiated tissue-specialized cells, and are responsible for the regenerative capacities of tissue. AS cells present a more limited range of differentiation lineages.

Mesenchymal stem cells are stem cells residing in variety of adult mesenchymal tissues, and can be isolated from bone marrow, or other hematopoietic and non-hematopoietic tissues.

Mesenchymal stem cells derived from non-hematopoietic tissues, such as adipose tissues are very attractive future area of tissue engineering.

Embryonic Germ Cells

Embryonic germ cells are derived from the cells in the ridge of an embryo or a fetus, which give rise to eggs or sperm. They are able to rise to virtually all cell types. This potential makes pluripotent cells very attractive candidates for the development of unprecedental medical treatments.

Fig. 1 The main source of stem cells

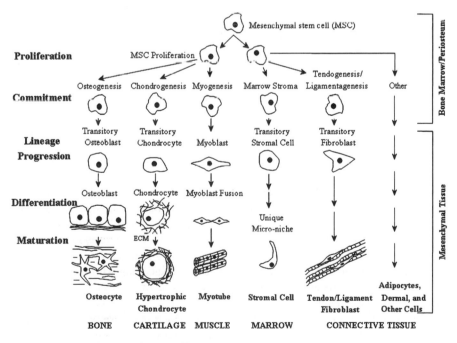

Fig. 2 Differentiation tree of stem cells

the development of biological substitutes that may replace lost tissue function (8). Over the past decade, scientists have applied the principles of tissue engineering in the fabrication of a wide variety of tissues, including both structural and complex tissues.

Tissue Engineering Principles in Stem Cells Technology

Regenerative medicine is a new field of science using stem cells to generate biological tissues and improve tissue functions. The application of MSCs has rapidly improved through research to evaluate their therapeutic applications (9). Tissue engineering is an interdisciplinary field that applies principles and methods of engineering toward the development of biological substitutes to improve the function of damaged tissue and organs (10, 11). The motivation of using tissue engineering in regenerative medicine centers around several factors:

1. Since 1970s, organ transplantation has become a common therapeutic approach for end-stage organ failure patients.
2. Demand is greater than supply; for example: 19,095 patients (1989), 80,766 patients (December 2002) on the UNOS National Patient Waiting List.
3. Cost of organ replacement therapy: $305 billion (US$, 2000).
4. The interdisciplinary approach of tissue engineering is a combinational technology, requiring the use of molecular biology, materials engineering, and reconstructive surgery.

For successful tissue regeneration, the cells constituting the tissue to be regenerated, such as matured, progenitor, or precursor cells, are necessary. Considering the proliferation activity and differentiation potential of cells, stem cells are most promising. Among them, MSC have been widely investigated for use by themselves or combined with scaffolds if necessary for promotion of cell proliferation and differentiation. It was found that MSC have an inherent nature to differentiate into not only osteogenic linage cells but also chondrogenic, myogenic, adipogenic, and neurogenic lineages (12–17). MSC have been experimentally used to demonstrate their in vivo potential to induce the regeneration of mesenchymal tissues (17–20). Since it is reported that the cells are effective in inducing the regeneration of tissues other than mesenchym, their feasibility in the cell source for regenerative medicine is highly expected. They are practically isolated from patients themselves (21–24). Material design of scaffold for cell proliferation and differentiation is one of the key technologies for tissue engineering. In conventional cell culture, such as static tissue culture dish (2D), the initial rate of cell growth is higher, but the proliferation stops once the cells have reached confluence. Porous materials with various dimensional (3D) structures have been investigated for the cell scaffold because they have a larger surface for cell attachment and proliferation than 2D materials and are preferable to assist the formation of 3D cell constructs, which may resemble the structure and function of body tissues. In addition, the 3D scaffold also plays an important role in the substrate for in vitro cell culture to increase the number of cells as high as clinical application. Three-dimensional scaffolds, through their ability to regenerate or restore tissue and/or organs, have begun to revolutionize medicine and biomedical science. Scaffolds have been used to support and promote the regeneration of tissues. Different processing techniques have been developed to design and fabricate 3D scaffolds for tissue engineering implants. However, there is neither a simple nor an inexpensive method for producing the main characteristics that a scaffold should have for application in tissue engineering.

Because the proliferation of cells in the 3D scaffold needs oxygen and nutrition supply, the 3D scaffold materials should provide such an environment. Diffusion of nutrients, bioactive factors, and oxygen through 3D scaffolds is sufficient for survival of large numbers of cells for extended periods of time. A major constraint in the use of biodegradable polymer scaffolds for vascular tissue engineering is poor cell adhesion and lack of signals for new tissue generation. The presence of extracellular matrix (ECM) within the scaffold is desirable for growth of stem cells and in vitro formation of remodeled vascular conduit *(25)*.

Tissue engineering is designed to regenerate natural tissues or to create biological substitutes for defective or lost organs by making use of cells. Considering the usage of cells in the body, it is no doubt that a sufficient supply of nutrients and oxygen to the transplanted cells is vital for their survival and functional maintenance *(26)*. Without a sufficient supply, only a small number of cells that have been preseeded in the scaffold or migrated into the scaffold from the surrounding tissue would survive. Rapid formation of a vascular network at the transplanted site of cells must be a promising way to provide cells with the vital supply. This process of generating new microvasculature, termed neovascularization, is a process observed physiologically in development and wound healing *(27)*. It is recognized that basic fibroblast growth factor (bFGF) functions to promote such an angiogenesis process *(27, 28)*. The growth factor stimulates the appropriate cells (e.g., endothelial cells), already present in the body, to migrate from the surrounding tissue, proliferate, and finally differentiate into blood vessels *(27)*. However, these proteins cannot always reach the sustained angiogenesis activity if they are only injected in solution form, most likely because of their rapid diffusional excretion from the injection site. One possible way for enhancing the in vivo efficacy is to achieve its controlled release over an extended time period by incorporating the growth factor in a polymer carrier. If this carrier is biodegraded or harmonized with tissue growth, it will work as a scaffold for tissue regeneration in addition to a carrier matrix for the growth factor release. The use of angiogenic factors is a popular approach to induce neovascularization. Among them, bFGF plays a multifunctional role in stimulation of cell growth and tissue repair. However, it has a very short half-life when injected and it is unstable in solution. To overcome these problems, bFGF was encapsulated within alginate, gelatin, agarose/heparin, collagen, and poly(ethylene-co-vinyl acetate) carriers *(29–31)*. According to the results of these studies, it is conceivable to incorporate the angiogenic factor into a sustained releasing system and use it prior to the implantation.

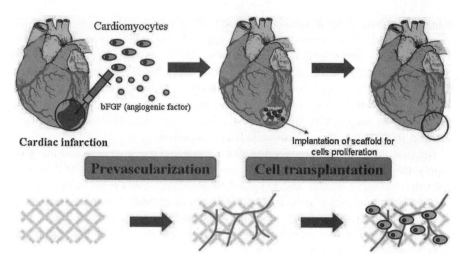

Fig. 3 Schematic illustration of tissue regeneration based on the principle of tissue engineering

Some studies have demonstrated that bFGF achieved promoted angiogenesis when used in combination with delivery matrices and scaffold *(31–36)*. Figure 3 indicates tissue regeneration based on the principle of tissue engineering *(37)*.

There are other growth factors currently used in tissue regeneration. Hepatocyte growth factor (HGF) was originally discovered as a protein factor to accelerate hepatocyte proliferation *(38)*. Previous studies have demonstrated that HGF has great potential for proliferation, differentiation, mitogenesis, and morphogenesis of various cells *(39–41)*. Therefore, HGF can be used for various tissue engineering applications where angiogenesis is needed. However, since growth factors such as HGF have a very short half-life, when they inject into the body, they lose their biological activities due to rapid digestion. Sustained-release technology has been used widely for different drugs and proteins to overcome this problem. Previous studies using growth factors such as bFGF, bone morphogenetic protein (BMP), and transforming growth factor (TGF) have demonstrated that their expected biological activities could be achieved when incorporated in carrier matrices *(28, 42, 43)*. It has been shown that bFGF and vascular endothelial growth factor (VEGF) exhibited properties to promote the angiogenesis process *(27, 31, 44)*. Osteogenic growth factors such as TGF-β, BMP, and bFGF can induce bone formation in both ectopic and orthotopic sites in vivo *(45–47)*. Table 1 summarizes the characteristics of the growth factors used in tissue engineering.

Tissue-Engineered Scaffolds

The proposed technique of cell culture in 3D cell scaffold constructs is based on the use of 3D fibrous scaffold to guide cell organization. In comparison with conventional culture, cells maintained in 3D culture more closely resemble the in vivo environment with regard to cell shape and cellular environment that can influence the behavior of cells.

Table 1 Characteristics of the growth factors used in tissue engineering

Growth factor	Isoelectric point (IEP)	Molecular weight (kDa)	Biological substances for growth factor binding	Functions of growth factor
Basic fibroblast growth factor (bFGF)	9.6	16	Heparin or heparan sulfate	Stimulating the cells involved in the healing process (bone, cartilage, nerve, etc.). Angiogenesis
Transforming growth factor β1 (TGF-β1)	9.5	25	Heparin or heparan sulfate Collagen type IV Latency associated protein Latent TGF-β1 binding protein	Enhancing the wound healing, stimulating the osteoblast proliferation to enhance bone formation
Bone morphogenetic protein-2 (BMP-2)	8.5	32	Collagen type IV	Stimulating the mesenchymal stem cells to osteoblast lineage and inducing the bone formation both at bone and ectopic sites
Vascular endothelial growth factor (VEGF)	8.5	38	Heparin or heparan sulfate	Stimulating the endothelial cell growth, angiogenesis, and capillary permeability
Hepatocyte growth factor (HGF)	5.5	100	Heparin or heparan sulfate	Stimulating of matrix remodeling and epithelial regeneration (liver, spleen, kidney, etc.)

It has been recognized that induction of tissue regeneration based on tissue engineering can be achieved through three key steps: the proliferation of cells, the seeding of cells and proliferation in a suitable scaffold, and the maintenance of the differentiation phenotype of the engineered tissues *(48)*. The property of scaffold material for cell attachment is one of the major factors contributing its morphology, proliferation, functions, and the subsequent tissue organization *(49)*. At first, cells attach to the material surface of scaffold, then spread, and proliferate. The 3D scaffold can provide a larger surface area available for cell attachment and spreading than a 2D scaffold could (i.e., tissue culture plate). Xie et al. *(50)* have reported that the initial rate of cells growth was higher for the 2D culture, but once the cells reached confluence, their proliferation stopped. However, the cells' growth in the 3D scaffold was continued for longer time periods than that of the 2D scaffold. Other reports have demonstrated that cell proliferation was superior in the 3D scaffold versus the 2D scaffold *(51–55)*.

Regenerative medicine is an interdisciplinary field that combines engineering and live sciences in order to develop techniques that enable the restoration, maintenance, or enhancement of living tissues and organs. Its fundamental aim is the creation of natural tissue with the ability to restore missing organ or tissue function, which the organism has not been able to regenerate in physiological conditions. As a result, it aspires to improve the health and quality of life for millions of people worldwide and to provide a solution to the present limitations of rejections, low quantity of donors, and so forth *(56)*. Tissue engineering needs a scaffold to serve as a substrate for seeding cells and as a physical support in order to guide the formation of the new tissue *(56–61)*. The majority of researchers use techniques that utilize 3D polymeric scaffolds that are composed of natural or synthetic polymers. Synthetic materials are attractive because their chemical and physical properties (e.g., porosity, mechanical strength) can be specifically optimized for a particular application. The polymeric scaffolds structures are endowed with a complex internal architecture, channels, and porosity that provide sites for cell attachment and maintenance of differentiated function without hindering proliferation *(56)*. Ideally, a polymeric scaffold for tissue engineering should have the following characteristics: (a) appropriate surface properties promoting cell adhesion, proliferation, and differentiation, (b) biocompatibility, (c) highly porous, with a high surface area to volume ratio, with an interconnected pore network for cell growth and flow transport of nutrients and metabolic waste, and (d) mechanical properties sufficient to withstand any in vivo stresses *(56, 57, 61–65)*. The last requirement is difficult to combine with the high porosity in a large volume of material. That is why it is necessary to use polymeric matrices with special or reinforced properties, especially if the polymer is a hydrogel.

Among many materials currently used as cell scaffolds, collagen has been the most widely used. Its in vivo safety has been proven through the long-term applications in clinical medicine, cosmetics, and foods. The collagen sponge fabricated by freeze-drying method, followed by crosslinking of combined dehydrothermal, glutaraldehyde, and ultraviolet (UV), is highly porous with an interconnected pore structure. This method is effective in the infiltration of cells and for supplying oxygen and nutrients to the cells or excluding the cells wastes, while the shape and bioresorbability can be readily regulated by changing the formulation conditions. However, as shown in Fig. 4, the drawback of using a collagen sponge as a scaffold for cell proliferation and differentiation is its poor mechanical strength. To overcome the inherent material problem of the sponge, the combination with other materials has been attempted. Considering implantation, the materials to be combined should also be bioabsorbable. From the viewpoint of clinical application, it is preferable to select the material that has been clinically used. Several biodegradable synthetic polymers, such as poly(glycolic acid) (PGA) and its copolymers with L-lactic acid, DL-lactic acid, and β-caprolactone, have been fabricated into the cell scaffolds of nonwoven fabric and sponge shapes for tissue engineering. The mechanical resistance of the scaffolds to compression is practically acceptable for the tissue engineering applications because of their hydrophobic nature. However, the cell attachment to the surface of synthetic polymer scaffolds is poor compared with that of collagen. PGA has been approved by the U.S. Food and Drug Administration for clinical applications. Our previous study revealed that incorporation of PGA fiber enabled a collagen sponge to increase the resistance to compression in vitro and in vivo *(66)*. The in vitro culture experiment revealed that the number of MSC attached increased with the incorporation of PGA fiber to a significantly higher extent compared with that of the original collagen sponge *(66)*. It is key for the present technology to fabricate mechanically strong collagen sponges

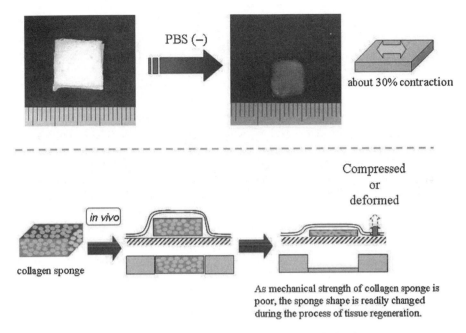

PBS (−)

about 30% contraction

Compressed
or
deformed

in vivo

collagen sponge

As mechanical strength of collagen sponge is
poor, the sponge shape is readily changed
during the process of tissue regeneration.

Fig. 4 Contraction of collagen sponge by in the presence of biological substances

by incorporating the PGA fiber of which the amount is as low as possible. Since collagen is more compatible to cells than PGA, at a higher amount of PGA fiber incorporated, the fiber may cause inflammation response to the sponge. Moreover, the collagen sponge does not become strong enough to resist the compressed deformation only by increasing the extent of crosslinking. Because the PGA fiber incorporation also suppressed the shrinkage of collagen sponge, it is possible that the volume available for cell attachment was larger, resulting in a higher number of cells attached. We have shown that mouse fibroblast L929 cells infiltrated into the collagen sponge incorporated PGA fiber more deeply than the collagen sponge alone (Figs. 5 and 6). This phenomenon also can be explained in terms of suppressed shrinkage of sponge by PGA fiber incorporation. The collagen sponge mechanically reinforced by PGA fiber incorporation is a promising scaffold for tissue regeneration. The incorporation of PGA fiber enabled the sponge to increase the resistance to compression. In comparing in vivo degradability, the collagen scaffold is generally digested faster than the PGA fabric. This degradation profile greatly depends on the crosslinking extent of collagen sponge and the molecular weight of PGA and the formulation shape. In our study, a combined crosslinking method of dehydrothermal, glutaraldehyde, and UV was used to prepare collagen sponges with or without PGA fiber incorporation. Weadock et al. *(67)* have evaluated the physical, mechanical, and biological behaviors of collagen sponge crosslinked by physical (UV irradiation and dehydrothermal) and chemical (carbodiimide and glutaraldehyde) or a combination of physical (dehydrothermal) and chemical (carbodiimide) crosslinking.

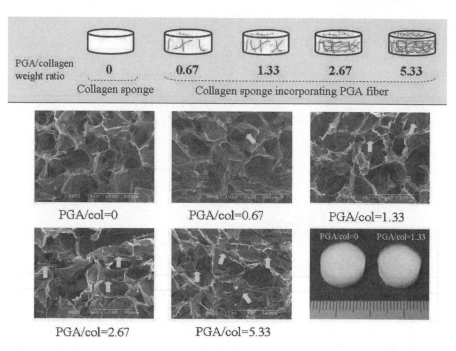

Fig. 5 Cross-sectional scanning electron microscopy (SEM) photographs of the structural morphology of collagen sponge with different poly(glycolic acid) (PGA) fiber incorporation. *Arrows* shows the location of PGA fibers inside the collagen sponge. The *right panel* in bottom *brown color* shows light microscopic photographs of a collagen sponge without PGA fiber incorporation and a collagen sponge incorporating PGA fiber

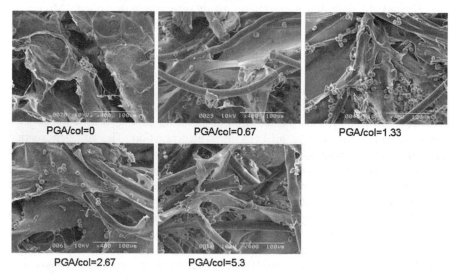

Fig. 6 Cross-sectional scanning electron microscopy (SEM) photographs of the structural morphology of collagen sponge with different poly(glycolic acid) (PGA) fiber incorporation 2 days after culturing the mouse fibroblast L929 cells

The results revealed that combination of physical (dehydrothermal) and chemical (carbodiimide) crosslinking of collagen reduced significantly the swelling ratio and increased the collagenase resistance time and low- and high-strain modulus compared with a single crosslinking of UV, dehydrothermal, and carbodiimide. The glutaraldehyde crosslinking itself showed the same physical and mechanical properties as the combination of physical (dehydrothermal) and chemical (carbodiimide) crosslinking.

The polymeric scaffold design depends on its anticipated application, but in any application it must achieve a structure with the aforementioned characteristics, which are necessaries to their correct function. Successfully achieving this is conditional on two factors: the materials used, both the porogen, and the reticulate polymer, which is infiltrated in the porogen to become a scaffold; and the structural architecture, both external and internal, basically shown by its porosity (high surface area to volume ratio), geometry, and pore size, keeping in mind that the structures must be easily processed into three dimensional format. On the basis of the extensive range of polymeric materials, different processing techniques have been developed to design and fabricate 3D scaffolds for tissue engineering implants (56, 57, 61, 68–74). They include (a) phase separation, (b) gas foaming, (c) fiber bonding, (d) photolithography, (e) solid free form (SFF), and (f) solvent casting in combination with particle leaching.

However, none of the techniques have achieved a suitable model of 3D architecture so that the scaffolds can fulfill their purpose in the desired way using high-cost equipment, for the reasons discussed below. When using phase separation, a porous structure can be easily obtained by adjusting thermodynamic and kinetic parameters. However, because of the complexity of the processing variables involved in the phase-separation technique, the pore structure cannot be easily controlled. Moreover, it is difficult to obtain large pores, which may exhibit a lack of interconnectivity (57, 58, 71). Gas foaming has the advantage of room temperature processing but produces a largely nonporous outer skin layer and a mixture of open and closed pores within the center, leaving incomplete interconnectivity. The main disadvantage of the gas foaming method is that it often results in a nonconnected cellular structure within the scaffold (58, 71). Fiber bonding provides a large surface area for cell attachment and a rapid diffusion of nutrients in favor of cell survival and growth. However, these scaffolds, as the ones used to construct a network of bonded PGA, lacked the structural stability necessary for in vivo use. In addition, the technique does not lend itself to easy and independent control of porosity and pore size (56–58). Photolithography has also been employed for patterning and obtaining structures with high resolution, although this resolution may be unnecessary for many applications of patterning in cell biology. In any case, the disadvantage of this technique is the high cost of the equipment needed, which limits its applicability (75). SFF scaffold manufacturing methods provide excellent control over scaffold external shape, internal pore interconnectivity, and geometry, but it offers limited microscale resolution. Also, it is important to consider the following items: (1) the minimum size of global pores is 100 μm, (2) SFF requires complex correction of scaffold design for anisotropic shrinkage during fabrication; and (3) it requires high-cost equipment (63).

Finally, solvent casting in combination with particulate leaching method, which involves the casting of a mixture of monomers and initiator solution and a porogen in a mold, termed polymerization, followed by leaching-out of the porogen with the proper solvent to generate the pores, is inexpensive but still has to overcome some disadvantages in order to find engineering applications, namely the problem of residual porogen remains, irregular shaped pores, and insufficient interconnectivity *(61, 76)*. The proposed scaffolds may find application as structures that facilitate either tissue regeneration or repair during reconstructive operations *(56, 77, 78)*. The new structure could also find application in other areas in which the pore morphology might play an essential role, such as membranes *(79)* and filters *(80)*. In the United States alone, each year over 10,000 newly injured people are added to the total of more than 250,000 who are confined to a wheelchair *(81)*. A major limitation in treating nerve injury, central nervous system (CNS), and peripheral nervous system (PNS) is the failure of current therapies to induce nerve regeneration. Unfortunately, for CNS injury, and particularly spinal cord injury, there is currently no treatment available to restore nerve function *(82)*. One possible avenue to remedy this situation is to artificially engineer nerve tissue. It is commonly accepted that physical guidance of axons is a vital component of nerve repair. Many materials have been used in an attempt to physically guide the regeneration of damaged nerves *(82)*. Kang et al. *(83)* have concluded that preferential alignment of channel pores may provide a unique advantage in certain medical applications, such as nerve regeneration. In another research work, Blacher et al. *(78)* fabricated a highly oriented poly-lactic acid (PLA) scaffold for spinal cord regeneration and demonstrated that highly oriented macroporous have efficiency in axonal regeneration both in the PNS and CNS. Cell migration and angiogenesis were observed, as well as the expected orientation of axonal growth. The axons were perfectly aligned along the pore direction, which confirmed the crucial role of 3D polymer structures. Plant et al. *(84)* have demonstrated that 3D sponges of poly-hydroxy ethyl methacrylate (PHEMA) sponges are able to house a purified population of glial cells and provide a scaffold for regenerative growth of axons in the lesioned rat optic tract and may be a candidate for use as prosthetic bridges in the repair of the damaged CNS. However, they deduce that further work is necessary to optimize their procedure, such as providing a more oriented trabecular network within the hydrogel scaffold. In the research carried out by Shugens et al. *(85)*, macroporous foams of 100 μm were produced in the form of channels by the solid–liquid phase separation technique for nerve regeneration. They concluded that nerve regeneration can only occur through a structure of interconnected pores of ideal diameter in the range of 10–100 μm. In the study developed by Maquet et al. *(86)*, poly(D,L-lactide) foams with macroporous of 100 μm organized longitudinally were prepared by freeze-drying technique for spinal cord regeneration. They showed that the parallel assembly of rods of porous (diameter ~100 μm) containing an amphophilic copolymer was a promising strategy to bridge a defect in the spinal cord of adult rats, and they confirmed a high density of cells in the surface of porous interconnected structures as well.

Porosity of fabricated scaffolds can be determined through the measurement of the apparent density of the scaffold. For this, distilled water is used as a filler of the porous structure. The dried scaffold is weighted and placed in a glass tube connected to a vacuum pump then filled with distilled water before breaking the vacuum. The scaffolds filled with water are weighed again and the porosity is calculated as:

$$\Pi(\%) = \frac{V_p}{V_t} \times 100 = \frac{m_l / \rho_l}{\frac{m_l}{\rho_l} + \frac{m_m}{\rho_m}} \times 100,$$

where Π is the porosity, V_p and V_t are the volume occupied by the pores and the volume of the scaffolds, respectively, ρ_l is the density of the filler liquid and ρ_m is the bulk density of the scaffolding material, m_l and are m_m the liquid mass and dried scaffold mass, respectively.

Tissue Engineered Nanoscaffolds

The design of materials that can regulate cell behavior, such as proliferation and differentiation, is a key component for the fabrication of tissue engineering scaffolds. From the viewpoint of immune system response of the body, the implanted biomaterials should mimic the structure and biological function of native ECM, both in terms of chemical composition and physical structure as reported by Ma et al. *(87)*. Therefore, in order to mimic the biological function of ECM proteins, the scaffold materials used in tissue engineering need to be chemically functionalized to promote tissue regeneration as ECM does. It has been reported that collagen and elastin as ECM proteins are made from fibers in dimensions smaller than micrometers *(87)*. It seems that artificial nanoscaled fibers have great potential application in the field of biomaterials and tissue engineering.

The initial report showed that nanoscaled features influenced cell behaviors *(88)*. Nanoscaled surface topography has been found to promote osteoblast adhesions *(89)*. It has been demonstrated that osteoblast adhesion, proliferation, alkaline phosphatase activity, and ECM secretion on carbon nanofibers increased with decreasing fiber diameter in the range of 60–200 nm, whereas the adhesion of other kinds of cells such as chondrocytes, fibroblasts, and smooth muscle cells was not influenced *(90, 91)*. It has been supposed that the nanoscaled surface affects the conformation of adsorbed adhesion proteins such as vitronectin, thus affecting the cell behaviors *(92)*. In addition, the nanoscaled dimensions of cell membrane receptors such as integrins should also be considered. It has been reported that there are three different approaches toward the formation of nanofibrous materials: phase separation, electrospinning, and self-assembly *(93)*. Phase separation and self-assembling of biomolecules can generate smaller diameter nanofibers in the same range of natural ECM, while electrospinning generates large diameter nanofibers on the upper end of the range of natural ECM *(93)*. Electrospinning is a common technique used to fabricate tissue engineering scaffolds *(94)*. It is an easy technique and extremely inexpensive and can be applied for many different types of polymers. The authors' recent study demonstrated that fabricated PGA/collagen nanofibers through electrospinning significantly enhanced cell adhesion compared with PGA/ collagen microfibers *(95)*. Figure 7 shows a schematic illustration of an electrospinning device for the fabrication of nanofibers, while Fig. 8 shows cross-sectional scanning electron microscopy (SEM) photographs of the structural morphology of PGA/collagen nanofibers fabricated by electrospinning before and after culturing the MSC.

One of the most common approaches to produce fibers similar to ECM proteins such as collagen is self-assembly. It has been shown that peptide amphiphile (PA) contains a carbon alkyl tail and several other functional peptide formed nanofibers through self-assembly by mixing cell suspensions in media with dilute aqueous solutions of the peptide *(96)*. Another type of peptide containing 16 alternating hydrophobic and hydrophilic amino acids was fabricated to self-assemble into nanofibers under appropriate pH values *(97)*. Nanoscaled fibers produced by self-assembly of PA may be a promising approach in designing the next generation of biomaterials for drug delivery and tissue engineering.

It would be beneficial for biomedical applications if scaffold materials could promote the adhesion and growth of cells on their surfaces. The sequence of arginine–glycine–aspartic acid (RGD) has been discovered as a cell attachment sequence in various adhesive proteins present in the ECM and found in many proteins, such as fibronectin, collagen type 1, vitronectin, fibrin, and von Willebrand factor *(98)*. It has been well recognized that the sequence of RGD interacts with various types of integrin receptors of mammalian cells. Since the discovery of the

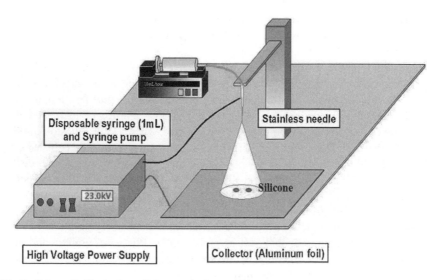

Fig. 7 Schematic illustration of electrospinning machine

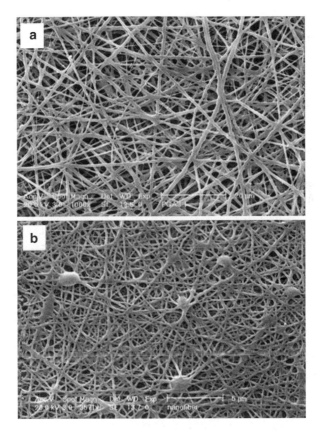

Fig. 8 Cross-sectional scanning electron microscopy (SEM) photographs of the structural morphology of poly(glycolic acid) (PGA)/collagen nanofibers fabricated by (a) electrospinning and (b) 2 days after culturing the mesenchymal stem cells (MSC) on nanofibers

RGD sequence as a cell attachment sequence in adhesive proteins of the ECM, there have been several efforts to synthesize bioactive peptides incorporating RGD for therapeutic purpose (99). Micro- and nanopatterned scaffolds have been less well investigated in regard to stem cells, although two recent studies highlight their attractiveness (100). In one study, Silva et al. (101) included a five amino acid, laminin-specific, cell-binding domain (which binds to specific integrins on cell surface) at the hydrophilic head of their amphiphiles and showed that neural stem cells could be induced to differentiate into neurons when cultured within a peptide gel. In contrast, cells grown in control scaffolds without the laminin-specific domain or on 2D tissue culture plastic coated with laminin solution differentiated much less. This was hypothesized to be largely a result of the density of the cells' binding ligands to which the cells were exposed, indicating clearly the importance of ECM in influencing cell function. Our recent studies have indicated that when the laminin-specific domain in the amphiphilic molecule was replaced with the amino acid sequence, RGD, a common cell-binding domain in many ECM proteins, especially collagen, differentiation of MSC to osteoblasts was significantly enhanced compared with amphiphilic nanofibers without this sequence or to 2D controls *(105)*. This is because the interaction of MSC integrins receptors with RGD of the peptide enhanced cell attachment on peptide nanofibers. The artificial scaffolds formed by self-assembling molecules not only provides a suitable support for cell proliferation but also serves as a medium through which diffusion of soluble factors and migration of cells can occur. The result of the cell attachment and proliferation revealed that diffusion of nutrients, bioactive factors, and oxygen through these highly hydrated networks is sufficient for survival of large numbers of cells for extended periods of time.

As understood from the findings, proteins and peptides can self-assemble into various structures like nanotubes, nanovesicles, and 3D peptide matrices with interwoven nanofibers. Macroscopic 3D peptide matrices can be engineered to form various shapes by changing the peptide sequence. Self-assembled peptide materials encouraged cell proliferation and differentiation. In regenerative medicine, these peptide matrices were used to cultivate chondrocyte ECM that can be used to repair cartilage tissue. Thus cartilage tissue engineering has been done by placing the primary chondrocytes and MSCs into these self-assembled peptide hydrogels to produce collagen and glycosaminoglycans. These peptide matrices can also be used in regeneration of bone by incorporating a phosphorylated serine that can attract and organize calcium ions to form hydroxyapatite crystals and functionalize them with a cell adhesion motif like RGD acid complex. The research studies have not been limited only to natural amphiphilic peptides. There are many research trials that have focused on synthesizing complex amphiphilic peptides by joining hydrophilic peptides into long alkyl chains. The peptide end of the molecule was designed to function and regulate biomineralization. Bone is produced as a result of deposition of calcium and phosphate ions to form hydroxyapatite crystals. This process is known as mineralization. Serine is a nonessential amino acid. When a phosphorylated serine was incorporated with the synthetic amphiphilic peptide complex, it served to attract and organize calcium and phosphate ions to form hydroxyapatite crystals. Furthermore, the synthetic amphiphilic peptides have been functionalized by adding a cell-adhesion motif. It was the RGD that was attached to the C-terminus of the peptide. This can be used to study the ability of bone cells to differentiate, proliferate, and adhere to a biomaterial surface like titanium. Titanium is the most widely used biomaterial surface to produce orthopedic implants, dental implants, and hip replacements. In spite of its excellent biocompatibility, titanium implants still fail. Most orthopedic implants have a lifetime of 15 years to the maximum. In order to produce a newer version of titanium implants that can stay in the body for a longer period of time, its surface has to be modified with nano-sized surface patterns so that bone cells (osteoblasts) differentiate and migrate into these patterns for better bone-implant adhesion. For such a purpose, these synthetic amphiphiles can be used to regulate and control the osteoblasts *(71, 102)*.

Our recent study has indicated that a 3D networks of self-assembled nanofibers was formed by mixing a bFGF suspension with aqueous solution of PA as an injectable carrier for controlled release of growth factors and was then used for feasibility of

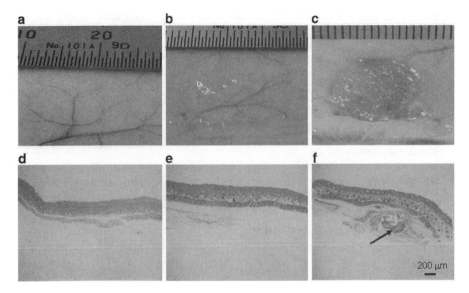

Fig. 9 Representative of tissue appearance (**a–c**) and histological cross-sections (**d–f**) of ectopically formed bone after subcutaneous injection of (**a, d**) transforming growth factor β (TGF-β), (**b, e**) peptide solution, and (**c, f**) peptide solution with TGF-β. The concentration of TGF-β is 10 µg. Each specimen subjected to hematoxylin and eosin staining. *Arrow* indicates the newly formed bone

prevascularization by the bFGF release from the 3D networks of nanofibers in improving efficiency of tissue regeneration *(103)*. The bFGF incorporated in alginate, gelatin, agarose/heparin, collagen, and poly(ethylene-co-vinyl acetate) releasing system *(28–31)* requires surgery for implantation, which is not always a welcomed option. In contrast, the bFGF incorporated in self-assembled peptide could be delivered to living tissues by simply injecting a liquid (i.e., PA solutions) and bFGF solution. The injected solutions would form a solid scaffold at the injected site of tissue and the release bFGF would induce significant angiogenesis around the injected site, in marked contrast to bFGF injection alone or PA injection alone *(102)*. This release system also was able to induce significant bone formation when PA solutions and TGF-β were subcutaneously injected to the back of rat (Fig. 9). The injected solutions of peptide and TGF-β formed a solid gel, and the sustained release of TGF-β induced significant ectopic bone compared with TGF-β injection. As a flexible delivery system, these scaffolds can be adapted for sustained release of many different growth factors and biomolecules.

References

1. Caplan AI. Mesenchymal stem cells. J Orthop Res 1991;9:641–50.
2. Owen M. Lineage of osteogenic cells and their relationship to the stromal system. Excerpta Med 1985;3:340–8.

3. Prockop DJ. Marrow stromal cells as stem cells for nonhematopoietic tissues. Science 1997;276:71–4.
4. Yamanaka S, Li J, Kania G, et al. Pluripotency of embryonic stem cells. Cell Tissue Res 2008;331:5–22.
5. Lensch MW, Daheron L, Schlaeger TM. Pluripotent stem cells and their niches. Stem Cell Rev 2006;2:185–201.
6. Dawson E, Mapili G, Erickson K, et al. Biomaterials for stem cell differentiation. Adv Drug Deliv Rev 2007;273:320–9.
7. Atala A. Engineering tissues, organs and cells. J Tissue Eng Regen Med 2007;1:83–96.
8. Kim SS, Vacanti JP. The current status of tissue engineering as potential therapy. Semin Pediatr Surg 1999;8:119–23.
9. Dazzi F, Horwood NJ. Potential of mesenchymal stem cell therapy. Curr Opin Oncol 2007;19:650–5.
10. Langer RS, Vacanti JP. Tissue engineering: the challenges ahead. Sci Am 1999;280:86–9.
11. Langer R, Vacanti JP. Tissue engineering. Science 1993;260:920–6.
12. Vandenburgh H, DelTatto M, Shansky J, et al. Tissue-engineered skeletal muscle organoids for reversible gene therapy. Hum Gene Ther 1996;7:2195–9.
13. Petite H, Viateau V, Bensaid W, et al. Tissue-engineered bone regeneration. Nat Biotechnol 2000;18:959–65.
14. Peretti GM, Randolph MA, Villa MT, et al. Cell-based tissue-engineered allogeneic implant for cartilage repair. Tissue Eng 2000;6:567–73.
15. Woerly S, Plant GW, Harvey AR. Neural tissue engineering from polymer to biohybrid organs. Biomaterials 1996;17:301–10.
16. Ishaug SL, Crane GM, Miller MJ, Yasko AW, Yaszmski MJ, Mikos AG. Bone formation by three dimensional stromal osteoblast culture in biodegradable polymer scaffold. J Biomed Mater Res 1997;36:17–28.
17. Freed LE, Vunjak-Novakovic G. Culture of organized cell communities. Adv Drug Deliv Rev 1998;33:15–30.
18. Freed LE, Vunjak-Novakovic G, Langer R. Cultivation of cell-polymer cartilage implants in bioreactors. J Cell Biochem 1993;51:257–64.
19. Niklason LE, Gao J, Abbott WM, et al. Functional arteries grow in vivo. Science 1999;284:489–93.
20. Ma PX, Langer R. Morphology and mechanical function of long-term *in vitro* engineered cartilage. J Biomed Mater Res 1999;44:217–21.
21. Schneider AI, Maier-Reif K, Graeve T. Constructing an in vitro cornea from cultures of the three specific corneal cell types. In Vitro Cell Dev Biol Anim 1999;35:515–26.
22. Papas KK, Long RC, Sambanis A, Constantinidis I. Development of a bioartificial pancreas: I. Long-term propagation and basal and induced secretion from entrapped TC3 cell cultures. Biotechnol Bioeng 1999;66:219–30.
23. Griffith M, Osborne R, Munger R, et al. Functional human corneal equivalents constructed from cell lines. Science 1999;286:69–72.
24. Kloth S, Ebenbeck C, Kubitza M. Stimulation of renal microvascular development under organotypic culture conditions. FASEB J 1995;9:963–7.
25. Pankajakshan D, Krishnan VK, Krishnan LK. Vascular tissue generation in response to signaling molecules integrated with a novel poly(epsilon-caprolactone)-fibrin hybrid scaffold. J Tissue Eng Regen Med 2007;1:389–97.
26. Colton CK. Implantable biohybrid artificial organs. Cell Transplant 1995;4:415–36.
27. Polverini PJ. The pathophysiology of angiogenesis. Crit Rev Oral Biol Med 1995;6:230–47.
28. Downs EC, Robertson NE, Riss TL, Plunkett ML. Calcium alginate beads as a slow release system for delivering angiogenic molecules in vivo and in vitro. J Cell Physiol 1992;152:422–9.
29. Iwakura A, Tabata Y, Tamura N, et al. Gelatin sheets incorporating basic fibroblast growth factor enhances healing of devascularized sternum in diabetic rats. Circulation 2001;104:1325–9.
30. Edelman ER, Mathiowitz E, Langer R, Klagsbrun M. Controlled and modulated release of basic fibroblast growth factor. Biomaterials 1991;12:619–26.

31. Ware JA, Simons M. Angiogenesis in ischemic heart disease. Nat Med 1997;3:158–64.
32. Thompson JA, Anderson KD, Dipietro JM, et al. Site-directed neovessel formation in vivo. Science 1988;241:1349–52.
33. Tabata Y, Nagano A, Ikada Y. Biodegradation of hydrogel carrier incorporating fibroblast growth factor. Tissue Eng 1999;5:127–38.
34. Tabata Y, Ikada Y. Vascularization effect of basic fibroblast growth factor released from gelatin hydrogels with different biodegradabilities. Biomaterials 1999;20:2169–75.
35. Cai S, Liu Y, Shu XZ, Prestwich GD. Injectable glycosaminoglycan hydrogels for controlled release of human basic fibroblast growth factor. Biomaterials 2005;26:6054–67.
36. Dogan AK, Gumusderelioglu M, Aksoz E. Controlled release of EGF and bFGF from dextran hydrogels in vitro and in vivo. J Biomed Mater Res B Appl Biomater 2005;74:504–10.
37. Tabata Y, Ikada Y. Protein release from gelatin matrices. Adv Drug Deliv Rev 1998;31:287–301.
38. Nakamura T, Nishikawa T, Hagiya M, et al. Molecular cloning and expression of human hepatocyte growth factor. Nature 1989;342:440–3.
39. Rubin RS, Chan AM, Bottaro DP, et al. A broad-spectrum human lung fibroblast-derived mitogen is a variant of hepatocyte growth factor. Proc Natl Acad Sci U S A 1991;88:415–9.
40. Weidner KM, Arakaki N, Vandekerckhove J, et al. Evidence for the identity of human scatter factor and human hepatocyte growth factor. Proc Natl Acad Sci U S A 1991;88:7001–5.
41. Montesano R, Matsumoto K, Nakamura T, Orci L. Identification of a fibroblast-derived epithelial morphogen as hepatocyte growth factor. Cell 1991;67:901–8.
42. Miyamoto S, Takaoka K, Okada T, et al. Evaluation of polylactic acid homopolymers as carriers for bone morphogenetic protein. Clin Orthop 1992;294:333–43.
43. Gombotz WR, Pankey SC, Bouchard LS, Ranchalis J, Puolakkainen P. Controlled release of TGF-beta 1 from a biodegradable matrix for bone regeneration. J Biomater Sci Polym Ed 1993;5:49–63.
44. Su Y, Cui Z, Li Z, Block ER. Calpain-2 regulation of VEGF-mediated angiogenesis FASEB J 2006;20:1443–51.
45. Ueda H, Hong L, Yamamoto M, et al. Use of collagen sponge incorporating transforming growth factor-beta1 to promote bone repair in skull defects in rabbits. Biomaterials 2002;23:1003–9.
46. Nagai H, Tsukuda R, Mayahara H. Effects of basic fibroblast growth factor (bFGF) on bone formation in growing rats. Bone 1995;16:367–74.
47. Wang EA, Rosen V, D'Alessandro JS, et al. Recombinant human bone morphogenetic protein induces bone formation. Proc Natl Acad Sci U S A 1990;87:2220–7.
48. Minuth WW, Sittinger M, Kloth S. Tissue engineering: generation of differentiated artificial tissues for biomedical applications. Cell Tissue Res 1998;291:1–11.
49. Ingber DE, Prusty D, Sun Z, Betensky H, Wang N. Cell shape, cytoskeletal mechanics, and cell cycle control in angiogenesis. J Biomech 1995;28:1471–84.
50. Xie Y, Yang ST, Kniss DA. Three-dimensional cell-scaffold constructs promote efficient gene transfection: implications for cell-based gene therapy. Tissue Eng 2001;7:585–98.
51. Yoshikawa T, Ohgushi H, Akahane M, et al. Analysis of gene expression in osteogenic cultured marrow/hydroxyapatite construct implanted at ectopic sites: a comparison with the osteogenic ability of cancellous bone. J Biomed Mater Res 1998;41:568–73.
52. Noshi T, Yoshikawa T, Ikeuchi M, et al. Enhancement of the in vivo osteogenic potential of marrow/hydroxyapatite composites by bovine bone morphogenetic protein. J Biomed Mater Res 2000;52:621–30.
53. Goldstein AS, Juarez TM, Helmke CD, Gustin MS, Mikos AG. Effect of convection on osteoblastic cell growth and function in biodegradable polymer foam scaffolds. Biomaterials 2001;22:1279–88.
54. Ma T, Li Y, Yang ST, Kniss DA. Effects of pore size in 3-D fibrous matrix on human trophoblast tissue development. Biotechnol Bioeng 2000;70:606–18.
55. Mueller-Klieser W. Three-dimensional cell cultures: from molecular mechanisms to clinical applications. Am J Physiol 1997;273:C1109–23.

56. Thomson RC, Wake MC, Yaszemski MJ, Mikos AG. Biodegradable polymer scaffolds to regenerate organs. Adv Polym Sci (Biopolymers II) 1995;122:245–73.
57. Chen G, Ushida T, Tateishi T. Development of biodegradable porous scaffolds for tissue engineering. Mater Sci Eng C 2001;17:63–9.
58. Chen G, Ushida T, Tateishi T. Preparation of poly(l-lactic acid) and poly(dl-lactic-co-glycolic acid) foams by use of ice microparticules. Biomaterials 2001;22:2563–7.
59. Wintermantel E, Mayer J, Blum J, Eckert KL, Lüscher P, Mathey M. Tissue engineering scaffolds using superstructures. Biomaterials 1996;17:83–91.
60. Zhao K, Deng Y, Chen JC, Chen GQ. Polyhydroxyalkanoate (PHA) scaffolds with good mechanical properties and biocompatibility. Biomaterials 2003;24:1041–5.
61. Hutmatcher DW. Scaffolds in tissue engineering bone and cartilage. Biomaterials 2000;21:2529–43.
62. Whang K, Thomas CH, Healy KE, Nuber G. A novel method to fabricate bioabsorbable scaffolds. Polymer 1995;36:837–42.
63. Taboas JM, Maddox RD, Krebsbach PH, Hollister SJ. Indirect solid free form fabrication of local and global porous, biomimetic and composite 3D polymer-ceramic scaffolds. Biomaterials 2003;24:181–94.
64. Gomes ME, Ribeiro AS, Malafaya PB, Reis RL, Cunha AM. A new approach based on injection moulding to produce biodegradable starch-based polymeric scaffolds: morphology, mechanical and degradation behaviour. Biomaterials 2001;22:883–9.
65. Hutmatcher DW, Schantz T, Zein I, Ng KW, Teoh SH, Tan KC. Mechanical properties and cell cultural response of polycaprolactone scaffolds designed and fabricated via fused deposition modeling. Mechanical properties and cell cultural response. J Biomed Mater Res 2001;5:203–16.
66. Hosseinkhani H, Hosseinkhani M, Tian F, Kobayashi H, Tabata Y. Bone regeneration on a collagen sponge-self assembled peptide-amphiphile nanofibers hybrid scaffold. Tissue Eng 2007;13:1–9.
67. Weadock K, Olson RM, Silver FH. Evaluation of collagen crosslinking techniques. Biomater Med Devices Artif Organs 1983;11:293–318.
68. Cai Z, Cheng G. Novel method to produce poly (3-hydroxybutyrate) scaffolds with controlled multi-pore size. J Mater Sci Lett 2003;22:153–5.
69. Cai Q, Yang J, Bei J, Wang S. A novel porous cells scaffold made of polylatic-dextran blend by combining phase-separation and particle-leaching techniques. Biomaterials 2002;23:4483–92.
70. Zein I, Hutmacher DW, Tan KC, Teoh SH. Fused deposition modeling of novel architectures for tissue applications. Biomaterials 2002;23:1169–85.
71. Lin ASP, Borrows TH, Cartmell SH, Guldberg RE. Microarchitectural and mechanical characterization of oriented porous polymer scaffolds. Biomaterials 2003;24:481–9.
72. Oxley HR, Corkhill PH, Fitton JH, Tighe BJ. Macroporous hydrogels for biomedical applications: methodology and morphology. Biomaterials 1993;14:1064–72.
73. Chirila TV, Constable IJ, Crawford GJ, et al. Poly(2-hydroxyethyl methacrylate) sponges as implant materials: in vivo and in vitro evaluation. Biomaterials 1993;14:26–38.
74. Zhang X, Jiang XN, Sun C. Micro-stereolithography of polymeric and ceramic microstructures. Sens Actuators A 1997;77:149–56.
75. Kane RS, Takayama S, Ostuni E, Ingber DE, Whitesides GM. Patterning proteins and cells using soft lithography. Biomaterials 1999;20:2363–76.
76. Mikos AG, Sakarinos G, Vacanti JP, Langer RS, Cima LG. Biocompatible polymer membranes and methods of preparation of three dimensional membrane structures. Patent number 5514378, issue date May 7, 1996.
77. Hutmacher DW, Ng KW, Kaps C, Sittinger M, Klaring S. Elastic cartilage engineering using novel scaffold architectures in combination with a biomimetic cell carrier. Biomaterials 2003;24:4445–58.
78. Blacher S, Maquet V, Schils F, et al. Image analysis of the axonal ingrowth into poly(d,l-lactide) porous scaffolds in relation to the 3-D porous structure, Biomaterials 2003;24:1033–40.
79. Nikpour M, Chaouk H, Mau A, et al. Porous conducting membranes based on polypyrrole-PMMA composites. Synth Met 1999;99:121–6.

80. Cavallini A, Notarnicola M, Berloco P, Lippolis A, De Leo A. Use of macroporous polypropylene filter to allow identification of bacteria by PCR in human fecal simples. J Microbiol Methods 2000;39:265–70.
81. Pastist CM, Mulder MB, Gautier SE, Maquet V, Jerome R, Oudega M. Freeze-dried poly (D,l-lactic acid) macroporous guidance scaffolds impregnated with brain-derived neurotrophic factor in the transected adult rat thoracic spinal cord. Biomaterials 2004;25:1569–82.
82. Schmidt CE, Leach JB. Neural tissue engineering: strategies for repair and regeneration. Annu Rev Biomed Eng 2003;5:293–347.
83. Kang HW, Tabata Y, Ikada Y. Fabrication of porous gelatine scaffolds for tissue engineering. Biomaterials 1999;20:1339–44.
84. Plant GW, Harvey AR, Chirila TV, et al. Axonal growth within poly (2-hydroxyethyl methacrylate) sponges infiltrated with Schwann cells and implanted into the lesioned rat optic tract. Brain Res 1995;671:119–30.
85. Shugens CH, Maquet V, Grandfils C, et al. Biodegradable and macroporous polylactide implants for cell transplantation: 1. Preparation of macroporous polylactide supports by solid–liquid phase separation. Polymer 1996;37:1027–38.
86. Maquet V, Martin D, Scholtes F, et al. Poly(D,l-lactide) foams modified by poly(ethylene oxide)-block-poly(D,l-lactide) copolymers and a-FGF: in vitro and in vivo evaluation for spinal cord regeneration. Biomaterials 2001;22:1137–46.
87. Ma Z, Kotaki M, Inai R, Ramakrishna S. Potential of nanofiber matrix as tissue engineering scaffolds. Tissue Eng 2005;11:101–9.
88. Rosenberg MD. Cell guidance by alterations in monomolecular films. Science 1963;139:411–2.
89. Webster TJ, Siegel RW, Bizios R. Osteoblast adhesion on nanophase ceramics. Biomaterials 1999;20:1221–7.
90. Price RL, Waidh MC, Haberstroha KM, et al. Selective bone cell adhesion on formulations containing carbon nanofibers. Biomaterials 2003;24:1877–87.
91. Elias KL, Price RL, Webster TJ. Enhanced functions of osteoblasts on nanometer diameter carbon fibers. Biomaterials 2002;23:3279–87.
92. Webster TJ, Schalder LS, Siegel RW, et al. Mechanisms of enhanced osteoblast adhesion on nanophase alumina involve vitronectin. Tissue Eng 2001;7:291–301.
93. Smith LA, Ma PX. Nano-fibrous scaffolds for tissue engineering. Colloids Surf B Biointerfaces 2004;10:125–31.
94. Kameoka J, Verbridge SS, Liu H, et al. Fabrication of suspended silica glass nanofibers from polymeric materials using a scanned electrospinning source. Nano Lett 2004;4:2105–8.
95. Tian F, Hosseinkhani H, Hosseinkhani M, et al. Quantitative analysis of cell adhesion on aligned micro- and nanofibers. J Biomed Mater Res A 2008;84:291–9.
96. Hartgerink JD, Beniash E, Stupp SI. Self-assembly and mineralization of peptide-amphiphile nanofibers. Science 2001;294:1684–8.
97. Hong Y, Legge RL, Zhang S, Chen P. Effect of amino acid sequence and pH on nanofiber formation of self-assembling peptides EAK16-II and EAK16-IV. Biomacromolecules 2003;4:1433–42.
98. Ruoslahti E, Pierschbacher MD. New perspectives in cell adhesion: RGD and integrins. Science 1987;238:491–7.
99. Williams JA. Disintegrins: RGD-containing proteins which inhibit cell/matrix interactions (adhesion) and cell/cell interactions (aggregation) via the integrin receptors. Pathol Biol 1992;40:813–21.
100. Evans ND, Gentleman E, Polak JM. Scaffolds for stem cells. Materials Today 2006;12:26–33.
101. Silva GA, Czeisler C, Niece KL, et al. Selective differentiation of neural progenitor cells by high-epitope density nanofibers. Science 2004;303:1352–5.
102. Hosseinkhani H, Hosseinkhani M, Tian F, Kobayashi H, Tabata Y. Osteogenic differentiation of mesenchymal stem cells in self-assembled. Biomaterials 2006;27:4079–86.
103. Hosseinkhani H, Hosseinkhani M, Tian F, et al. Enhanced angiogenesis through controlled release of basic fibroblast growth factor from peptide amphiphile for tissue regeneration. Biomaterials 2006;27:5836–44.

Mechanotransduction and Its Role in Stem Cell Biology

Christopher B. Wolf and Mohammad R.K. Mofrad

Abstract Most cells can sense and actively respond to mechanical stimuli. Yet the exact processes that convert mechanical signals into a cascade of biochemical signals that affect the phenotype of the cell (a process called cellular mechanotransduction) remain elusive. However, such active response promises a large potential in stem cells for future work in therapeutics, tissue engineering, and synthetic bioengineering. Stem cells are highly responsive undifferentiated cells in the biological environment that are able to adapt and differentiate into an appropriate cell type based on the microenvironment within which they reside. Mechanotransduction, in combination with other experimental techniques, may provide new insights into the operations that occur at the cellular level. Understanding cellular mechanotransduction can also prove useful in understanding the overall effect on biological systems resulting from a change in just a few small variables. To elucidate the particular roles that stem cells play in healing during the adult stages, a role for stem cells that is still poorly understood as compared to what is known about them in an embryonic environment, experimental approaches must combine both mechanical and biochemical observations.

Keywords Mechanotransduction • Mechanics • Stem cells

Introduction

What defines a given cell? This is a crucial question that any stem cell needs to resolve before it can choose to differentiate into a particular cell type. The answer to this important question deals with a broad range of characteristics and states that

M.R.K. Mofrad (✉)
Molecular Cell Biomechanics Laboratory, Department of Bioengineering,
University of California, Berkeley, CA 94720, USA
e-mail: mofrad@berkeley.edu

H. Bharavand (ed.), *Trende in Stem Cell Biology and Technology*,
DOI 10.1007/978-1-60327-905-5_20

comprise the overall operation of the cell in its environment: from its specific biological role and functions to the morphology and physiology of the cell itself. Yet,
as a whole, these characteristics are a mix of the initial defining factors, such as cell
type, and the results that the cell has arrived at in the process of adapting to the
environment, such as cytoskeletal, nucleoskeletal, mechanical response, and adhesion to a substrate. Yet a major part that defines the characteristics of a cell lies in
both the biophysical response to its environment as well as the inherent biophysical
properties of that particular type of cell *(1)*. If, for example, the differentiated cell
must be part of the endothelial vasculature, it must be able to respond to the cyclic
shear strains that are encountered from the blood flow in the proper manner by
lengthening along the direction of flow, stiffening, and changing its shape and compliance to pack better with other cells in the vasculature. This response represents
an adaptation to the environment around the cell as well as describing yet another
parameter that can be used to precisely define, in the case of an undifferentiated
stem cell, the final differentiated cell type it should result in. In this manner, the
biochemistry and biology of the cell plus the environment combine with the biophysical forces experienced to determine the "coordinate" of the final differentiated
state. Hence, a better understanding is needed of the roles that mechanotransduction, the biological response to biomechanical forces that are experienced by the
cell, plays in stem cell development, maturation, differentiation, and final biology.
This would also prove to be of great use in the development of efficient and practical clinical or medical therapeutic techniques and laboratory practices. Currently,
there is an incomplete understanding of the methods that trigger undifferentiated
stem cells to differentiate into various resultant cell types. This, as a matter for
practical concern, results in undesirable premature differentiation of stem cell cultures that are intended to be kept for a given period of time until a particular cell
type is required. In practice, there exists a large number of factors and parameters
that affect the differentiation events that stem cells undergo, both in vivo and in
vitro (Fig. 1). The stem cell goes through a process that dictates its response to a
variety of stimuli and, based on whichever set of stimuli is present as the dominant
one, that causes the stem cell to up-regulate certain genes that respond to stimuli.
If the major part of this stimuli is, for example, cyclic loading (due to blood flow
or an artificial imposition), then the stem cell response will eventually grow to
adapt to the biophysical forces. This includes the start of the differentiation process
and the appropriate cell type results. In real biological systems, it is not just one or
two factors that determine the fate of a given portion of stem cells, but rather a collection of such factors that are applied to the cells at the right time to induce the
progenitor stem cells to decide a particular fate. It is of importance to note that for
a functional organ or tissue to be developed in vivo or in vitro, there should exist
the proper combination of these parameters that act in a concerted fashion so as to
not only instruct the desired fate path to be acted upon, but also to form the particulars of the resultant cell itself such as stiffness, directionality, and other biophysical
properties. For example, it is relatively easy to find stem cells that differentiate into
endothelial vasculature or cardiomyocytes, but these differentiated stem cell cultures usually have random ordering among the cells with regard to their directions,

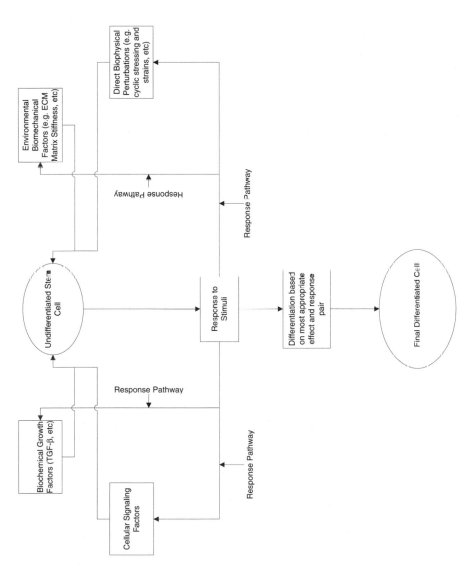

Fig. 1 A systematic flow chart that describes the input and responses that can cause a given stem cell to differentiate into a cell type whose morphology is defined by the microenvironment

as well as a wide heterogeneity in their biomechanical response. For these cells to be properly used in vivo, such as an artery or vein replacement, it is imperative that they can be made to grow and form with the proper directionality and homogeneity in all their properties within a reasonable threshold, as an area that is too stiff or too weak could cause major problems later. Therefore, one of the goals of development of stem cell–based therapeutics is to understand how the interplay of biomechanical and biochemical factors leads to the proper differentiation and modeling of the resultant cell type. Furthermore, accurate and precise knowledge of the response of undifferentiated stem cells, or even adult stem cells, to biomechanical stimuli would be of great help in developing minimally invasive, self-originating regenerative therapeutics where, through modification of the diseased or wounded site, they would encourage progenitor cells to develop in controllable ways to better accelerate healing and reconstruct the damaged area as opposed to applying the progenitor cells plus the growth factors and hoping that the majority of the cells take their cues from the surrounding, undamaged cells that border the region of injury to form up properly.

Mechanotransduction and the Primary Cellular Senses

From the point of view of the cell, the primary sensory transduction is via biochemical sensing and biomechanical transduction. Hence, these are the most important factors whenever a cell wants to adapt and respond to its environment appropriately. Stem cells are no exception to this, and these cellular "senses" take on an even larger role for stem cells since these cells have to coordinate and form tissues and organs out of large groups of themselves in a highly organized manner. It has been shown in recent work that stem cell lineages and fates can be directed by the elasticity of the extracellular matrix (ECM) within which the cells are situated (2). This provides a link in understanding that mechanotransduction processes can provide a large impetus sufficient to provoke a change in the fates and, subsequently, the resulting differentiated cells. Yet to understand the entire process, it is first important to look at the initial response of the stem cell to an applied, mechanically transduced force and see how the cell remodels on the molecular level to account for such impositions. A useful stem cell to look at for such activities are the mesenchymal stem cells (MSCs), which are marrow-derived stem cells able to differentiate into various cell types including neurons, myoblasts, osteoblasts, and fibroblasts. Setting aside the role of the biochemical growth and signaling factors at this point, the first biomechanical transduction change that is noted in the differentiation of these cells into the various aforementioned cell types is the anchorage of the MSC versus the ECM or substrate that it is deposited within or upon. How does the cell determine the material and mechanical properties of this sort of environment? The answer lies in one of the two major cellular senses mentioned previously, namely biophysical mechanotransduction and biochemical signaling, as illustrated in Fig. 2. In this case, the cell applies a force to measure the mechanical

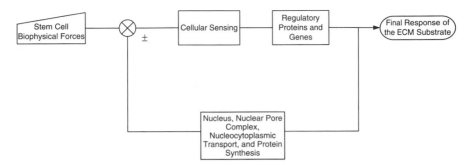

Fig. 2 A control feedback system that describes the adaptation of the biophysical properties of the stem cell to the microenvironment of the extracellular matrix or substrate. It is also possible to see how a small change in one of the mechanotransduction pathways, particularly in the error feedback pathway seen at the *bottom*, can lead to a larger change in the final result of the differentiation, adaption, and growth processes

and material properties of the ECM or substrate around it and adjusts appropriately. But it remains to be defined what is "appropriately" in this case. During development, it is crucial that, once a structure prototype is formed that is later going to become an important part of the organism, the cells around it pick up on the precise formation of this construct by sensing the ECM and the cells that comprise it. Once these values have been ascertained via active mechanosensing from the cell, it is possible to match the properties of the cell, both biomechanical and biochemical, to that of the surrounding biology, thereby continuing the formation and maintenance of the structure until some predetermined point where the macroscale modeling of the structure stops. Careful coordination of such an effort practically requires the tight integration and coupling of the biomechanical sensing pathways and biochemical sensing pathways. As such, this comprises one of the first steps in the early differentiation of a stem cell, in this case an MSC, via biomechanical sensing. It has been shown that, within an early time period, this type of differentiation can be reversed *(2)*. It has also been observed that if the MSCs are continually exposed to the same environment for an extended period of time, they become permanently fated to the particular cell type deemed appropriate for that environment based primarily on mechanosensing of the environment *(3)*. A variety of substrates, ranging from glass to reconstructed ECM, have been used to typify the response of these cells to their environment and the role this plays in developing biological constructs, usually in the form of tissues or organs, on the macroscopic level *(4)*.

Not only do the cells respond to the properties that they themselves sense in their environment, but they also respond to forces that are transduced upon them, via the environment, from external origins. This may be flow over a recently developed ECM, as in the case of developing endothelial vasculature, or high stress-bearing regions of bone tissue where their osteoblasts are capable of remodeling the macroscopic structure of the bone tissues to optimize loading. In a growing organism, it is crucial for the normal physiological development of the tissues, especially the load-bearing or force-generating tissues and organs, to be able to continuously function and undergo many loading cycles at the same time as tissue growth. This imposes a requirement on all the tissues that are growing in these regions to be able to remodel dynamically when required to accommodate larger clusters of the same cell type, form a new interface to, say, new vasculature that is bifurcating, and growing into the area for higher nutrient delivery and waste disposal for better performance. The previous example is a case where the cells have to respond directly to mechanical loading, trigger the release of growth factors, as well as send a signal to the surrounding groups of cells, both mechanically transduced as well as biochemically, to increase energy production, growth rates, and prepare for remodeling that may be required. Indeed, such reactions to increased matrix loading have been seen elsewhere as well, wherein either increased loading on the matrix or increased stiffness of the matrix trigger an up-regulation in protein phosphorylation, stress fiber assembly, and cellular stiffness to nearly match that of the matrix itself *(3)*. In this way, the cells that are undergoing some form of activity, be it on the macroscopic or microscopic scale, are constantly attuned to their environment and have the ability to rapidly transduce a change in the input to other cells in the region and, subsequently, remodel their behavior and structure to suit the change more optimally. Including MSCs in the cells that are applied to such loading adds another degree of freedom in that the MSCs are pluripotent and can derive a large quantity of connective and supportive tissue cell types, by definition. This adds yet another degree of freedom to tissues undergoing extensive loading where there is the requirement of dynamic growth during the course of continued biological function, such as heart muscle, vasculature, bone, and neural tissues.

Application of Mechanotransduction and Techniques

Returning to the example given of vasculature, a particular application of the aforementioned processes arises in the genesis of new, branching vasculature to either supply a new tissue or organ growth or to resupply a site that is undergoing the healing process and needs both nutrients and waste disposal. It can be shown that, biomechanically, the leading edge of the growing vasculature causes one of two effects of major consequence: first, the force applied to the cells that are in the path of the vasculature growth, most likely expressing a growth factor that attracts the growing vasculature, and, second, the parting of two cells so that this pushing will immediately result in a separation force between the cells at the apex of the forming vasculature, following from a casual analysis of the mechanical forces. As the apex is forcing aside cells, the cells respond by stiffening and increasing their surface area of contact with the vasculature. This effect propagates through the ECM or medium that the vasculature is penetrating due to signal mechanotransduction as well as biochemical signaling. After it has developed for a while, the vasculature matures and forms into the final, desired shape. One of the effects on the MSCs that are forming a large part of the developing vascular growth as they develop into smooth muscle cells (SMCs) is that they are subject to a variety of biomechanical forces in this condition *(5)*. It has been shown that SMC markers on MSCs undergoing the differentiation process have been up-regulated in cells experiencing strain where the cells additionally attempted to align themselves perpendicularly to the direction of the strain *(4–7)*, as seen in Fig. 3. There are also large, widespread changes throughout the entire cell once the vasculature has formed, and there is operational blood flow through it, subjecting other cells to the same shear strains; whereby additional MSCs that are recruited to the vasculature have been shown to require such mechanical strains to undergo proper differentiation changes *(5)*. In another example, such as healing of vasculature after injury, it is possible for the MSCs to encounter the ECM around the damaged site, begin a period of rapid growth triggered by the release of growth factors, such as transforming growth factor b, that are released from the regular vasculature that surrounds the injured site due to higher loading on these cells, and finally differentiate perpendicularly to minimize shear stress on the cell, to the shear strains experienced from blood flow,

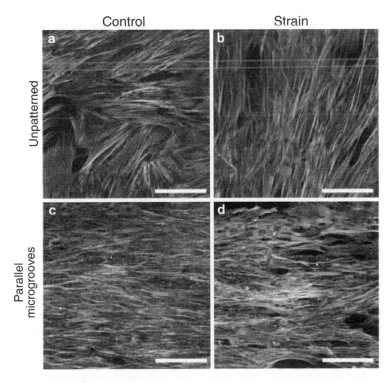

Fig. 3 A set of images, that illustrate the tendency of the cells, visualized via F-actinin staining, to align perpendicular to the direction of strain, which was left to right; the same alignment effect is seen when the cells are placed on a surface whose microstructure topology consists of parallel microgroves; however, they align parallel to the direction of shear strain in the case of the patterned microgroves on a polydimethylsiloxane membrane *(5)*

as well as the stiffness of the surrounding undamaged vasculature. Because of the requirement that MSCs need to react to their environment via mechanotransduction in developing organisms, it is possible to affect the biological and material properties of tissues simply by applying mechanical loads. This is of great importance when considering that it is often difficult to obtain the desired results, such as a large, intricate three-dimensional scaffold comprised of multiple types of cells, without a complex treatment of biochemical signaling and system biology on the molecular level of the cell. But we can include the ability to mechanically affect the cells via both mechanotransduction effectors and direct biochemical approaches. Continuing to use the vasculature as an example, we can envision a large scaffold that can be biocompatibly synthesized outside the body and implanted to aid in the reconstruction of large sections of damaged tissue. Typical issues with such approaches may consist of the cells randomly arranging themselves and not forming a proper cohesive matrix and connections strong enough for in vivo implantation. Furthermore, upon the degradation of the scaffolding, the construct would lose a large part of its mechanical integrity and become unsuitable for the typical biomechanical loading that would be encountered in its physiological role(s). If such scaffolds were to be used for therapeutic tissue engineering purposes on vasculature, then it is of great importance that the resulting tissue section be not only biologically sound, but biomechanically sound as well, since a failure in the mechanical properties of a section of vasculature will have catastrophic consequences. A similar example would be bone implants, particular structures that receive a great deal of loading such as the organization around hip and knee joints, where the implant material must be matched to the mechanical properties of bone. The main reason for this is that bone is an actively remodeling tissue that responds to changes in the stresses, sensed by the osteoblasts mechanically transducing strain throughout the bone, and responds by making regions that experience a significant amount of stress stronger by adding more bone tissue and ECM secretions by the osteoblasts and decreasing the amount of bone tissue in areas that appear to have a minimal amount of loading as determined by the osteoblasts. If an implant that has a modulus that is far too large is used in, for example, a hip joint replacement, then it takes the majority of stresses that would previously have been applied uniformly along the structure of the femur and/or the ilium of the pelvis and the cotyloid cavity, or acetabulum, that forms the socket, which the head of the femur fits and actuates within. It was discovered in short order that, while implants, made out of stainless steel designed to withstand very large loads, were able to be used well enough as implants, they loosened over the course of use, sometimes very dangerously as a loosening and mechanical failure of the implant embedded into the bone would cause serious injury once the mechanical junction between the bone and implant fully decoupled. The problem in cases such as the ones mentioned come down to the fact that, while care was taken to make sure the bone would grow around the implant and attach, thereby providing traction to grip the implant well, the bone would still loosen because the implant was too stiff and handling a very large part of the load that section of the femur, usually the upper shaft and parts of the lower neck, would otherwise experience. However, the osteoblasts within the bone tissue ECM that are removed from the implant or bone interface have no way of sensing this, since they measure the stiffness of the environment via active forcing on attachments and mechanotransduction of the response. Because the cells would have no way to determine the fact that an implant, in an otherwise biological and biomechanically sound interface with the bone tissue around it, was assuming the "missing" loading, they begin a response that is similar to muscle atrophy in that they started to digest the surrounding ECM and bone tissue. This in turn would lead to a mechanical failure whereby the circumferential, and to some extent the depth, of the "grip" around the implant lessened to the point of the implant coming loose from the bone. Taking this analogy to the vasculature scaffold, it is important for the scaffold, in cases such as bypass or grafting where an entirely new section is needed, to provide a framework for the cells to grow, differentiate, and adapt themselves to the flow and other biomechanical factors within a reasonable amount of time. The scaffold must not last too long, be too stiff, or have the microenvironment present as too stiff to the cells (Fig. 4). The latter most part is a concern that must be taken into account not only when choosing a material, but also a microstructure to adopt for the cells; a good example of a substrate that gives the cell little

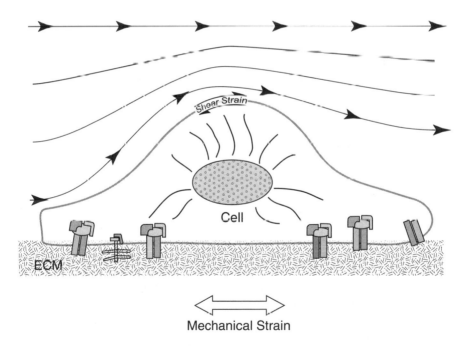

Fig. 4 An illustration of a cell being exposed to a flow that is causing shear strain to be experienced on the surface of the cellular membrane. The nucleus, nucleoskeleton, and cytoskeleton take a large deal of the load and are connected to the local adhesions to the extracellular matrix (ECM), which also experiences a mechanical strain. Depending on the difference between the response of the cell versus the response of the ECM, the tissue surface that is experiencing the shearing could deform in different ways macroscopically, affecting the physiology and the biomechanical effect on the flow and that locale of the organism

to no information about its compliance and modulus is glass, where cells can be cultured on glass in layers, but do not form proper alignment or attachments that would otherwise be seen in proper ECMs in vivo since, during the mechanosensing that the cell applies to determine the properties of the ECM or substrate, the glass will not yield due to a high stiffness. Conversely, it is equally as detrimental to choose a matrix that is too pliable for the cells such that, as much as they push or pull on the matrix microstructure, there is neither a stiff enough location to anchor nor efficient motility, both of which are very important for migration of cells to a wound site to promote healing and repair. These factors are even more important to consider when culturing MSCs since there is an extra degree of freedom added in that the MSCs can choose to differentiate into a cell type that they deem to be appropriate for the given environment, even though the desire of the investigators or clinicians would be to hold them in the undifferentiated state for an extended period of time until they either have enough substance or the time comes to perform the therapeutic procedure.

Challenges and Potential Applied Solutions

A major issue that has to be resolved in proceeding toward the goal of realizing the practical applications of mechanotransduction in stem cells and therapeutic tissue engineering is how to get the progenitor cells to accurately and precisely emulate the developmental progression that the cells to be replaced from the injured area have already undergone from the beginning. In some cases, especially neural tissue in the CNS and PNS, the cellular age is comparable to the age of the organism itself. How, then, would it be possible to repair a large injury to one of these areas by the exogenous application of progenitor cells, usually in a scaffold of ECM or substrate of some form, to differentiate, grow, and make the proper connections? This still remains one of the primary hurdles that must be overcome for these types of applications to become practical. One of the ways that this hurdle may be overcome is the use of mechanical factors as a stimulus for the progenitor cells that have yet to develop. It is well known that spatial configuration, structure, topology, and biophysical forces play a role in the development of multicellular organisms beginning from the oocyte and egg then proceeding to the birthing event; the development of the neural crest and various asymmetric polarities in embryos is a good example of this. Here we see the integration of a multitude of biochemical signals, and biophysical forces help guide and shape the macroscopic structures that are formed from a variety of cells, each of which serves a particular set of functions and roles. This type of development proceeds over time to form a developing organism. Moreover, this type of development provides us with an opportunity for crucial insight into the particulars of mechanotransduction and associated biochemical signaling. With such an understanding, it would be possible to reverse engineer the developmental process and develop "shortcuts" that would help us apply the same mechanotransductive and biochemical triggers to stem cells for therapeutic and laboratory use. For example, when neurons are developing, the terminal of one neuron is attracted to the branching on the body of the closest neuron initially by biochemical signals, but must then balance the biomechanics of the end-state result such as the consistency of the synapse gap, which is important for proper summation and temporal convolution of signals sent to a neuron for integration, as well as the, as of yet unknown, mechanisms of targeting particular dendrites for neural connections *(8)*. After the dynamics and behaviors of the stem cells at the early developmental level have been ascertained, it is possible to apply that knowledge to therapeutics in the resolution of complex injuries, where the cell interface microstructure has to be rebuilt, such as neural connections, by using mechanotransduction, as in the form of nerve growth guides to micromanipulation of growing neural bodies, similar to the micropatterned grooves in Fig. 3 *(5)*. To guide the development toward the desired outcome, the cell is given the impression that it is in an early stage developmental environment so the proper connections are made and there is a minimum of invasive intervention that is both costly and risky. Experiments have been performed recently where mechanotransductive control has been exerted over the development of vascular smooth muscle using a variety of techniques *(9)*. Similar techniques can be applied to neural developmental control and nerve growth guides in conjunction with mechanotransduction to yield similar results in the neurological areas. Such applications help us circumvent the hurdle of getting the stem cells to differentiate and "redevelop" as if they were in the early stages of development of that area. However, this is not only important for therapeutics, but also for laboratory persistence of viable stem cell cultures free of the typical mouse feeder layers they are currently raised upon. Because of such advances, it is important to also understand the pathways, both biochemical and biomechanical, within the cell and how they interact to produce results that have macroscopic consequences. In the case of feeder layers, which could be replaced by a network of microchannels or microtubes *(10)* as the BioMEMs tissue engineering technology for these already exists and is practically applicable, the expected mechanical and material properties that deliver the required nutrients to the stem cells without the associated risks and special conditions are required to maintain a feeder layer. Beyond this, it is also possible to use mechanotransduction in the development of new techniques in understanding the response of stem cells to acute biophysical forcing, thereby allowing investigators to determine particular responses to biomechanical forces in vivo.

Tools and Experimental Vehicles

The ability to control and modify the developmental progression of stem cells via mechanotransduction has large ramifications for clinical or medical therapies as well as future laboratory techniques. Along these lines, there are many methods and tools available to the investigator for analyzing stem cells undergoing developmental changes in real time via mechanotransduction. As an example, BioMEMS technologies can be used to sequester a small assembly of cells within a deep groove and shear them using flow that can be controlled. Furthermore, the flow contents, which can contain soluble growth and signaling factors, can be mixed in rapidly and the effects observed in real time. Another potential method would use atomic force microscopy or optical traps to tug on a stem cell that is situated within an ECM for a period of time and determine what the overall response of the cell would be to the forces. Inversely, this can be used to measure forces that occur at the local adhesion sites to the matrix and determine what the threshold values are that trigger a change in the morphology of the stem cell. Going a step further, fluorophores can be attached to a few cells at specified locations that are inserted into a given ECM or deposited on a substrate. From detailed knowledge of the material properties of the substrate as well as those of the cell, the movement of the beads over time will yield displacement data that can then be used to generate a biophysical model of the mechanotransduction pathways of primary importance or simply for comparison to an already existing theoretical model for the predicted cellular response. Knowing the biophysical responses of the stem cell and the biological developments that are coupled with the morphological or physiological and biomechanical changes in the cell, a very comprehensive model of the cell can be developed that can account for not only purely biochemical factors affecting development, but also mechanotransduction and various other physical factors exerted by the environment. A typical example of this is the development of differentiated structures within embryos during the early developmental stages. A particular example consists of the migration of neurons across the midline during embryogenesis. This has been an extensively studied, yet never fully understood, system, and the neurons have to migrate long distances that are either independent or have no homogenous uniform concentration of attractive or repulsive factor cues except at key junctures that include the floor plate and alar plate *(11)*. From a theoretical point of view, since the growth cone needs to travel a long distance, it is reasonable to assume there is some mechanosensing going on as the growth cone travels through the environment, since this method of cellular sensing causes it to maintain a stable growth fashion and directionality. One of the ways of quantitatively analyzing this migratory behavior is by observing and tracking the growth cone as it works its way across the midline by using its filopodia as a guide to sense changes in the anchorage in an immediate arc extending out from the forward region of the growth cone. Research suggests that the growth cone selectively enhances the protrusion and further extension of filopodia in the preferred direction or orientation *(12)* depending on the amount of tension that is generated by the filopodia; the greater the tension, the larger the ability of the filopodia to adhere and gain traction, and the more preferentially it is further extended in that direction versus the entire arbor of filopodia of the growth cone (Fig. 5). The growth cone itself appears to use many filopodia that are extended out into the path of determined locomotion to sample the microenvironment in those areas and determine the next suitable and appropriate course of action. Further indications that support this hypothesis come from the disruption of the growth cone when it was being prepared for mounting and imaging, as it would extend and retract the filopodia in equal measure over a period of time, resulting in no net gain of growth. An exploratory response, such as this one, would tend to suggest that the growth cone, being mechanically perturbed, is now resensing and

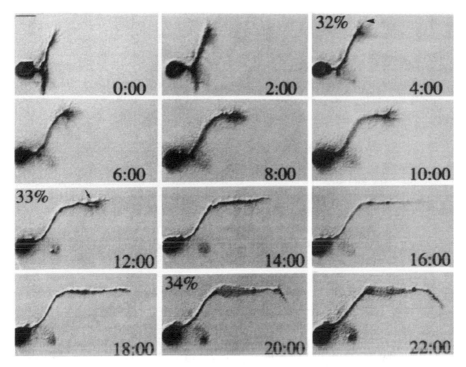

Fig. 5 A time lapse series of images where the Q1 growth cone can be seen; at first it continues upward and to the right, until it encounters the MP1 soma, where it turns and then eventually traverses the midline after some time, and reflects the behavior on the originating side by this time turning downward, with respect to this frame of visualization *(12)*

resampling an environment, by mostly random exploratory extensions, that appears to have changed. After a given period of time, however, this behavior stops and the growth cone resumes the typically observed migratory behavior. This type of motility behavior has also been seen in cells that participate in healing response and have to migrate to a single wound site or multiple wound sites to form a cohesive and confluent layer for the next stages of the healing and repair process. Further experiments of worth would be to biomechanically perturb multiple parts of, for instance, a growth cone and examine its transient response as well as the final developmental outcome of the migration. Further examination of this type of behavior will lead to methods that can manipulate the responses of key cells in development through easily applied, mechanical forces.

Conclusions, Questions, and the Outlook for the Future

The large potential that stem cells contain, especially for future work in therapeutics, tissue engineering, and synthetic bioengineering, is tempered by the difficulty with which they is accessed, controlled, and applied to current problems in both the clinical and medical fields as well as experimentally in a laboratory setting. One of the ways whereby we can learn a great deal more about stem cells is by understanding that they are, at the basic level, highly responsive undifferentiated cells in the biological environment that are able to adapt and differentiate into an appropriate cell type based on the microenvironment with which they reside. It bears mention that not only can using mechanotransduction, in combination with other experimental techniques, provide new insights into the operations that occur at the cellular level, but it can also prove useful in understanding the overall effect on biological systems resulting from a change in just a few small variables. To elucidate the particular roles that stem cells play in healing during the adult stages, a role for stem cells that is still poorly understood as compared to what is known about them in an embryonic environment, experimental approaches must combine both mechanical and biochemical observations. In development, and indeed throughout life, this plays a major role in the activation and reactivation plus recruitment of stem cells in all their forms for development and repair or healing processes that occur in the body; the latter occurs with adult stem cells and, while not yet fully understood, would play a key role in future therapeutics that revolve around in vivo tissue engineering or reengineering and healing without the need for complicated invasive surgeries or applications of pharmaceuticals that, in higher doses, would present with adverse side effects. However, before such research can progress toward the ultimate goal of the clinical or medical and laboratory realizations, answers are needed to basic questions that relate to the response dynamics of stem cells and their associated mechanotransduction pathways while they are being exposed to a multitude of applied biophysical forces and biochemical signals at the same time. One of these questions deals with what constitutes a threshold stimulus whereupon the cell decides to react against the force. Tracking down the exact threshold is a very tricky process as the response the cell provides can be broken down into two parts: the immediate portion where the deformations are simply a biomechanical reaction to the force, and a longer-term response, e.g., cytoskeletal remodeling, that formulates a reactive force initiated by the cell itself. These two effects combined comprise the mechanotransduced cellular response to the force. To observe the separate components of this dual-part response, one has to first understand what in the cell, given a particular configuration, physically responds to the force immediately and then what in the cell construes a directed, cell-initiated response to the force. Such requirements demand that experiments and models be developed where the two processes can be accurately and precisely described as well as predicted. After the development of such models, it would be possible to observe a wider range of experiments that place the stem cells in conditions that would typically be encountered in the body of the given organism as well as perturbations in variables to measure the integrated response of the stem cells under study within the biological system. The next question would then be how these threshold conditions are reached and, in certain cases, what can be done to promote or inhibit the predicted response. Still other questions would comprise the efficacy of a unit change, the sensitivity of the biological system, of which the stem cells would be a part of, to the change, and the long-term physiological impacts of the change desired, depending on the rate of response of the cells.

It must be said before concluding that mechanotransduction and the internal cellular processes play a crucial role in practically all aspects of any cell's biological function, role, and morphology. The cell, as defined, is a complex set of systems that combines physical changes, chemical reactions, mechanical processes, and overall biological responses to form an organism that can perform all the required functions of life. A thorough understanding and modeling of such a system will come not just from analyzing a specific subset for a long period of time and concluding that all there is to be learned about that subset's role in the cell has been learned, but to understand and realize the overall role that the process is a key component of and how that role changes as the process changes or is changed. Since both therapeutic and laboratory procedures operate on the basis of altering how a given process functions, understanding the changes and dynamics within those processes, instead of just the overall results, will yield key clues that allow for more and better knowledge to be discovered about those very systems. These will, in the end, prove to be highly applicable when they make the transition from the bench to the bedside as well as other cross-discipline interactions where not only will the effect be beneficial from the point of view of therapeutics and novel laboratory techniques, but also the stimulation of a deeper intuitive understanding of complex systems in other fields.

References

1. Mofrad MRK, Kamm RD (Eds.). Cellular mechanotransduction: diverse perspectives from molecules to tissues. Cambridge University Press, Cambridge, 2009 (in press).
2. Engler AJ, Sen S, Sweeney HL, Discher DE. Matrix elasticity directs stem cell lineage specification. Cell 2006;126:677–89.
3. Wells RG, Discher DE. Matrix elasticity, cytoskeletal tension, and TGF-b: the insoluble and soluble meet. Sci Signal 2008;1:pe13.
4. Li C, XuQ. Mechanical stress-initiated signal transductions in vascular smooth muscle cells. Cell Signal 2000;12:435–45.
5. Kurpinski K, Chu J, Hashi C, Li S. Anisotropic mechanosensing by mesenchymal stem cells. Proc Natl Acad Sci U S A 2006;103:16095–100.
6. Birukov KG, Shirinsky VP, Stepanova OV, et al. Stretch affects phenotype and proliferation of vascular smooth muscle cells. Mol Cell Biochem 1995;144:131–9.
7. Tock J, Van Putten V, Stenmark KR, Nemenoff RA. Induction of SM-alpha-actin expression by mechanical strain in adult vascular smooth muscle cells is mediated through activation of JNK and p38 MAP kinase. Biochem Biophys Res Commun 2003;301:1116–21.
8. Korte M. Neuroscience. A protoplasmic kiss to remember. Science 2008;319:1627–8.
9. Jeong SI, Kwon JH, Lim JI, et al. Mechano-active tissue engineering of vascular smooth muscle using pulsatile perfusion bioreactors and elastic PLCL scaffolds. Biomaterials 2005;26:1405–11.
10. Huang NF, Patel S, Thakar RG, et al. Myotube assembly on nanofibrous and micropatterned polymers. Nano Lett 2006;6:537–42.
11. Taniguchi H, Tamada A, Kennedy TE, Murakami F. Crossing the ventral midline causes neurons to change their response to floor plate and alar plate attractive cues during transmedian migration. Dev Biol 2002;249:321–32.
12. Myers PZ, Bastiani MJ. Growth cone dynamics during the migration of an identified commissural growth cone. J Neurosci 1993;13:127–43.

Index

Printed in the United States of America